T0250902

Odyssey of Light in Nonlinear Optical Fibers

THEORY AND APPLICATIONS

Odyssey of Light in Nonlinear Optical Fibers

THEORY AND APPLICATIONS

EDITED BY

Kuppuswamy **PORSEZIAN**
Ramanathan **GANAPATHY**

CRC Press
Taylor & Francis Group
Boca Raton London New York

CRC Press is an imprint of the
Taylor & Francis Group, an **informa** business

Cover painting: Jungle Gym (Tuner)© by Molly Barker

CRC Press
Taylor & Francis Group
6000 Broken Sound Parkway NW, Suite 300
Boca Raton, FL 33487-2742

First issued in paperback 2017

© 2016 by Taylor & Francis Group, LLC
CRC Press is an imprint of Taylor & Francis Group, an Informa business

No claim to original U.S. Government works

ISBN-13: 978-1-4822-3613-2 (hbk)
ISBN-13: 978-1-138-74958-0 (pbk)

This book contains information obtained from authentic and highly regarded sources. Reasonable efforts have been made to publish reliable data and information, but the author and publisher cannot assume responsibility for the validity of all materials or the consequences of their use. The authors and publishers have attempted to trace the copyright holders of all material reproduced in this publication and apologize to copyright holders if permission to publish in this form has not been obtained. If any copyright material has not been acknowledged please write and let us know so we may rectify in any future reprint.

Except as permitted under U.S. Copyright Law, no part of this book may be reprinted, reproduced, transmitted, or utilized in any form by any electronic, mechanical, or other means, now known or hereafter invented, including photocopying, microfilming, and recording, or in any information storage or retrieval system, without written permission from the publishers.

For permission to photocopy or use material electronically from this work, please access www.copyright.com (http://www.copyright.com/) or contact the Copyright Clearance Center, Inc. (CCC), 222 Rosewood Drive, Danvers, MA 01923, 978-750-8400. CCC is a not-for-profit organization that provides licenses and registration for a variety of users. For organizations that have been granted a photocopy license by the CCC, a separate system of payment has been arranged.

Trademark Notice: Product or corporate names may be trademarks or registered trademarks, and are used only for identification and explanation without intent to infringe.

Library of Congress Cataloging-in-Publication Data

Odyssey of light in nonlinear optical fibers : theory and applications / editors, Kuppuswamy Porsezian, Ramanathan Ganapathy.
    pages cm
Includes bibliographical references and index.
ISBN 978-1-4822-3613-2 (hardcover : alk. paper) 1. Fiber optics. I. Porsezian, K. (Kuppuswamy), 1963- II. Ganapathy, Ramanathan.
TA1800.O29 2015
621.36'92--dc23   2015013881

Visit the Taylor & Francis Web site at
http://www.taylorandfrancis.com

and the CRC Press Web site at
http://www.crcpress.com

Dr. K. Porsezian dedicates this book to his wife, P. Senthamizhselvi, and his sons, P. Gokul and P. Ragul.

Dr. R. Ganapathy dedicates this book to his parents, G. Ramanathan and Chithra Ramanathan, and his better half, Radhika Ganapathy.

Contents

Foreword xix

Preface xxi

Acknowledgments xxv

Editors xxvii

Contributors xxix

Chapter 1   Basic nonlinear fiber optics 1
K. Thyagarajan and Ajoy Ghatak

    1.1   Introduction 1
    1.2   Modes of a step-index fiber 1
    1.3   Guided modes of a step-index fiber 3
    1.4   Single-mode fiber 5
         1.4.1   Spot size of the fundamental mode 6
    1.5   Pulse dispersion in optical fibers 7
         1.5.1   Material dispersion 7
         1.5.2   Waveguide dispersion 8
         1.5.3   Total dispersion 9
         1.5.4   Dispersion and maximum bit rate in
                  single-mode fibers 11
         1.5.5   Dispersion-compensating fibers 12
    1.6   Nonlinear effects in optical fibers 13
         1.6.1   Self-phase modulation 13
    1.7   Nonlinear Schrödinger equation 16
    1.8   Spectral broadening due to SPM 18
         1.8.1   Cross-phase modulation 18
         1.8.2   Four-wave mixing 21
         1.8.3   Fiber optic parametric amplifier 25
         1.8.4   Supercontinuum 27

Chapter 2   Waveguide electromagnetic pulse dynamics: Projecting
operators method 31
Mateusz Kuszner and Sergey Leble

    2.1   Introduction 31
    2.2   Theory of initialization of a pulse propagation 33

2.2.1   Cauchy problem formulation in one spatial
          dimension 33
2.2.2   Projection method for the 1D Cauchy
          problem 34
2.2.3   Dispersion and nonlinearity with
          polarization interaction, general relations 38
2.3   Comparison of results obtained with the multiple
      scale method 39
2.4   Projection method for boundary regime propagation .. 41
2.5   Cylindrical waveguide 45
      2.5.1   On transversal fiber modes........................... 45
      2.5.2   Solution of the linear problem, a way to
                the projecting procedure 45
2.6   Including nonlinearity 54
2.7   Conclusion... 56

Chapter 3   Coupled-mode dynamics in continuous and discrete
              nonlinear optical systems 61

Alejandro Aceves

3.1   Coupled-mode dynamics in nonlinear optical systems . 61
3.2   Parity-time optical coupled systems 64
3.3   Binary arrays 68
3.4   Dual-core photonic crystal fiber.......................... 73
3.5   Fiber amplifiers 75
3.6   Future directions and conclusions 75

Chapter 4   Continuous-discrete duality of the nonlinear Schrödinger
              and Ablowitz–Ladik rogue wave hierarchies 79

A. Ankiewicz, D. J. Kedziora, and N. Akhmediev

4.1   Introduction 79
4.2   Theory.. 80
      4.2.1   Derivation of discrete result 80
      4.2.2   Darboux scheme 83
4.3   Rogue wave triplet 83
4.4   Discretization effects 85
4.5   Ablowitz–Ladik rogue wave hierarchy........................ 89
4.6   Conclusion... 93

Chapter 5   A theoretical study on modulational instability in relaxing
              saturable nonlinear optical media 97

Kuppuswamy Porsezian and K. Nithyanandan

5.1   Introduction 97
5.2   Scalar MI in the relaxing saturable nonlinearity
      (SNL) system 99

      5.2.1   Theoretical framework.................................99
      5.2.2   Linear stability analysis............................100
      5.2.3   Results and discussion..............................100
  5.3 Vector MI in a relaxing system with the effect of walk-off and higher-order dispersion 105
      5.3.1   Theoretical framework..............................105
      5.3.2   Results and discussion..............................105
  5.4 MI in a two-core nonlinear directional coupler with relaxing nonlinearity 110
      5.4.1   Theoretical framework..............................110
      5.4.2   Results and discussion..............................110
  5.5 MI in a two-core fiber with the effects of saturable nonlinearity and coupling coefficient dispersion.........114
      5.5.1   Theoretical framework..............................114
      5.5.2   Results and discussion..............................114
  5.6 Two-state behavior in the instability spectrum of a saturable nonlinear system..122
      5.6.1   Theoretical framework..............................122
  5.7 MI in a semiconductor doped dispersion decreasing fiber..125
      5.7.1   Theoretical framework..............................125
      5.7.2   Results and discussion..............................125
  5.8 Summary and conclusion 128

Chapter 6 Modulational instabilities in a system of four coupled, nonlinear Schrödinger equations with the effect of a coupling coefficient...133

H. Tagwo, S. Abdoulkary, A. Mohamadou, C. G. Latchio Tiofack, and T. C. Kofane

  6.1 Introduction 133
  6.2 Model..135
  6.3 Linear stability analysis...............................136
  6.4 Modulational instability gain.......................137
  6.5 Propagation of waves through the system 140
  6.6 Conclusion...142

Chapter 7 Hidden symmetry reductions and the Ablowitz–Kaup–Newell–Segur hierarchies for nonautonomous solitons 145

V. N. Serkin, A. Hasegawa, and T. L. Belyaeva

  7.1 Introduction 145
  7.2 Husimi–Taniuti and Talanov transformations in quantum mechanics and the soliton theory................147
      7.2.1   Darwin wave packet and Chen and Liu accelerating soliton 149

7.2.2   Airy and Hermite accelerating wave packets in free space 151

7.2.3   Coherent states, squeezed states, and squeezions 153

7.3   Lax operator method and exact integrability of nonautonomous nonlinear and dispersive models with external potentials 157

7.4   Nonautonomous nonlinear evolution equations 160

7.4.1   Nonautonomous Hirota equation 160

7.4.2   Solitons of the nonautonomous Korteweg–de Vries equation 162

7.4.3   Nonautonomous modified Korteweg–de Vries equation 164

7.5   Generalized NLSE and nonautonomous solitons 165

7.6   Soliton adaptation law to external potentials 168

7.7   Bright and dark NLSE nonautonomous solitons 169

7.8   Colored nonautonomous solitons 172

7.9   Conclusion 176

Chapter 8   Hot solitons, cold solitons, and hybrid solitons in fiber optic waveguides 189

P. Tchofo Dinda, E. Tchomgo Felenou, and C. M. Ngabireng

8.1   Introduction 189

8.2   Isothermic solitons 192

8.3   Hyperthermic solitons 192

8.4   Hypothermic solitons 193

8.5   Hybrid solitons 193

Chapter 9   Optical solitary modes pumped by localized gain 199

Boris A. Malomed

9.1   Introduction and models 199

9.2   Dissipative solitons pinned to hot spots in the ordinary waveguide 205

9.2.1   Analytical considerations 205

9.2.2   Exact results 205

9.2.3   Exact results for $\gamma = 0$ (no linear background loss) 207

9.2.4   Perturbative results for the self-defocusing medium 207

9.2.5   Perturbative results for the self-focusing medium 208

9.2.6   Stability of the zero solution and its relation to the existence of pinned solitons 209

9.2.7   Numerical results.. 210
9.2.8   Self-trapping and stability of pinned
         solitons 210
9.2.9   Model with a double hot spot 211
9.2.10   Related models.. 213
9.3   Solitons pinned to the $\mathcal{PT}$-symmetric dipole............ 215
9.3.1   Analytical results 215
9.3.2   Numerical findings.. 216
9.4   Gap solitons supported by a hot spot in the Bragg
      grating.. 219
9.4.1   Zero-order approximation............................. 220
9.4.2   First-order approximation 220
9.4.3   Stability of the zero solution 223
9.4.4   Numerical results.. 223
9.5   Discrete solitons pinned to the hot spot in the lossy
      lattice... 225
9.5.1   Analytical results 225
9.5.2   Numerical results.. 226
9.5.3   Self-defocusing regime ($B = -1$)................... 226
9.5.4   Self-focusing regime ($B = +1$) 228
9.6   Conclusion.. 229

Chapter 10   Exploring the frontiers of mode locking with fiber lasers .. 235

Philippe Grelu

10.1 Introduction 235
10.1.1   The wonder of mode locking 235
10.1.2   Exploring the mode-locking frontier............. 236
10.1.3   Partially mode-locked regimes...................... 238
10.2 Soliton rain dynamics 238
10.2.1   Introduction 238
10.2.2   Fiber laser setup.. 239
10.2.3   First-order mode-locking transition............. 240
10.2.4   Soliton rain dynamics.................................. 242
10.3 Chaotic pulse bunches.. 243
10.3.1   Extended chaotic bunches: dissipative rogue
         waves 243
10.3.2   Compact chaotic bunches: Noise-like pulses. 245
10.3.3   Spectral rogue waves 246
10.4 Conclusion.. 249

Chapter 11   Matter wave solitons and other localized excitations in
             Bose–Einstein condensates in atom optics........................ 253

P. Muruganandam and M. Lakshmanan

11.1 Introduction 253

11.2 Gross–Pitaevskii equation 254
11.3 Matter wave bright and dark solitons 256
    11.3.1   One-soliton dynamics 256
    11.3.2   Time-independent trap 259
    11.3.3   Time-dependent trap 260
    11.3.4   Dark and bright multi-soliton dynamics.......260
    11.3.5 N-dark soliton solution 265
11.4 Matter wave solitons in multi-component BECs........266
    11.4.1   Dark–bright solitons 268
    11.4.2   Bright–bright solitons 270
11.5 Summary 275

Chapter 12 $\mathcal{PT}$-symmetric solitons 279

Chandroth P. Jisha and Alessandro Alberucci

12.1 Introduction 279
12.2 Ruling equation 280
12.3 $\mathcal{PT}$ linear modes 281
    12.3.1   Particle conservation 283
12.4 Nonlinear modes 284
    12.4.1   Gaussian potential 285
    12.4.2   Periodic potential 287
12.5 Variational approach for periodic potential and
    defocusing nonlinearity 291
12.6 Stability analysis 294
    12.6.1   Gaussian potential 295
    12.6.2   Periodic potential 297
12.7 Dynamical evolution of the soliton 298
    12.7.1   Gaussian potential 298
    12.7.2   Periodic potential 299
12.8 Conclusion 300

Chapter 13 Suspended core photonic crystal fibers and generation
of dual radiation 305

Samudra Roy, Debashri Ghosh, and Shyamal K. Bhadra

13.1 Introduction 305
13.2 Solid core photonic crystal fiber: A brief outline 308
13.3 Group velocity dispersion 309
13.4 Fabrication of suspended core PCFs 311
13.5 Characteristics of suspended core PCF 313
13.6 Dual-resonant radiation 317

Chapter 14   Parabolic similaritons in optical fibers 323

Finot Christophe and Boscolo Sonia

14.1 Introduction 323
14.2 Short-pulse dynamics in normally dispersive fibers ... 324
     14.2.1   Model and situation under investigation 324
     14.2.2   Parabolic waveform as a transient stage of
           evolution in a passive fiber 325
     14.2.3   Parabolic waveform as an asymptotic
           attracting state of pulse evolution in a fiber
           amplifier 327
14.3 Properties of self-similar pulses and extension to
     other configurations 329
     14.3.1   Parabolic similariton properties 329
     14.3.2   Impact of higher-order effects 331
     14.3.3   Extension to other configurations 332
           14.3.3.1 In dispersion-tailored fibers 332
           14.3.3.2 In fiber-based cavities 332
           14.3.3.3 In nonlinear waveguides 333
14.4 Experimental generation of parabolic pulse shape 334
     14.4.1   In passive segments 334
     14.4.2   In fiber amplifiers 334
     14.4.3   Through linear pulse shaping 337
14.5 Applications of parabolic pulses 337
     14.5.1   High-power pulse amplification and
           ultrashort pulse generation 337
     14.5.2   Highly coherent continuums for optical
           communications 339
     14.5.3   Ultrafast all-optical signal processing 339
     14.5.4   Spectral compression 340
14.6 Conclusion 341

Chapter 15   Brillouin scattering: From characterization to novel
applications 351

**Victor Lambin Iezzi, Sébastien Loranger, and
Raman Kashyap**

15.1 Introduction 351
15.2 Basic concepts 352
     15.2.1   Generalities 352
     15.2.2   Spontaneous Brillouin scattering 352
     15.2.3   Stimulated Brillouin scattering 355
     15.2.4   Brillouin gain 357
     15.2.5   Power threshold 360
     15.2.6   Brillouin strain and temperature
           dependence 363

15.2.7   Brillouin mitigation 367
15.2.8   Pump laser modulation 368
15.2.9   Segmented fibers.. 368
15.2.10   Modulation via temperature or strain 368
15.2.11   Polarization and self-induced effects............ 369
15.3   Brillouin fiber laser 371
15.3.1   Continuous wave SBS lasers 371
15.3.2   DFB Brillouin laser 373
15.3.3   Multi-Stokes order comb 374
15.3.4   Mode-locked Brillouin laser 378
15.3.5   Self-phase-locked Brillouin laser 384
15.4   Brillouin scattering for sensors............................. 388
15.4.1   BOTDR... 390
15.4.2   BOTDA... 392
15.4.3   Advances in distributed sensing 394
15.4.4   Strain sensing vs. temperature sensing......... 395
15.4.5   Improving sensitivity.................................. 396
15.4.6   Other techniques and limitations 397
15.5   Conclusion.. 400

Chapter 16   Nonlinear waves in metamaterials—Forward and
backward wave interaction 409

Andrei I. Maimistov

16.1   Introduction 409
16.2   Forward and backward waves................................ 410
16.2.1   Discrete linear models 410
16.2.2   Discrete nonlinear models for backward
          waves 411
16.3   Resonant interaction of forward and backward
       waves.. 412
16.4   Parametric interaction 415
16.4.1   Second-harmonic generation....................... 415
16.4.2   Third-harmonic generation 416
16.5   Waveguide systems: Couplers, arrays and bundles 417
16.5.1   Alternating nonlinear optical waveguide
          zigzag array 417
16.5.2   Linear properties of the alternating
          waveguide zigzag array............................... 419
16.5.3   Nonlinear waves in ANOWZA...................... 419
16.5.4   Alternating nonlinear optical waveguide
          array 421
16.5.5   Linear properties of the alternating
          waveguide array... 422
16.5.6   Nonlinear waves in ANOWA 422

16.5.7   Spatial discrete solitons in ANOWA 423

16.5.8   Bundles of waveguides 424

16.5.9   Linear modes of an alternating waveguide
bundle 424

16.5.10 Linear modes of the twisted alternating
waveguide bundle 425

16.5.11 Nonlinear solitary waves in ANOWB 426

16.5.12 Oppositely directional couplers 427

16.5.13 Gap solitons 428

16.5.14 Bistability 428

16.5.15 Modulation instability 429

16.5.16 Waveguide amplifier based on ODC 429

Chapter 17 Optical back propagation for compensation of dispersion and
nonlinearity in fiber optic transmission systems 435

Xiaojun Liang, Jing Shao, and Shiva Kumar

17.1 Introduction 435

17.2 Optical back propagation using optical phase
conjugation 437

17.3 Optical back propagation with optimal step size 442

17.4 Ideal optical backpropagation using
dispersion-decreasing fiber 450

17.5 Conclusion 456

Chapter 18 Eigenvalue communications in nonlinear fiber channels 459

Jaroslaw E. Prilepsky and Sergei K. Turitsyn

18.1 Introduction and main model description 461

18.2 Nonlinear Fourier transform associated with NLSE ... 464

18.2.1   Forward nonlinear Fourier transform
(Zakharov–Shabat direct scattering prob-
lem) for the focusing NLSE 465

18.2.2   Modification of the FNFT for the normal
dispersion case 467

18.2.3   Backward nonlinear Fourier transform
(Gelfand–Levitan–Marchenko equation) 467

18.2.4   Some remarks on numerical methods for
computing NFT and associated complexity . 468

18.3 Transmission using continuous nonlinear spectrum
— Normal dispersion case 469

18.4 Method of nonlinear and linear spectra equalization
for low energy signals. Anomalous dispersion 472

18.4.1   Nonlinear spectrum expansions for low
signal amplitude 473

18.4.2   Linear and nonlinear spectra equalization
using signal pre-distortion 474
18.4.3   Illustration of the method 475
18.4.3.1 Optical frequency division
multiplexing (OFDM) modulation 475
18.4.3.2 Spectra equalization for OFDM
input signals 476
18.5 Nonlinear inverse synthesis (NIS)
method—Anomalous dispersion 478
18.5.1   General idea of the method 478
18.5.2   Illustration of the method 480
18.5.2.1 Synthesis of profiles from some
characteristic shapes in the
nonlinear spectral domain 480
18.5.2.2 NIS for high-efficiency OFDM
transmission—Comparison with
digital backpropagation 480
18.6 Conclusion.. 486

Chapter 19   Digital coherent technology-based eigenvalue modulated
optical fiber transmission system 491

**Akihiro Maruta, Yuki Matsuda, Hiroki Terauchi,
and Akifumi Toyota**

19.1 Introduction 491
19.2 Principle of eigenvalue demodulation......................... 492
19.3 Numerical demonstration of eigenvalue modulated
transmission 493
19.4 Experimental demonstration of eigenvalue modu-
lated transmission 493
19.5 Noise tolerance of eigenvalues 498
19.6 Conclusion.. 504

Chapter 20   Quantum field theory analog effects in nonlinear
photonic waveguides.. 507

Andrea Marini and Fabio Biancalana

20.1 Optical analog of relativistic Dirac solitons in
binary waveguide arrays... 507
20.1.1   Introduction 507
20.1.2   Analytical soliton solutions 508
20.1.3   Soliton propagation and generation............. 512
20.1.4   Dirac solitons 515
20.1.5   Conclusion 516
20.2 Optical analog of spontaneous symmetry breaking
and tachyon condensation in plasmonic arrays.......... 516

20.2.1   Vacuum expectation value and nonlinear tachyon-like Dirac equation............................518

20.2.2   Spontaneous symmetry breaking..................519

20.3   Optical analog of neutrino oscillations in binary waveguide arrays..520

20.3.1   Dirac limit: Neutrinos 522

20.3.2   Neutrino oscillations.....................................523

20.4   Negative frequencies in nonlinear optics 525

20.4.1   Introduction 525

20.4.2   Existence and reality of negative frequencies in optics...527

20.4.3   Derivation of the envelope equation for the analytic signal from the unidirectional pulse propagation equation 529

20.4.4   Phase-matching conditions between soliton and radiation...533

20.4.5   Numerical simulations 535

20.4.6   Discussion and conclusion 537

Index.. 545

Foreword

Optical fiber waveguides provide very rich media for research and applications in modern optics because of the combination of small (group) dispersion and matching nonlinearity coupled with impressively small dissipation. Light waves in fiber waveguides propagate three-dimensionally but the energy or information represented by the envelope of the light waves propagates one-dimensionally in the direction of the fiber. Thus the dynamical evolution of information in fiber waveguides can be treated in one space

Dr. Akira Hasegawa *wearing the hat sent by Dr. V. N. Serkin for his 80th birthday.*

dimension and time coordinates. The evolution equation including cubic nonlinearity and group dispersion (without loss), called the nonlinear Schrödinger equation, is integrable and the characteristic solution is given by (optical) solitons. In the more than 40 years since the discovery of optical solitons in fibers by Hasegawa and Tappert, impressive progress has been made both in theory and applications of solitons and related phenomena, much of which is summarized in this book.

The primary nonlinear response in fibers is the (electronic) Kerr effect, which increases the index of refraction with instantaneous response. However, its effect is very small; it induces a change in the dielectric constant on the order of 10^{-22} (V/m)^2, resulting in an increase of the index of refraction on the order of 10^{-12} for a lightwave with a power of 1 mW. Because the effect is so small, few people noticed the importance of this nonlinear effect. However, it should be recognized that this small change in the index of refraction appears in every wavelength of propagation, which is about 10^{-6} m. Thus it begins to become significant for a practical distance of propagation of a few kilometers even for a lightwave with a milliwatt level of power. Coupled with this, fiber (group) dispersion has been reduced and combined effects of nonlinearity and dispersion have become comparable. Furthermore, development of optical amplifiers has made fiber waveguides practically lossless. Thus the nonlinear Schrödinger equation is now regarded as the master equation for transmission of information in fiber waveguides. Since the nonlinear Schrödinger equation is integrable based on the inverse scattering transform of Zakharov and Shabat, information carried in nonlinear and dispersive optical fibers is robust like Fourier modes in linear systems. In a linear transmission system, the output signal in Fourier space is given by the product of the Fourier transform of the input signal and of the transfer function; thus the input signal can be precisely recovered by inverse Fourier transformation of the product. The integrability of the nonlinear Schrödinger equation in principle warrants the same procedure and all the information should be recoverable, even when the transmission system is nonlinear, where Fourier information is destroyed. More concretely, the complex eigenvalues (both amplitude and phase) in the inverse scattering transform are preserved in the transmission, like Fourier components in a linear system, and thus should in principle be recoverable with proper reverse transformation of the output signal. In this regard coherent transmission is possible in nonlinear optical fiber; namely, using a soliton for one bit of information, although it is robust, is a waste of the channel capacity. Only recently this fact began to be recognized and used for ultra-high speed transmission. This will truly open a new era of nonlinear communication systems where the Fourier transform becomes irrelevant. In this regard, the soliton-based transmission concept in optical waveguides has just begun its new paradigm.

Soliton-related phenomena in optical fiber waveguides have wide applications, including modulational instability, pulse compression and generation of a continuous spectrum and the like. New theoretical developments of integrability of the modified nonlinear Schrödinger equation, including nonautonomous cases by Serkin et al.,* have opened up a new paradigm of the soliton concept to nonlinear quantum field theory in addition to a new information transfer concept. In this respect, the soliton concept in fiber waveguides is still on the way to full blossom.

Dr. Akira Hasegawa
Professor Emeritus
Osaka University Suita, Osaka, Japan

---

*V. N. Serkin, A. Hasegawa, and T. L. Belyaeva. Phys. Rev. Lett. 85 (2007)

Preface

Optical fiber technology has emerged as one of the prominent technologies of the century as it encompasses a wide range of applications—from medical optics to all optical information processing. Extensive research carried out in the field of optical fiber communications has revolutionized the fiber optic industry by paving the way for the discovery of new laser materials and optical fibers with new designs and configurations. In particular, after the invention of lasers, the role of nonlinear optical effects and their applications in different nonlinear materials have been extensively used by the scientific community as well as industry. Among different nonlinear materials, optical fiber is a widely used medium for many important applications, including ultra-high-speed communication networks, pulse compression, erbium-doped fiber amplifiers, etc. Further, the study of optical solitons in nonlinear materials has also attracted a lot of attention due to its potential applications in many branches of science and technology, including the generation of supercontinuum sources through soliton fission and modulational instability. The previously mentioned ventures, in turn, have opened the door to new technical possibilities and exploration of new opportunities for the commercialization of fiber optic deliverables. As an example, appropriate dispersion engineering of highly nonlinear photonic crystal fibers has resulted in the development of a special class of dispersion compensating fibers that exhibit interesting phenomena such as supercontinuum generation. As the supercontinuum light source is characterized by its ability to provide ultra-wide band and high coherence, its applications have spread over a wide range of fields, encompassing telecommunication, medical imaging, spectroscopy, optical frequency metrology, gas sensors, etc.

Conventional silica core photonic crystal fibers are the best contenders for telecom applications as the carrier wavelength can be shifted to 1.55 micrometers where loss is at a minimum. Some important applications, such as industrial welding, surgery, sensing, etc., demand higher operating wavelengths. In these cases, photonic crystal fibers having core materials that provide minimum losses at higher operating wavelengths are usually preferred. This has opened the gateway to study the performance of potential core materials such as chalcogenide glasses, heavy metal fluoride glasses, polycrystalline materials, etc., in order to explore the possibilities of obtaining optimal higher operating wavelengths. Special classes of photonic crystal fibers that have hollow cores are most sought for potential applications in the fields of medicine, telecommunication, gas cells, remote sensing, etc., as they provide much smaller absorption, nonlinearity, and material dispersion than their solid-core counterparts. Due to the high performance characteristics of hollow-core photonic band gap fibers, a plethora of opportunities have been opened up for minimally invasive laser surgery. Other than the conventional microhole photonic crystal fibers, macrohole microstructured fibers, whose air holes or other cladding structures that extend in the axial direction have been studied extensively as they can be manufactured easily, thereby leading to cost-sensitive applications. No wonder that multinational companies such as Alcatel-Lucent, ATT, etc., have invested heavily in optical fiber technology. This calls for a conglomeration of the core concepts and applications of nonlinear optical fiber technology where the unique optical properties, ranging from optical solitons to optical rogue waves, are contemplated, thereby acting as a tribute to the odyssey of light.

This is a collection of breakthrough research work that portrays the odyssey of light from optical solitons to optical rogue waves in nonlinear optical fibers, exploring the very frontiers of light-wave technology. Furthermore, this book also provides a simple yet holistic view on the theoretical and application-oriented aspects of light, with special focus on the underlying nonlinear phenomena. The reader is exposed to some of the latest advances in nonlinear optical fiber technology, ranging from optical solitons to optical rogue waves. In a nutshell, the contents of this book are as follows: The basics of nonlinear fiber optics are discussed at length in Chapter 1. In Chapter 2, the dynamics of electromagnetic pulse propagation in nonlinear waveguides are discussed in detail. Employing the coupled mode dynamics approach, nonlinear propagation in various continuous and discrete systems encompassing parity-time optical coupled systems, binary arrays, dual-core photonic crystal fiber and fiber amplifiers are discussed in detail in Chapter 3. The possibility of the existence of Ablowitz–Ladik rogue wave hierarchies is investigated at length in Chapter 4. In Chapter 5, a detailed theoretical study on modulational instability in relaxing saturable nonlinear optical media is carried out. The conditions for the occurrence of various types of modulational instabilities in two twin-core fibers that are governed by coupling coefficient dispersion are discussed in Chapter 6. By employing the generalized non-isospectral Ablowitz–Kaup–Newell–Segur (AKNS) hierarchies, the nonlinear dynamics of solitons in various nonautonomous nonlinear and dispersive physical systems, whose properties are completely determined by hidden symmetry parameters, are studied extensively in Chapter 7. Various types of soliton propagation nonlinear fiber optics that encompass isothermic-, hypothermic-, hyperthermic- and hybrid solitons are discussed in detail in Chapter 8. In Chapter 9, a compact review of theoretical results that throw light on the prediction of a generic method in supporting stable spatial solitons in dissipative optical media is carried out. The frontiers of mode locking using fiber lasers are explored in detail in Chapter 10, mainly aided by the development of advanced statistical tools on one hand and the technological improvements of the characterization methods and devices on the other hand. An extensive study on matter wave solitons and other localized excitations pertaining to Bose–Einstein condensates in atom optics is carried out in Chapter 11. An in-depth study based on symmetric soliton propagation in nonlocal nonlinear media where pertaining to linear Gaussian and periodic potentials is carried out in Chapter 12. The theoretical and experimental procedures for the generation of dual radiation in suspended photonic crystal fibers are studied extensively in Chapter 13. In Chapter 14, the theory and the experiment of parabolic similariton propagation in nonlinear optical fibers and their subsequent applications that encompass pulse amplification, coherent continuum for optical communications, ultrafast all-optical signal processing and spectral compression are studied extensively. Chapter 15 discusses the theoretical concepts, experimental techniques, and applications of Brillouin scattering. Various aspects of nonlinear effects and their ensuing propagation occurring in optical metamaterials owing to forward and backward wave interactions are discussed in detail in Chapter 16. Backpropagation, an effective technique to compensate for deterministic dispersive and nonlinear effects in optical fibers, is discussed in detail in Chapter 17. Chapter 18 dwells on the concept of eigenvalue communication, a powerful nonlinear digital signal processing technique that paves the way to overcome current limitations of traditional communications methods in nonlinear fiber channels. The feasibility of the eigenvalue demodulation scheme based on digital coherent technology by throwing light on the experimental study of the noise tolerance of the demodulated eigenvalues is the subject of Chapter 19. Finally, Chapter 20 discusses in an extensive manner the various quantum field theory analogue effects occurring in binary waveguide arrays, plasmonic arrays, etc., and their ensuing nonlinear wave propagation.

We hope that this book serves as a commemoration of the International Year of Light (2015).

Kuppuswamy Porsezian
Ramanathan Ganapathy

Acknowledgments

First and foremost, the editors acknowledge Dr. Akira Hasegawa, the father of soliton communication systems, for his valuable support and advice throughout the preparation of this book. We are thankful to Dr. M. Lakshmanan, the father of nonlinear dynamics in India, for his constant encouragement. We also express our sincere gratitude to all the contributing authors of this book for having charted the road map for *Odyssey of Light in Nonlinear Optical Fibers*, and without whose valuable contributions this book would not have materialized. The editors acknowledge Dr. Ajoy Ghatak and Dr. K. Thyagarajan for providing permission for the front cover from the Fianium, United Kingdom. We also acknowledge the Fianium, United Kingdom, for the same. CRC Press deserves a special acknowledgment of their editorial and production departments for the valuable support rendered by their capable team.

Dr. K. Porsezian fondly remembers the valuable support of his students Dr. T. Uthayakumar, K. Nithyanandan, T. Mithun, S. Sabari, A. K. Shafeeque Ali, Tamil Thiruvalluvar, Monisha Kumar and Ishfaq Ahmad Bhat toward the consolidation of this book.

Dr. R. Ganapathy acknowledges his spiritual guru, Dr. K. Kamakshi, for showing him the ways of life. He acknowledges Dr. R. Sethuraman, the hon. vice chancellor of SASTRA University, for providing valuable support in his research endeavor. He is very much indebted to Dr. K. Porsezian for giving him the opportunity to be one of the editors of this book. Dr. R. Ganapathy greatfully acknowledges the valuable help rendered by his students Ganapathy Joshva Raj, Balasubramanian Sai Venkatesh, Bommepalli Madhava Reddy, and Thadigotla Venkata Subba Reddy, and his colleague M. Easwaran.

Last but not least, the editors acknowledge the Almighty for clearing all the obstacles in their odyssey in the preparation of this book.

Editors

Kuppuswamy Porsezian, PhD, earned his MSc degree in physics from the University of Madras, Chennai, India, in 1985, and his PhD degree in physics from Bharathidasan University, Tiruchirapalli, India, in 1991. After working as a research scientist with the SERC, Department of Science and Technology (DST), Government of India Project, in 1993, he joined the Department of Physics, Anna University, Chennai, India as a lecturer. He is

Dr. K. Porsezian

currently a professor with the Department of Physics, Pondicherry University, Puducherry, India. He has published more than 175 papers in international journals and edited five books on optical solitons. His current research interests include solitons and modulational instability in nonlinear fiber optics, self-induced transparency solitons, nonlinear pulse propagation in periodic structures, metamaterials and photonic crystal fibers, and integrability aspects of nonlinear partial differential equations.

Dr. Porsezian received the Indian National Science Academy (INSA) Young Scientists Award of the Year 1995, the Deutscher Akademischer Austauschdienst (DAAD) Post-Doctoral Fellowship from 1995 to 1997, the Anil Kumar Bose Memorial Award by the INSA in 1998, the Sathya Murthy Memorial Award of the Indian Physics Association for the Year 1998, Junior and Regular Associateship Award from 1995 to 2002, the All-India Council of Technical Education Career Award for Young Teachers from 1998 to 2001, the University Grants Commission Research Award from 2004 to 2007, and the Department of Science and Technology Ramanna Fellowships from 2006 to 2009. He was a program advisory committee member (2004–2012) and is a member of the Fund for Improvement of Science and Technology Infrastructure (FIST) (2008–present) of the DST, Government of India. He was also a member of the New Millennium Indian Technology Leadership Initiative committee of CSIR, received an incoming fellowship from the European Union to visit Gdansk University, Poland (2010). He was elected a fellow of the Indian Academy of Sciences Bangalore (2012) and the National Academy of Sciences, Ahmadabad (2013).

Ramanathan Ganapathy, PhD, earned a MSc degree in physics from the University of Hyderabad in 1993 and the M.Phil. in physics from Cochin University of Science and Technology in 1996. He pursued his research at the same university and earned his PhD degree in physics in 2003. He worked as a post-doctoral fellow for three years in the CSIR-sponsored project Nonlinear Dynamics of Femtosecond Pulse Propagation in Nonlin-

Dr. R. Ganapathy

ear Fibers at Pondicherry University. Presently he is a senior assistant professor at the Centre for Nonlinear Science and Engineering, School of Electrical and Electronics Engineering, SASTRA University, Tamilnadu, India.

Contributors

S. Abdoulkary
Département des Sciences Physiques
Ecole Normale Supérieure
Université de Maroua
Maroua, Cameroon

Alejandro Aceves
Department of Mathematics
Southern Methodist University
Dallas, Texas

N. Akhmediev
Optical Sciences Group
Research School of Physics and
    Engineering
The Australian National University
Canberra, Australia

Alessandro Alberucci
Nonlinear Optics and
OptoElectronics Lab (NooEL)
Rome, Italy

A. Ankiewicz
Optical Sciences Group
Research School of Physics and
    Engineering
The Australian National University
Canberra, Australia

T. L. Belyaeva
Benemerita Universidad Autonoma
    de Puebla
Instituto de Ciencias
Puebla, Mexico

Shyamal K. Bhadra
CSIR-Central Glass & Ceramic
Research Institute
Kolkata, India

Fabio Biancalana
School of Engineering and Physical
    Sciences
Heriot–Watt University
Edinburgh, Scotland, United
    Kingdom

Finot Christophe
Laboratoire de Physique de
l'Université de Bourgogne
Dijon, France

P. Tchofo Dinda
Laboratoire Interdisciplinaire Carnot
    de Bourgogne (ICB)
Dijon, France

E. Tchomgo Felenou
Laboratoire Interdisciplinaire Carnot
    de Bourgogne (ICB)
Dijon, France

Ajoy Ghatak
Department of Physics
Indian Institute of Technology Delhi
New Delhi, India

Debashri Ghosh
XLIM Research Institute
University of Limoges
Limoges, France

Philippe Grelu
Laboratoire Interdisciplinaire Carnot
    de Bourgogne (ICB)
Dijon, France

A. Hasegawa
Osaka University Suita
Osaka, Japan

Victor Lambin Iezzi
Department of Engineering Physics
École Polytechnique Montreal
Montreal, Canada

Chandroth P. Jisha
Centro de Física do Porto
Faculdade de Ciências
Universidade do Porto
Porto, Portugal

Raman Kashyap
Department of Engineering Physics
École Polytechnique de Montreál
Montreal, Canada

D. J. Kedziora
Optical Sciences Group
Research School of Physics and
    Engineering
The Australian National University
Canberra, Australia

T. G. Kofane
Laboratory of Mechanics
Department of Physics
Faculty of Science
University of Yaounde I
Yaounde, Cameroon

Shiva Kumar
Department of Electrical and
    Computer Engineering
McMaster University
Ontario, Canada

Mateusz Kuszner
Gdansk University of Technology
Gdansk, Poland

M. Lakshmanan
Centre for Nonlinear Dynamics
Bharathidasan University
Tamilnadu, India

Sergey Leble
Gdansk University of Technology
Gdansk, Poland

Xiaojun Liang
Department of Electrical and
    Computer Engineering
McMaster University
Ontario, Canada

Sébastien Loranger
Department of Engineering Physics
École Polytechnique de Montreal
Montreal, Canada

Andrei I. Maimistov
National Nuclear Research
University MEPh I
Moscow Institute for Physics and
    Technology
Moscow, Russia

Boris A. Malomed
Department of Physical Electronics
School of Electrical Engineering
Faculty of Engineering
Tel Aviv University
Tel Aviv, Israel

Andrea Marini
Max Planck Institute for the Science
    of Light
Erlangen, Germany

Akihiro Maruta
Osaka University Suita
Osaka, Japan

Yuki Matsuda
Osaka University Suita
Osaka, Japan

A. Mohamadou
Department of Physics
Faculty of Science
University of Maroua
Maroua, Cameroon

P. Muruganandam
Department of Physics
Bharathidasan University
Tamilnadu, India

C. M. Ngabireng
Laboratoire Interdisciplinaire Carnot
 de Bourgogne (ICB)
Dijon, France

K. Nithyanandan
Department of Physics
Pondicherry University
Puducherry, India

Kuppuswamy Porsezian
Department of Physics
Pondicherry University
Puducherry, India

Jaroslaw E. Prilepsky
Aston Institute of Photonic
 Technologies
Aston University Birmingham
United Kingdom

Samudra Roy
Department of Physics
Indian Institute of Technology
Kharagpur, India

V. N. Serkin
Benemerita Universidad Autonoma
 de Puebla
Instituto de Ciencias
Puebla, Mexico

Jing Shao
Department of Electrical and
 Computer Engineering
McMaster University
Ontario, Canada

Boscolo Sonia
School of Engineering and Applied
 Science
Aston University
Birmingham, United Kingdom

H. Tagwo
Laboratory of Mechanics
Department of Physics
Faculty of Science
University of Yaounde I
Yaounde, Cameroon

Hiroki Terauchi
Osaka University Suita
Osaka, Japan

K. Thyagarajan
Department of Physics
Indian Institute of Technology Delhi
New Delhi, India

C. G. Latchio Tiofack
Laboratory of Mechanics
Department of Physics
Faculty of Science
University of Yaounde I
Yaounde, Cameroon

Akifumi Toyota
Osaka University Suita
Osaka, Japan

Sergei K. Turitsyn
Aston Institute of Photonic
 Technologies
Aston University
Birmingham, United Kingdom

1 Basic nonlinear fiber optics

K. Thyagarajan and *Ajoy Ghatak*

1.1 INTRODUCTION

With the development of extremely low loss optical fibers and their application to communication systems, a revolution has taken place over the last 35
years. With technological developments in the area of low loss optical fibers
with optimum characteristics, optical fiber amplifiers and wavelength division
multiplexing techniques in 2001, fiber optic communication at a rate of more
than 1 terabit per second (which is roughly equivalent to the transmission of
about 15 million simultaneous telephone conversations) through one hair-thin
optical fiber was achieved. With further advances in modulation formats and
other multiplexing schemes, transmission at the rate of 14 terabits per second
over a 160 km long single fiber was demonstrated in 2006, which is equivalent
to sending 140 digital high definition movies in one second.

With increased optical power propagating through the optical fibers, the
optical intensities within the fiber cause nonlinear optical effects, which cause
problems in communication systems. Understanding the nonlinear effects, developing techniques to overcome their limitations, and also using some of the
nonlinear effects advantageously for amplification or signal distortion correction, have become extremely important. In this chapter we discuss the propagation characteristics of single-mode optical fibers and the various nonlinear
optical effects that take place within the fibers.

1.2 MODES OF A STEP-INDEX FIBER

The propagation of light through an optical fiber can be analyzed using modal
analysis. Here we describe light propagation in terms of modes of propagation
that propagate through the optical fiber without any change in their transverse
field distribution.

In order to understand the concept of modes, we consider the simplest
refractive index distribution, namely, the step index optical fiber, which is
characterized by the following refractive index distribution:

$$
\begin{aligned}
n(r) &= n_1 \quad 0 < r < a \quad \text{core,} \\
&= n_2 \quad r > a \quad \text{cladding,}
\end{aligned}
\tag{1.1}
$$

where we are using the cylindrical system of coordinates (r, ϕ, z), z being
along the axis of the optical fiber, which is also the direction of propagation.

In actual fibers the fractional index difference between the core and cladding is much less than unity:

$$\Delta \equiv \frac{n_1 - n_2}{n_2} \leq 0.01, \tag{1.2}$$

and this allows the use of the so-called scalar wave approximation (also known as the weakly guiding approximation. For more details about the weakly guiding approximation, see, e.g., [1] and [2]). In this approximation, the modes of propagation are assumed to be nearly transverse and having an arbitrary state of polarization. Thus, the two independent sets of modes can be assumed to be x-polarized and y-polarized, and in the weakly guiding approximation they have the same propagation constants. These are usually referred to as LP modes (linearly polarized modes).

In the weakly guiding approximation, the transverse component of the electric field (E_x or E_y) is given by

$$\psi(r, \phi, z, t) = \psi(r, \phi)e^{i(\omega t - \beta z)}, \tag{1.3}$$

where ω is the angular frequency and β is known as the propagation constant. The above equation defines the modes of the system. Since $\psi(r, \phi)$ depends only on the transverse coordinates r and ϕ, *the modes represent transverse field configurations that do not change as they propagate through the optical fiber except for a phase change.*

The dependence of the transverse field on the spatial coordinates and time is given by

$$\psi(r, \phi, z, t) = R(r)e^{i(\omega t - \beta z)} \left\{ \begin{array}{c} \cos l\phi \\ \sin l\phi \end{array} \right\} ; l = 0, 1, 2 \ldots, \tag{1.4}$$

where $R(r)$ satisfies the radial part of the equation

$$r^2 \frac{d^2 R}{dr^2} + r\frac{dR}{dr} + \{[k_0^2 n^2(r) - \beta^2]r^2 - l^2\}R = 0, \tag{1.5}$$

where

$$k_0 = \omega/c, \tag{1.6}$$

is the free space wave number and

$$c = \left(\frac{1}{\sqrt{\epsilon_0 \mu_0}}\right) \approx 3 \times 10^8 \text{m/s} \tag{1.7}$$

is the speed of light in free space. In fact, Eqs. (1.3–1.5) are valid as long as n^2 depends only on the cylindrical coordinate r, which is true for most practical fibers. In fact, for an arbitrary cylindrically symmetric profile having a refractive index that decreases monotonically from a value n_1 on the axis to a constant value n_2 beyond the core-cladding interface $r = a$, we can make the following general observations about the solutions of Eq. (1.5):

Since for each value of l there can be two independent states of polarization, modes with $l > 1$ are four-fold degenerate (corresponding to two orthogonal polarization states and to the ϕ dependence being $\cos l\phi$ or $\sin l\phi$). Modes with $l = 0$ are ϕ independent and have two-fold degeneracy. Further, the solutions of Eq. (1.5) can be divided into two distinct classes: The first class of solutions corresponds to

$$n_2^2 < \frac{\beta^2}{k_0^2} < n_1^2 \quad \text{Guided modes.} \tag{1.8}$$

For β^2 lying in the above range, the field $R(r)$ is oscillatory in the core and decay in the cladding and β^2 assumes only discrete values; these are known as the *guided* modes of the waveguide. For a given value of l, there will be a finite number of guided modes; these are designated as LP_{lm} modes ($m = 1, 2, 3, $). The second class of solutions corresponds to

$$\beta^2 < k_0^2 n_2^2, \quad \text{Radiation modes.} \tag{1.9}$$

For such β values, the field is oscillatory even in the cladding and β can assume a continuum of values. These are known as the *radiation modes*. We will give below the complete modal solutions for a step index fiber.

1.3   GUIDED MODES OF A STEP-INDEX FIBER

For a step index fiber, an analytical solution to the wave equation in the core and cladding regions can be obtained in terms of the Bessel functions, and the modal field is given by (using continuity of fields at the core-cladding interface)

$$\psi(r,\phi) = \begin{cases} \dfrac{A}{J_l(U)} J_l\left(\dfrac{Ur}{a}\right) \left(\begin{matrix} \cos l\phi \\ \sin l\phi \end{matrix} \right) ; r < a, \\[4mm] \dfrac{A}{K_1(W)} K_l\left(\dfrac{Wr}{a}\right) \left(\begin{matrix} \cos l\phi \\ \sin l\phi \end{matrix} \right) ; r > a, \end{cases} \tag{1.10}$$

where

$$U \equiv a\sqrt{k_0^2 n_1^2 - \beta^2}, \tag{1.11}$$

and

$$W \equiv a\sqrt{\beta^2 - k_0^2 n_2^2}, \tag{1.12}$$

$$V = \sqrt{U^2 + W^2} = k_0 a\sqrt{n_1^2 - n_2^2} = \frac{2\pi}{\lambda_0} a\sqrt{n_1^2 - n_2^2}. \tag{1.13}$$

The waveguide parameter V is an extremely important quantity characterizing an optical fiber. It is convenient to define the normalized propagation constant

$$b = \frac{\frac{\beta^2}{k_0^2} - n^2}{n_1^2 - n_2^2} = \frac{W^2}{V^2}. \tag{1.14}$$

Thus

$$W = V\sqrt{b}, \tag{1.15}$$

and

$$U = V\sqrt{1 - b}. \tag{1.16}$$

For guided modes, $0 < b < 1$. Continuity of $\partial\psi/\partial r$ at $r = a$ and use of identities involving Bessel functions (see, e.g., [1, 3, 4]) give us the following transcendental equations which determine the allowed discrete values of the normalized propagation constant b of the guided LP_{lm} modes:

$$V(1 - b)^{1/2}\frac{J_{l-1}[V(1-b)^{1/2}]}{J_l[V(1-b)^{1/2}]} = -Vb^{1/2}\frac{K_{l-1}[Vb^{1/2}]}{K_l[Vb^{1/2}]}; l \geq 1, \tag{1.17}$$

and

$$V(1 - b)^{1/2}\frac{J_1[V(1-b)^{1/2}]}{J_0[V(1-b)^{1/2}]} = Vb^{1/2}\frac{K_1[Vb^{1/2}]}{K_0[Vb^{1/2}]}; l = 0. \tag{1.18}$$

For a given value of l and V, the above two equations represent transcendental equations the solutions of which will give allowed values of b. thus we will obtain universal curves describing the dependence of b(and therefore of U and W) on V. For a given value of l,there will be a finite number of solutions and the mth solution ($m = 1,2,3,...$) is referred to as the LP_{lm} mode. The variation of b with V form a set of universal curves, which are plotted in Fig. 1.1. The field patterns for some low order modes are shown in Fig. 1.2. The phase and group velocities of the mode are given by

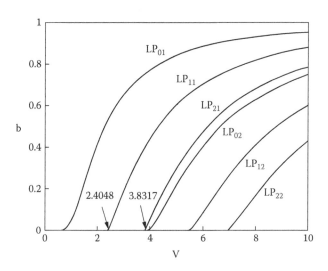

FIGURE 1.1: Variation of normalized propagation constant b with V for a step index fiber.

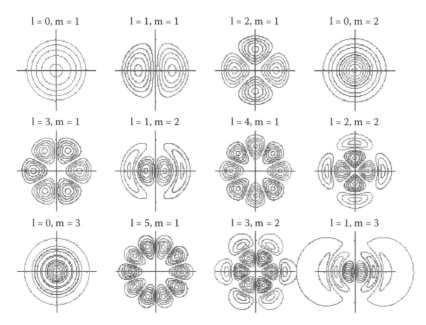

FIGURE 1.2: Transverse electric field patterns of some low-order guided modes (after http://www.rp-photonics.com/fibers.html, Rutger Paschotta, Date of last access: January 5, 2007).

$$v_p = \frac{\omega}{\beta} = \frac{c}{n_{eff}(\omega)} \quad and \quad v_g = \left(\frac{d\beta}{d\omega}\right)^{-1}, \qquad (1.19)$$

where

$$n_{eff} = \frac{\beta}{k_0}, \qquad (1.20)$$

is referred to as the effective index of the mode and k_0 represents the free space propagation constant defined earlier.

Solutions of the eigenvalue equations (1.17) and (1.18) would give us the variation of the effective index with free space wavelength from which we can obtain the wavelength dependence of the phase velocity and group velocity of the various modes. Using the value of the effective index in Eq. (1.10) we can obtain the exact transverse field profile of the modes. All these parameters are extremely important for understanding and optimizing the propagation characteristics of fibers.

1.4 SINGLE-MODE FIBER

As is obvious from Fig. 1.1, for a step index fiber with $0 < V < 2.4048$ we will have only one guided mode namely, the LP_{01}, mode, which is the fundamental mode. Such a fiber is referred to as a single-mode fiber and is of tremendous importance in optical fiber communication systems. *From now on, we will consider only single-mode fibers.*

As an example, we consider a step index fiber with $n_2 = 1.447$, $\Delta = 0.003$ and $a = 4.2$ μm, giving $V = 2.958/\lambda_0$, where λ_0 is measured in μm. Thus for $\lambda_0 > 1.23$ μm the fiber will be of single-mode. The wavelength for which $V = 2.4045$ is known as the *cut-off wavelength* and is denoted by λ_c. In this example, $\lambda_c = 1.23$ μm.

Although we can obtain the exact value of b for any given V by solving the eigenvalue equation, many times it is advantageous to use the following convenient empirical formula $b(V)$ for single-mode step index fibers:

$$b(V) = \left(A - \frac{B}{V} \right)^2 \; ; 1.5 \leq V \leq 2.5, \tag{1.21}$$

where $A \approx 1.1428$ and $B \approx 0.996$. Similar empirical formulae can be created for graded index fibers.

1.4.1 SPOT SIZE OF THE FUNDAMENTAL MODE

The transverse field distribution associated with the fundamental mode of a single-mode fiber is an extremely important quantity and it determines various important parameters like splice loss at joints between fibers, launching efficiencies from sources, bending loss, etc. For a step index fiber one has an analytical expression for the fundamental field distribution in terms of Bessel functions. For most single-mode fibers with a general transverse refractive index profile, the fundamental mode field distributions can be well approximated by a Gaussian function, which may be written in the form

$$\psi(x,y) = A \, e^{-\frac{x^2+y^2}{w^2}} = A \, e^{-\frac{r^2}{w^2}}, \tag{1.22}$$

where w is referred to as the spot size of the mode field pattern and $2w$ is called the mode field diameter (MFD). MFD is a very important characteristic of a single-mode optical fiber. For a step index fiber one has the following empirical expression for w (see [5]):

$$\frac{w}{a} \approx 0.65 + \frac{1.619}{V^{3/2}} + \frac{2.879}{V^6} ; 0.8 \leq V \leq 2.5, \tag{1.23}$$

where a is the core radius.

As an example, for the step index fiber considered earlier and operating at 1300 nm we have $V \approx 2.28$, giving $w = 4.8$ μm. Note that the spot size is larger than the core radius of the fiber; this is due to the penetration of the modal field into the cladding of the fiber. The same fiber will have a V value of 1.908 at $\lambda_0 = 1550$ nm, giving a value of the spot size ≈ 5.5 μm. *Thus, in general, the spot size increases with wavelength.* The standard single-mode fiber designated as G.652 fiber for operation at 1310 nm has an MFD of $9.2 \pm 0.4\mu$m and an MFD of 10.4 ± 0.8 μm at 1550 nm.

For $V \geq 10$, the number of modes (for a step index fiber) is approximately $V^2 2$ and the fiber is said to be a multimoded fiber. Different modes (in a multimoded fiber) travel with different group velocities, leading to what is known as intermodal dispersion; in the language of ray optics, this is known as ray dispersion arising due to the fact that different rays take different amounts of time in propagating through the fiber.

1.5 PULSE DISPERSION IN OPTICAL FIBERS

Most fiber optic communication systems use digital communication in which information to be sent is first coded in the form of pulses and then these pulses of light are transmitted from the transmitter to the receiver where the information is decoded. The bit rate is the number of pulses being sent per second and the larger the bit rate the larger the transmission capacity of the system. Although the pulses may be resolvable at the input of the fiber optic communication link, due to the phenomenon of pulse dispersion, the pulses of light broaden in the time domain as they propagate through the optical fiber. This pulse dispersion would limit the bit rate at which information can be transmitted through a fiber optic link of given length. In single-mode fibers pulse dispersion occurs primarily because of the following mechanisms:

1. Any given light source emits over a range of wavelengths and, because of the dependence of the refractive index on wavelength, different wavelengths take different amounts of time to propagate along the same path. This is known as material dispersion.
2. Apart from material dispersion, there is another important phenomenon referred to as waveguide dispersion, which arises due to the fact that the propagation constant (of the fundamental mode) has an explicit wavelength dependence.
3. A single-mode fiber can support two orthogonally polarized LP_{01} modes. In a perfectly circular core fiber laid along a perfectly straight path, the two polarizations propagate with the same velocity. However, due to small random deviations from circularity of the core or due to random bends and twists present in the fiber, the orthogonal polarizations travel with slightly different velocities and get coupled randomly along the length of the fiber. This phenomenon leads to polarization mode dispersion (PMD), which becomes important for high speed communication systems operating at 40 Gb/s and higher.

We will now discuss these sources of pulse dispersion.

1.5.1 MATERIAL DISPERSION

Every source of light has a certain wavelength spread which is often referred to as the *spectral width of the source*. A light emitting diode (LED) would have a spectral width of about 25 nm and a typical laser diode (LD) operating at 1300 nm would have a spectral width of about 2 nm or less. The pulse broadening (due to wavelength dependence of the refractive index) is given in terms of the material dispersion coefficient D_m (which is measured in ps/km.nm) and is defined by

$$D_m = -\frac{10^4}{3\lambda_0} \left[\lambda_0^2 \frac{d^2 n}{d\lambda_0^2} \right] \text{ ps/km nm.} \tag{1.24}$$

λ_0 is measured in μm and the quantity inside the square brackets is dimensionless. Thus D_m represents the material dispersion in picoseconds per kilometer length of the fiber per nanometer spectral width of the source.

Conventional optical fibers are made of silica and such a material is characterized by a specific dependence of refractive index on wavelength, referred to as the Sellmeier equation. Doping does alter the wavelength dependence of the refractive index, but the value of D_m is (almost) the same for *all* silica fibers. When D_m is negative, it implies that longer wavelengths travel faster than shorter wavelengths; this is referred to as normal group velocity dispersion (GVD). Similarly, a positive value of D_m implies that shorter wavelengths travel faster than longer wavelengths; this is referred to as anomalous GVD.

The specific wavelength dependence of silica is such that D_m passes through a zero at about 1270 nm, being negative below this wavelength and positive above this wavelength. The wavelength at which D_m becomes zero is referred to as the zero material dispersion wavelength.

It is because of the very low material dispersion value that optical communication systems based on silica fibers shifted their operation to around $\lambda_0 \approx 1300$ nm.

Optical communication systems in operation today use laser diodes (Usually abbreviated as LDs) operating at $\lambda_0 \approx 1550$ nm having a spectral width of about 2 nm. At this wavelength, the material dispersion coefficient is ≈ 21.5 ps/km.nm and the material dispersion $\Delta\tau_m$ would be 43 ps/km.

1.5.2 WAVEGUIDE DISPERSION

The effective index $n_{eff} = \dfrac{\beta}{k_0}$ of the mode of a single-mode fiber depends on the core and cladding refractive indices as well as the waveguide parameters (refractive index profile shape and radii of various regions). Hence n_{eff} would vary with wavelength even if the core and cladding media were assumed to be dispersionless (i.e., the refractive indices of core and cladding are assumed to be independent of wavelength). This dependence of effective index on wavelength is due to the waveguidance mechanism and gives rise to *waveguide dispersion*. Waveguide dispersion can be understood from the fact that the effective index of the mode depends on the fraction of power in the core and the cladding at a particular wavelength. As the wavelength changes, this fraction also changes. Thus, even if the refractive indices of the core and the cladding are assumed to be independent of wavelength, the effective index will change with wavelength. It is this dependence of n_{eff} that leads to waveguide dispersion.

The waveguide contribution for a step index fiber is given by (see, e.g., [4]):

$$D_w = -\frac{n_2\Delta}{c\lambda_0}V\left(\frac{d^2bV}{dV^2}\right). \tag{1.25}$$

A simple empirical expression for waveguide dispersion for step index fibers is

$$D_w = -10^7\frac{n_2\Delta}{3\lambda_0}\left[0.080 + 0.549\,(2.834 - V)^2\right] \text{ ps/km.nm}, \tag{1.26}$$

where λ_0 is measured in nanometers.

1.5.3 TOTAL DISPERSION

The total dispersion in the case of a single-mode optical fiber occurs due to a combination of material dispersion and waveguide dispersion. Indeed, it can be shown that the total dispersion coefficient D is given to a good accuracy by the sum of material D_m and waveguide D_w dispersion coefficients. The material contribution is given by Eq. (1.24) while the waveguide dispersion is given by Eq. (1.26).

In the single-mode regime, the quantity within the bracket in Eq. (1.26) is usually positive; hence the waveguide dispersion is negative. Since the sign of material dispersion depends on the operating wavelength region, it is possible that the two effects, namely, material and waveguide dispersions, cancel each other at a certain wavelength. Such a wavelength, which is a very important parameter of single-mode fibers, is referred to as the zero-dispersion wavelength, λ_{ZD}. For typical step index fibers the zero dispersion wavelength falls in the 1310 nm wavelength window.

Since the lowest loss in an optical fiber occurs at a wavelength of 1550 nm and optical amplifiers are available in the 1550 nm window, fiber designs can be modified to shift the zero dispersion wavelength to the 1550 nm wavelength window. This is possible since the waveguide dispersion is determined by the refractive index profile of the single-mode fiber and by appropriately choosing the transverse refractive index profile, it is possible to shift the zero dispersion wavelength close to 1550 nm; such fibers are referred to as dispersion shifted fibers (with zero dispersion around 1550 nm) or non-zero dispersion shifted fibers (with finite but small dispersion around 1550 nm). With proper fiber refractive index profile design it is also possible to have a flat dispersion spectrum, leading to dispersion flattened designs. Figure 1.3 gives the spectral variations of a standard single mode fiber (SMF) (with zero dispersion close to 1310 nm) while Fig. 1.4 shows the total dispersion in three standard types of fibers, namely, G.652, G.653 and G.655 fibers. The G.655 fibers have a small but finite dispersion around the 1550 nm wavelength. The small dispersion is required to avoid four wave mixing.

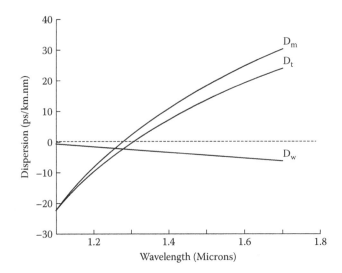

FIGURE 1.3: Variation of material, waveguide and total dispersion with wavelength for a standard single-mode fiber.

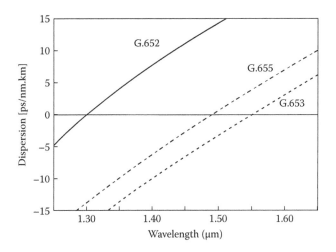

FIGURE 1.4: Variation of total dispersion for three types of single-mode fibers.

It appears that when an optical fiber is operated at the zero dispersion wavelength, the pulses will not suffer any dispersion at all. In fact, zero dispersion only signifies that the second order dispersive effects are absent. In this case the next higher order dispersion, namely, third order dispersion characterized by $d^3\beta/d\omega^3$, will become the dominating term in determining the dispersion. Thus, in the absence of second order dispersion, we can write for

TABLE 1.1

Value of dispersion and dispersion slope for some standard fibers at 1550 nm. After [6].

Fiber Type	D (ps/km-nm)	S (ps/km-nm^2)
Standard SMF (G.652)	17	0.058
LEAF (Corning)	4.2	0.085
Truewave-Reduced slope (OFS)	4.5	0.045
TeraLight (Alcatel)	8.0	0.057

dispersion suffered by a pulse

$$\Delta\tau = \frac{L(\Delta\lambda_0)^2}{2}\frac{dD}{d\lambda_0}, \tag{1.27}$$

where $S = dD/d\lambda_0$ represents the dispersion slope at zero dispersion wavelength and is measured in units of ps/km $-$ nm^2. Third order dispersion becomes important when operating close to the zero-dispersion wavelength. In the presence of only third order dispersion, the pulse does not remain symmetric. Table 1.1 lists values of D and S for some standard fibers at 1550 nm.

1.5.4   DISPERSION AND MAXIMUM BIT RATE IN SINGLE-MODE FIBERS

In a digital communication system employing light pulses, pulse broadening would result in an overlap of adjacent pulses, resulting in intersymbol interference, leading to errors in detection. Apart from this, since the energy in the pulse gets reduced within the time slot, the corresponding signal to noise ratio (SNR) will decrease. One can offset this by increasing the power in the pulses. This additional power requirement is termed the *dispersion power penalty*. Increased dispersion would imply an increased power penalty.

In order to keep the interference between adjacent bits below a specified level, the root mean square width of the dispersed pulse needs to be kept below a certain fraction of the bit period. For a 2 dB power penalty, $\epsilon \sim 0.491$ (see, e.g., [7]). Using this condition we can estimate the maximum bit rate B for a given link length L and dispersion coefficient D operating at 1550 nm as

$$B^2DL < 1.9 \times 10^5 Gb^2 \text{ ps/nm}, \tag{1.28}$$

where B is measured in Gbps, D in ps/km.nm and L in km. Thus for a bit rate of 2.5 Gb/s the maximum allowed dispersion (D.L) is approximately 30,400 ps/nm, while for a bit rate of 10 Gb/s the maximum allowed dispersion is 1900 ps/nm.

1.5.5   DISPERSION-COMPENSATING FIBERS

Most single-mode optical fibers laid in the early years were the conventional single-mode fibers operating at 1310 nm and these fibers have very low dispersion at the operating wavelength. One could significantly increase the transmission capacity of these system by operating these fibers at 1550 nm (where the loss is extremely small) and we can have the added advantage of using EDFA (erbium doped fiber amplifiers) for optical amplification in this wavelength range. However, if we operate the conventional single-mode fibers at 1550 nm, we will have a significant residual dispersion of about. Such a large dispersion would result in a significant decrease in the information carrying capacity of the communication system. On the other hand, replacing the existing conventional single-mode fibers with lower dispersion fibers would involve huge costs. As such, in recent years there has been a considerable amount of work in upgrading the installed 1310 nm optimized optical fiber links for operation at 1550 nm. This is achieved by developing fibers with very large negative dispersion coefficients, a few hundred meters to a kilometer of which can be used to compensate for dispersion accumulated over tens of kilometers of the fiber in the link.

By changing the refractive index profile, we can alter the waveguide dispersion and hence the total dispersion. Indeed, it is possible to have specially designed fibers whose dispersion coefficient (D) is large and negative at 1550 nm. These types of fibers are known as dispersion compensating fibers (DCFs). A short length of DCF can be used in conjunction with the 1310 nm optimized fiber link so as to have a small total dispersion value at the end of the link. There are a number of different fiber designs with optimized refractive index profiles which have extremely large negative dispersion coefficients and a small length of which can compensate for the accumulated dispersion of a link. Some of the important ones include depressed clad designs, W-type fiber designs, dual core coaxial designs, etc. (see [7–10]).

If $D_T(\lambda_n)$ and L_T represent the dispersion coefficient and length of the transmission fiber and $D_C(\lambda_n)$ and L_C represent the corresponding quantities of the DCF, then for achieving zero net dispersion at a chosen wavelength λ_n we must have

$$D_T(\lambda_n)L_T + D_C(\lambda_n)L_C = 0. \tag{1.29}$$

Hence, for given values of $D_T(\lambda_n)$, L_T and $D_C(\lambda_n)$, the length of the DCF required is given by Eq. (1.29), showing that $D_C(\lambda_n)$ and $D_T(\lambda_n)$ should have opposite signs. Also, the larger the value of $D_C(\lambda_n)$ the smaller would be the length of the required DCF. Since the wavelength dependence of the dispersion of the link fiber and the dispersion compensating fiber are in general different, in general the DCF would compensate for dispersion only at the design wavelength. However, in a wavelength division multiplexed system it is necessary to compensate for the accumulated dispersion of all the wavelength channels simultaneously. For this to happen, the relative dispersion slope (RDS), which is the ratio of the dispersion slope (S) and dispersion coefficient

(D) of the two fibers evaluated at the wavelength λ_n must be equal. Typically, the RDS of G.652 fiber is about 0.0034 nm^{-1} while that of a large effective area (LEAF) fiber from Corning is about 0.0202 nm^{-1}. DCFs with similar RDS values are commercially available.

1.6 NONLINEAR EFFECTS IN OPTICAL FIBERS

When a light beam with a power of 100 mW propagates through an optical fiber having an effective mode area of 50 μm^2, the corresponding optical intensity within the fiber core is about 2×10^9 W/m^2. At such high intensities, nonlinear effects in optical fibers start to influence the propagation of the light beam. The most important nonlinear effects that affect optical fiber communication systems [11] include self-phase modulation (SPM), cross-phase modulation (XPM) and four wave mixing (FWM). Stimulated Raman scattering (SRS) and stimulated Brillouin scattering (SBS) are also important nonlinear phenomena. In this section, we will discuss mainly SPM, XPM and FWM, which affect pulse propagation through optical fibers.

At high optical intensities (which correspond to high electric fields), all media behave in a nonlinear fashion. The electric polarization generated in the medium is given by the following equation:

$$P = \epsilon_0 \left(\chi \mathcal{E} + \chi^{(2)} \mathcal{E}^2 + \chi^{(3)} \mathcal{E}^3 + \cdots \right), \tag{1.30}$$

where χ represents the linear susceptibility of the medium, $\chi^{(2)}, \chi^{(3)} \dots$ are higher order susceptibilities giving rise to the nonlinear terms and $\mathcal{E}$ represents the total electric field of the propagating light waves. The second term on the right hand side is responsible for second harmonic generation (SHG), sum and difference frequency generation, parametric interactions, etc., while the third term is responsible for third harmonic generation, intensity dependent refractive index, self-phase modulation, cross-phase modulation, four wave mixing, etc. For media possessing an inversion symmetry, $\chi^{(2)}$ is zero and there is no second order nonlinear effect. Thus silica optical fibers do not possess second order nonlinearity and the first nonlinear term is the third order nonlinearity.

1.6.1 SELF-PHASE MODULATION

When a single electromagnetic wave propagates through the medium, the nonlinearity due to $\chi^{(3)}$ results in an intensity dependent refractive index given by

$$n = n_0 + n_2 I, \tag{1.31}$$

where $n_2 = 3\chi^{(3)}/4c\epsilon_0 n_0$ and n_0 is the refractive index of the medium at low intensities. It is this intensity dependent refractive index that gives rise to self-phase modulation (SPM). Due to the intensity dependent refractive index, the propagation constant of a mode becomes intensity dependent and can be written as

$$\beta_{NL} = \beta + \gamma P, \tag{1.32}$$

where

$$\gamma = \frac{k_0 n_2}{A_{eff}}; A_{eff} = 2\pi \frac{\left(\int \psi^2(r) r dr\right)^2}{\int \psi^4(r) r dr}, \tag{1.33}$$

represent the nonlinear coefficient and the nonlinear mode effective area, respectively, and β is the propagation constant of the mode at low powers.

If we assume the mode to be described by a Gaussian function, $A_{eff} = \pi w_0^2$, where w_0 is the Gaussian mode spot size. Note that the nonlinear coefficient γ of the fiber depends on the effective area of the mode; the larger the effective mode area, the smaller are the nonlinear effects. Table 1.2 gives the mode effective area of some common fiber types.

TABLE 1.2

Mode effective area of typical commercially available fibers

Fiber Type	Effective Area (μm^2)
Single-mode fiber (SMF) G652	≈ 85
Dispersion shifted fiber (DSF)	≈ 46
Non-Zero DSF (NZDSF)	≈ 52 ($D > 0$), 56 ($D < 0$) and 73
Dispersion compensating fiber (DCF)	≈ 23
Photonic crystal fiber/holey fiber	≈ 3

Since nonlinear effects depend on the power propagating in the fiber, attenuation of the lightwave plays a major role in the overall nonlinear effects suffered by the propagating light. Thus, if α represents the attenuation coefficient of the optical fiber, then the power propagating through the fiber decreases exponentially as $P(z) = P_0 e^{-\alpha z}$ where P_0 is the input power. In such a case, the phase shift suffered by an optical beam in propagating through a length L of the optical fiber is given by

$$\Phi = \int_0^L \beta_{NL} dz = \beta L + \gamma P_0 L_{eff}, \tag{1.34}$$

where

$$L_{eff} = \frac{1 - e^{-\alpha L}}{\alpha} \tag{1.35}$$

is called the effective length of the fiber. If $\alpha L \gg 1$, then $L_{eff} \sim 1/\alpha$, and if $\alpha L \ll 1$, then $L_{eff} \sim L$. For single-mode fibers operating at 1550 nm, with loss 0.25 dB/km we get $L_{eff} \sim L$ for $L << 20$ km and $L_{eff} \approx 20$ km for $L >> 20$ km.

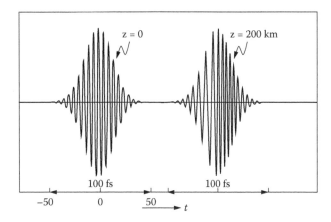

FIGURE 1.5: An unchirped pulse at the input after propagating through an optical fiber gets chirped due to self-phase modulation.

Since the propagation constant β_{NL} of the mode depends on the power carried by the mode, the phase Φ of the emergent wave depends on its power, and hence this is referred to as *self-phase modulation* (SPM).

For a light pulse P_0 in Eq.(1.34) becomes time dependent and this leads to an additional time dependent phase apart from $\omega_0 t$. Thus the output pulse is chirped and the instantaneous frequency of the output pulse is given by

$$\omega(t) = \frac{d}{dt}(\omega_0(t) - \gamma P_0 L_{eff}) = \omega_0 - \gamma L_{eff}\frac{dP_0}{dT}. \tag{1.36}$$

The leading edge of the pulse corresponds to the positive values of dP_0/dt and the trailing edge to negative values of dP_0/dt. Thus, in the presence of SPM, the leading edge gets downshifted in frequency while the trailing edge gets upshifted in frequency. The frequency at the center of the pulse remains unchanged from ω_0. Figure 1.5 shows an input unchirped and the output chirped pulse generated due to SPM. The output chirped pulse with the same temporal width has a larger frequency spectrum. These new frequencies have been generated by the nonlinear process within the fiber. By Fourier transform theory, an increased spectral width implies that the pulse can now be compressed in the temporal domain by passing it through a device with the proper sign of dispersion. This is indeed one of the standard techniques to realize ultrashort femtosecond optical pulses.

The chirping due to nonlinearity without any corresponding increase in pulse width leads to increased spectral broadening of the pulse. This spectral broadening coupled with the dispersion in the fiber leads to modified dispersive propagation of the pulse in the presence of nonlinearity. In the normal dispersion region the chirping due to dispersion is to downshift the leading edge and upshift the trailing edge of the pulse. This is of the same sign as that due to SPM. Thus, in the normal dispersion regime (wavelength less than the zero dispersion wavelength), the chirping due to dispersion and nonlinearity add. Thus, at high powers, where the nonlinear effects are not negligible, the pulse will suffer additional dispersion as compared to the dispersion of the same pulse at low powers. On the other hand, in the anomalous dispersion region (wavelength greater than the zero dispersion wavelength), the chirping due to dispersion is opposite to that due to nonlinearity, and thus, in this wavelength region, nonlinearity and dispersion induced chirpings can partially or even totally cancel each other. When total cancellation takes place, the pulse neither broadens in time nor in its spectrum and such a pulse is called a *soliton*. Such solitons can hence be used for dispersionless propagation of pulses to realize very high bit rate systems.

1.7 NONLINEAR SCHRÖDINGER EQUATION

Let $E(x, y, z, t)$ represent the electric field variation of an optical pulse. It is usual to express E in the following way:

$$E(x, y, z, t) = \frac{1}{2}[A(z, t)\psi(x, y)e^{i(\omega_0 t - \beta_0 z)} + c.c],\tag{1.37}$$

where $A(z, t)$ represents the slowly varying complex envelope of the pulse, $\psi(x, y)$ represents the transverse electric field distribution, ω_0 represents the center frequency and β_0 represents the propagation constant at ω_0 and c.c represents complex conjugate.

In the presence of attenuation, second order dispersion and third order nonlinearity, the complex envelope $A(z, t)$ can be shown to satisfy the following equation (see, e.g., [13]):

$$\frac{\partial A}{\partial z} = -\frac{\alpha}{2}A - \beta_1 \frac{\partial A}{\partial z} + i\frac{\beta_2}{2}\frac{\partial^2 A}{\partial t^2} - i\gamma|A|^2 A.\tag{1.38}$$

Here

$$\beta_1 = \left.\frac{d\beta}{d\omega}\right|_{\omega=\omega_0} = \frac{1}{v_g}\tag{1.39}$$

represents the inverse of the group velocity of the pulse, and

$$\beta_2 = \left.\frac{d^2\beta}{d\omega^2}\right|_{\omega=\omega_0} = \frac{-\lambda_0^2}{2\pi c}D,\tag{1.40}$$

where D represents the group velocity dispersion (measured in ps/km.nm). The various terms on the right hand side of Eq. (1.38) represent the following:
I term: attenuation
II term: group velocity term
III term: second order dispersion
IV term: second order nonlinearity

If we change to a moving frame defined by coordinates $T = t - \beta_1 z$, Eq. (1.38) becomes

$$\frac{\partial A}{\partial z} = -\frac{\alpha}{2}A + i\frac{\beta_2}{2}\frac{\partial^2 A}{\partial t^2} - i\gamma|A|^2. \qquad (1.41)$$

If we neglect the attenuation term, we obtain the following equation, which is also referred to as the nonlinear Schrödinger equation:

$$\frac{\partial A}{\partial z} = i\frac{\beta_2}{2}\frac{\partial^2 A}{\partial T^2} - i\gamma|A|^2 A. \qquad (1.42)$$

The above equation has a solution given by

$$A(z,t) = A_0 \mathrm{sech}(\sigma T)\, e^{-igz}, \qquad (1.43)$$

with

$$A_0^2 = -\frac{\beta_2}{\gamma}\sigma^2, g = \frac{-\sigma^2}{2}\beta_2. \qquad (1.44)$$

Equation (1.43) represents an envelope soliton and has the property that it propagates undispersed through the medium. The full width at half maximum (FWHM) of the pulse envelope will be given by $\tau_f = 2\tau_0$ where $\mathrm{sech}^2\sigma\tau_0 = \frac{1}{2}$, which gives the FWHM τ_f

$$\tau_f = 2\tau_0 = \frac{2}{\sigma}ln(1 + \sqrt{2}) \approx \frac{1.7627}{\sigma}. \qquad (1.45)$$

The peak power of the pulse is

$$P_0 = |A_0|^2 = \frac{\beta_2}{\gamma}\sigma^2. \qquad (1.46)$$

Replacing σ by τ_f, we obtain

$$P_0\tau_f^2 \approx \frac{\lambda_0^2}{2\pi c}D. \qquad (1.47)$$

The above equation gives the required peak for a given f for the formation of a soliton pulse. A heuristic derivation of the required power for the soliton formation can be found in [4].

As an example, we have $\tau_f = 10$ ps, $\gamma = 2.4\mathrm{W}^{-1}$ km^{-1}, $\lambda_0 = 1.55\ \mu$m, $D = 2$ ps/km.nm and the required peak power will be $P_0 = 33$ mW.

Soliton pulses are being extensively studied for application to long distance optical fiber communication. In actual systems, the pulses have to be optically amplified at regular intervals to compensate for the loss suffered by the pulses. The amplification could be carried out using erbium doped fiber amplifiers (EDFAs) or fiber Raman amplifiers.

1.8   SPECTRAL BROADENING DUE TO SPM

In the presence of only nonlinearity, Eq. (1.42) becomes

$$\frac{dA}{dz} = -i\gamma|A|^2)A \tag{1.48}$$

whose solution is given by

$$A(z,t) = A(z=0,t)\exp(-i\gamma P z) \tag{1.49}$$

where $P = |A|^2$ is the power in the pulse. If P is the function of time, then the time dependent phase term at $z = L$ becomes

$$exp(i\phi(t)) = \exp(i(\omega_0 t - \gamma P(t)L)). \tag{1.50}$$

We can define an instantaneous frequency as

$$w(t) = \frac{d\phi}{dt} = \omega_0 - \gamma L \frac{dP}{dt}. \tag{1.51}$$

For a Gaussian pulse,
$$P = P_0 \exp(-2T^2/\tau_0^2), \tag{1.52}$$

giving
$$w(t) = \omega_0 + \frac{4\gamma L T P_0 \exp(-2T^2/\tau_0^2)}{\tau_0^2}. \tag{1.53}$$

Thus the instantaneous frequency within the pulse changes with time, leading to chirping of the pulse (see Fig. 1.5). Note that since the pulse width has not changed, but the pulse is chirped, the frequency spectrum of the pulse has increased. Thus SPM leads to the generation of new frequencies. By Fourier transform theory, an increased spectral width implies that the pulse can now be compressed in the temporal domain by passing it through a medium with the proper sign of dispersion. This is indeed one of the standard techniques to realize ultrashort femtosecond pulses.

1.8.1   CROSS-PHASE MODULATION

Like SPM, cross-phase modulation also arises due to the intensity dependence of the refractive index, leading to spectral broadening. Consider the propagation of two lightwaves of frequencies ω_1 and ω_2. In such a case the propagating lightwave at ω_1 will lead to a change in the refractive index due to nonlinearity and this change in the refractive index will in turn affect the lightwave at frequency ω_2. Thus the lightwave at ω_1 alters the phase of the lightwave at ω_2; this is termed cross-phase modulation. The reverse process will also take place at the same time. If the signals at both frequencies are pulses, then due to the difference in the group velocities of the pulses, there is a walk off between the two pulses, i.e., if they start together, they will separate as they propagate through the medium. Nonlinear interaction takes place as long as they physically overlap in the medium. The smaller the dispersion, the smaller will be the difference in the group velocities (assuming closely spaced wavelengths) and their overlap will be over a longer length of the fiber. This would lead to stronger XPM effects. At the same time, if two pulses pass through each other, then since one pulse will interact with both the leading and the trailing edge of the other pulse, XPM effects will be nil provided there is no attenuation. In the presence of attenuation in the medium, the pulse will still get modified due to XPM.

To study XPM, we assume simultaneous propagation of two waves at two different frequencies through the medium. If ω_1 and ω_2 represent the two frequencies, then one obtains for the variation of the amplitude A_1 of the frequency ω_1 as

$$\frac{dA_1}{dz} = -i\gamma(\tilde{P}_1 + 2\tilde{P}_2)A_1, \tag{1.54}$$

where $\tilde{P}_1$ and $\tilde{P}_2$ represent the powers at frequencies ω_1 and ω_2, respectively. The first term in Eq. (1.54) represents SPM, while the second term corresponds to XPM. If the powers are assumed to attenuate at the same rate, i.e.,

$$\tilde{P}_1 = P_1 e^{-\alpha z}, \tilde{P}_2 = P_2 e^{-\alpha z}, \tag{1.55}$$

then the solution of Eq. (1.54) is

$$A_1(L) = A_1(0)e^{-i\gamma(P_1 + 2P_2)L_{eff}}, \tag{1.56}$$

where, as before, L_{eff} represents the effective length of the medium. When we are studying the effect of power at ω_2 on the light beam at frequency ω_1, we will refer to the wave at frequency ω_2 as a pump and the wave at frequency ω_1 as a probe or signal. From Eq. (1.56) it is apparent that the phase of the signal at frequency ω_1 is modified by the power at another frequency. This is referred to as XPM. Note also that XPM is twice as effective as SPM.

Similar to the case of SPM, we can now write for the instantaneous frequency in the presence of XPM

$$\omega(t) = \omega_0 - 2\gamma L_{eff}\frac{dP_2}{dt}. \tag{1.57}$$

Hence the part of the signal that is influenced by the leading edge of the pump will be down shifted in frequency (since in the leading edge $dP_2/dt > 0$) and the part overlapping the trailing edge will be up shifted in frequency (since $dP_2/dt < 0$). This leads to a frequency chirping of the signal pulse just as in the case of SPM.

If the probe and pump beams are pulses, then XPM can lead to induced frequency shifts depending on whether the probe pulse interacts only with the leading edge or the trailing edge or both as they both propagate through the medium. Let us consider a case when the group velocity of a pump pulse is greater than that of the probe pulse. Thus, if both pulses enter the medium together, then, since the pump pulse travels faster, the probe pulse will interact only with the trailing edge of the pump. Since in this case dP_2/dt is negative, the probe pulse suffers a blue-induced frequency shift. Similarly, if the pulses enter at difference instants but completely overlap at the end of the medium, then $dP_2/dt > 0$ and the probe pulse would suffer a red-induced frequency shift. Indeed, if the two pulses start separately and walk through each other, then there is no induced shift due to cancellation of shifts induced by the leading and trailing edges of the pump.

When pulses of light at two different wavelengths propagate through an optical fiber, due to different group velocities of the pulses, they pass through each other, resulting in what could be termed a collision. In the linear case, the pulses will pass through without affecting each other, but when intensity levels are high, XPM induces phase shifts in both pulses. We can define a parameter termed walk off length L_{wo}, which is the length of the fiber required for the interacting pulses to walk off relative to each other. The walk off length is given by

$$L_{wo} = \frac{\Delta\tau}{D\Delta\lambda}, \tag{1.58}$$

where D represents the dispersion coefficient and $\Delta\lambda$ represents the wavelength separation between the interacting pulses. For return to zero (RZ) pulses, $\Delta\tau$ represents the pulse duration while for non-return to zero (NRZ) pulses, $\Delta\tau$ represents the rise term or fall time of the pulse. Closely spaced channels will thus interact over longer fiber lengths, thus leading to greater XPM effects. A larger dispersion coefficient will reduce L_{wo} and thus the effects of XPM. Since the medium is attenuating, the power carried by the pulses decreases as they propagate, thus leading to a reduced XPM effect. The characteristic length for attenuation is the effective length L_{eff} defined by Eq. (1.35). If $L_{wo} << L_{eff}$, then over the length of interaction of the pulses, the intensity levels do not change appreciably and the magnitude of the XPM induced effects will be proportional to the wavelength spacing $\Delta\lambda$. For small $\Delta\lambda$âĂŹs, $L_{wo} >> L_{eff}$ and the interaction length is now determined by the fiber losses (rather than by walk off) and the XPM induced effects become almost independent of $\Delta\lambda$. Indeed, if we consider XPM effects between a continuous wave (cw) probe beam and a sinusoidally industry modulated pump beam, then the amplitude of the XPM induced phase shift $(\Delta\Phi_{pr})$ in the probe beam is given by [12]

$$\begin{aligned}
\Delta\Phi_{pr} &\approx 2\gamma P_{2m} L_{eff} \text{ for } L_{wo} \gg L_{eff}, \\
\Delta\Phi_{pr} &\approx 2\gamma P_{2m} L_{wo} \text{ for } L_{wo} \ll L_{eff}.
\end{aligned} \tag{1.59}$$

Here P_{2m} is the amplitude of the sinusoidal power modulation of the pump beam.

XPM induced intensity interference can be studied by simultaneously propagating an intensity modulated pump signal and a continuous wave probe

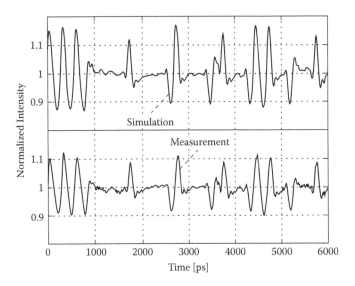

FIGURE 1.6: Intensity fluctuations induced by cross—phase modulation on a probe signal at 1550 nm by a modulated pump for a channel separation of 0.6 nm. After [14].

signal at a different wavelength. The intensity modulated signal will induce phase modulation on the cw probe signal and the dispersion of the medium will convert the phase modulation to intensity modulation of the probe. Thus the magnitude of the intensity fluctuation of the probe signal serves as an estimate of the XPM induced interference. Figure 1.6 shows the intensity fluctuations on a probe signal at 1550 nm induced by a modulated pump for a channel separation of 0.6 nm. Figure 1.7 shows the variation of the RMS value of probe intensity modulation with the wavelength separation between the intensity modulated signal and the probe. The experiment has been performed over four amplified spans of 80 km of standard single-mode fiber (SMF) and non-zero dispersion shifted fiber (NZDSF). The large dispersion in SMF has been compensated using dispersion compensating chirped gratings. The probe modulation in the case of SMF decreases approximately linearly with $1/\Delta\lambda$ for all $\Delta\lambda$s; the modulation is independent of $\Delta\lambda$. This is consistent with the earlier discussion in terms of L_{wo} and L_{eff}.

1.8.2   FOUR-WAVE MIXING

Four-wave mixing (usually abbreviated FWM) is a nonlinear interaction that occurs in the presence of multiple wavelengths in a medium, leading to the generation of new frequencies. Thus, if lightwaves at three different frequencies $\omega_2, \omega_3, \omega_4$ are launched simultaneously into a medium, the same nonlinear polarization that led to the intensity dependent refractive index leads to a

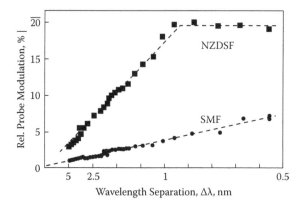

FIGURE 1.7: Variation of the RMS value of probe intensity modulation with the wavelength separation between the intensity modulated signal and the probe. After [15].

nonlinear polarization component at a frequency

$$\omega_1 = \omega_3 + \omega_4 - \omega_2. \tag{1.60}$$

This nonlinear polarization, under certain conditions, leads to the generation of electromagnetic waves at ω_1. This process is referred to as four-wave mixing due to the interaction between four different frequencies. Degenerate four wave mixing corresponds to the case when two of the input waves have the same frequency. Thus, if $\omega_3 = \omega_4$, then inputting waves at ω_2 and ω_3 leads to the generation of waves at the frequency

$$\omega_1 = 2\omega_3 - \omega_2. \tag{1.61}$$

During the FWM process, there are four different frequencies present at any point in the medium. We write the electric field of the waves as

$$E_i = \frac{1}{2}[A_i(z)\psi_i(x,y)e^{i(\omega_i\, t - \beta_i\, z)} + cc], \quad i = 1,2,3,4, \tag{1.62}$$

where as before $A_i(z)$ represents the amplitude of the wave, $\psi_i(x,y)$ the transverse field distribution and β_i the propagation constant of the wave. The total electric field is given by

$$E = E_1 + E_2 + E_3 + E_4. \tag{1.63}$$

Substituting for the total electric field in the equation for nonlinear polarization, the term with frequency ω_1 comes out to be

$$\mathcal{P}_{NL}^{(\omega_1)} = \frac{1}{2}[P_{NL}^{\omega_1}e^{i(\omega_1 t - \beta_1 z)} + cc], \tag{1.64}$$

where

$$P_{NL}^{(\omega_1)} = \frac{3\epsilon_0}{2}\chi^{(3)}A_2^*A_3A_4\psi_2\psi_3\psi_4 e^{-i\Delta\beta z}, \tag{1.65}$$

and

$$\Delta\beta = \beta_3 + \beta_4 - \beta_2 - \beta_1, \tag{1.66}$$

assuming $\omega_3 + \omega_4 = \omega_1 + \omega_2$. In writing Eq. (1.65), we have only considered the FWM term, neglecting the SPM and XPM terms. Substituting the expression for $P_{NL}^{\omega_1}$ in the wave equation for ω_1 and making the slowly varying approximation (in a manner similar to that employed in the case of SPM and XPM), we obtain the following equation for $A_1(z)$:

$$\frac{dA_1}{dz} = -2i\gamma A_2^*A_3A_4 e^{-i\Delta\beta z}, \tag{1.67}$$

where γ has been defined earlier and A_{eff} is the average effective area of the modes.

Assuming all waves to have the same attenuation coefficient α and neglecting depletion of waves at frequencies ω_2, ω_3 and ω_4, due to nonlinear conversion we obtain for the power in the frequency ω_1

$$P_1(L) = 4\gamma^2 P_2 P_3 P_4 L_{eff}^2 \eta e^{-\alpha L}, \tag{1.68}$$

where

$$\eta = \frac{\alpha^2}{\alpha^2 + \Delta\beta^2}\left[1 + \frac{4e^{-\alpha L}\sin^2\frac{\Delta\beta L}{2}}{(1 - e^{-\alpha L})^2}\right], \tag{1.69}$$

and L_{eff} is the effective length. Maximum four wave mixing takes place when $\Delta\beta = 0$, since in such a case $\eta = 1$. Now

$$\Delta\beta = \beta(\omega_3) + \beta(\omega_4) - \beta(\omega_2) - \beta(\omega_1). \tag{1.70}$$

Assuming that the frequencies lie close to each other (as happens in a WDM system), we can make a Taylor series expansion about any frequency, say ω_2. In such a case, we obtain

$$\Delta\beta = (\omega_3 - \omega_2)(\omega_3 - \omega_1)\frac{d^2\beta}{d\omega^2}|_{\omega=\omega_2}. \tag{1.71}$$

In optical fiber communication systems, the channels are usually equally spaced. Thus we assume the frequencies to be given by $\omega_4 = \omega_2 + \Delta\omega$, $\omega_3 = \omega_2 - 2\Delta\omega$ and $\omega_1 = \omega_2 - \Delta\omega$.

Using these frequencies, Eq. (1.71) gives us

$$\Delta\beta = -\frac{4\pi D\lambda^2}{c}(\Delta v)^2, \tag{1.72}$$

where $\Delta\omega = 2\pi\Delta\nu$. Thus maximum FWM takes place when $D = 0$, when operating at the zero dispersion wavelength of the fiber.

This is the main problem in using wavelength division multiplexing (WDM) in dispersion shifted fibers which are characterized by zero dispersion at the operating wavelength of 1550 nm, as FWM will then lead to cross talk among various channels. FWM efficiency can be reduced by using fibers with non-zero dispersion. This has led to the development of non-zero dispersion shifted fibers (NZDSF) which have a finite non-zero dispersion of about ± 2 ps/km-nm at the operating wavelength.

From Eq. (1.71), we notice that for a given dispersion coefficient D, FWM efficiency will reduce as $\Delta\nu$ increases.

In order to get a numerical appreciation, we consider a case with $D = 0$, i.e., $\Delta\beta = 0$. For such a case $\eta = 1$. If all channels were launched with equal power P_{in}, then

$$P_1(L) = 4\gamma^2 P_{in}^3 L_{eff}^2 e^{-\alpha L}. \tag{1.73}$$

Thus the ratio of power generated at ω_1 due to FWM and that existing at the same frequency is

$$\frac{P_g}{P_{out}} = \frac{P_1(L)}{P_{in}e^{-\alpha L}} = 4\gamma^2 P_{in}^2 L_{eff}^2. \tag{1.74}$$

Typical values are $L_{eff} = 20$ km, $\gamma = 2.4$ W^{-1} km^{-1}. Thus

$$\frac{P_g}{P_{out}} \approx 0.01 P_{in}^2 (\text{mW}^2). \tag{1.75}$$

Figure 1.8 shows the output spectrum measured at the output of a 25 km long dispersion shifted fiber ($D = -0.2$ ps/km-nm) when three 3 mW wavelengths are launched simultaneously. Notice the generation of many new frequencies by FWM. Figure 1.8 also shows the ratio of generated power to the output as a function of channel spacing $\Delta\lambda$ for different dispersion coefficients. It can be seen that by choosing a non-zero value of dispersion, the four wave mixing efficiency can be reduced. The larger the dispersion coefficient, the smaller can be the channel spacing for the same cross talk.

Since dispersion leads to increased bit error rates in fiber optic communication systems, it is important to have low dispersion. On the other hand, lower dispersion leads to cross talk due to FWM. This problem can be resolved by noting that FWM depends on the local dispersion value in the fiber while the pulse spreading at the end of a link depends on the overall dispersion in the fiber link. If one chooses a link made up of positive and negative dispersion coefficients, then by an appropriate choice of the lengths of the positive and negative dispersion fibers, it would be possible to achieve a zero total link dispersion while at the same time maintaining a large local dispersion. This is referred to as dispersion management in fiber optic systems.

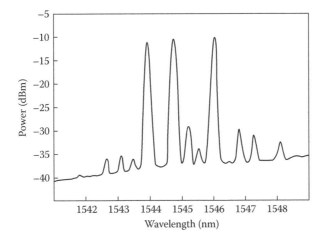

FIGURE 1.8: New frequency components generated due to four wave mixing in an optical fiber operating close to the zero dispersion wavelength. The input consists of the three large amplitude frequencies. After [16].

Although FWM leads to cross talk among different wavelength channels in an optical fiber communication system, it can be used for various optical processing functions such as wavelength conversion, high speed time division multiplexing, pulse compression, parametric amplification, etc. For such applications, there is a concerted worldwide effort to develop highly nonlinear fibers with much smaller mode areas and higher nonlinear coefficients. Some of the very novel fibers that have been developed recently include holey fibers, photonic bandgap fibers or photonic crystal fibers, which are very interesting since they possess extremely small mode effective areas (2.5 μm^2 at 1550 nm) and can be designed to have zero dispersion even in the visible region of the spectrum. This is expected to revolutionize nonlinear fiber optics by providing new geometries to achieve highly efficient nonlinear optical processing at lower powers.

1.8.3 FIBER OPTIC PARAMETRIC AMPLIFIER

In this section we will show that we can use FWM for amplifying a light-wave and such an amplifier is referred to as a fiber optic parametric amplifier (FOPA). In order to show this we consider degenerate FWM in which a strong light-wave at frequency ω_p (subscript p stands for pump) and a low amplitude wave at frequency ω_s (subscript s stands for signal) are incident simultaneously into an optical fiber. Four wave mixing among these waves can produce a new wave at frequency $\omega_i = 2\omega_p - \omega_s$; subscript i refers to idler.

Using an analysis similar to the one carried out for FWM we can obtain the following two equations describing the changes in the amplitude of the signal and idler waves:

$$\frac{dA_s}{dz} = -i\gamma A_p^2 A_i^* e^{-i\Delta\beta z},$$

$$\frac{dA_i}{dz} = -i\gamma A_p^2 A_s^* e^{-i\Delta\beta z},$$

where, to keep the analysis simple, we have neglected the pump depletion due to conversion to other wavelengths and also the attenuation of the pump. Here

$$\Delta\beta = 2\beta(\omega_p) - \beta(\omega_s) = \beta(\omega_i). \qquad (1.76)$$

The above two equations can be solved to obtain the evolution of the signal and idler waves as they propagate through the fiber.

As an example, we assume $\Delta\beta = 0$, i.e., we operate at the zero dispersion wavelength of the fiber and neglect pump depletion. We also assume that at the input apart from the pump we have a signal wave with amplitude $A_s(0)$. Under these conditions, the solution to the above two equations gives us

$$P_s(z) = P_s(0)\cosh^2(gz),$$
$$P_i(z) = P_s(0)\sinh^2(gz), \qquad (1.77)$$

where $g = \gamma P_p$.

As can be seen from the above equations, the signal gets amplified as it propagates through the fiber and the gain coefficient is given by g, which increases with an increase in pump power. It is also interesting to note that the signal and idler powers satisfy the following equation:

$$P_s(z) - P_s(0) = P_i(z). \qquad (1.78)$$

This is a consequence of the fact that in the process pairs of parametric amplification, each photon at the pump frequency generates a pair of signal and idler photons as denoted by the equation $2\omega_p = \omega_s + \omega_i$. Thus, along with signal amplification, the process also generates a wave at the idler frequency. If the pump and signal frequencies are close to each other, the idler frequency can be small and thus this process can be used for the generation of light at infrared frequencies.

If we assume a pump power of 1 W and a fiber with $\gamma = 20$ W^{-1} km^{-1}, which is a highly nonlinear fiber, then the amplification of the signal over a length of 100 m is about 17 dB, which is indeed significant. It may be mentioned that at such high pump powers we should not neglect the self-phase modulation of the pump itself, which gives an additional phase variation to the pump, resulting in a changed condition for achieving maximum gain. This condition is given by

$$2\beta(\omega_p) - \beta(\omega_s) - \beta(\omega_i) - 2\gamma P_p = 0. \qquad (1.79)$$

In the above discussion the FOPA operates as a phase insensitive amplifier, which implies that the amplification is independent of the phase of the input signal. FOPAs can also operate in the phase sensitive regime where, apart from the pump, both signal and idler waves are input. In such a case it can be shown that, depending on the relative phases of the pump, signal and idler, the signal and idler can be amplified or attenuated. Such phase sensitive amplifiers can be shown to add much less noise during the amplification process compared to amplifiers operating in the phase insensitive regime or EDFAs. In view of lower noise, such amplifiers are expected to find applications in optical fiber communication systems and other areas where noise can be an important consideration.

Fiber optic parametric amplifiers are finding a number of applications and use the nonlinear effect in the optical fiber for amplification ([17,18] and [19]). Unlike other amplifiers such as EDFA based on energy levels of atoms, using nonlinearity for amplification can provide for amplifiers at any wavelength. In fact, recent experiments have demonstrated applications of FOPAs in optical fiber communication at rates above 1 Tb/s (see [17]). The process of degenerate FWM is also being used for the generation of nonclassical states of light such as squeezed light, entangled photon pairs, etc.; these are expected to find applications in the upcoming field of quantum information processing.

1.8.4 SUPERCONTINUUM

Supercontinuum (SC) generation is the phenomenon in which a nearly continuous spectrally broadened output is produced through nonlinear effects on high peak power picosecond and subpicosecond pulses. Such broadened spectra find applications in spectroscopy, wavelength characterization of optical components such as a broadband source from which many wavelength channels can be obtained by slicing the spectrum.

Supercontinuum generation in an optical fiber is a very convenient technique since it provides a very broad bandwidth (>200 nm), the intensity levels can be maintained high over long interaction lengths by choosing small mode areas and the dispersion profile of the fiber can be appropriately designed by varying the transverse refractive index profile of the fiber.

The spectral broadening that takes place in the fiber is attributed to a combination of various third order effects such as SPM, XPM, FWM and Raman scattering. Since dispersion plays a significant role in the temporal evolution of the pulse, different dispersion profiles have been used in the literature to achieve broadband SC.

Some studies have used dispersion decreasing fibers or dispersion flattened fibers, while others have used a constant anomalous dispersion fiber followed by a normal dispersion fiber.

Figure 1.9 shows a supercontinuum source based on nonlinear optical effects in an optical fiber. The entire visible spectrum gets generated through nonlinear effects; the spectrum is dispersed using a diffraction grating.

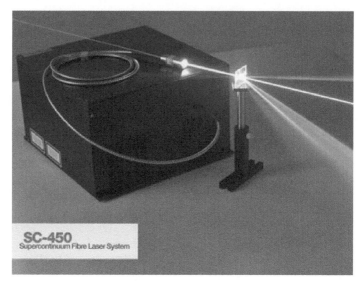

FIGURE 1.9: Supercontinuum white light source. Laser pulses of 6ps duration are incident on a special optical fiber characterized by a very small mode field diameter which leads to very high intensities. Because of the high intensities, SPM (Self-Phase Modulation) and other non-linear effects can be observed; these nonlinear effects result in the generation of new frequencies. In this experiment, the entire visible spectrum gets generated which can be observed by passing the light coming out of the optical fiber through a grating. The repetition rate of the laser pulses is 20 MHz. The wavelengths generated range from 460 nm to 2200 nm. Photograph courtesy Fianium, UK.

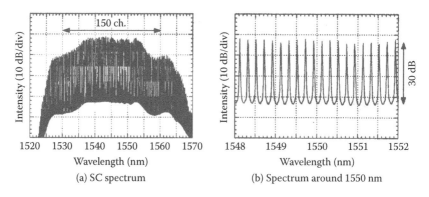

(a) SC spectrum   (b) Spectrum around 1550 nm

FIGURE 1.10: Supercontinuum spectrum generated by passing a 25 GHz optical pulse train at 1544 nm. After [20].

Figure 1.10 shows the supercontinuum spectrum generated by passing a 25 GHz optical pulse train at 1544 nm generated by a mode locked laser diode and amplified by an erbium doped fiber. As can be seen, the output spectrum is very broad and contains more than 150 spectral components at a spacing of 25 GHz with a flat top spectrum over 18 nm. Such sources are being investigated as attractive solutions for dense WDM (DWDM) optical fiber communication systems.

REFERENCES

1. A. W. Snyder and J. D. Love, Optical Waveguide Theory, Chapman and Hall, London (1987).
2. D. Gloge, Weakly guiding fibers, Appl. Opt. 10, 2252 (1971).
3. J. Irving and N. Mullineux, Mathematics in Physics and Engineering, Academic Press, New York (1959).
4. A. Ghatak and K. Thyagarajan, Introduction to Fiber Optics, Cambridge University Press, Cambridge (1998).
5. D. Marcuse, Gaussian approximation of the fundamental modes of graded index fibers, J. Opt. Soc. Am. 68, 103 (1978).
6. M.J. Li, Recent progress in fiber dispersion compensators, Proc ECOC 2001, Amsterdam, Paper ThM1.1,4, 486, London (1983).
7. Y. Nagasawa, K. Aikawa, N. Shamoto, A. Wada, Y. Sugimasa, I. Suzuki and Y. Kikuchi, High performance dispersion compensating fiber module, Fujikura Tech. Review 1 (2001).
8. J. L. Auguste, R. Jindal, J. M. Blondy, M. J. Clapeau, B. Dussardier, G. Monnom, D. B. Ostrowsky, B. P. Pal and K. Thyagarajan, Electron. Lett. 36, 1689, (2000).
9. S. Ramachandran (Ed.), Fiber-based dispersion compensation, Springer Verlag, New York (2008).
10. K. Thyagarajan, R. K. Varshney, P. Palai, A. Ghatak and I. C. Goyal, A novel design of a dispersion compensating fiber, Photon. Tech. Letts., 8, 1510 (1996).
11. A. R. Chraplyvy, Limitations on lightwave communications imposed by optical-fiber nonlinearities, J. Lightwave Tech. 8, 1548 (1990).
12. T. K. Chiang, N. Kagi, M. E. Marhic, and L. G. Kazovsky, Cross-phase modulation in fiber links with multiple optical amplifiers and dispersion compensators, J. Lightwave Tech. 14, 249 (1996).
13. G. P. Agrawal, Nonlinear Fiber Optics, Academic Press, Boston (2012).
14. L. Rapp, Experimental investigation of signal distortions induced by cross-phase modulation combined with dispersion, IEEE Photon Tech. Letts. 9, 1592 (1997).
15. M. Shtaif, M. Eiselt and L. D. Garret, Cross-phase modulation distortion measurements in multispan WDM systems, IEEE Photon. Tech. Letts. 12, 88 (2000).
16. R. W. Tkach, A. R. Chraplyvy, F. Forghiari, A. H. Gnanck and R. M. Derosier, Four-photon mixing and high speed WDM systems. J. Lightwave Tech. 13, 841 (1995).
17. N. El. Dahdah, D. S. Govan, M. Jamshidifar, N. J. Doran, and M. E. Marhic, Fiber optical parametric amplifier performance in a 1-Tb/s DWDM communication system, IEEE J. Sel. Topics Quant. Electron. 18, 950 (2012).

18. J. Hansryd, P. A. Andrekson, M. Westlund, J. Li, and P. Hedekvist, Fiber-based optical parametric amplifiers and their applications, IEEE J. Sel. Topics Quant. Electron. 8, 506 (2002).

19. Z. Tong, C. Lundström, P. A. Andrekson, M. Karlsson and A. Bogris, Ultralow noise, broadband phase-sensitive optical amplifiers, and their applications, IEEE J. Sel. Topics Quant. Electron. 18, 1016 (2011).

20. E. Yamada, H. Takara, T. Ohara, K. Sato, T. Morioka, K. Jinguji, M. Itoh and M. Ishi, 150 channel supercontinuum CW optical source with high SNR and precise 25 GHz spacing for 10 Gbit/s DWDM systems, Electronics Letters 37, 304 (2001).

2 Waveguide electromagnetic pulse dynamics: Projecting operators method

Mateusz Kuszner and *Sergey Leble*

2.1 INTRODUCTION

In this review we discuss applications of the dynamical projecting method (DPM) [1] to problems of short pulse propagation in a dispersive medium. We concentrate on two basic mathematical formulations: the Cauchy problem and boundary regime propagation along a waveguide. The DPM method is a convenient way to specify the problems of a wave initialization for directed modes and for a given polarization account. A minimal version of such a study appears when one integrates across a waveguide with respect to transversal coordinates (e.g., [2]). More advanced theories include one-mode fiber description [3] and multimode waves projecting to transversal eigenfunction subspaces [4]. In such a case we have only four possibilities for electromagnetic mode specifications that in linear descriptions correspond to roots of the dispersion relation inside a transversal mode. It is division with respect to polarization and direction of propagation. All such modes and their interaction were intensely studied during the last decades [5].

Assuming the unidirectional propagation of a polarized light pulses in a 1D medium with cubic nonlinearity it is conventional to use the nonlinear Schrödinger equation (NS) which was derived by Zkharov [6]. Interaction of such pulses is described by the coupled nonlinear Schrödinger equation (CNS) [7].

Here we study pulses whose length is shorter than a few cycles of the center frequency. Propagation of such pulses was investigated by others [8–11]. In the derivation of equations describing the propagation of ultra-short pulses there are two basic approximations for electromagnetic fields, which are described in the review by Maimistov [13], namely, the slowly varying envelope approximation and the unidirectional wave approximation. For ultra-short pulses whose length does not exceed a few picoseconds, the first approximations lead to inaccurate results. Therefore, the validity of the application of either NS or CNS to describe the propagation of ultra-short pulses of light has been questioned by different groups [14–16]. Such approaches to the problem use slowly evolving wave approximation (SEWA) [15] or the application of Bloch

31

equations [17]. Another approach is to introduce new variables of the form

$$\psi^{\pm} = \epsilon \frac{1}{2} E_i \pm \mu \frac{1}{2} H_j,$$

as did Fleck [8], Kinsler [11, 18, 19] and Amiranashvili [12] in their work. It is a method in a sense similar to the projection operator method which was described in Leble's book [1]. The projection operator method is a powerful tool which was initially set for the Cauchy problem and which allows for detailed analysis of the field and its modes. Applying the projection operators to the one-dimensional model of light pulse propagation has led to the generalized short pulse (SP) equation [20] where the physical form of cofactors has been shown and interaction between waves in two directions has been taken into consideration. The case of the interaction of two counterpropagating waves was the topic of experimental studies [21] in which the authors have presented proof of the interaction of counterpropagating pulse polarizations. In the other case the authors [22] have proven the existence of the wave propagating backwards in optical resonators. The most fitting case is a paper about backward pulse generation in a transmission line [23], which lacks a theoretical description of the phenomenon. The aim of the paper [24] was to show how, starting from Maxwell's equations

$$
\begin{aligned}
\nabla \cdot \mathbf{D} &= 0 \\
\nabla \cdot \mathbf{B} &= 0 \\
\nabla \times \mathbf{E} &= -\frac{1}{c}\frac{\partial \mathbf{B}}{\partial t} \\
\nabla \times \mathbf{H} &= \frac{1}{c}\frac{\partial \mathbf{D}}{\partial t}
\end{aligned}
\tag{2.1}
$$

for electric and magnetic fields $\mathbf{E}, \mathbf{H}$ and taking into account material relations for inductions $\mathbf{D}, \mathbf{B}$, one can derive projection operators and by this means [1] derive equations of pulse propagation with an account of polarization and nonlinear effects presented by new variables for right Π_i and left Λ_i waves which can lead to the answer to the question of the generation of backward waves as in the terahertz left-handed transmission line ([23] Section 2). Moreover, the results were compared with the works of Kinsler [11, 18, 19], Schäfer and Wayne ([10] Section 3) and, as a novel step in the dynamic projection operators theory, were derived for boundary regime propagation. The acquired results were compared with the works of Schäfer and Chung [25] et al. and U. Bandelow ([26] Section 4).

In [27] we go to the complete multimode problem for pulse propagation in a dielectric fiber. As a three-dimensional problem, we continue to study cylindrical dielectric waveguides [4], e.g., optic fibers, using the Hondros–Debye basis, containing exponential in polar coordinate and Bessel ones in radial coordinate. The transverse (Hondros–Debye) expansion leads to equations that could be used for both initial (Cauchy) and boundary regime problems. A

Cauchy problem for a Maxwell system with Kerr nonlinearity and waveguide dispersion leads to a system of equations for interacted modes. Constants of interaction, the transverse and polarization direction modes, are expressed in the form of integrals of the Hondros–Debye basic functions. Here, besides reviewing, we essentially develop the technique of dynamical projecting, constructing the explicit form of the operators. The idempotent operators and corresponding equations were derived by the projecting procedure of the guide modes interaction, including the effects of the unidirectional pulses. We return to the general method of integrable evolution equations implementation adjusted to the problems to be considered. This method links the unidirectional pulse propagation approach with the derivation of the nonlinear Schrödinger (NS), Manakov [28,29], coupled NS (CNS) and SW (spectral width) [10] equations and their generalizations (e.g., coupled SW [20]) in conditions of waveguides. The results also account for opposite directed waves with all possible polarization propagations, interaction and, as a new element, the matter dispersion account. The CNS equations that appear in the envelope of optic pulse theory are proven [30] to be numerically integrable (N-integrable) in the sense of the existence of stable and convergent in L_2 finite difference representation of the equations. In Sec. 2.5 we outline an idea and implementation of the projection operators method that specify evolution operator subspaces. Further, we present an application of this approach to the derivation of a bidirectional wave system for hybrid elds [20, 4]. Then we account for the dispersion terms and show in which conditions a transition to equations of earlier works [6] is accomplished. In Sec. 2.6 we present applications of the projecting technique to cylindric waveguide modes and CNS-type equation derivation. The present review is mainly based on our recent publications [20, 24, 27].

2.2 THEORY OF INITIALIZATION OF A PULSE PROPAGATION

2.2.1 CAUCHY PROBLEM FORMULATION IN ONE SPATIAL DIMENSION

To create a description of a one-dimensional polarized pulse propagation in both directions [11,18,26], developing results of [20] we start with the standard linearized Maxwell equation setup

$$
\begin{aligned}
\frac{1}{c}\frac{\partial D_y}{\partial t} &= -\frac{\partial H_z}{\partial x}, \\
\frac{1}{c}\frac{\partial D_z}{\partial t} &= \frac{\partial H_y}{\partial x}, \\
\frac{1}{c}\frac{\partial B_z}{\partial t} &= -\frac{\partial E_y}{\partial x}, \\
\frac{1}{c}\frac{\partial B_y}{\partial t} &= \frac{\partial E_z}{\partial x},
\end{aligned}
\tag{2.2}
$$

and material equations in operator form $D_y = \hat{\epsilon} E_y$ and $D_z = \hat{\epsilon} E_z$, $\hat{\mu} H_y = B_y$, $\hat{\mu} H_z = B_z$ end the description.

From the point of view of the initial problem to be considered, the operators may by introduced on a base of Fourier transform on the axis $x \in (-\infty, \infty)$ to be provided for the variable $k \in (-\infty, \infty)$ as

$$D_y = \int_{-\infty}^{\infty} \int_{-\infty}^{\infty} \epsilon(k) e^{-ik(x-x')} dk\, E_y(x') dx' \quad = \quad \hat{\epsilon} E_y \qquad (2.3)$$

$$B_y = \int_{-\infty}^{\infty} \int_{-\infty}^{\infty} \mu(k) e^{-ik(x-x')} dk\, H_y(x') dx' \quad = \quad \hat{\mu} H_y. \qquad (2.4)$$

Its Fourier transforms

$$\tilde{D}_y(k) = \frac{1}{\sqrt{2\pi}} \int_{-\infty}^{+\infty} D_y e^{-ikx} dx \qquad (2.5)$$

and one for E_y

$$\tilde{E}_y(k) = \frac{1}{\sqrt{2\pi}} \int_{-\infty}^{+\infty} E_y e^{-ikx} dx \qquad (2.6)$$

are linked as follows:

$$\tilde{D}_y(k) = \epsilon(k) \tilde{E}_y(k). \qquad (2.7)$$

The relation characterizes material properties of a medium of propagation used, for example, in the seminal work [10]. The magnetic fields are connected by similar integral operators; we further traditionally choose magnetic induction as a basic one, excluding $H_y = \hat{\mu}^{-1} B_y$ and $H_z = \hat{\mu}^{-1} B_z$.

The Cauchy problem needs four initial conditions

$$\begin{aligned}
E_y(x,0) &= \eta_1(x), \quad E_z(x,0) = \eta_2(x), \\
B_y(x,0) &= \xi_1(x), \quad B_z(x,0) = \xi_2(x)
\end{aligned} \qquad (2.8)$$

to make the solution of Eq. (2.2) unique.

2.2.2   PROJECTION METHOD FOR THE 1D CAUCHY PROBLEM

From the point of view of the initial problem to be considered, we transform the system (2.2) to the more convenient split form (mode representation [1]) and specify the correspondent initial conditions for such mode variables. We introduce projection operators in x-space that have to fulfill the standard properties of orthogonality and completeness:

$$P_i * P_j = 0, \quad P_i^2 = P_i, \quad \sum_i P_i = I, \qquad (2.9)$$

$i = 1, 2, 3, 4$. To construct projection operators for a specific problem like Eq. (2.2) it is necessary to find eigenvectors for the linear evolution operator L [31] that enter an evolution equation of the general form

$$\psi_t = L\psi.$$

In the case we consider, Eq. (2.2) can be rewritten as follows:

$$
\psi = \begin{pmatrix} E_z \\ B_y \\ E_y \\ B_z \end{pmatrix}, L = \begin{pmatrix} 0 & c\hat{e}^{-1}\hat{\mu}^{-1}\partial_x & 0 & 0 \\ c\partial_x & 0 & 0 & 0 \\ 0 & 0 & 0 & -c\hat{e}^{-1}\hat{\mu}^{-1}\partial_x \\ 0 & 0 & -c\partial_x & 0 \end{pmatrix}. \quad (2.10)
$$

Entering elements from Eq. (2.10) to the matrix operator equation gives the formula

$$
\begin{pmatrix} (E_z)_t \\ (B_y)_t \\ (E_y)_t \\ (B_z)_t \end{pmatrix} = \begin{pmatrix} 0 & c\hat{e}^{-1}\hat{\mu}^{-1}\partial_x & 0 & 0 \\ c\partial_x & 0 & 0 & 0 \\ 0 & 0 & 0 & -c\hat{e}^{-1}\hat{\mu}^{-1}\partial_x \\ 0 & 0 & -c\partial_x & 0 \end{pmatrix} \begin{pmatrix} E_z \\ B_y \\ E_y \\ B_z \end{pmatrix}.
$$
$$(2.11)$$

Let us apply the Fourier transformation by x in the form of Eq. (2.6) to Eq. (2.11), which yields the system of ordinary differential equations with material constants ϵ, ν as functions of wave vector k.

$$
\begin{pmatrix} (\tilde{E}_z(k,t))_t \\ (\tilde{B}_y(k,t))_t \\ (\tilde{E}_y(k,t))_t \\ (\tilde{B}_z(k,t))_t \end{pmatrix} = \begin{pmatrix} ck\epsilon^{-1}\mu^{-1}\tilde{B}_y(k,t) \\ ck\tilde{E}_z(k,t) \\ -ck\epsilon^{-1}\mu^{-1}\tilde{B}_z(k,t) \\ -ck\tilde{E}_y(k,t) \end{pmatrix}, \quad (2.12)
$$

of which particular solutions have the form

$$
\tilde{E}_i = \breve{E}_i \exp{(i\omega t)} \quad \tilde{B}_j = \breve{B}_j \exp{(i\omega t)}. \quad (2.13)
$$

Plugging the solutions into the system (2.12) yields a system of equations

$$
\begin{pmatrix} \omega \breve{E}_z(k,\omega) \\ \omega \breve{B}_y(k,\omega) \\ \omega \breve{E}_y(k,\omega) \\ \omega \breve{B}_z(k,\omega) \end{pmatrix} = \begin{pmatrix} -ck\epsilon^{-1}\mu^{-1}\breve{B}_y(k,\omega) \\ -ck\breve{E}_z(k,\omega) \\ ck\epsilon^{-1}\mu^{-1}\breve{B}_z(k,\omega) \\ ck\breve{E}_y(k,\omega) \end{pmatrix} \quad (2.14)
$$

that have a sense of the eigenvalue problem under solvability condition

$$
\omega^2 = k^2 c^2 \mu^{-1}\epsilon^{-1}. \quad (2.15)
$$

There is an important dependency hidden in the above equation. The dispersion relation is described by ϵ, which in this case is a function of k. Strictly from those equations one can construct projection operators. Hence we can present a direct relation between $\breve{E}_z$ and $\breve{H}_y$ and between $\breve{E}_y$ and $\breve{H}_z$,

$$
P_{11} = \frac{1}{2}\begin{pmatrix} 1 & -\mu^{-\frac{1}{2}}\epsilon^{-\frac{1}{2}} & 0 & 0 \\ -\epsilon^{\frac{1}{2}}\mu^{\frac{1}{2}} & 1 & 0 & 0 \\ 0 & 0 & 0 & 0 \\ 0 & 0 & 0 & 0 \end{pmatrix}, \quad P_{12} = \frac{1}{2}\begin{pmatrix} 0 & 0 & 0 & 0 \\ 0 & 0 & 0 & 0 \\ 0 & 0 & 1 & \mu^{-\frac{1}{2}}\epsilon^{-\frac{1}{2}} \\ 0 & 0 & \epsilon^{\frac{1}{2}}\mu^{\frac{1}{2}} & 1 \end{pmatrix},
$$
$$(2.16)$$

$$P_{21} = \frac{1}{2} \begin{pmatrix} 1 & \mu^{-\frac{1}{2}}\epsilon^{-\frac{1}{2}} & 0 & 0 \\ \epsilon^{\frac{1}{2}}\mu^{\frac{1}{2}} & 1 & 0 & 0 \\ 0 & 0 & 0 & 0 \\ 0 & 0 & 0 & 0 \end{pmatrix}, \quad P_{22} = \frac{1}{2} \begin{pmatrix} 0 & 0 & 0 & 0 \\ 0 & 0 & 0 & 0 \\ 0 & 0 & 1 & -\mu^{-\frac{1}{2}}\epsilon^{-\frac{1}{2}} \\ 0 & 0 & -\epsilon^{\frac{1}{2}}\mu^{\frac{1}{2}} & 1 \end{pmatrix},$$

$$(2.17)$$

which, acting on ψ, gives us four eigenstates:

$$\breve{\psi}_{11} = \begin{pmatrix} 1 \\ -\epsilon^{\frac{1}{2}}\mu^{\frac{1}{2}} \\ 0 \\ 0 \end{pmatrix} \breve{E}_z, \quad \breve{\psi}_{12} = \begin{pmatrix} 0 \\ 0 \\ 1 \\ \epsilon^{\frac{1}{2}}\mu^{\frac{1}{2}} \end{pmatrix} \breve{E}_y, \tag{2.18}$$

and

$$\breve{\psi}_{21} = \begin{pmatrix} 1 \\ \epsilon^{\frac{1}{2}}\mu^{\frac{1}{2}} \\ 0 \\ 0 \end{pmatrix} \breve{E}_z, \quad \breve{\psi}_{22} = \begin{pmatrix} 0 \\ 0 \\ 1 \\ -\epsilon^{\frac{1}{2}}\mu^{\frac{1}{2}} \end{pmatrix} \breve{E}_y \tag{2.19}$$

of the evolution matrix.

After the inverse Fourier transformation of Eqs. (2.16) and (2.17) the projection operators obtain the form of matrix-integral ones:

$$\hat{P}_{11} = \frac{1}{2} \begin{pmatrix} 1 & -\hat{\mu}^{-\frac{1}{2}}\hat{\epsilon}^{-\frac{1}{2}} & 0 & 0 \\ -\hat{\epsilon}^{\frac{1}{2}}\hat{\mu}^{\frac{1}{2}} & 1 & 0 & 0 \\ 0 & 0 & 0 & 0 \\ 0 & 0 & 0 & 0 \end{pmatrix}, \quad \hat{P}_{12} = \frac{1}{2} \begin{pmatrix} 0 & 0 & 0 & 0 \\ 0 & 0 & 0 & 0 \\ 0 & 0 & 1 & \hat{\mu}^{-\frac{1}{2}}\hat{\epsilon}^{-\frac{1}{2}} \\ 0 & 0 & \hat{\epsilon}^{\frac{1}{2}}\hat{\mu}^{\frac{1}{2}} & 1 \end{pmatrix},$$

$$(2.20)$$

$$\hat{P}_{21} = \frac{1}{2} \begin{pmatrix} 1 & -\hat{\mu}^{-\frac{1}{2}}\hat{\epsilon}^{-\frac{1}{2}} & 0 & 0 \\ \hat{\epsilon}^{\frac{1}{2}}\hat{\mu}^{\frac{1}{2}} & 1 & 0 & 0 \\ 0 & 0 & 0 & 0 \\ 0 & 0 & 0 & 0 \end{pmatrix}, \quad \hat{P}_{22} = \frac{1}{2} \begin{pmatrix} 0 & 0 & 0 & 0 \\ 0 & 0 & 0 & 0 \\ 0 & 0 & 1 & -\hat{\mu}^{-\frac{1}{2}}\hat{\epsilon}^{-\frac{1}{2}} \\ 0 & 0 & -\hat{\epsilon}^{\frac{1}{2}}\hat{\mu}^{\frac{1}{2}} & 1 \end{pmatrix},$$

$$(2.21)$$

where the integral operator $\hat{\epsilon}$ is given by Eq. (2.3). The operators acting on ψ give us four subspaces, generated by the vectors

$$\psi_{11} = \begin{pmatrix} 1 \\ -\hat{\epsilon}^{\frac{1}{2}}\hat{\mu}^{\frac{1}{2}} \\ 0 \\ 0 \end{pmatrix} E_z, \quad \psi_{12} = \begin{pmatrix} 0 \\ 0 \\ 1 \\ \hat{\epsilon}^{\frac{1}{2}}\hat{\mu}^{\frac{1}{2}} \end{pmatrix} E_y,$$

and

$$\psi_{21} = \begin{pmatrix} 1 \\ \hat{\epsilon}^{\frac{1}{2}}\hat{\mu}^{\frac{1}{2}} \\ 0 \\ 0 \end{pmatrix} E_z, \quad \psi_{22} = \begin{pmatrix} 0 \\ 0 \\ 1 \\ -\hat{\epsilon}^{\frac{1}{2}}\hat{\mu}^{\frac{1}{2}} \end{pmatrix} E_y. \tag{2.22}$$

The result of identity operator $I = \hat{P}_{11} + \hat{P}_{12} + \hat{P}_{21} + \hat{P}_{22}$ action on the vector ψ defines the following transition to new variables:

$$\Pi_1 = \frac{1}{2}E_z - \frac{1}{2}\hat{\mu}^{-\frac{1}{2}}\hat{\epsilon}^{-\frac{1}{2}}B_y, \qquad \Lambda_1 = \frac{1}{2}E_z + \frac{1}{2}\hat{\mu}^{-\frac{1}{2}}\hat{\epsilon}^{-\frac{1}{2}}B_y, \quad (2.23)$$

$$\Pi_2 = \frac{1}{2}E_y - \frac{1}{2}\hat{\mu}^{-\frac{1}{2}}\hat{\epsilon}^{-\frac{1}{2}}B_z, \qquad \Lambda_2 = \frac{1}{2}E_y + \frac{1}{2}\hat{\mu}^{-\frac{1}{2}}\hat{\epsilon}^{-\frac{1}{2}}B_z, \quad (2.24)$$

$$\Lambda_1 + \Pi_1 = E_z, \qquad \Lambda_2 + \Pi_2 = E_y, \quad (2.25)$$

$$\hat{\epsilon}^{\frac{1}{2}}\hat{\mu}^{\frac{1}{2}}(\Lambda_1 - \Pi_1) = B_y, \qquad \hat{\epsilon}^{\frac{1}{2}}\hat{\mu}^{\frac{1}{2}}(\Lambda_2 - \Pi_2) = B_z. \quad (2.26)$$

From the definition of the projectors it follows that $[\mathbb{L}, P_i] = 0$. The result of operator $\hat{P}_{12}$ application to Eq. (2.14) is:

$$\hat{P}_{11}\begin{pmatrix} (E_z)_t \\ (B_y)_t \\ (E_y)_t \\ (B_z)_t \end{pmatrix} - \mathbb{L}\hat{P}_{11}\begin{pmatrix} E_z \\ B_y \\ E_y \\ B_z \end{pmatrix} = 0. \quad (2.27)$$

Further evaluation yields

$$\begin{pmatrix} (\Pi_1)_t \\ (-\hat{\epsilon}^{\frac{1}{2}}\hat{\mu}^{\frac{1}{2}}\Pi_1)_t \\ 0 \\ 0 \end{pmatrix}$$

$$-\begin{pmatrix} 0 & c\hat{\epsilon}^{-1}\hat{\mu}^{-1}\partial_x & 0 & 0 \\ c\partial_x & 0 & 0 & 0 \\ 0 & 0 & 0 & -c\hat{\epsilon}^{-1}\hat{\mu}^{-1}\partial_x \\ 0 & 0 & -c\partial_x & 0 \end{pmatrix}\begin{pmatrix} \Pi_1 \\ -\hat{\epsilon}^{\frac{1}{2}}\hat{\mu}^{\frac{1}{2}}\Pi_1 \\ 0 \\ 0 \end{pmatrix} = 0,$$

$$(2.28)$$

$$\begin{pmatrix} (\Pi_1)_t \\ (-\hat{\epsilon}^{\frac{1}{2}}\hat{\mu}^{\frac{1}{2}}\Pi_1)_t \\ 0 \\ 0 \end{pmatrix} - \begin{pmatrix} -c\hat{\epsilon}^{-\frac{1}{2}}\hat{\mu}^{-\frac{1}{2}}(\Pi_1)_x \\ c(\Pi_1)_x \\ 0 \\ 0 \end{pmatrix} = 0, \quad (2.29)$$

which reads as one of the main integro-differential equations:

$$(-\hat{\epsilon}^{\frac{1}{2}}\hat{\mu}^{\frac{1}{2}}\Pi_1)_t + c(\Pi_1)_x = 0, \quad (2.30)$$

with initial condition, specified by Eq. (2.23),

$$\Pi_1(x,0) = \frac{1}{2}\eta_2(x) - \frac{1}{2}\hat{\mu}^{-\frac{1}{2}}\hat{\epsilon}^{-\frac{1}{2}}\xi_1(x). \quad (2.31)$$

One obtains similar results by action on the Eq. (2.14) by the rest of operators. The resulting equations define independent linear mode propagations that correspond to left and right waves, each with two fixed orthogonal polarizations and initial conditions as in Eq. (2.31).

2.2.3 DISPERSION AND NONLINEARITY WITH POLARIZATION INTERACTION, GENERAL RELATIONS

Following the general scheme of [31]), we have shown in previous work [20] that new variables, Eqs. (2.23) and (2.24), may be introduced into Eq. (2.11) by acting with the projection operators of Eq. (2.17):

$$\frac{\partial}{\partial t}\hat{P}_{11}\Psi - L\hat{P}_{11}\Psi = \hat{P}_{11}\mathbb{N}(E). \tag{2.32}$$

Stepping to a more precise model, we look closer at a description of propagation of electromagnetic pulses in silica fiber. According to our main target, we restrict ourselves to $\mu = 1$ because we would ignore the contribution of dispersion and nonlinear effects from the magnetic field.

As it was obvious how to present $(P_{NL})_i$ in the one-dimensional case [20], we analyze third-order optical susceptibility $\chi^{(3)}$ as Kielich [32] did earlier:

$$\chi_{ijkl} = \chi_{xxyy}\delta_{ij}\delta_{kl} + \chi_{xyxy}\delta_{ik}\delta_{jl} + \chi_{xyyx}\delta_{il}\delta_{jk}, \tag{2.33}$$

$$\chi_{xxxx} = \chi_{yyyy} = \chi_{zzzz} = \chi_{xxyy} + \chi_{xyyx} + \chi_{xyxy}. \tag{2.34}$$

We have based our approach on the interaction of two polarization modes, described by the terms $(P_{NL})_y$ and $(P_{NL})_z$, taking into account directions of propagation.

Next we show give additional information about nonlinearity of a Kerr type:

$$\mathbb{N}(E_y, E_z) = -4\pi\frac{\partial}{\partial t}\begin{pmatrix} \hat{\epsilon}^{-1}\left(3\chi_{yyyy}(E_y)^3 + (\chi_{yyzz} + \chi_{yzzy} + \chi_{yzyz})E_y(E_z)^2\right) \\ 0 \\ \hat{\epsilon}^{-1}\left(3\chi_{zzzz}(E_z)^3 + (\chi_{zzyy} + \chi_{zyyz} + \chi_{zyzy})E_z(E_y)^2\right) \\ 0 \end{pmatrix}. \tag{2.35}$$

With the use of Eq. (2.33) we have arrived at

$$\mathbb{N}(E_y, E_z) = -4\pi\frac{\partial}{\partial t}\begin{pmatrix} \hat{\epsilon}^{-1}\chi_{yyyy}(3(E_y)^3 + E_y(E_z)^2) \\ 0 \\ \hat{\epsilon}^{-1}\chi_{zzzz}(3(E_z)^3 + E_z(E_y)^2) \\ 0 \end{pmatrix}. \tag{2.36}$$

In the frame of weak nonlinearity theory one applies the (linear!) projection operators to third-order nonlinearities, which are so common in many materials, for example, in the silica used to make optical fibers. Acting with Eq. (2.16) on Eq. (2.35), transition to new variables via Eq. (2.25) yields

$$P_{11}\mathbb{N}(E_y, E_z) = \mathbb{N}(\Lambda_1, \Lambda_2, \Pi_1, \Pi_2)$$

$$= -2\pi\frac{\partial}{\partial t}\begin{pmatrix} \hat{\epsilon}^{-1}\chi_{yyyy}\left(3(\Pi_1 - \Lambda_1)^3 + (\Pi_1 + \Lambda_1)(\Pi_2 - \Lambda_2)^2\right) \\ -\hat{\epsilon}^{-\frac{1}{2}}\chi_{yyyy}\left(3(\Pi_1 - \Lambda_1)^3 + (\Pi_1 + \Lambda_1)(\Pi_2 + \Lambda_2)^2\right) \\ 0 \\ 0 \end{pmatrix} \tag{2.37}$$

the third-order nonlinearity has been projected on the directed polarization mode. We obtain an equation for bidirectional pulse propagation with a Kerr effect.

Equation (2.37) is just one part of our result. The second part we can obtain by acting with projection operator P_{12} from Eq. (2.17) on Eq. (2.35):

$$P_{12}N(E_y, E_z) = -2\pi \frac{\partial}{\partial t} \begin{pmatrix} 0 \\ 0 \\ \hat{e}^{-1}\chi_{yyyy}\big(3(\Pi_2 - \Lambda_2)^3 + (\Pi_2 + \Lambda_2)(\Pi_1 - \Lambda_1)^2\big) \\ \hat{e}^{-\frac{1}{2}}\chi_{yyyy}\big(3(\Pi_2 - \Lambda_2)^3 + (\Pi_2 + \Lambda_2)(\Pi_1 + \Lambda_1)^2\big) \end{pmatrix}.$$
(2.38)

With the above result, the full formula of Eq. (2.32) can be presented as

$$\hat{e}(\Pi_1)_t + c\hat{e}^{\frac{1}{2}}(\Pi_1)_x = 2\pi \frac{\partial}{\partial t}\chi_{yyyy}\big(3(\Pi_1 - \Lambda_1)^3 + (\Pi_1 + \Lambda_1)(\Pi_2 - \Lambda_2)^2\big). \quad (2.39)$$

If we consider a unidirectional approach to our result, neglecting opposite waves (supposed to be absent in the initial disturbance), we assume that $\Lambda_1 = 0$, $\Lambda_2 = 0$ and that implies that

$$\hat{e}(\Pi_1)_t + c\hat{e}^{\frac{1}{2}}(\Pi_1)_x = -2\pi \frac{\partial}{\partial t}\big(\chi_{yyyy}(\Pi_1(3\Pi_1^2 + \Pi_2^2))\big), \quad (2.40)$$

having a directed electromagnetic pulse.

2.3 COMPARISON OF RESULTS OBTAINED WITH THE MULTIPLE SCALE METHOD

To compare our general system with those of [10, 26] we develop a method of nonsingular perturbation theory (see also [31]). The sense of the method is an expansion of the evolution operator (e.g., in Eq. (2.32)) with respect to a small dimensionless parameter. In our case such a parameter may be chosen as a unified dispersion nonlinear one.

The resulting equation type depends on the dispersion form. The dielectric susceptibility coefficient $\epsilon(k)$ originated from [10], where the authors focused on the propagation of light in the infrared range with $\lambda = 1600-3000$ nm. Such pulses correspond to femtosecond pulses. In this range it can be approximated as

$$\epsilon = 1 + 4\pi\chi^{(1)}(\lambda) \approx 1 + 4\pi(\chi_0^{(1)} + \chi_2^{(1)}\lambda^2). \quad (2.41)$$

where λ is equal to $\frac{2\pi}{k}$. In further calculations it has been treated as a parameter, as we consider propagation in dielectric media. An alternative assumption will open further investigation for materials qualified as metamaterials. At this point we consider permittivity as a function of k:

$$\epsilon = 1 + 4\pi\chi^{(1)}(\lambda) \approx 1 + 4\pi(\chi_0^{(1)} + \chi_2^{(1)}4\pi^2\frac{1}{k^2}), \quad (2.42)$$

which, in fact is the Taylor series in the vicinity of $k = \infty$. Having in mind that the calculations of integral operators as in Eq. (2.3) and their functions as in Eq. (2.10) will be continued in the x-representation, it is crucial to present the expansion of ϵ^α in the same vicinity.

$$\epsilon(k)^{\frac{1}{2}} \approx \sqrt{1 + 4\pi\chi_0} + \frac{8\pi^3\chi_2}{\sqrt{1 + 4\pi\chi_0}k^2}, \tag{2.43}$$

$$\epsilon(k)^{-1} \approx \frac{1}{1 + 4\pi\chi_0} - \frac{16\left(\pi^3\chi_2\right)}{\left(1 + 4\pi\chi_0\right)^2 k^2}, \tag{2.44}$$

$$\epsilon(k)^{-\frac{1}{2}} \approx \frac{1}{\sqrt{1 + 4\pi\chi_0}} - \frac{8\left(\pi^3\chi_2\right)}{\left(1 + 4\pi\chi_0\right)^{3/2} k^2}. \tag{2.45}$$

After inverse Fourier transformation,

$$\hat{\epsilon}(x)^{\frac{1}{2}} \approx \sqrt{1 + 4\pi\chi_0} + \frac{8\pi^3\chi_2}{\sqrt{1 + 4\pi\chi_0}}(i\partial_x)^{-2}, \tag{2.46}$$

because the factor k^{-1} is represented by $(i\partial_x)^{-1} = i\int dx$,

$$\hat{\epsilon}(x)^{-1} \approx \frac{1}{1 + 4\pi\chi_0} - \frac{16\left(\pi^3\chi_2\right)}{\left(1 + 4\pi\chi_0\right)^2}(i\partial_x)^{-2}. \tag{2.47}$$

Similarly,

$$\hat{\epsilon}(x)^{-\frac{1}{2}} \approx \frac{1}{\sqrt{1 + 4\pi\chi_0}} - \frac{8\left(\pi^3\chi_2\right)}{\left(1 + 4\pi\chi_0\right)^{3/2}}(i\partial_x)^{-2}. \tag{2.48}$$

With projection operators in correspondent approximation, we rewrite relations Eq. (2.40) with new conditions:

$$\left(1 + 4\pi\chi_0^{(1)} + 4\pi\chi_2^{(1)}(i\partial_x)^{-2}\right)\frac{\partial}{\partial t}\Pi_1$$
$$- c\left(\sqrt{1 + 4\pi\chi_0} + \frac{8\pi^3\chi_2}{\sqrt{1 + 4\pi\chi_0}}(i\partial x)^{-2}\right)\frac{\partial}{\partial x}\Pi_1 = -2\pi\frac{\partial}{\partial t}\left(\chi_{yyyy}\Pi_1(3\Pi_1^2 + \Pi_2^2)\right). \tag{2.49}$$

With analogy to [10] we can substitute a multiple scales ansatz. The ansatz, by construction, accounts for only one of the directed waves, hence we simply do not take the opposite waves into account, regarding excitation of the correspondent mode.

$$\Pi_1(x,t) = \left(\kappa A_0(\phi, x_1, x_2) + \kappa^2 A_1(\phi, x_1, x_2) + \dots\right),$$

$$\Pi_2(x,t) = \left(\kappa B_0(\phi, x_1, x_2) + \kappa^2 B_1(\phi, x_1, x_2) + \dots\right),$$

where $\phi = \dfrac{ct - x}{\kappa}$ and $x_n = \kappa^n x$. Then we get

$$
\left(2 + 8\pi\chi_0^{(1)} - 3\mu^{-\frac{1}{2}}\sqrt{1 + 4\pi\chi_0^{(1)}}\right)\left(\frac{\partial^2}{\partial\phi\partial x_1}\right)A_0
$$

$$
-\left(4\pi\chi_2^{(1)} - \frac{8\pi^3\chi_2}{\sqrt{1 + 4\pi\chi_0}}\right)A_0 = 2\pi\left(\frac{\partial^2}{\partial\phi^2}\right)\chi^{(3)}A_0\left(3A_0^2 + B_0^2\right),
$$

$$
\left(2 + 8\pi\chi_0^{(1)} - 3\mu^{-\frac{1}{2}}\sqrt{1 + 4\pi\chi_0^{(1)}}\right)\left(\frac{\partial^2}{\partial\phi\partial x_1}\right)B_0
$$

$$
-\left(4\pi\chi_2^{(1)} - \frac{8\pi^3\chi_2}{\sqrt{1 + 4\pi\chi_0}}\right)B_0 = 2\pi\left(\frac{\partial^2}{\partial\phi^2}\right)\chi^{(3)}B_0\left(3B_0^2 + A_0^2\right).
$$

$$(2.50)$$

Comparing our result to the one obtained by et al. and Bandelow [26],

$$
\frac{\partial^2}{\partial x_1\partial\phi}A_0 = A_0 + \frac{1}{6}\frac{\partial^2}{\partial\phi^2}\left(A_0^3 + 3A_0B_0^2\right), \tag{2.51}
$$

we conclude the coincidence of the result.

2.4 PROJECTION METHOD FOR BOUNDARY REGIME PROPAGATION

From the point of a waveguide propagation the conventional mathematical problem is formulated as a boundary one. An application of the projection operator method to the time dependent boundary problem needs significant modification. From the point of view of the boundary problem to be considered, we use a version of the antisymmetric in t basic functions E_y, H_y, E_z, H_z, defined on $t \in (-\infty, \infty)$ [33], which guarantee the zero value of the field at $t = 0$. Such continuation of the field to the range $t < 0$ gives a possibility for defining its Fourier transforms as

$$
E_y = \frac{1}{\sqrt{2\pi}}\int_{-\infty}^{\infty}\tilde{E}_y(\omega)e^{i\omega t}d\omega, \quad H_z = \frac{1}{\sqrt{2\pi}}\int_{-\infty}^{\infty}\tilde{H}_z(\omega)e^{i\omega t}d\omega. \tag{2.52}
$$

The consequent procedure is supposed to be provided for the variable $\omega \in (-\infty, \infty)$ with correspondent continuation.

It is essential to stress that $\hat{\varepsilon}$ and $\hat{\mu}$ are integral operators, whose form differs from those of Sec. 2.2:

$$
D_y = \int_{-\infty}^{\infty}\int_{-\infty}^{\infty}\varepsilon(\omega)e^{-i\omega(t-t')}d\omega E_y(t')dt' = \hat{\varepsilon}E_y, \tag{2.53}
$$

$$
B_z = \int_{-\infty}^{\infty}\int_{-\infty}^{\infty}\mu(\omega)e^{-i\omega(t-t')}d\omega H_z(t')dt' = \hat{\mu}H_z. \tag{2.54}
$$

where

$$D_y = \frac{1}{\sqrt{2\pi}} \int_{-\infty}^{\infty} \tilde{D}_y(\omega) e^{i\omega t} d\omega, \quad B_z = \frac{1}{\sqrt{2\pi}} \int_{-\infty}^{\infty} \tilde{B}_z(\omega) e^{i\omega t} d\omega. \quad (2.55)$$

This gives a simple link between transforms

$$\tilde{D}_y(\omega) = \varepsilon(\omega) \tilde{E}_y(\omega), \quad \tilde{B}_z(\omega) = \mu(\omega) \tilde{H}_z(\omega), \quad (2.56)$$

which describes material properties by the medium of propagation. The dielectric susceptibility coefficient $\varepsilon(\omega)$ either originated from the quantum version of the Lorentz formula (see, e.g., [33, 34]) or directly from phenomenology [10, 35], for example, approximated as a Taylor expansion at $\omega = \infty$ by Bandelow et al. [26].

$$\varepsilon(\omega) \approx 1 + 4\pi\chi(\omega) \approx 1 - 4\pi\chi_0\omega^{-2}, \quad (2.57)$$

$$\varepsilon^{-1}(\omega) \approx 1 + 4\pi + 4\pi\frac{\chi_0}{\omega^2}. \quad (2.58)$$

The problem is accomplished by the boundary conditions

$$E_y(0,t) = \phi_y(t), \quad E_z(0,t) = \phi_z(t),$$
$$H_y(0,t) = \theta_y(t), \quad H_z(0,t) = \theta_z(t), \quad (2.59)$$

at the time points $t > 0$.

Exchanging the x and t interpretation, we present a new operator, which is similar to the evolution one L in Eq. (2.10):

$$\frac{\partial \psi'}{\partial x} = L'\psi'. \quad (2.60)$$

To construct the shift operator L' it is necessary to use $\hat{\varepsilon}$ and $\hat{\varepsilon}^{-1}$ in time representation in Eq. (2.53):

$$\frac{\partial}{\partial x}\begin{pmatrix} E_z \\ B_y \\ E_y \\ B_z \end{pmatrix} = \begin{pmatrix} 0 & \frac{1}{c}\frac{\partial}{\partial t} & 0 & 0 \\ \frac{\mu\hat{\varepsilon}}{c}\frac{\partial}{\partial t} & 0 & 0 & 0 \\ 0 & 0 & 0 & -\frac{1}{c}\frac{\partial}{\partial t} \\ 0 & 0 & -\frac{\mu\hat{\varepsilon}}{c}\frac{\partial}{\partial t} & 0 \end{pmatrix}\begin{pmatrix} E_z \\ B_y \\ E_y \\ B_z \end{pmatrix}. \quad (2.61)$$

System (2.61) can be treated with a procedure similar to the one presented in Eqs. (2.11)–(2.14). This leads us to the dispersion relation, which is similar to Eq. (2.15)

$$k^2 = \mu\varepsilon\frac{\omega^2}{c^2}, \quad (2.62)$$

but ε is a function of ω.

Strictly from those equations one can construct projection operators. Hence we can present a direct relation between $\check{E}_z$ and $\check{H}_y$ and between $\check{E}_y$ and $\check{H}_z$,

$$\varepsilon^{\frac{1}{2}}\mu^{\frac{1}{2}}\check{E}_z = -\check{B}_y, \quad \varepsilon^{\frac{1}{2}}\mu^{\frac{1}{2}}\check{E}_y = \check{B}_z, \quad (2.63)$$

as $k = \frac{1}{c}\omega\mu^{\frac{1}{2}}\varepsilon^{\frac{1}{2}}$. For one of direction propagation for both polarizations, otherwise

$$\varepsilon^{\frac{1}{2}}\mu^{\frac{1}{2}}\breve{E}_z = \breve{B}_y, \quad \varepsilon^{\frac{1}{2}}\mu^{\frac{1}{2}}\breve{E}_y = -\breve{B}_z, \tag{2.64}$$

as $k = -\frac{1}{c}\omega\mu^{\frac{1}{2}}\varepsilon^{\frac{1}{2}}$. with eigenvectors

$$\Psi_{11} = \begin{pmatrix} \tilde{E}_z \\ \varepsilon^{\frac{1}{2}}\mu^{\frac{1}{2}}\tilde{E}_z \\ 0 \\ 0 \end{pmatrix}, \quad \Psi_{12} = \begin{pmatrix} 0 \\ 0 \\ -\tilde{E}_y \\ -\varepsilon^{\frac{1}{2}}\mu^{\frac{1}{2}}\tilde{E}_y \end{pmatrix},$$

$$\Psi_{21} = \begin{pmatrix} -\tilde{E}_z \\ -\varepsilon^{\frac{1}{2}}\mu^{\frac{1}{2}}\tilde{E}_z \\ 0 \\ 0 \end{pmatrix}, \quad \Psi_{22} = \begin{pmatrix} 0 \\ 0 \\ \tilde{E}_y \\ \varepsilon^{\frac{1}{2}}\mu^{\frac{1}{2}}\tilde{E}_y \end{pmatrix} \tag{2.65}$$

that specify four modes. As a starting point we know that they have to fulfill the standard properties of orthogonal projecting operators (Eq. ((2.9))).

Let us show the form of two corresponding projection operators:

$$\mathcal{P}_{11} = \frac{1}{2}\begin{pmatrix} 1 & -\mu^{-\frac{1}{2}}\varepsilon^{-\frac{1}{2}} & 0 & 0 \\ -\varepsilon^{\frac{1}{2}}\mu^{\frac{1}{2}} & 1 & 0 & 0 \\ 0 & 0 & 0 & 0 \\ 0 & 0 & 0 & 0 \end{pmatrix}, \quad \mathcal{P}_{12} = \frac{1}{2}\begin{pmatrix} 0 & 0 & 0 & 0 \\ 0 & 0 & 0 & 0 \\ 0 & 0 & 1 & \mu^{-\frac{1}{2}}\varepsilon^{-\frac{1}{2}} \\ 0 & 0 & \varepsilon^{\frac{1}{2}}\mu^{\frac{1}{2}} & 1 \end{pmatrix}. \tag{2.66}$$

Projection operators for this model have the same form as projection operators generated in the Cauchy problem, but with a significant difference hidden in the form of $\hat{\varepsilon}$. After inverse Fourier transformation one contains its explicit form. The application of the first one yields the transition to new variables in the space of solutions

$$\hat{\mathcal{P}}_{11}\psi' = \frac{1}{2}\begin{pmatrix} 1 & -\mu^{-\frac{1}{2}}\hat{\varepsilon}^{-\frac{1}{2}} & 0 & 0 \\ -\hat{\varepsilon}^{\frac{1}{2}}\mu^{\frac{1}{2}} & 1 & 0 & 0 \\ 0 & 0 & 0 & 0 \\ 0 & 0 & 0 & 0 \end{pmatrix}\begin{pmatrix} E_z \\ B_y \\ E_y \\ B_z \end{pmatrix} = \frac{1}{2}\begin{pmatrix} E_z - \mu^{-\frac{1}{2}}\hat{\varepsilon}^{-\frac{1}{2}}B_y \\ -\hat{\varepsilon}^{\frac{1}{2}}\mu^{\frac{1}{2}}E_z + B_y \\ 0 \\ 0 \end{pmatrix}, \tag{2.67}$$

$$\frac{1}{2}\begin{pmatrix} E_z - \mu^{-\frac{1}{2}}\hat{\varepsilon}^{-\frac{1}{2}}B_y \\ -\hat{\varepsilon}^{\frac{1}{2}}\mu^{\frac{1}{2}}E_z + B_y \\ 0 \\ 0 \end{pmatrix} = \begin{pmatrix} \Pi_1 \\ -\mu^{\frac{1}{2}}\hat{\varepsilon}^{\frac{1}{2}}\Pi_1 \\ 0 \\ 0 \end{pmatrix}. \tag{2.68}$$

The analogous actions allow us to define a system of new variables

$$\begin{aligned} \Pi'_1 &= E_z - \mu^{-\frac{1}{2}}\hat{\varepsilon}^{-\frac{1}{2}}B_y \\ \Lambda'_1 &= E_z + \mu^{-\frac{1}{2}}\hat{\varepsilon}^{-\frac{1}{2}}B_y \\ \Pi'_2 &= E_y - \mu^{-\frac{1}{2}}\hat{\varepsilon}^{-\frac{1}{2}}B_z \\ \Lambda'_2 &= E_y + \mu^{-\frac{1}{2}}\hat{\varepsilon}^{-\frac{1}{2}}B_z. \end{aligned} \tag{2.69}$$

Taking nonlinearity into consideration, the system is described by the equation

$$\partial_x \psi' - L' \psi' = N(\psi'). \tag{2.70}$$

With the use of new projection operators in Eq. (2.70) we get

$$\partial_x \hat{P}_{ij} \psi' - L' \hat{P}_{ij} \psi' = \hat{P}_{ij} N(\psi'). \tag{2.71}$$

This action will generate the system of equations which describes pulse propagation in the right and left directions:

$$\partial_x \hat{P}_{11} \psi' - L' \hat{P}_{11} \psi' = \hat{P}_{11} N(\psi'), \tag{2.72}$$
$$\partial_x \hat{P}_{12} \psi' - L' \hat{P}_{12} \psi' = \hat{P}_{12} N(\psi'), \tag{2.73}$$
$$\partial_x \hat{P}_{21} \psi' - L' \hat{P}_{21} \psi' = \hat{P}_{21} N(\psi'), \tag{2.74}$$
$$\partial_x \hat{P}_{22} \psi' - L' \hat{P}_{22} \psi' = \hat{P}_{22} N(\psi'), \tag{2.75}$$

taking account of both polarizations. Let's take a closer look at Eq. (2.72):

$$\frac{\partial}{\partial_x} \begin{pmatrix} \Pi_1 \\ -\mu^{\frac{1}{2}} \hat{\varepsilon}^{\frac{1}{2}} \Pi_1 \\ 0 \\ 0 \end{pmatrix} - \begin{pmatrix} 0 & \frac{1}{c}\frac{\partial}{\partial t} & 0 & 0 \\ \frac{\mu \hat{\varepsilon}}{c}\frac{\partial}{\partial t} & 0 & 0 & 0 \\ 0 & 0 & 0 & -\frac{1}{c}\frac{\partial}{\partial t} \\ 0 & 0 & -\frac{\mu \hat{\varepsilon}}{c}\frac{\partial}{\partial t} & 0 \end{pmatrix} \begin{pmatrix} \Pi_1 \\ -\mu^{\frac{1}{2}} \hat{\varepsilon}^{\frac{1}{2}} \Pi_1 \\ 0 \\ 0 \end{pmatrix}$$
$$= \hat{P}_{11} N(\psi'). \tag{2.76}$$

As the projection operators have a similar form, the relation given by Eq. (2.37) can be used.

$$\frac{\partial}{\partial x} \hat{\varepsilon} \Pi_1 + \frac{1}{c}\frac{\partial}{\partial t} \mu^{\frac{1}{2}} \hat{\varepsilon}^{\frac{3}{2}} \Pi_1 = -2\pi \frac{\partial}{\partial t} \chi_{yyyy} \left(3(\Pi_1 - \Lambda_1)^3 + (\Pi_1 + \Lambda_1)(\Pi_2 - \Lambda_2)^2 \right). \tag{2.77}$$

After Taylor series expansion and inverse Fourier transformation of Eq. (2.37), we get the formulas

$$\varepsilon^{\frac{1}{2}}(\omega) \approx 1 - 2\left(\pi \chi_0\right) (i\partial_t)^{-2}, \tag{2.78}$$
$$\varepsilon^{\frac{3}{2}}(\omega) \approx 1 - 6\left(\pi \chi_0\right) (i\partial_t)^{-2}, \tag{2.79}$$
$$\varepsilon^{-1}(\omega) \approx 1 + 4\pi \chi_0 (i\partial_t)^{-2}, \tag{2.80}$$

which we apply to Eq. (2.77).

Differentiating twice, both sides over time, we get the equation

$$\frac{\partial}{\partial x}\frac{\partial^2}{\partial t^2}\Pi_1 - 4\pi \chi_0 \frac{\partial}{\partial x}\Pi_1 + \frac{1}{c}\mu^{\frac{1}{2}}\frac{\partial^3}{\partial t^3}\Pi_1$$
$$+ \frac{1}{c}\mu^{\frac{1}{2}} 6\pi \chi_0 \frac{\partial}{\partial t}\Pi_1 = 2\pi \frac{\partial^3}{\partial t^3}\chi_{yyyy}\left(3(\Pi_1 - \Lambda_1)^3 + (\Pi_1 + \Lambda_1)(\Pi_2 - \Lambda_2)^2\right). \tag{2.81}$$

With analogy to [25, 26] we can substitute a multiple scales ansatz, as we have done in Sec. 2.3. The ansatz, by construction, accounts for only one of the directed waves, hence we simply do not take the opposite waves into account, regarding excitation of the correspondent mode which is specified by the projection operators application to boundary conditions vector.

$$\Pi_1(x,t) = \left(\kappa A_0(\phi, x_1, x_2) + \kappa^2 A_1(\phi, x_1, x_2) + \ldots\right),$$

$$\Pi_2(x,t) = \left(\kappa B_0(\phi, x_1, x_2) + \kappa^2 B_1(\phi, x_1, x_2) + \ldots\right),$$

where $\phi = \frac{ct-x}{\kappa}$ and $x_n = \kappa^n x$. Then we get $\Lambda_i = 0$,

$$\left(\frac{\partial^2}{\partial x_1 \partial \phi}\right) A_0 = \frac{4\pi\chi_0}{c^2}\left(1 - \frac{3}{2}\mu^{\frac{1}{2}}\right) A_0 + 2\pi\chi_{yyyy}\frac{\partial^2}{\partial \phi^2}\left(3A_0^3 + A_0 B_0^2\right). \quad (2.82)$$

Comparing our result to one obtained by Bandelow et al. [26],

$$\frac{\partial^2}{\partial x_1 \partial \phi} A_0 = A_0 + \frac{1}{6}\frac{\partial^2}{\partial \phi^2}\left(A_0^3 + 3A_0 B_0^2\right), \quad (2.83)$$

we state that the system (2.82) realizes the result of [26, 36] in the frame of the accepted approximations. The system describes collisions of oppositely propagated short pulses as in [21].

2.5   CYLINDRICAL WAVEGUIDE

2.5.1   ON TRANSVERSAL FIBER MODES

In this part we extend our consideration to a dielectric cylindric waveguide with nonlinearities arising from third-order dielectric susceptibility. The dielectric susceptibility for silica as an isotropic medium is represented only by one component $\chi^{(3)}_{xxxx}$ of the whole susceptibility tensor. This assumption restricts the area of time of the pulse to pulses longer than 0.1 ps. This means an immediate response of the medium and it is the standard model for communication fibers in this range.

Starting from the linearized problem, we saw solutions of the transversal Hondros–Debye problem (in polar variables r, ϕ) inside and outside a infinite waveguide at boundary $r = r_0$, getting the equation for the eigenvalues α_{ln}. We use the complete field description [30] apart from the standard mode distribution of this field, producing the basis of separated transversal and longitudinal z variables, implying a standard Sturm–Liuoville problem for the transverse variables.

2.5.2   SOLUTION OF THE LINEAR PROBLEM, A WAY TO THE PROJECTING PROCEDURE

Projecting in a three-dimensional waveguide needs two-step procedure: first is waveguide transversal mode expansion, which reduces the problem to one dimension for each mode. The second is similar to the ones used in previous sections, but may include extra dynamical variables (fields) and dispersion at the first stage. A similar but simpler problem for rectangular metal waveguides was studied in [1].

The procedure of projecting follows [1]. First we expand basic fields in series with respect to the Hondros–Debye basis $N_{nl}^{-1} J_l(\alpha_{nl}r)e^{il\phi}$ (see [4] expansion):

$$E_z^{\pm}(r,\varphi,z,t) = \sum_{p,l,n} A_{ln}^p(z,t) J_l(\alpha_{nl}r)e^{il\varphi} + c.c., \tag{2.84a}$$

$$E_r^{\pm}(r,\varphi,z,t) = \sum_{p,l,n} B_{ln}^p(z,t)\frac{1}{r} J_l(\alpha_{ln}r)e^{il\varphi}$$

$$+ C_{ln}^p(z,t)\frac{\partial}{\partial r} J_l(\alpha_{ln}r)e^{il\varphi} + c.c., \tag{2.84b}$$

$$E_\varphi^{\pm}(r,\varphi,z,t) = \sum_{p,l,n} D_{ln}^p(z,t)\frac{1}{r} J_l(\alpha_{ln}r)e^{il\varphi}$$

$$+ \sum_{p,l,n} F_{ln}^p(z,t)\frac{\partial}{\partial r} J_l(\alpha_{ln}r)e^{il\varphi} + c.c., \tag{2.84c}$$

$$B_z^{\pm}(r,\varphi,z,t) = \sum_{p,l,n} K_{ln}^p(z,t) J_l(\alpha_{nl}r)e^{il\varphi} + c.c., \tag{2.84d}$$

$$B_r^{\pm}(r,\varphi,z,t) = \sum_{p,l,n} L_{ln}^p(z,t)\frac{1}{r} J_l(\alpha_{ln}r)e^{il\varphi}$$

$$+ \sum_{p,l,n} M_{ln}^p(z,t)\frac{\partial}{\partial r} J_l(\alpha_{ln}r)e^{il\varphi} + c.c., \tag{2.84e}$$

$$B_\varphi^{\pm}(r,\varphi,z,t) = \sum_{p,l,n} R_{ln}^p(z,t)\frac{1}{r} J_l(\alpha_{ln}r)e^{il\varphi}$$

$$+ \sum_{p,l,n} S_{ln}^p(z,t)\frac{\partial}{\partial r} J_l(\alpha_{ln}r)e^{il\varphi} + c.c., \tag{2.84f}$$

and substitute the expansion into linear equations. The Maxwell equations in cylindrical coordinates have the form

$$\frac{\partial}{r\partial r}(rD_r) + \frac{\partial D_\varphi}{r\partial\varphi} + \frac{\partial D_z}{\partial z} = 0, \tag{2.85}$$

$$\frac{\partial}{r\partial r}(rB_r) + \frac{\partial B_\varphi}{r\partial\varphi} + \frac{\partial B_z}{\partial z} = 0, \tag{2.86}$$

$$\left(\frac{\partial E_z}{r\partial\varphi} - \frac{\partial E_\varphi}{\partial z}\right) = -\frac{1}{c}\frac{\partial B_r}{\partial t}, \tag{2.87}$$

$$\left(\frac{\partial E_r}{\partial z} - \frac{\partial E_z}{\partial r}\right) = -\frac{1}{c}\frac{\partial B_\varphi}{\partial t}, \tag{2.88}$$

$$\frac{1}{r}\left(\frac{\partial (rE_\varphi)}{\partial r} - \frac{\partial E_r}{\partial \varphi}\right) = -\frac{1}{c}\frac{\partial B_z}{\partial t}, \tag{2.89}$$

$$\left(\frac{\partial H_z}{r\partial \varphi} - \frac{\partial H_\varphi}{\partial z}\right) = \frac{1}{c}\frac{\partial D_r}{\partial t}, \tag{2.90}$$

$$\left(\frac{\partial H_r}{\partial z} - \frac{\partial H_z}{\partial r}\right) = \frac{1}{c}\frac{\partial D_\varphi}{\partial t}, \tag{2.91}$$

$$\frac{1}{r}\left(\frac{\partial (rH_\varphi)}{\partial r} - \frac{\partial H_r}{\partial \varphi}\right) = \frac{1}{c}\frac{\partial D_z}{\partial t}. \tag{2.92}$$

Now it is possible to present the first two Maxwell equations in a transverse function basis. Bessel functions satisfy orthogonality conditions and hence form a convenient basis. To complete the statement of the problem, we define an initial value problem with conditions at the boundary $z = 0$ that fix the transversal components (polarizations) of the electric field.

$$\frac{\partial}{r\partial r}r\left(\sum_{p,l,n}\mathcal{B}^p_{ln}(z,t)\frac{1}{r}J_l(\alpha_{ln}r)e^{il\varphi} + \sum_{p,l,n}\mathcal{C}^p_{ln}(z,t)\frac{\partial}{\partial r}J_l(\alpha_{ln}r)e^{il\varphi}\right)$$

$$+ \frac{\partial}{r\partial\varphi}\left(\sum_{p,l,n}\mathcal{D}^p_{ln}(z,t)\frac{1}{r}J_l(\alpha_{ln}r)e^{il\varphi} + \sum_{p,l,n}\mathcal{F}^p_{ln}(z,t)\frac{\partial}{\partial r}J_l(\alpha_{ln}r)e^{il\varphi}\right)$$

$$+ \frac{\partial}{\partial z}\left(\sum_{p,l,n}\mathcal{A}^p_{ln}(z,t)J_l(\alpha_{nl}r)e^{il\varphi}\right) = 0, \tag{2.93}$$

$$\frac{\partial}{r\partial r}\left(r(\sum_{p,l,n}\mathcal{L}^p_{ln}(z,t)\frac{1}{r}J_l(\alpha_{ln}r)e^{il\varphi} + \sum_{p,l,n}\mathcal{M}^p_{ln}(z,t)\frac{\partial}{\partial r}J_l(\alpha_{ln}r)e^{il\varphi})\right)$$

$$+ \frac{\partial}{r\partial\varphi}\left(\sum_{p,l,n}\mathcal{R}^p_{ln}(z,t)\frac{1}{r}J_l(\alpha_{ln}r)e^{il\varphi} + \sum_{p,l,n}\mathcal{S}^p_{ln}(z,t)\frac{\partial}{\partial r}J_l(\alpha_{ln}r)e^{il\varphi}\right)$$

$$+ \frac{\partial}{\partial z}\sum_{p,l,n}\mathcal{K}^p_{ln}(z,t)J_l(\alpha_{nl}r)e^{il\varphi} = 0. \tag{2.94}$$

For further calculations the Bessel function properties will be necessary.

$$\frac{\partial}{\partial r}J_l(r) = \frac{1}{2}\left(J_{l-1}(r) - J_{l+1}(r)\right), \tag{2.95}$$

$$\frac{l}{r}J_l(r) = \frac{1}{2}\left(J_{l-1}(r) + J_{l+1}(r)\right). \tag{2.96}$$

As we will solve the first equation from the above system (2.94), it will be possible to expand the result to the second one.

$$\mathcal{B}_{ln}^{p}(z,t)\frac{\partial}{r\partial r}J_{l}(\alpha_{ln}r) + \mathcal{C}_{ln}^{p}(z,t)\frac{\partial}{r\partial r}r\frac{\partial}{\partial r}J_{l}(\alpha_{ln}r) + \tag{2.97}$$

$$\mathcal{D}_{ln}^{p}(z,t)\frac{\partial}{r\partial\varphi}\frac{1}{r}J_{l}(\alpha_{ln}r) + \mathcal{F}_{ln}^{p}(z,t)\frac{\partial}{r\partial\varphi}\frac{\partial}{\partial r}J_{l}(\alpha_{ln}r) +$$

$$\frac{\partial}{\partial z}\mathcal{A}_{ln}^{p}(z,t)J_{l}(\alpha_{nl}r) = 0.$$

This allows us to rearrange the coefficients by C_{ln}. After transformations, one can apply the result to Eq. (2.97).

$$(\mathcal{B}_{ln}^{p}(z,t) + il\mathcal{F}_{ln}^{p}(z,t))\frac{\partial}{r\partial r}J_{l}(\alpha_{ln}r)$$

$$+ (\mathcal{C}_{ln}^{p}(z,t)l^{2} + il\mathcal{D}_{ln}^{p}(z,t))\frac{1}{r^{2}}J_{l}(\alpha_{ln}r)$$

$$+ \left(\frac{\partial}{\partial z}\mathcal{A}_{ln}^{p}(z,t) - \mathcal{C}_{ln}^{p}(z,t)\alpha_{ln}^{2}\right)J_{l}(\alpha_{nl}r) = 0.$$

Hence, with the use of the independence of $\frac{\partial}{r\partial r}J_{l}(\alpha_{ln}r)$, $\frac{1}{r^{2}}J_{l}(\alpha_{ln}r)$ and $J_{l}(\alpha_{nl}r)$, we can write the equations for coefficients:

$$\mathcal{B}_{ln}^{p}(z,t) + il\mathcal{F}_{ln}^{p}(z,t) = 0, \tag{2.98}$$

$$\mathcal{C}_{ln}^{p}(z,t)l^{2} + il\mathcal{D}_{ln}^{p}(z,t) = 0, \tag{2.99}$$

$$\frac{\partial}{\partial z}\mathcal{A}_{ln}^{p}(z,t) - \mathcal{C}_{ln}^{p}(z,t)\alpha_{ln}^{2} = 0. \tag{2.100}$$

Based on analogy between Eq. (2.93) and Eq. (2.94), the relation between coefficients can be presented by the system of equations

$$\mathcal{L}_{ln}^{p}(z,t) + il\mathcal{S}_{ln}^{p}(z,t) = 0, \tag{2.101}$$

$$\mathcal{M}_{ln}^{p}(z,t)l^{2} + il\mathcal{R}_{ln}^{p}(z,t) = 0, \tag{2.102}$$

$$\frac{\partial}{\partial z}\mathcal{K}_{ln}^{p}(z,t) - \mathcal{M}_{ln}^{p}(z,t)\alpha_{ln}^{2} = 0. \tag{2.103}$$

We repeat the procedure for the rest of the Maxwell equations in a Bessel function s basis. The results obtained for the system directly are complex. The relations between coefficients and algebraic relations have been used to simplify the results to gain a more adoptable one. The result obtained can be presented as a system of six equations:

$$\frac{\varepsilon\mu}{c}\frac{\partial}{\partial t}\mathcal{A}_{ln}^p(z,t) = \frac{i\alpha_{ln}^2}{l}\mathcal{L}_{ln}^p(z,t),$$

$$\frac{\varepsilon\mu}{c}\frac{\partial}{\partial t}\mathcal{B}_{ln}^p(z,t) = il\mathcal{K}_{ln}^p(z,t) - il\frac{\partial}{\partial z}\mathcal{M}_{ln}^p(z,t),$$

$$\frac{\varepsilon\mu}{c}\frac{\partial}{\partial t}\mathcal{C}_{ln}^p(z,t) = -\frac{i}{l}\frac{\partial}{\partial z}\mathcal{L}_{ln}^p(z,t),$$

$$\frac{1}{c}\frac{\partial}{\partial t}\mathcal{K}_{ln}^p(z,t) = -\frac{i\alpha_{ln}^2}{l}\mathcal{B}_{ln}^p(z,t),\qquad (2.104)$$

$$\frac{1}{c}\frac{\partial}{\partial t}\mathcal{L}_{ln}^p(z,t) = -il\mathcal{A}_{ln}^p(z,t) + \frac{\partial}{\partial z}il\mathcal{C}_{ln}^p(z,t),$$

$$\frac{1}{c}\frac{\partial}{\partial t}\mathcal{M}_{ln}^p(z,t) = \frac{i}{l}\frac{\partial}{\partial z}\mathcal{B}_{ln}^p(z,t),$$

which is a basis for an application of the projection operator method.

From now on the differential operator $\frac{\partial}{\partial x_i}$ will be written as ∂_{x_i}. The system of equations can be rewritten in the matrix form

$$\Psi = \begin{pmatrix} \mathcal{A}_{ln}^p(z,t) \\ \mathcal{B}_{ln}^p(z,t) \\ \mathcal{C}_{ln}^p(z,t) \\ \mathcal{K}_{ln}^p(z,t) \\ \mathcal{L}_{ln}^p(z,t) \\ \mathcal{M}_{ln}^p(z,t) \end{pmatrix} \quad L$$

$$= \begin{pmatrix} 0 & 0 & 0 & 0 & ic\alpha_{ln}^2(l\varepsilon\mu)^{-1} & 0 \\ 0 & 0 & 0 & ilc(\varepsilon\mu)^{-1} & 0 & -ilc(\varepsilon\mu)^{-1}\partial_z \\ 0 & 0 & 0 & 0 & -ic(l\varepsilon\mu)^{-1}\partial_z & 0 \\ 0 & -ic\alpha_{ln}^2(l)^{-1} & 0 & 0 & 0 & 0 \\ -ilc & 0 & ilc\partial_z & 0 & 0 & 0 \\ 0 & ic(l)^{-1}\partial_z & 0 & 0 & 0 & 0 \end{pmatrix}.$$

$$(2.105)$$

To construct projection operators for our problem (2.104) it is convenient to find eigenvectors for the linear evolution operator $\hat{L}$ [31] which will be applied to an evolution equation $\partial_t \Psi(z,t) - \hat{L}\Psi = 0$. On the other hand, the system can be formulated as a boundary regime (along z) problem.

$$\frac{i}{l}\frac{\partial}{\partial z}\mathcal{L}_{ln}^p(z,t) = -\frac{\varepsilon\mu}{c}\frac{\partial}{\partial t}\mathcal{C}_{ln}^p(z,t), \qquad (2.106a)$$

$$il\frac{\partial}{\partial z}\mathcal{M}_{ln}^p(z,t) = -\frac{\varepsilon\mu}{c}\frac{\partial}{\partial t}\mathcal{B}_{ln}^p(z,t) + il\mathcal{K}_{ln}^p(z,t), \qquad (2.106b)$$

$$\frac{\partial}{\partial z}il\mathcal{C}_{ln}^p(z,t) = \frac{1}{c}\frac{\partial}{\partial t}\mathcal{L}_{ln}^p(z,t) + il\mathcal{A}_{ln}^p(z,t), \qquad (2.106c)$$

$$\frac{i}{l}\frac{\partial}{\partial z}\mathcal{B}_{ln}^p(z,t) = \frac{1}{c}\frac{\partial}{\partial t}\mathcal{M}_{ln}^p(z,t), \qquad (2.106d)$$

where two variables $\mathcal{L}_{ln}^{p}(z,t)$, $\mathcal{B}_{ln}^{p}(z,t)$ may be excluded, by

$$i\frac{\alpha_{ln}^{2}c}{l\varepsilon\mu}\mathcal{L}_{ln}^{p}(z,t) = \frac{\partial}{\partial t}\mathcal{A}_{ln}^{p}(z,t), \tag{2.107a}$$

$$-\frac{ic\alpha_{ln}^{2}}{l}\mathcal{B}_{ln}^{p}(z,t) = \frac{\partial}{\partial t}\mathcal{K}_{ln}^{p}(z,t). \tag{2.107b}$$

$$\tag{2.107c}$$

From the point of view of boundary problem consideration we use an anti-symmetric version in t basis functions $\mathcal{A}_{ln}^{p}$, $\mathcal{B}_{ln}^{p}$, $\mathcal{C}_{ln}^{p}$, $\mathcal{K}_{ln}^{p}$, $\mathcal{L}_{ln}^{p}$, $\mathcal{M}_{ln}^{p}$ defined for $t \in (-\infty;\infty)$. Such an approach guarantees a zero field amplitude at $t = 0$. Moreover, with such a continuation of the field, its Fourier transform is

$$\mathcal{A}_{ln}^{p} = \tilde{A}_{ln}^{p}e^{i\omega t}. \tag{2.108}$$

After some algebra, we obtain

$$\tilde{A}_{ln}^{p}(z,\omega) = \frac{\alpha_{ln}^{2}c}{l\omega\varepsilon\mu}\tilde{L}_{ln}^{p}(z,\omega), \tag{2.109a}$$

$$\tilde{K}_{ln}^{p}(z,\omega) = -\frac{c\alpha_{ln}^{2}}{l\omega}\tilde{B}_{ln}^{p}(z,\omega), \tag{2.109b}$$

$$\frac{\partial}{\partial z}\tilde{B}_{ln}^{p}(z,\omega) = \frac{l\omega}{c}\tilde{M}_{ln}^{p}(z,\omega), \tag{2.109c}$$

$$\frac{\partial}{\partial z}\tilde{C}_{ln}^{p}(z,\omega) = \left(l\frac{\omega}{c} + \frac{\alpha_{ln}^{2}c}{l\omega\varepsilon\mu}\right)\tilde{L}_{ln}^{p}(z,\omega), \tag{2.109d}$$

$$\frac{\partial}{\partial z}\tilde{L}_{ln}^{p}(z,\omega) = -l\omega\frac{\varepsilon\mu}{c}\tilde{C}_{ln}^{p}(z,\omega), \tag{2.109e}$$

$$\frac{\partial}{\partial z}\tilde{M}_{ln}^{p}(z,\omega) = \left(-\varepsilon\mu l\frac{\omega}{c} - \frac{c\alpha_{ln}^{2}}{l\omega}\right)\tilde{B}_{ln}^{p}(z,\omega). \tag{2.109f}$$

In matrix form it is equivalent to

$$\partial_{z}\begin{pmatrix} \tilde{B}_{ln}^{p}(z,\omega) \\ \tilde{C}_{ln}^{p}(z,\omega) \\ \tilde{L}_{ln}^{p}(z,\omega) \\ \tilde{M}_{ln}^{p}(z,\omega) \end{pmatrix} = \begin{pmatrix} 0 & 0 & 0 & \frac{l\omega}{c} \\ 0 & 0 & l\frac{\omega}{c} + \frac{\alpha_{ln}^{2}c}{l\omega\varepsilon\mu} & 0 \\ 0 & -l\omega\frac{\varepsilon\mu}{c} & 0 & 0 \\ -\varepsilon\mu l\frac{\omega}{c} - \frac{c\alpha_{ln}^{2}}{l\omega} & 0 & 0 & 0 \end{pmatrix}\begin{pmatrix} \tilde{B}_{ln}^{p}(z,\omega) \\ \tilde{C}_{ln}^{p}(z,\omega) \\ \tilde{L}_{ln}^{p}(z,\omega) \\ \tilde{M}_{ln}^{p}(z,\omega) \end{pmatrix}. \tag{2.110}$$

Or as the system of ordinary differential equations

$$\partial_{z}\begin{pmatrix} \tilde{B}_{ln}^{p}(z,\omega) \\ \tilde{C}_{ln}^{p}(z,\omega) \\ \tilde{L}_{ln}^{p}(z,\omega) \\ \tilde{M}_{ln}^{p}(z,\omega) \end{pmatrix} = \begin{pmatrix} \frac{l\omega}{c}\tilde{M}_{ln}^{p}(z,\omega) \\ (l\frac{\omega}{c} + \frac{\alpha_{ln}^{2}c}{l\omega\varepsilon\mu})\tilde{L}_{ln}^{p}(z,\omega) \\ -l\omega\frac{\varepsilon\mu}{c}\tilde{C}_{ln}^{p}(z,\omega) \\ -(l\varepsilon\mu\frac{\omega}{c} + \frac{c\alpha_{ln}^{2}}{l\omega})\tilde{B}_{ln}^{p}(z,\omega) \end{pmatrix}. \tag{2.111}$$

Going to the corresponding eigenvalue matrix problem by $\tilde{B}^p_{ln}(z,\omega) = \check{B}^p_{ln}(k,\omega)e^{ikz}$,

$$
\begin{pmatrix}
ik\check{B}^p_{ln}(k,\omega) \\
ik\check{C}^p_{ln}(k,\omega) \\
ik\check{L}^p_{ln}(k,\omega) \\
ik\check{M}^p_{ln}(k,\omega)
\end{pmatrix}
=
\begin{pmatrix}
\frac{l\omega}{c}\check{M}^p_{ln}(k,\omega) \\
(l\frac{\omega}{c} + \frac{\alpha^2_{ln}c}{l\omega\varepsilon\mu})\check{L}^p_{ln}(k,\omega) \\
-l\omega\frac{\varepsilon\mu}{c}\check{C}^p_{ln}(k,\omega) \\
-(l\varepsilon\mu\frac{\omega}{c} + \frac{c\alpha^2_{ln}}{l\omega})\check{B}^p_{ln}(k,\omega)
\end{pmatrix} .
\tag{2.112}
$$

The condition of the nonzero solution is obtained by expansion of the determinant.

$$
\det
\begin{pmatrix}
ik & 0 & 0 & \frac{l\omega}{c} \\
0 & ik & (l\frac{\omega}{c} + \frac{\alpha^2_{ln}c}{l\omega\varepsilon\mu}) & 0 \\
0 & -l\omega\frac{\varepsilon\mu}{c} & ik & 0 \\
-(l\varepsilon\mu\frac{\omega}{c} + \frac{c\alpha^2_{ln}}{l\omega}) & 0 & 0 & ik
\end{pmatrix}
= 0.
\tag{2.113}
$$

$$
ik\det
\begin{pmatrix}
ik & (l\frac{\omega}{c} + \frac{\alpha^2_{ln}c}{l\omega\varepsilon\mu}) & 0 \\
-l\omega\frac{\varepsilon\mu}{c} & ik & 0 \\
0 & 0 & ik
\end{pmatrix}
$$

$$
-\frac{l\omega}{c}\det
\begin{pmatrix}
0 & ik & (l\frac{\omega}{c} + \frac{\alpha^2_{ln}c}{l\omega\varepsilon\mu}) \\
0 & -l\omega\frac{\varepsilon\mu}{c} & ik \\
-(l\varepsilon\mu\frac{\omega}{c} + \frac{c\alpha^2_{ln}}{l\omega}) & 0 & 0
\end{pmatrix}
= 0.
\tag{2.114}
$$

It yields the dispersion relation that accounts for both waveguide and material dispersion hidden in ϵ, μ.

Solving it gives, for positive k,

$$
\left(k^2 - \alpha^2_{ln} - l^2\varepsilon\mu\frac{\omega^2}{c^2}\right) = 0.
\tag{2.115}
$$

Finally, each number α_{nl} defines a branch of the dispersion relation via

$$
k^2 = (l^2\omega^2\frac{\varepsilon\mu}{c^2} + \alpha^2_{nl}),
\tag{2.116}
$$

where ε is the dielectric constant of the fiber. The differentiation over ω gives the group velocity

$$
k'(\omega) = \frac{2\omega c^2\varepsilon + \omega^2 c^2\varepsilon'}{2\sqrt{\omega^2 c^2\varepsilon + \alpha^2_{nl}}},
\tag{2.117}
$$

where ε is a function over ω and $\varepsilon' = \frac{\varepsilon}{\omega}$.

Now we obtain the group velocity dependent on a small parameter ε which is proportional to $\Delta k/2\alpha \ll 1$ and is related to the wave packet width.

$$
\begin{pmatrix}
i\sqrt{l^2\omega^2\frac{\varepsilon\mu}{c^2}+\alpha_{nl}^2}\,\breve{B}_{ln}^p(k,\omega) \\
i\sqrt{l^2\omega^2\frac{\varepsilon\mu}{c^2}+\alpha_{nl}^2}\,\breve{C}_{ln}^p(k,\omega) \\
i\sqrt{l^2\omega^2\frac{\varepsilon\mu}{c^2}+\alpha_{nl}^2}\,\breve{L}_{ln}^p(k,\omega) \\
i\sqrt{l^2\omega^2\frac{\varepsilon\mu}{c^2}+\alpha_{nl}^2}\,\breve{M}_{ln}^p(k,\omega)
\end{pmatrix}
=
\begin{pmatrix}
\frac{l\omega}{c}\breve{M}_{ln}^p(k,\omega) \\
(l\frac{\omega}{c}+\frac{\alpha_{ln}^2 c}{l\omega\varepsilon\mu})\breve{L}_{ln}^p(k,\omega) \\
-l\omega\frac{\varepsilon\mu}{c}\breve{C}_{ln}^p(k,\omega) \\
-(l\varepsilon\mu\frac{\omega}{c}+\frac{c\alpha_{ln}^2}{l\omega})\breve{B}_{ln}^p(k,\omega)
\end{pmatrix}.
\tag{2.118}
$$

$$
\breve{B}_{ln}^p(k,\omega) = -i\frac{l\omega}{c\sqrt{l^2\omega^2\frac{\varepsilon\mu}{c^2}+\alpha_{nl}^2}}\breve{M}_{ln}^p(k,\omega),
\tag{2.119}
$$

$$
\breve{C}_{ln}^p(k,\omega) = -i\frac{c}{l\omega}(\sqrt{l^2\omega^2\frac{\varepsilon\mu}{c^2}+\alpha_{nl}^2})\breve{L}_{ln}^p(k,\omega),
\tag{2.120}
$$

$$
\breve{L}_{ln}^p(k,\omega) = i\varepsilon\mu\frac{l\omega}{c\sqrt{l^2\omega^2\frac{\varepsilon\mu}{c^2}+\alpha_{nl}^2}}\breve{C}_{ln}^p(k,\omega),
\tag{2.121}
$$

$$
\breve{M}_{ln}^p(k,\omega) = i\frac{c}{l\omega}(\sqrt{l^2\omega^2\frac{\varepsilon\mu}{c^2}+\alpha_{nl}^2})\breve{B}_{ln}^p(k,\omega).
\tag{2.122}
$$

$$
P_{11} = \frac{1}{2}
\begin{pmatrix}
1 & 0 & 0 & -i\frac{l\omega}{c(\sqrt{l^2\omega^2\frac{\varepsilon\mu}{c^2}+\alpha_{nl}^2})} \\
0 & 0 & 0 & 0 \\
0 & 0 & 0 & 0 \\
i\frac{c}{l\omega}(\sqrt{l^2\omega^2\frac{\varepsilon\mu}{c^2}+\alpha_{nl}^2}) & 0 & 0 & 1
\end{pmatrix},
\tag{2.123}
$$

$$
P_{21} = \frac{1}{2}
\begin{pmatrix}
1 & 0 & 0 & i\frac{l\omega}{c(\sqrt{l^2\omega^2\frac{\varepsilon\mu}{c^2}+\alpha_{nl}^2})} \\
0 & 0 & 0 & 0 \\
0 & 0 & 0 & 0 \\
-i\frac{c}{l\omega}(\sqrt{l^2\omega^2\frac{\varepsilon\mu}{c^2}+\alpha_{nl}^2}) & 0 & 0 & 1
\end{pmatrix},
\tag{2.124}
$$

$$
P_{12} = \frac{1}{2}
\begin{pmatrix}
0 & 0 & 0 & 0 \\
0 & 1 & -i\frac{c}{l\omega}(\sqrt{l^2\omega^2\frac{\varepsilon\mu}{c^2}+\alpha_{nl}^2}) & 0 \\
0 & i\frac{l\omega}{c(\sqrt{l^2\omega^2\frac{\varepsilon\mu}{c^2}+\alpha_{nl}^2})} & 1 & 0 \\
0 & 0 & 0 & 0
\end{pmatrix},
\tag{2.125}
$$

$$
P_{22} = \frac{1}{2}
\begin{pmatrix}
0 & 0 & 0 & 0 \\
0 & 1 & i\frac{c}{l\omega}(\sqrt{l^2\omega^2\frac{\varepsilon\mu}{c^2}+\alpha_{nl}^2}) & 0 \\
0 & -i\frac{l\omega}{c(\sqrt{l^2\omega^2\frac{\varepsilon\mu}{c^2}+\alpha_{nl}^2})} & 1 & 0 \\
0 & 0 & 0 & 0
\end{pmatrix}.
\tag{2.126}
$$

$$\breve{\Psi}_{11} = \frac{1}{2} \begin{pmatrix} 1 \\ 0 \\ 0 \\ i\frac{c}{l\omega}(\sqrt{l^2\omega^2\frac{\varepsilon\mu}{c^2} + \alpha_{nl}^2}) \end{pmatrix} \breve{B}_{ln}^p(k,\omega), \qquad (2.127)$$

$$\breve{\Psi}_{21} = \frac{1}{2} \begin{pmatrix} 0 \\ 1 \\ i\frac{c}{l\omega}(\sqrt{l^2\omega^2\frac{\varepsilon\mu}{c^2} + \alpha_{nl}^2}) \\ 0 \end{pmatrix} \breve{C}_{ln}^p(k,\omega), \qquad (2.128)$$

$$\breve{\Psi}_{12} = \frac{1}{2} \begin{pmatrix} 1 \\ 0 \\ 0 \\ -i\frac{c}{l\omega}(\sqrt{l^2\omega^2\frac{\varepsilon\mu}{c^2} + \alpha_{nl}^2}) \end{pmatrix} \breve{B}_{ln}^p(k,\omega), \qquad (2.129)$$

$$\breve{\Psi}_{22} = \frac{1}{2} \begin{pmatrix} 0 \\ 1 \\ -\frac{c}{l\omega}(\sqrt{l^2\omega^2\frac{\varepsilon\mu}{c^2} + \alpha_{nl}^2}) \\ 0 \end{pmatrix} \breve{C}_{ln}^p(k,\omega). \qquad (2.130)$$

$$\partial_z P_{11}\Psi(z,\omega) - \hat{L}P_{11}\Psi(z,\omega) = 0. \qquad (2.131)$$

$$\partial_z \begin{pmatrix} -i\frac{l\omega}{c\hat{k}_{nl}}\Lambda(z,\omega) \\ 0 \\ 0 \\ \Lambda(z,\omega) \end{pmatrix}$$

$$- \begin{pmatrix} 0 & 0 & 0 & \frac{l\omega}{c} \\ 0 & 0 & l\frac{\omega}{c} + \frac{\alpha_{ln}^2 c}{l\omega\varepsilon\mu} & 0 \\ 0 & -l\omega\frac{\varepsilon\mu}{c} & 0 & 0 \\ -\frac{c}{l\omega}(\varepsilon\mu l^2\frac{\omega^2}{c^2} - \alpha_{ln}^2) & 0 & 0 & 0 \end{pmatrix} \begin{pmatrix} -i\frac{l\omega}{c\hat{k}_{nl}}\Lambda(z,\omega) \\ 0 \\ 0 \\ \Lambda(z,\omega) \end{pmatrix}$$

$$= 0. \qquad (2.132)$$

$$\partial_z \begin{pmatrix} -i\frac{l\omega}{c\hat{k}_{nl}}\Lambda(z,\omega) \\ 0 \\ 0 \\ \Lambda(z,\omega) \end{pmatrix} - \begin{pmatrix} \frac{l\omega}{c}\Lambda(z,\omega) \\ 0 \\ 0 \\ i\hat{k}_{nl}\Lambda(z,\omega) \end{pmatrix} = 0. \qquad (2.133)$$

$$\partial_z\Lambda(z,\omega) - i\hat{k}_{nl}\Lambda(z,\omega) = 0. \qquad (2.134)$$

The Taylor series expansion of $\frac{\hat{k}}{\omega} = \sqrt{l^2 \frac{\epsilon\mu}{c^2} + \frac{\alpha_{nl}^2}{\omega^2}}$ implies the modified form of the equation

$$\partial_z \Lambda(z,\omega) - i(\frac{l}{c}(\epsilon)^{\frac{1}{2}}(\mu)^{\frac{1}{2}}\omega + \frac{c\alpha^2(\epsilon)^{-\frac{1}{2}}(\mu)^{-\frac{1}{2}}}{2l}\frac{1}{\omega})\Lambda(z,\omega) = 0. \qquad (2.135)$$

After inverse Fourier transformation,

$$\partial_z \Lambda(z,\omega) - i(\sqrt{\frac{l^2\epsilon\mu}{c^2}}\partial_t + \frac{c^2\alpha^2\sqrt{\frac{l^2\epsilon\mu}{c^2}}}{2l^2\epsilon\mu}\partial_t^{-1})\Lambda(z,\omega) = 0. \qquad (2.136)$$

The dielectric susceptibility coefficient $\varepsilon(\omega)$ either originated from the quantum version of the Lorentz formula (see, e.g., [33, 34]) or directly from phenomenology [10, 35], for example, approximated as a Taylor expansion at $\omega = \infty$ by Bandelow et al. [26].

$$\varepsilon(\omega) \approx 1 + 4\pi\chi(\omega) \approx 1 - 4\pi\chi_0\omega^{-2}. \qquad (2.137)$$

However, including the ϵ dependency over ω (Eq. (2.137)), it is necessary to rewrite Eq (2.135).

$$\partial_z \Lambda(z,\omega) - i\omega\sqrt{l^2\frac{(1 - 4\pi\chi_0\omega^{-2})\mu}{c^2} + \frac{\alpha_{nl}^2}{\omega^2}}\Lambda(z,\omega) = 0. \qquad (2.138)$$

$$\partial_z \Lambda(z,\omega) - i\omega\sqrt{l^2\frac{\mu}{c^2} + \frac{\alpha^2 - l^2 * \mu c^{-2}4\pi\chi_0}{\omega^2}}\Lambda(z,\omega) = 0. \qquad (2.139)$$

$$\partial_z \Lambda(z,\omega) - i\omega(\sqrt{\alpha^2 - \frac{4l^2\pi\mu\chi_0}{c^2}}\omega^{-1} - \frac{l^2\mu}{2c^2\sqrt{\alpha^2 - \frac{4l^2\pi\mu\chi_0}{c^2}}}\omega)\Lambda(z,\omega) = 0. \qquad (2.140)$$

After inverse Fourier transformation,

$$\partial_z \hat{\Lambda}(z,t) - i\partial_t(\sqrt{\alpha^2 - \frac{4l^2\pi\mu\chi_0}{c^2}}\partial_t^{-1} + \frac{l^2\mu}{2c^2\sqrt{\alpha^2 - \frac{4l^2\pi\mu\chi_0}{c^2}}}\partial_t)\hat{\Lambda}(z,t) = 0. \qquad (2.141)$$

2.6   INCLUDING NONLINEARITY

Allow us to present now the equation for the nl-mode amplitude A of the electric field z component:

$$\left(\Box_z + \alpha_{ln}^2\right) A_{ln}^p = \frac{2}{\pi c^2 N_{nl}} \int\limits_0^{r_0} \int\limits_0^{2\pi} r J_l(\alpha_{ln} r) e^{-il\varphi} \frac{\partial^2}{\partial t^2} \sum_{klm} \chi_{zklm} E_k E_l E_m d\varphi dr.$$

$$(2.142)$$

The coefficients of expansion (functions of z, t) form a system to which we apply the procedure explained in previous sections, and arriving at 4×4 matrix projectors which define mode subspaces that differ by polarization and propagation direction. Next, acting by projecting operators on nonlinear equations, we obtain a system defining the mode interaction.

Introducing scaling (applied in the context of experiments on the propagation of pulses with a slowly changing envelope $K^\pm$ [37] as the simplest way to pass from the dimensionless description to the experimental units), we obtain

$$\xi = \sigma z, \quad \tau = (t - k'z)\epsilon, \quad A_{ln}^\pm(t, z) = \sigma X_{ln}^\pm(\tau, \xi) e^{i(\omega t - k^\pm z)}$$
$$= \sigma X e^{i\omega\tau/\epsilon + i(k'-k)\xi/\sigma} + c.c..$$

$$(2.143)$$

We now evaluate the right-hand side of Eq. (2.142). We first substitute Eq. (2.84) in the right-hand side of Eq. (2.142) and integrate over r and ϕ. We then take into account that a fiber carries only a few modes and emits the higher modes. For example, a fiber can exist carrying only one mode. After a scaling procedure, the new derivations have been introduced.

$$A_{ln\,t}^\pm = A_{ln\,\tau}^\pm \epsilon \quad A_{ln\,tt}^\pm = A_{ln\,\tau\tau}^\pm \epsilon^2, \tag{2.144}$$

$$A_{ln\,z}^\pm = A_{ln\,\xi}^\pm \sigma - k' A_{ln\,\tau}^\pm \epsilon \quad A_{ln\,zz}^\pm$$

$$= A_{ln\,\xi\xi}^\pm \sigma^2 - 2k' A_{ln\,\xi\tau}^\pm \sigma\epsilon + k'^2 A_{ln\,\tau\tau}^\pm \epsilon^2. \tag{2.145}$$

To present a final equation, we will apply the solution (2.143) which can be applied to the left-hand side of Eqs. (2.142)–(2.144).

$$\frac{\hat{\epsilon}}{c^2} A_{ln\,tt}^\pm - A_{ln\,zz}^\pm = [\frac{\hat{\epsilon}}{c^2} - k'^2]\epsilon^2 A_{ln\,\tau\tau}^\pm - A_{ln\,\xi\xi}^\pm \sigma^2 + 2k' A_{ln\,\xi\tau}^\pm \sigma\epsilon, \tag{2.146}$$

with new variables. The vector field can by presented with the use of solution (2.143).

$$A_{ln\,\tau}^\pm = [\sigma X e^{i\omega\tau/\epsilon + i(k'-k)\xi/\sigma} + c.c.]_\tau = \sigma X_\tau e^{i\omega\tau/\epsilon + i(k'-k)\xi/\sigma} +$$
$$i\sigma \frac{\omega}{\epsilon} X e^{i\omega\tau/\epsilon + i(k'-k)\xi/\sigma} + c.c.$$

$$A_{ln\,\tau\tau}^\pm = \sigma[X_{\tau\tau} + i\frac{\omega}{\epsilon} X_\tau - \frac{\omega^2}{\epsilon^2} X] e^{i\omega\tau/\epsilon + i(k'-k)\xi/\sigma} + c.c.$$

$$A_{ln\,\xi\tau}^\pm = [\sigma X_\tau e^{i\omega\tau/\epsilon + i(k'-k)\xi/\sigma} + i\frac{\omega}{\epsilon}\sigma X e^{i\omega\tau/\epsilon + i(k'-k)\xi/\sigma} + c.c.]_\xi = \sigma[X_{\tau\xi} +$$
$$i\frac{k'-k}{\sigma} X_\tau / + i\frac{\omega}{\epsilon} X_\xi] e^{i\omega\tau/\epsilon + i(k'-k)\xi/\sigma} + c.c.$$

$$A_{ln\,\xi}^\pm = [\sigma X e^{i\omega\tau/\epsilon + i(k'-k)\xi/\sigma} + c.c.]_\xi = \sigma[X_\xi + i\frac{k'-k}{\sigma} X] e^{i\omega\tau/\epsilon + i(k'-k)\xi/\sigma} + c.c.$$

$$\mathcal{A}^{\pm}_{ln\,\xi\xi} = [\sigma(X_\xi + i\tfrac{k'-k}{\sigma}X)e^{i\omega\tau/\epsilon + i(k'-k)\xi/\sigma} + c.c.]_\xi = \sigma[X_{\xi\xi} - (\tfrac{k'-k}{\sigma})^2 X +$$
$$i2\tfrac{k'-k}{\sigma}X_\xi]e^{i\omega\tau/\epsilon + i(k'-k)\xi/\sigma} + c.c.$$

Applying our result to Eq. (2.146),

$$[\frac{\hat{\varepsilon}}{c^2} - k'^2\epsilon^2]\mathcal{A}^{\pm}_{ln\,\tau\tau} - \mathcal{A}^{\pm}_{ln\,\xi\xi}\sigma^2 + 2k'\mathcal{A}^{\pm}_{ln\,\xi\tau}\sigma\epsilon =$$

$$\sigma\Big([\frac{\hat{\varepsilon}}{c^2} - k'^2]\epsilon^2\big(X_{\tau\tau} + i\frac{\omega}{\epsilon}X_\tau - \frac{\omega^2}{\epsilon^2}X\big) - \big(X_{\xi\xi} - (\frac{k'-k}{\sigma})^2 X \hspace{2cm} (2.147)$$

$$+ i2\frac{k'-k}{\sigma}X_\xi\big)\sigma^2 + 2k'\sigma\epsilon\big(X_{\tau\xi} + i\frac{k'-k}{\sigma}X_\tau + i\frac{\omega}{\epsilon}X_\xi\big)\Big)e^{i\omega\tau/\epsilon + i(k'-k)\xi/\sigma} + c.c.,$$

the left-hand side of the final equation is obtained. Keeping terms of the expansion up to the third power of the amplitude parameter ϵ leads to

$$[\frac{\hat{\varepsilon}}{c^2} - k'^2]\epsilon^2\sigma X_{\tau\tau} - 2i(k'-k)X_\xi\sigma^2 + 2k'\sigma^2\epsilon X_{\tau\xi}. \hspace{1.5cm} (2.148)$$

The right-hand side of the equation can be presented as

$$\frac{2}{\pi c^2 N_{nl}}\int\limits_0^{r_0}\int\limits_0^{2\pi} rJ_l(\alpha_{ln}r)e^{-il\varphi}\frac{\partial^2}{\partial t^2}\sum_{klm}\chi_{zklm}E_kE_lE_m\,d\varphi dr = \sigma^3 q_1|X|^2 X$$
$$\hspace{10cm} (2.149)$$

where q_1 is the nonlinear coefficient. Thus the final equation has the form

$$[\frac{\hat{\varepsilon}_1}{c^2} - k'^2]\epsilon^2 X_{\tau\tau} + 2k'\sigma\epsilon X_{\tau\xi} - 2i\sigma(k'-k)X_\xi = \sigma^2 q_1|X|^2 X. \hspace{1cm} (2.150)$$

2.7 CONCLUSION

The problem we considered related to the derivation of waveguide mode interaction equations. We have derived the system of equations that describe the interaction of waves with fixed polarizations and directions of propagations in the most appropriate form. The form implies clear definitions of all four modes by means of projection operators in two standard mathematical statements of problems: the Cauchy problem (given initial conditions) and the boundary regime problem (given boundary regime dependence of the fields on time at a boundary x = 0). The account of nonlinear terms introduces into the physical picture the important phenomenon of mode interaction, which allows us to investigate energy and momentum transfer in a physically well-defined context. Again the projector applications extract a mode contribution in arbitrary time for the Cauchy problem and in a space point for a boundary regime. Application of asymptotic expansion with respect to a small parameter (e.g., nonlinearity or amplitude) in the evolution operator allows us simplify significantly the equations for a specific initial (boundary) problem (e.g., specific mode excitation).

The case of 3D space in cylindrical variables adapted to an optical fiber problem also results in a mode interaction system. The main difference is an appearance of waveguide dispersion terms that drastically change pulse evolution even in a linear description. The dispersion account in the nonlinear pulse evolution case stabilizes the wave breaking while in 1D systems such a phenomenon restricts applicability by a time on wave packet collapse.

So, we have solved the problem of mode definition together with its linear and nonlinear evolution and interaction. An application of the theory means an essential step toward the system solution. While in the case of the linearized problem the answer after the transformation to mode equations is almost trivial (resulting in differential equations for the separated modal variables being of the first order), the nonlinear system solution looks essentially nontrivial and needs effort. There is an approach to developing a finite-difference scheme with proved convergence and stability ([30]), but its adaptation to the considered problem of the short-pulse equations stands as a challenge. Namely, the numeric solution of the problem of opposite moving pulse (with arbitrary polarization) collisions is an open question. An answer to this question would open a way to qualitative comparison with the experiments of [22, 23, 38] and to very interesting applications (optical control, gates, etc.).

REFERENCES

1. S. B. Leble. *Nonlinear Waves in Waveguides.* Springer, 1990.
2. C. R. Menyuk. Nonlinear pulse propagation in birefringent optical fibers. *IEEE J. Quantum Electron,* 23:174–176, 1987.
3. B. S. Marks and C. R. Menyuk. Interaction of polarization mode dispersion and nonlinearity in optical fiber transmission systems. *Journal of Lightwave Technology,* 23(7), 2006.
4. B. Reichel and S. Leble. The equations for interactions of polarization modes in optical fibers including the Kerr effect. *Journal of Modern Optics,* 55(11):3653–3666, 2008.
5. G. P. Agrawal. *Nonlinear Fiber Optics.* Academic, San Diego, 1989.
6. V. E. Zakharov. Stability of periodic waves of finite amplitude on the surface of a deep fluid. *Journal of Applied Mechanics and Technical Physics,* 9(2):190, 1968.
7. V. E. Zakharov and A. L. Berkhoer. Self-excitation of waves with different polarizations in nonlinear media. *ZhETP,* 58:901–911, 1970.
8. J. A. Fleck. Ultrashort-pulse generation by q-switched lasers. *Phys. Rev. B,* 84 (1), 1970.
9. S. V. Chenikov and P. V. Mamyshev. Ultrashort-pulse propagation in optical fibers. *Optics Letters,* 15 (19), 1990.
10. G. E. Wayne and T. Schafer. Propagation of ultra-short optical pulses in cubic nonlinear media. *Physica D,* 196:90–105, 2004.
11. P. Kinsler. Unidirectional optical pulse propagation equation for materials with both electric and magnetic responses. *Phys. Rev. A,* 81 (023808), 2010.

12. Sh. Amiranashvili and A. Demircan. Ultrashort optical pulse propagation in terms of analytic signal. *Advances in Optical Technologies*, vol. 2011, ID 989515, 2011.

13. A. I. Maimistov and J.-G. Caputo. Unidirectional propagation of an ultrashort electromagnetic pulse in a resonant medium with high frequency stark shift. *Phys. Lett. A*, 296 (34), 2002.

14. J. E. Rothenberg. Space-time focusing: breakdown of the slowly varying envelope approximation in the self-focusing of femtosecond pulses. *Opt. Lett.*, 17, 123001, 1992.

15. T. Brabec and F. Krausz. Nonlinear optical pulse propagation in the single-cycle regime. *Phys. Rev. Lett.*, 78, 3282, 1997.

16. T. Brabec and F. Krausz. Nonlinear optical pulse propagation in the single-cycleregime. *Phys. Rev. Lett.*, 78, 3282, 1997.

17. A. M. Zheltikov and A. A. Voronin. Soliton-number analysis of soliton-effect pulse compression to single-cycle pulse widths. *Phys. Rev. A*, 78, 063834, 2008.

18. G. H. C. New, P. Kinsler and B. P. Radnor. Theory of direction pulse propagation. *Phys. Rev. A*, 72, 063807, 2005.

19. P. Kinsler. Limits of the unidirectional pulse propagation approximation. *J. Opt. Soc. Am.*, B24, 2363, 2007.

20. S. Leble and M. Kuszner. Directed electromagnetic pulse dynamics: Projecting operators method. *J. Phys. Soc. Jpn.*, 80, 024002, 2011.

21. S. Wabnitz, S. Pitois and G. Millot. Nonlinear polarization dynamics of counterpropagating waves in an isotropic optical fiber: Theory and experiments. *J. Opt. Soc. Am. B*, 18, 432, 2001.

22. C. Montes et al. *Without Bounds: A Scientific Canvas of Nonlinearity and Complex Dynamics.* Springer, 2013.

23. T. Crepin et al. Experimental evidence of backward waves on terahertz left-handed transmission lines. *Appl. Phys. Lett.*, 87, 104105, 2005.

24. S. Leble and M. Kuszner. Ultrashort opposite directed pulses dynamics with Kerr effect and polarization account. *J. Phys. Soc. Jpn.*, 83, 034005, 2014.

25. T. Schafer and Y. Chung. Stabilization of ultra-short pulses in cubic nonlinear media. *Phys. Lett. A*, 361(1–2):63–69, January 2007.

26. U. Bandelow, M. Pietrzyk and I. Kanattsikov. On the propagation of vector ultra-short pulses. *J. of Nonlin. Math. Phys.*, 15, 162, 2008.

27. S. Leble, M. Kuszner and B. Reichel. Multimode systems of nonlinear equations: Derivation, integrability, and numerical solutions. *Theoretical and Mathematical Physics*, 168(1):974–984, 2011.

28. S. Manakov. Remarks on the integrals of the Euler equations of the n-dimensional heavy top. *Functional Anal. Appl.*, vol. 10, 9394, 1976.

29. C. R. Menyuk. Application of multiple-length-scale methods to the study of optical fiber transmission *Journal of Engineering Mathematics* 36:113–136, 1999.

30. B. Reichel and S. Leble. On convergence and stability of a numerical scheme of coupled nonlinear Schrödinger equations. *Computers and Mathematics with Applications*, 55:745–759, 2008.

31. V. C. Kuriakose, S. B. Leble and K. Porsezian. *Nonlinear Waves in Optical Waveguides and Soliton Theory Applications. Optical Solitons, Theoretical and Experimental Challenges.* Springer, 2003.

32. S. Kielich. *Nonlinear Molecular Optics*. PWN, 1977.

33. R. W. Boyd. *Nonlinear Optics*. Academic Press, Boston, 1992.

34. V. A. Fock. *Fundamentals of Quantum Mechanics*. Mir Publishers, Moscow, 1978.

35. T. Schafer, Y. Chung, C. K. R. T. Jones and C. E. Wayne. Ultra-short pulses in linear and nonlinear media. *Nonlinearity*, 18:1354–74, 2005.

36. S. Sakovich. Integrability of the vector short pulse equation. *J. Phys. Soc. Jpn.*, 77, 2008.

37. Y. Kodama and A. Hasegawa. Signal transmission by optical solitons in monomode fiber. *Proc. IEEE*, 69:1145–1150, 1981.

38. Ayhan Demircan, Shalva Amiranashvili, Carsten Bre, Uwe Morgner and Gunter Steinmeyer. Supercontinuum generation by multiple scatterings at a group velocity horizon. *Optics Express*, vol. 22, no. 4, 3866, 2014.

3 Coupled-mode dynamics in continuous and discrete nonlinear optical systems

Alejandro Aceves

3.1 COUPLED-MODE DYNAMICS IN NONLINEAR OPTICAL SYSTEMS

In many instances, theoretical advances in nonlinear science result from universal descriptions from which knowledge can be transferable and even expanded into many fields. A good example closely connected to nonlinear optics and well supported by several chapters in this book and in the nonlinear optics literature is the case of quasi-monochromatic weakly nonlinear wave phenomena which is well described by the nonlinear Schrödinger equation (NLSE) and close "cousins." Similarly, when the dynamics is still weakly nonlinear, but instead energy flows between two or more modes coupled by some resonant mechanism, then one obtains a coupled system of equations. The simplest example of this is the presence of birefringence in an optical fiber, where the single guided mode can have two states of polarization. Here the resonant coupling mechanism is due to the Kerr effect and the system is then modeled by the coupled nonlinear Schrödinger equation. Instead, in the cases discussed in this chapter, the common feature is the presence at the leading order of linear resonant coupling, for which one obtains as envelope equations coupled mode equations (CME) similar to

$$i\frac{\partial u}{\partial t} \quad - \quad i\frac{\partial u}{\partial z} = \kappa v + c_{11}|u|^2 u + c_{12}|v|^2 u,$$

$$i\frac{\partial v}{\partial t} \quad + \quad i\frac{\partial v}{\partial z} = \kappa u + c_{21}|u|^2 v + c_{22}|v|^2 v.$$

$$(3.1)$$

In this particular form, this CME describes the propagation of the envelopes of counter-propagating waves in a fiber Bragg grating (FBG), where the period of the grating Λ satisfies the resonance condition $\Lambda = 2\beta$, where β is the propagation constant of the modes. As we will highlight with examples throughout this chapter, this system and its "cousins" are universal in that they describe multiple phenomena in nonlinear science. An equivalent statement can be made when we talk about discrete (as opposed to continuous) physical systems and their theoretical description.

The chart shown below, which by no means is exhaustive, summarizes what has just been said.

Physics: Wave propagation in a weakly nonlinear periodic medium. Examples: Light propagation in a *periodic* optical fiber *grating*; surface waves in deep water with periodic bottom; delay lines in a photonic crystal fiber	$\longrightarrow$	Equation (CME): $$i\frac{\partial u}{\partial t} - i\frac{\partial u}{\partial x} = \kappa v + (c_{11}|u|^2 + c_{12}|v|^2)u$$ $$i\frac{\partial v}{\partial t} + i\frac{\partial v}{\partial x} = \kappa u + (c_{21}|u|^2 + c_{22}|v|^2)v$$

Physical system: *Binary* array of 2-type coupled elements. Examples: Metallic-dielectric arrays	$\longrightarrow$	Equation (CME): $$i\frac{\partial u_n}{\partial z} = K(v_n - v_{n-1}) + |u_n|^2)u_n$$ $$i\frac{\partial v_n}{\partial z} = K(u_n - u_{n-1}) + |v_n|^2)v_n$$

Instead of presenting a summary of results and progress in a specific optical system, what we will attempt to do throughout this chapter is first briefly highlight some of the most relevant earlier theoretical and experimental work as it relates in particular to nonlinear fiber Bragg grating. We then follow up with a few examples of other nonlinear photonic systems that are the focus of ongoing research and whose models bear a resemblance to Eq. (3.1). In all instances we briefly describe the structure, objective and assumptions that lead to the model of study and present some analytical results. In the conclusions we attempt to suggest possible future directions. What we hope to accomplish is to establish how the theory first considered in the study of pulse propagation in FBG can be naturally extended to a wide variety of device oriented models; we do this at the expense of giving due credit to the vast amount of work centered on gap solitons and the CME. As such, an apology is due in advance for not referencing many of these important contributions.

In the last years before the turn of the century, there was a big effort to build and test the performance of FBG for intense pulse propagation. One of the main objectives was that as the capacity of fibers to transmit information at very high bit-rates increased, there was a necessity to build all optical devices to address bottle-neck problems arising from operations such as buffering, re-routing, and delaying of optically based information. In particular, the property of slowing light, first conceived by use of self-induced transparency, was considered and demonstrated experimentally in FBG. Two groups led these efforts: at Bell Laboratories/Lucent Technologies, Eggleton and Slusher demonstrated the existence of Bragg solitons [1] propagating at wavelengths corresponding to the edge of the bandgap corresponding to anomalous dispersion. The Bragg induced dispersion was orders of magnitude greater than the chromatic dispersion of the fiber, thus inducing a reduction in the group velocity of the pulse. These experiments were well described by a further reduction from the CME to the NLSE. Parallel experiments by Mohideen et al. produced nonlinear pulses close to the center of the gap where the CME predicts pulses at or near zero velocity [2]. Therefore, these pulses were true gap solitons. At the time, these exciting experiments triggered an intense period of research whose models at the core were identical or similar to the CME written above. By now, the theory of Bragg grating solitons is a mature field with a history close to 25 years old. The book *Nonlinear Photonic Crystals* [3] is still a good reference for fiber base Bragg or gap soliton systems.

On the theoretical side, while there is also a long history, beginning with the directional coupler, of time independent (valid for the treatment of the fields as continuous waves) CME, here we start from a model that in fact comes from quantum field theory known as the Massive Thirring Model (MTM), whose equations are like Eq. (3.1) with $c_{11} = c_{22} = 0$. While on one hand this condition limits direct applicability to nonlinear optics, it is an important building block since, as was shown in the early 1970s, the MTM as in the case of the NLSE is a completely integrable system and has soliton solutions. While Eq. (3.1) is not integrable, it maintains three key properties on the MTM: it is Hamiltonian, it conserves momentum and it is Lorenz invariant. The initial reports on the existence of Bragg and/or gap solitary waves in FBG can be found in [4,5]. Both exact solitary wave solutions at zero [4] and any velocity [5] were obtained. In particular, in [6] analytical results were obtained by first proving the property that any solution of the MTM such that the amplitude moduli are proportional to each other led to exact solutions of Eq. (3.1) by a proper scaling factor and a phase that can be computed by quadrature. The notion of the gap soliton was in fact originally proposed for a discrete system, a super-lattice [6]. This correspondence of discrete versus continuous periodic systems is central to this chapter. A nice intuitive explanation of the existence of such solutions is that the interference pattern of the field generated inside the envelope of the counter-propagating modes "erases" the Bragg grating, or equivalently, for the moving pulses this interference distorts the gap structure in such a way that it modes the location of the central frequency of the pulse outside the gap. In fact, the qualitative properties of these solitary waves are important to point out. The functional form of the fields satisfying Eq. (3.1) as given in [5] is

$$
\begin{aligned}
u(z,t) &= \frac{1}{2}[K_1 g_1(\xi) + iK_2 g_2(\xi)]e^{i\cos(Q)\psi}, \\
v(z,t) &= \frac{1}{2i}[K_1 g_1(\xi) - iK_2 g_2(\xi)]e^{i\cos(Q)\psi}, \\
\xi &= \frac{z + vt}{\sqrt{1 - v^2}}, \psi = \frac{vz + t}{\sqrt{1 - v^2}}, \\
K_1 &= \left(\frac{1 + v}{1 - v}\right)^{1/4}, K_2 = \left(\frac{1 - v}{1 + v}\right)^{1/4}.
\end{aligned}
\tag{3.2}
$$

Observe in particular that the solutions are Lorentz invariant and that the velocity at which they propagate depends on the ratio of peak pulse peak powers. True gap (zero velocity) solitons correspond to the case $v = 0$.

To see how the field of gap solitons in optics for continuous and discrete systems has evolved, we refer readers to references [7–9]. There are in fact several articles that discuss gap solitons outside optics (see, for example, [11,12]. Instead and for the remainder of the chapter we will present recent examples of optical or photonics systems whose models bear strong similarities with that of the fiber Bragg grating. The first example comes from a current area of intense research: Parity-Time symmetric optical systems.

3.2 PARITY-TIME OPTICAL COUPLED SYSTEMS

The discussion in the previous section dealt with conservative systems. In most instances losses are present, but if they are small compared to characteristic propagation lengths, they will only slightly perturb the type of nonlinear modes or solitary waves of the ideal system. On the other hand, a novel approach now considered in photonic systems includes losses which are properly balanced by gain. As in many other instances in the study of optics and photonics, these so-called Parity-Time (PT) photonic systems are based on a question initially posed in a quantum mechanical framework. The question posed by Bender et al. [13] is simple: If one allows the potential in Schrödinger's equation to be complex, under which condition the now non-Hermitian operator has a real spectrum. By use of additional symmetries, one finds that a necessary condition is for the potential $V(x)$ to have the property $V(x) = V^*(-x)$. In other words, the real part is an even function while the imaginary part is odd. Understanding that Helmholtz's equation in optics being analogous to Schrödinger's equation in quantum mechanics, the role of the potential is the square of the index of refraction n^2. The question of when does one have real spectrum here means having a real propagation constant under the condition of a complex refractive index. A complex index of refraction means having a medium with gain and/or loss, so the importance of the question is the possibility of having propagation modes in such a medium. The necessary condition, in particular on the imaginary part, is that the gain/loss property has to be odd. Intuitively, this states the possibility that losses are properly balanced by amplification, allowing for the system to behave as if it were conservative. Equally important to the observation of this PT symmetry condition is the fact that the condition is only necessary. In fact, if the imaginary part of the index is parametric dependent, one can find bifurcations or phase transitions, which, as we know well, can be used for applications. Before we go into recent examples of nonlinear photonic PT systems similar to those discussed above, it is useful to illustrate this behavior in the simplest of systems: a coupler where one waveguide is lossy and the other one is an amplifier.

$$\frac{du}{dz} = i\kappa v + \gamma u,$$

$$\frac{dv}{dz} = i\kappa u - \gamma u.$$

Seeking propagating modes, one finds $\beta = \pm\sqrt{\kappa^2 - \gamma^2}$; thus the condition β real happens if $\gamma < \gamma_{crit} = \kappa$. Below the critical value, the coupler acts like the usual one, where energy oscillates between the waveguides. Here the loss/gain parameter only modifies the period of oscillation. Above γ_{crit}, the oscillations stop and from the general solution, one can show that, regardless of the input conditions, the output energy flows out from the port with gain; thus the bifurcation can be viewed as an optical diode.

It is perhaps not surprising that a large portion of recent work on optical PT-systems evolved around the modified NLSE $i\frac{\partial A}{\partial z} + V(x)A + |A|^2 A = 0$, with $V(x) = V^*(-x)$, where again the model assumes a design of the index of refraction in a nonlinear medium that satisfies the PT gain/loss property in the index of refraction. What follows are examples of PT systems modeled by coupled mode equations. The first one is a natural extension of the fiber Bragg grating.

Recalling that a fiber Bragg grating is built by imposing a weak periodic modification of the index of refraction to the fiber $\Delta n(z) = C\cos(\Lambda z)$, $\frac{C}{n_0} \ll 1$ where Bragg resonance is given by the condition $\Lambda = 2\beta$, where β is the propagation constant of the coupled counter-propagating modes in the fiber. In [14] the authors studied propagation properties when the variation of the index is complex $\Delta n(z) = C\cos(\Lambda z) + iD\sin(\Lambda z)$, where one can observe how the real part is an even function and the imaginary part is odd. Although this is not quite the definition of PT systems in that the index variation is in the direction of propagation, this work shows interesting phase transitions. For this index, the CME at Bragg resonance read

$$i\frac{\partial u}{\partial z} + i\frac{\partial u}{\partial t} + (\kappa + g)v + c_{11}|u|^2 u + c_{12}|v|^2 u = 0,$$

$$-i\frac{\partial v}{\partial z} + i\frac{\partial v}{\partial t} + (\kappa - g)u + c_{21}|u|^2 v + c_{22}|v|^2 v = 0, \qquad (3.3)$$

where κ and g are proportional to the strength of, respectively, the real and imaginary parts of Δn. Similar to the simpler PT coupler, in the linear regime, there is a phase transition at $g_c = \kappa$, shown in the Fig. 3.1.

Notice that the frequency gap typical of fiber Bragg gratings (a,b) switch to a wavenumber gap for $g > \kappa$ (d). In terms of the full system, if $\kappa - g > 0$, solitary wave solutions similar to Eq. (3.2) exist [14]. They describe the co-existence of forward and backward propagating fields. On the other hand, if $\kappa > g$, an incident forward field propagates through an amplifying medium without generating a backward component; thus a phase transition occurs.

We now turn our attention to discrete PT systems where in particular much progress has been made in coupled electric systems (coupled Van der Pol oscillators) and on synthetic circuits. The hope is that they can provide a proof of principle that can be extended to optical devices. Here we illustrate PT properties in coupled arrays in two cases: coupled circuits and coupled nano-resonators.

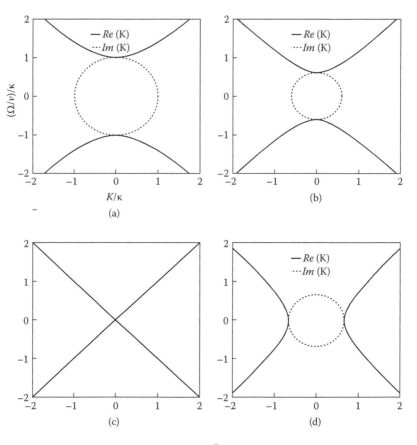

FIGURE 3.1: Band structure of a PT âĂŞsymmetric periodic grating (linear case) for different ratios of (a) 0, (b) 0.8, (c) 1, (d) 1.2.

We first discuss the coupled nonlinear circuits shown in Fig. 3.2. From Kirchhoff's laws, this dimer circuit proposed by Bender et al. [15] to study PT symmetry breaking dynamics is modeled by the coupled system written in dimensionless form [15],

$$\eta\frac{dI_{ND_L}}{d\tau} = \gamma(1 - V_{ND_L})^2\frac{dV_{ND_L}}{d\tau} + (1 + c)\frac{d^2V_{ND_L}}{d\tau^2} - c\frac{d^2V_{ND_R}}{d\tau^2},$$

$$\eta\frac{dI_{ND_R}}{d\tau} = \gamma(1 - V_{ND_R})^2\frac{dV_{ND_R}}{d\tau} - (1 + c)\frac{d^2V_{ND_L}}{d\tau^2} - c\frac{d^2V_{ND_R}}{d\tau^2}, \quad (3.4)$$

where η, γ and c are, respectively, dimensionless conductance, gain/loss parameter and capacitance. Here c is also the intra-dimer coupling coefficient. Analysis of this system [16] has shown the existence of a critical value γ_{crit} below which there is oscillatory behavior in the currents which is modulated

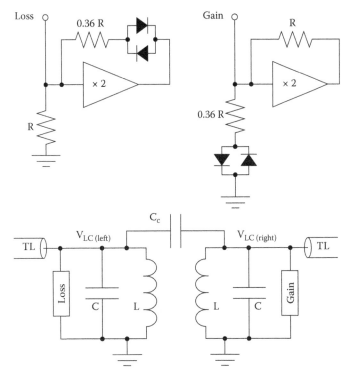

FIGURE 3.2: Schematics of two nonlinear circuits, one with gain and one with loss, that are coupled with a capacitor.

by the coupling strength and the value of γ. Instead, above the critical value, all current remains in the gain component, which behaves like the traditional Van der Pol oscillator. In fact, one can characterize in asymptotic form two regimes: $0 < \gamma \ll 1, \gamma \gg 1$. It is of interest to extend the behavior of the dimer to a large chain; this has not been explored yet. A similar system recently proposed is that of an array of single ring nano-resonators [17] (see Fig. 3.3).

The weakly nonlinear coupled mode system for such an array reads

$$\frac{d^2 A_m}{d\tau^2} + A_m = -\lambda_M \frac{d^2 B_m}{d\tau^2} - \lambda_M \frac{d^2 B_{m+1}}{d\tau^2} + \epsilon_0 \sin(\Omega \tau)$$
$$- \alpha A_m^2 - \beta A_m^3 - \gamma \frac{d A_m}{d\tau},$$
$$\frac{d^2 B_m}{d\tau^2} + B_m = -\lambda_M \frac{d^2 A_{m-1}}{d\tau^2} - \lambda_M \frac{d^2 A_m}{d\tau^2} + \epsilon_0 \sin(\Omega \tau)$$
$$- \alpha B_m^2 - \beta B_m^3 + \gamma \frac{d B_m}{d\tau},$$

where $A_m = q_{2n+1}$, $B_m = q_{2n}$ are the charges at the capacitors of the odd and

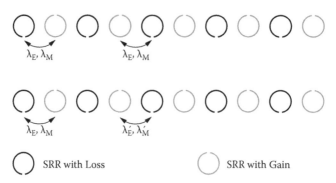

FIGURE 3.3: Schematic of a PT metamaterial (as in Fig. 1 in [7]). Upper panel: all the single ring resonators (SRRs) are equidistant. Lower panel: the separation between SRRs is modulated according to a binary pattern (PT dimer chain).

even resonators, respectively. A useful approach is to consider the long-wave (continuum) approximation $A_m(\tau) \to A(x, \tau)e^{iz} + cc.$, $B_m(\tau) \to B(x, \tau)e^{i\tau}$ where we restrict the discussion to the case where the external voltage frequency Ω is away from resonances. Then A, B satisfy

$$i\frac{\partial A}{\partial \tau} + i\frac{\partial B}{\partial x} = \frac{1}{\lambda_m}\left[-\alpha_M B + 3\beta|A|^2 A + i\gamma A\right],$$

$$-i\frac{\partial B}{\partial \tau} + i\frac{\partial A}{\partial x} = \frac{1}{\lambda_m}\left[\alpha_M B - 3\beta|B|^2 B + i\gamma B\right].$$

$A_M = \lambda_M + \lambda'_{M'}$.

This CME system differs from those in conservative models by the presence of the last terms representing the gain/loss component. However, one can demonstrate the presence of is suitable for phase transitions, as demonstrated in [18]. In particular, for the stationary case, it was shown that a homoclinic orbit which is an exact solitary wave for $\gamma = 0$ persists up to a critical value of γ. This solution represents an extended breather mode of the dimer chain which no longer persists if $\gamma > \gamma_c = |a_M|$. More examples of recent work on Pt symmetric systems can be found in [19].

3.3 BINARY ARRAYS

The work on light localization in uniform nonlinear waveguide arrays modeled by the discrete nonlinear Schrödinger equation (DNLSE) has a long history and continues to be an active area of research. When this uniformity is broken, in particular in a "binary" format, new phenomena are expected, including the presence of discrete gap solitons [20]. The particular example we discuss here consists of a binary array of alternating unequal (for example, metallic/dielectric) waveguides. This particular configuration brings novel features

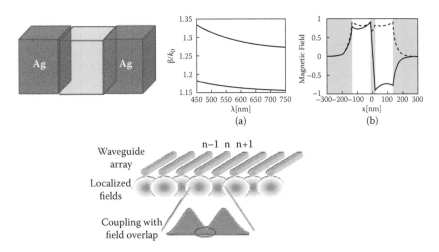

FIGURE 3.4: Schematics of a binary array. By alternating metallic (eg: Ag) and dielectric waveguides (top left), and odd (solid line top right) mode is formed. The overlap with the even (dashed) mode creates coupling constants of opposite signs. The propagation constants of each mode are displayed in the top middle figure. This profile is in contrast to the traditional uniform dielectric waveguide arrays (bottom) where the mode overlap produced coupling constants of equal signs.

that have only been recently studied [21, 22], the most important being the nearest neighbor coupling can be of opposite signs (see Fig. 3.4).

In this case the DNLSE no longer applies; instead the amplitude propagation at the nth guide satisfies

$$i\frac{\partial E_n}{\partial z} + \beta_n E_n + C_{n-1}E_{n-1} + C_{n+1}E_{n+1} + \chi_n E_n = 0$$
$$C_{n-1} \equiv C_1 > 0, C_{n+1} \equiv C_2 < 0, n = \text{even},$$
$$C_{n-1} \equiv C_1 < 0, C_{n+1} \equiv C_2 > 0, n = \text{odd}.$$

This system is transformed into a discrete CME syetem by the transformation $A_n = E_n, n = even, B_n = E_n, n = odd.$

$$i\frac{\partial A_n}{\partial z} + \frac{\Delta\beta}{2}A_n + C_1 B_n + C_2 B_{n+1} + \gamma_1|A_n|A_n = 0,$$
$$i\frac{\partial B_n}{\partial z} - \frac{\Delta\beta}{2}A_n + C_2 A_{n-1} + C_1 A_n + \gamma_2|B_n|B_n = 0,$$

where $\Delta\beta$ the difference between propagation constants, and γ's are the possible different values of the Kerr constants.

In close proximity of the band edge (i.e., around $k_x = 0$ for $C_1 < 0$) a very useful equivalent continuous model can be derived by performing a Taylor

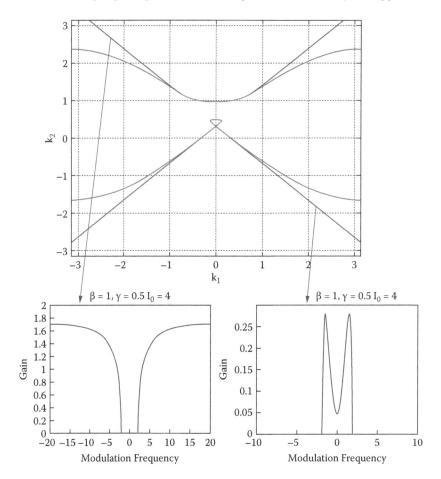

FIGURE 3.5: Top: Nonlinear dispersion relations for the discrete and continuous binary array models for and amplitudes of the continuous (plane) waves satisfying. Bottom: modulation instability plots for the continuous model and (left) for the upper branch and (right) for the lower branch.

expansion to obtain (as a first order approximation)

$$i\frac{\partial u}{\partial z} + \frac{\Delta\beta}{2}u + \frac{\partial w}{\partial x} + \epsilon w + \gamma_1|u|^2 u = 0,$$

$$i\frac{\partial w}{\partial z} - \frac{\Delta\beta}{2}w - \frac{\partial u}{\partial x} + \epsilon u + \gamma_2|w|^2 w = 0,$$

where the spatial variable x is scaled so that $C_2 = 1, C_1 = -1 + \epsilon$.

For comparison, Fig. 3.5 shows an example of nonlinear dispersion relations for both the discrete and the continuous models.

The case where $\epsilon = \Delta\beta$ is particularly interesting in that the lin-

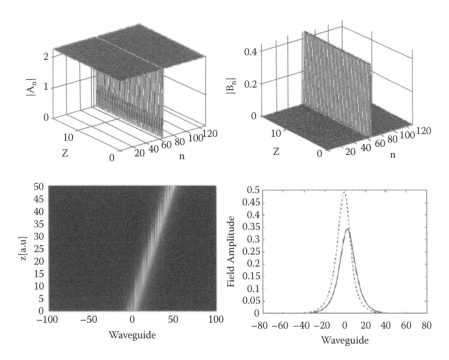

FIGURE 3.6: Top: A discrete dark/bright solitary wave mode, solution of the discrete binary array. Bottom: A traveling bright/bright solitary wave mode, solution of the continuous model.

ear dispersion relation for solutions of the form $A_n(z) = Ae^{i(k_z z - k_n n)}$, $B_n(z) = Be^{i(k_z z - k_n n)}(u(x,z) = Ae^{i(k_z z - k_x x)})$, $w(x,z) = Be^{i(k_z z - k_x x)}$ is $k_z = \pm 2|\sin(\frac{k_2}{2})|(k_z = \pm|k_x|$ does not have a gap, but instead a Dirac singular point at the origin. Similar binary systems and in particular honeycomb arrays share this behavior. In the presence of nonlinearity, a gap can open, as in the case shown in Fig. 3.5; it can also alter the behavior near the Dirac point. As a matter of fact, in some instances it can regularize it. Finally, both the discrete and the continuous models possess solitary wave solutions, both stationary, and, at least in the continuous model, traveling waves similar to Eq. (3.2) have been obtained. Figure 3.6 presents an example of solutions for each case

While it is generally true that it is impossible to find analytic solutions of the discrete system, there are instances where highly localized states can be obtained as an asymptotic expansion. Here we present a few examples that highlight analytical ways to find conditions for the existence and properties of localized solutions, including their asymptotic $|n| >>> 1$ features. We will do so for different sets of parameters but in all cases we assume $\Delta\beta = 0$ for simplicity of presentation. We begin by considering the ansatz $A_n = a_n e^{i\lambda z}$,

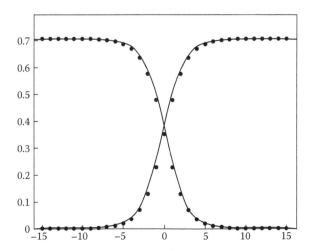

FIGURE 3.7: Intensity profile of the discrete (dots) and continuous (lines) solutions for odd and even sites. Here $\gamma_1 = \gamma_2 = 1$.

$B_n = b_n e^{i\lambda z}$. We first consider localized discrete stationary solutions for $\gamma_1 = \gamma_2 = \gamma$. These states satisfy the system

$$-\lambda a_n - b_n + b_{n+1} + \gamma a_n^3 = 0,$$
$$-\lambda b_n + a_{n-1} - a_n + \gamma b_n^3 = 0.$$

We first observe that these equations admit solutions satisfying the discrete symmetry relation $a_k = b_{-k}$ so that we need to solve only for $n \geq 0$. Looking at the asymptotic $n \gg 1$ tails, one can consider modes that are unstaggered for a_n, while they are staggered for b_n. These conditions translate into $a_{n-1} \approx a_n, b_n \approx \frac{\lambda}{y}, n \gg 1$. Finally, the combination of symmetry, that b_b is staggered and asymptotic analysis lead to the recurrence equation for the amplitude moduli

$$|b_n| = |b_{n-1}| + |a_{n-1}|(\lambda - \gamma|a_{n-1}|^2),$$
$$|a_n| = 2a_0 - |b_n|, a_0 = b_0 = \frac{1}{2}\sqrt{\cos(Q)/\gamma},$$

where $Q = Q(\lambda)$. Figure 3.7 shows the comparison of the amplitude moduli and the corresponding solution based on the continuum model. As one can see, the asymptotic analysis gives a good approximation. Most important, the figure shows the profiles of the discrete modes, which themselves cannot be approximated by a continuum model.

The same analysis strategy leads to unique staggered/unstaggered modes for different values of the nonlinear coefficient. For illustration, Figure 3.8 $\gamma 1 = \gamma 2 = 1$ (a self-focusing binary array) while Figs. 3.9 and 3.10 show the results for $\gamma 1 = 1, \gamma 2 = 0$ (a self-focusing/linear binary array). In the first

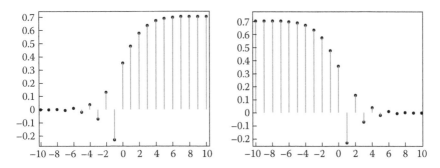

FIGURE 3.8: Amplitude profiles of the discrete solution for even (left) and odd (right) sites for $\gamma_1 = \gamma_2 = 1$.

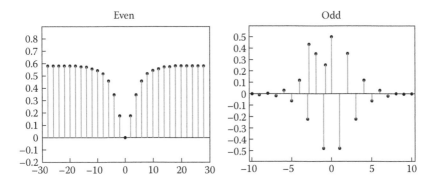

FIGURE 3.9: Amplitude profiles of the discrete solution for even (left) and odd (right) sites for $\gamma_1 = 1, \gamma_2 = 0$.

case (Fig 3.8) there is decay of the mode intensity at one end. Instead Fig. 3.9 shows a themode corresponding to the odd index which goes to zero as $|n| \to \infty$ (a discrete dark/bright soliton mode). Finally, in the case shown in Fig. 3.10, no such decay exists.

3.4 DUAL-CORE PHOTONIC CRYSTAL FIBER

The last two examples to be highlighted are based on photonic crystal fibers (PCFs). In the first case, an interesting possibility of having a delay line in a dual core PCF was recently proposed [23]. Figure 3.11 displays the cross section of the super-mode of an almost symmetric dual core fiber. A slight asymmetry in the features near each core is reflected by the fact that the super-modes look like two nearly identical modes, each supported by one core. This consideration makes the description of the field be approximated by a superposition $E(x, y, z, t) = [u(z, t)F_L(x - x_L)e^{i\beta_1 z} + v(z, t)F_R(x - x_R)e^{i\beta}]e^{-i\omega t}$, where $F_L \approx F_R$ thus $\beta_1 \approx \beta_2$. The envelopes written in a frame of reference

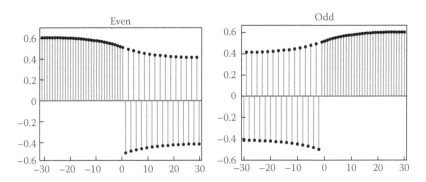

FIGURE 3.10: Amplitude profiles of the discrete solution for $\gamma_1 = 1, \gamma_2 = 0$ even (left) and odd (right) sites for non-zero asymptotic conditions for both components.

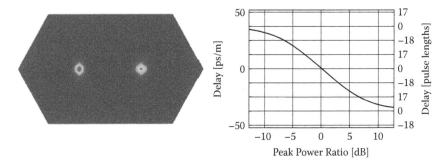

FIGURE 3.11: Left: Cross section of the super-mode profile in a nearly symmetric dual core PCF. Right: Time delay vs. power (amplitude) ratio numerically obtained for a particular PCF parameter value.

moving with the average group velocity of the two modes satisfy the CME

$$i\frac{\partial u}{\partial z} - i\frac{\partial u}{\partial t} = \kappa v + |u|^2 u,$$

$$i\frac{\partial v}{\partial z} + i\frac{\partial v}{\partial t} = \kappa u + |v|^2 v.$$

As with other systems discussed above, soliton solutions of the Thirring model can be modified to satisfy this system (see Eq. (3.2)). In this particular application, one envisions two co-propagating pulses, each propagating in a corresponding core, but with its proximity allowing them to "see" each other. As the solution form dictates, the ratio of amplitudes determines the speed of propagation. With this in mind, if one pulse carries information and the other one is a control pulse, by changing the power of the control pulse (therefore changing the amplitude ratio), the pulse velocity can be tuned; in other words, in this structure the dual core PCFs perform like a delay line (see Fig. 3.11)

3.5 FIBER AMPLIFIERS

The last example looks at the use of PCFs in the design of state of the art fiber amplifiers. In the search for building lasers that are compact and produce high powers, it is evident that lasers built from fiber amplifiers have the upper hand. State of the art fiber laser technology has reached power outputs in the kilowatt level. The threshold that limits more power output is that thermal distortions alter the index of refraction in such a way that it causes undesirable nonlinear mode coupling for modes at different frequencies, leading to modal instabilities. Mode dynamics in fiber amplifiers can be modeled by the CME

$$\frac{\partial c_1}{\partial z} + \frac{1}{v}\frac{\partial c_1}{\partial t} = (i\mu_1 + \Gamma_1)c_1 + (i\mu_{12} - \gamma_{12})|c_2|^2 c_1,$$

$$\frac{\partial c_2}{\partial z} - \frac{1}{v}\frac{\partial c_2}{\partial t} = (i\mu_2 + \Gamma_2)c_2 + (i\mu_{21} + \gamma_{12})|c_1|^2 c_2,$$

where c_1, c_2 are, respectively, the amplitudes of the fundamental and a higher order mode. The preferred performance corresponds to the sole amplification of the fundamental mode described by the solution $|c_1(z)| = Ae^{\Gamma_1 z}$, $c_2(z) = 0$. As one can easily see, once this mode is highly amplified, it becomes unstable due to the nonlinear gain $\gamma_{12}|c_1|^2$, which will amplify the other mode, leading to eventual spatio-temporal unstable dynamics. Other than putting a threshold on maximal output, this instability remains unless one adds new elements in the design of the amplifier. With this in mind, consider "drawing" a long period fiber Bragg grating, which leads to the modified CME

$$\frac{\partial c_1}{\partial z} + \frac{1}{v}\frac{\partial c_1}{\partial t} = (i\mu_1 + \Gamma_1)c_1 + i\Delta e^{i\delta z - \Omega_{12}t}c_2 + (i\mu_{12} - \gamma_{12})|c_2|^2 c_1,$$

$$\frac{\partial c_2}{\partial z} - \frac{1}{v}\frac{\partial c_2}{\partial t} = (i\mu_2 + \Gamma_2)c_2 + i\Delta e^{-i\delta z - \Omega_{12}t}c_1 + (i\mu_{21} + \gamma_{12})|c_1|^2 c_2,$$

where $\Omega_{12} = \omega_1 - \omega_2$ is the frequency offset between the modes and Δ, δ are, respectively, the strength and detuning from Bragg resonance of the grating. An interesting study currently in progress by the author and his collaborators is to search for stable solutions of this model, including those which can reflect improvement in amplifier performance.

3.6 FUTURE DIRECTIONS AND CONCLUSIONS

In this chapter by way of examples we attempted to highlight how universal the coupled mode equations are. With the fast and furious progress in optical technology, devices such as fiber Bragg gratings, dual or multi-core photonic crystal fibers and binary arrays are now a reality to test and implement applications, or for the study of fundamental physics using table top experiments [24]. In the theoretical aspect, it is clear many of these models are described by coupled mode equations like those presented here. Since others are better equipped to highlight future progress driven by experiments and new technologies, we want to conclude by doing this on the theoretical side. Clearly there is more work to be done on the study of models similar to those presented here. It is clear that dimers and in general binary arrays are good examples of devices that, for example, can apply to one of the current hot research topics, PT symmetry being one. Another hot topic of the day that goes along with both proof of principle in table top experiments and applications is the existence of rogue waves in CME based systems [25]. To our knowledge not much is known for these systems. Recently we obtained exact rational solutions of the integrable MTM model. There solutions sit on a continuous wave CW background and are spatially and temporarily localized, with a time/space location of high intensity. As in the general case in the theory of rogue waves, integrability is an essential component. A big challenge is if such important solutions exist for the optically relevant systems which resemble MTM but are nonintegrable. Take, for example, the theory of modulation of water waves the presence of a periodic bottom [25]. It is intuitively clear in this case, by Bragg-type resonance, energy on the surface of the ocean can be trapped; the hard question is if then nonlinear interactions of such trapped modes can create a rogue wave event. Finding such solutions for the CME as derived in [25] could be an important step toward answering this question. Finally, there is a large class of systems (semi-discrete [26], honeycomb lattices [27], ziz-zag arrays [28]) and modes (vortices, shock waves, discrete traveling waves) that were not presented here which by no means are less relevant. Simply stated, this chapter presents a sample of what we believe is an exciting theoretical and experimental area of research in nonlinear photonics.

ACKNOWLEDGMENTS

All of the work presented here to which the author has made contributions reflects collaborative efforts with too many colleagues to name here (you know who you are!). I want to thank them all for the many fruitful discussions and friendship. The author also acknowledges support from the US National Science Foundation.

REFERENCES

1. B. J. Eggleton, R. E. Slusher, C. M. de Sterke, P. A. Krug, J. E. Sipe, "Bragg grating solitons," *Physical Review Letters*, 76, 1627, 1996.
2. U. Mohideen, J. E Sipe, C. M. de Sterke, N. G. Broderick, R. E. Slusher, V. Mizrahi, P.J. Lemaire, "Gap soliton propagation in optical fiber gratings," *Optics Letters*, 20, 1674, 1995.
3. R. E. Slusher, B. J. Eggleton (Eds.). (2003). *Nonlinear Photonic Crystals* (Vol. 10). Springer.

4. D. N. Christodoulides, R. I. Joseph, "Slow Bragg solitons in nonlinear periodic structures," *Physical Review Letters*, 62, 1746, 1989.
5. A.B. Aceves, and S. Wabnitz, "Self-induced transparency solitons in nonlinear refractive periodic media," *Phys. Lett. A*, 141, 37, 1989.
6. W. Chen, D. L. Mills, "Gap solitons and the nonlinear optical response of superlattices," *Physical Review Letters*, 58, 160, 1987.
7. F. Lederer, G. I. Stegeman, D. N. Christodoulides, G. Assanto, M. Segev, Y. Silberberg, "Discrete solitons in optics," *Physics Reports*, 463, 1, 2008.
8. Y. S. Kivshar, G. Agrawal, *"Optical Solitons: From Fibers to Photonic Crystals."* Academic Press, 2003.
9. D. Mandelik, R. Morandotti, J. S. Aitchison, Y. Silberberg, "Gap solitons in waveguide arrays," *Physical Review Letters*, 92, 093904, 2004.
10. A. De Rossi, C. Conti, S. Trillo, "Stability, multistability, and wobbling of optical gap solitons," *Physical Review Letters*, 81, 85, 1998.
11. B. Eiermann, T. Anker, M. Albiez, M. Taglieber, P. Treutlein, K. P. Marzlin, M. K. Oberthaler, "Bright Bose-Einstein gap solitons of atoms with repulsive interaction," *Physical Review Letters*, 92, 230401, 2004; O. Zobay, S. Ptting, P. Meystre, E. M. Wright, "Creation of gap solitons in Bose-Einstein condensates," *Physical Review A*, 59, 643, 1999.
12. S. Longhi, "Gap solitons in metamaterials," *Waves in Random and Complex Media*, 15, 119, 2005.
13. C. M. Bender, S. Boettcher, "Real spectra in non-Hermitian Hamiltonians having P T symmetry," *Physical Review Letters* 80, 5243, 1998.
14. M.-A. Miri, A. B. Aceves, T. Kottos, V. Kovanis and D. N. Christodoulides, "Bragg solitons in nonlinear PT-symmetric periodic potentials," *Phys. Rev. A* 86, 033801, 2012.
15. N. Bender, S. Factor, J. D. Bodyfelt, H. Ramezani, F. M. Ellis, T. Kottos, "Observation of asymmetric transport in structures with active nonlinearities," *Phys. Rev. Lett* 110, 234101, 2013.
16. J. Schoendfeld and A. Aceves, Symmetry-breaking bifurcation in a PT-Van der Pol circuit. Private communication. 2014.
17. N. Lazarides and G. P. Tsironis, "Gain-driven discrete breathers in PT âĹŠsymmetric nonlinear metamaterials," *Phys. Rev. Lett.* 110, 053901, 2013.
18. D.Wang and A. B. Aceves, "Modulation theory in PT-symmetric magnetic metamaterial arrays in the continuum limit," *Phys. Rev. A.* 88, 043831, 2013.
19. Tsampikos Kottos and Alejandro B. Aceves, Synthetic Structures with Parity-Time Symmetry. *Contemporary Optoelectronics: From (Meta)Materials to Device Applications*, to be published by the Springer-Verlag series "Springer Series in Optical Sciences," 2015.
20. R. Morandotti, D. Mandelik, Y. Silberberg, J. S. Aitchison, M. Sorel, D. N. Christodoulides, Y. S. Kivshar, "Observation of discrete gap solitons in binary waveguide arrays," *Optics Letters*, 29, 2890, 2004.
21. M. Conforti, C. de Angelis, T. R. Akylas and A. B. Aceves, "Modulational instability and gap solitons of gapless systems: Continuous vs. discrete," *Phys. Rev. A* 85, 063836, 2012.
22. Aldo Auditore, Matteo Conforti, Costantino De Angelis, T. R. Akylas, Alejandro B. Aceves, "Dark-antidark solitons in waveguide arrays with alternating positive-negative couplings," *Optics Communications* 297, 125, 2013.

23. A. Tonello, M. Szpulak, J. Olszewski, S. Wabnitz, A. B. Aceves, W. Urbanczyk, "Nonlinear control of soliton pulse delay with asymmetric dual-core photonic crystal fibers," *Optics Letters*, 34, 920, 2009.

24. A. Marini, S. Longhi, F. Biancalana, "Optical simulation of neutrino oscillations in binary waveguide arrays," *arXiv:1405.1290v1*, 2014.

25. V. P. Ruban, "Water-wave gap solitons: An approximate theory and accurate numerical experiments," *arXiv:0810.1125v2 [physics.flu-dyn]*, 2013.

26. M. I. Molina, I. L. Garanovich, A. A. Sukhorukov, Y. S. Kivshar, "Discrete surface solitons in semi-infinite binary waveguide arrays," *Optics Letters*, 31, 2332, 2006.

27. O. Peleg, G. Bartal, B. Freedman, O. Manela, M. Segev, D. N. Christodoulides, "Conical diffraction and gap solitons in honeycomb photonic lattices," *Physical Review Letters*, 98, 103901, 2007.

28. N. K. Efremidis, D. N. Christodoulides, "Discrete solitons in nonlinear zigzag optical waveguide arrays with tailored diffraction properties," *Physical Review E*, 65, 056607, 2002.

4 Continuous-discrete duality of the nonlinear Schrödinger and Ablowitz–Ladik rogue wave hierarchies

A. Ankiewicz, D. J. Kedziora, and *N. Akhmediev*

4.1 INTRODUCTION

The nonlinear Schrödinger equation (NLSE) is a ubiquitous model for wave evolution in the presence of dispersion and self-phase modulation. It is applicable in fields ranging from oceanography [1, 2] and optics [3] to ionospheric wave beam propagation [4], and, more generally, describes the evolution of slowly varying quasi-monochromatic wave packets in weakly nonlinear media. Of course, the equation is only a first-order approximation for the physics that arise in many of these cases, yet it manages to depict real-world phenomena with surprising accuracy. Solitons [5], breathers [6], cnoidal structures [7] and rogue waves [8] are but some of the various solutions to this system that have inspired much investigation.

Amid numerous extensions and improvements of the model, the Ablowitz–Ladik equation (AL) takes the NLSE a step further in asking how nonlinear waves would evolve under the same effective physics but in a discontinuous periodic lattice, such as a fiber array or electronic circuit consisting of repeated elements (e.g., nonlinear capacitors) [9]. It is only one of numerous possible ways to spatially discretize the continuous NLSE [10] and can itself be extended into a family of equations [11, 12], but it has a major advantage in that it maintains integrability and thus enables various analytic methods that produce exact closed-form expressions [13].

Although there is debate on the practicality of the AL equation [14], it is certainly of great theoretical interest as it extends our understanding of the physics in the NLSE. Modulation instability, Fermi–Pasta–Ulam recurrence, first-order rogue waves and nonlinear phase shifts have all been previously studied within this context [15]. There has also been a recent drive to find correlations between NLSE and AL solutions of all types [16], to establish if structures can exist in one system but not the other.

This issue is the basis of this work, wherein we show that the rogue wave hierarchy is a feature of both the continuous NLSE and the discrete AL equations. Special emphasis is placed on the second-order rogue wave triplet, which we present in exact form in Sec. 4.3. This solution is inspired by similarities between the NLSE and AL versions of a first-order rogue wave, also known as either a Peregrine soliton or a Peregrine breather, and correspondingly involves the addition of small polynomials to the functional form of the NLSE triplet. We then use the Darboux scheme [17] to extend our investigation numerically to higher-order rogue waves, describing in Sec. 4.4 how discretization affects these structures. Particular attention is paid to the existence conditions of these higher-order waves in Sec. 4.5, where we also generate spectral results and display frequency-domain interference patterns of spatiotemporally localized peaks.

4.2   THEORY

The discrete Ablowitz–Ladik equation [18–20] is written as

$$i\frac{\partial \psi_n}{\partial t} \; + \; \frac{1}{2h^2}\left(\psi_{n-1} - 2\psi_n + \psi_{n+1}\right)$$
$$+ \; \frac{1}{2}\left(\psi_{n-1} + \psi_{n+1}\right)\left|\psi_n\right|^2 = 0, \tag{4.1}$$

where, for fixed t, the complex field ψ_n only exists at integer-valued gridpoints (i.e., $n \in \mathbb{Z}$). Any solution $\psi_n(t, h)$ of Eq. (4.1), with a unit background, for example, can be scaled in a simple manner. If such a solution does exist, then

$$q\psi_n(q^2 t, qh) \tag{4.2}$$

is also a solution, for arbitrary q [21].

In any case, it is possible to embed the discrete gridpoints on which the AL equation operates into a typical continuous spatial domain (defined by $x \in \mathbb{R}$), such that adjacent neighbors are a constant h units apart. In this way, ψ_n can be understood as the value of a wavefunction at $x = nh$. Naturally, for smaller h, the embedded gridpoints are closer to each other. In the $h \to 0$ limit, the discretized grid becomes indistinguishable from the continuous spatial domain, and Eq. (4.1) reduces to

$$i\frac{\partial \psi}{\partial t} + \frac{1}{2}\frac{\partial^2 \psi}{\partial x^2} + \psi\left|\psi\right|^2 = 0,$$

known as the nonlinear Schrödinger equation.

4.2.1   DERIVATION OF DISCRETE RESULT

This correspondence between the two systems is highly suggestive of a correlated set of solutions, differing only by terms dependent on grid spacing.

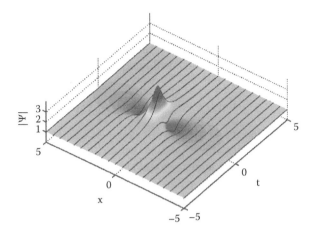

FIGURE 4.1: First-order AL rogue wave, represented by a periodic cross section of a continuous wavefunction, with background amplitude $q = 1$ and lattice spacing $h = 1/2$, so that $x = n/2$. Wavefunction shift is also set as $x_s = 2/3$. These parameters correspond to a lattice offset of $n_s = 4/3$ and the existence of 21 discrete gridpoints within the range $-5 \leq x \leq 5$. This rogue wave is degenerate with a solution using $n_s = 1/3$ and has a maximum amplitude of $7/2$ at $(n, t) = (1, 0)$. Given the values of q and h, the maximum possible amplitude for arbitrary n_s is 4.

In particular, given a solution of the AL equation for arbitary h, it should be possible to find a corresponding NLSE solution as a limiting case, with $h \to 0$ [16]. Indeed, such a concept is supported by the existence of an AL analog to the Peregrine soliton [22], expressed as

$$\psi_n(t) = q \left[\frac{4\left(h^2q^2 + 1\right)\left(1 + 2iq^2t\right)}{1 + 4n^2h^2q^2 + 4q^4t^2\left(h^2q^2 + 1\right)} - 1 \right] e^{iq^2t}. \qquad (4.3)$$

The wavefunction appears as a spatiotemporally localized event on a constant background level of q and, due to its peak amplitude evolving unexpectedly from an infinitesimal perturbation, this structure is considered to represent a first-order rogue wave.

As previously indicated, the solution in Eq. (4.3) is technically only defined for $n \in \mathbb{Z}$. However, by replacing n^2h^2 with x^2, it is clear that the AL rogue wave is nothing more than a periodic cross section of a continuous wavefunction, as the lattice overlay in Fig. 4.1 exemplifies. It is thus possible to reproduce many more non-degenerate AL solutions by simply shifting the continuous function with respect to the lattice. This is done by replacing nh with $(n - n_s)h$ in Eq. (4.3), where real-valued $0 \leq n_s < 1$ leads to unique AL rogue waves. Of course, n_s and corresponding $x_s = n_s h$ can have arbitrary values outside this range, with Fig. 4.1 showing the maximum of a shifted

$(q, h) = (1, 1/2)$ continuous structure arising between gridpoints $n = 1$ and $n = 2$.

It is evident that, for fixed q and h, this continuous function has a specific peak amplitude. However, the technical maximum of an AL rogue wave varies according to

$$\psi_0(0) = q\frac{3 + 4h^2q^2(1 - n_s^2)}{1 + 4h^2q^2n_s^2}, \tag{4.4}$$

within the non-degenerate $0 \leq n_s < 1$ range. Its highest possible value is $q(3 + 4h^2q^2)$ at $n_s = 0$. Its lowest possible value, $3q$, occurs when $n_s = 1/2$, in which case two neighboring nodes possess equal maximum amplitude. So, for unit spacing and background, $q = h = 1$, the maximum nodal value is between 3 and 7. Regardless, we shall henceforth display continuous wavefunctions as AL solutions, with the implicit understanding that they actually represent entire sets of "simply-shifted" discretized rogue waves.

Importantly, the AL wavefunction in Eq. (4.3) reduces to the NLSE Peregrine soliton for $h \to 0$, regardless of the fact that the limit compresses an infinite number of gridpoints n within any arbitrary interval of x. Both analytic solutions are quasi-rational functions, with polynomial numerators and denominators. An AL rogue wavefunction differs from its counterpart primarily in its general form (see Eq. (4.7) versus Eq. (4.5), both presented later) and the fact that the scaling of t is dependent on $\sqrt{1 + q^2h^2}$. Beyond that, any variation in closed-form expression arises from the presence of additional low-order terms proportional to even powers of hq in the discrete case. Assuming this trend for higher order, we can similarly investigate the general solution for the second-order NLSE rogue wave, given by

$$\psi(x, t) = q\left[4\frac{g_2(x, t) + i\, h_2(x, t)}{d_2(x, t)} + 1\right]e^{iq^2 t}, \tag{4.5}$$

with

$$
\begin{aligned}
g_2(x, t) &= -3(2\beta X - 2\gamma T + 6\left(X^2 + 3\right)T^2 \\
&\quad + 5T^4 + X^4 + 6X^2 - 3), \\
h_2(x, t) &= -3T(2\beta X + 2(X^2 + 1)T^2 \\
&\quad + T^4 + X^4 - 6X^2 - 15) \\
&\quad + 3\gamma(T^2 - X^2 - 1), \\
d_2(x, t) &= \beta^2 + 3(2\beta X + X^4 - 6X^2 + 33)T^2 \\
&\quad - 2\beta(X^2 - 3)X + \gamma^2 + 6\gamma(X^2 - 3)T \\
&\quad - 2\gamma T^3 + 3(X^2 + 9)T^4 + T^6 \\
&\quad + X^6 + 3X^4 + 27X^2 + 9, \tag{4.6}
\end{aligned}
$$

where $X = 2qx$, $T = 2q^2t$ and both β and γ represent arbitrary parameters. We note that this was presented in our previous work with $x \leftrightarrow t$ for both the solution and the NLSE [23]. Given all this, our discrete derivation effectively amounts to scaling appropriately and inserting new terms proportional to h^2q^2 and h^4q^4 in Eq. (4.6) before solving Eq. (4.1) and thus determining the required coefficients. The resulting exact solution is displayed and investigated in Sec. 4.3.

4.2.2   DARBOUX SCHEME

Continuous NLSE solutions of third order and beyond are hard to find in simple analytic form due to their complexity, rendering the intuitive direct method limited in its use. However, due to the integrability of the AL equation (see details in Sec. 4.6), we can establish a method involving Darboux transformations to recursively generate the whole rogue wave hierarchy [24]. We stress that this scheme can produce closed-form expressions for all solutions via a "zero modulation frequency" limit [25], but it is visually more convenient to apply the process numerically, with parameters as close to the computational limit as possible.

Much of the general scheme is extensively described throughout the literature [11,13] and the AL equation particulars that we have developed for rogue waves are displayed in sections 4.6, 4.7, and 4.8. We will not present discrete periodically modulating "breathers" here, but the NLSE analog of the rogue wave production process, along with helpful diagrams, are provided elsewhere [26]. To summarize, an AL breather solution of order m is a nonlinear combination of m "first-order" components, indexed by j, all superimposed on a plane wave background. Each component is solely dependent on three parameters, namely, an eigenvalue λ_j and two axial shifts (x_j, t_j). As mentioned in Sec. 4.2.1, the spatial shift can be converted from units of the continuous domain to a number of discrete gridpoints via the simple relation $n_j = x_j/h$.

Regardless, it is the eigenvalue that has the greatest effect on the shape of a breather, as it describes the relevant physics of modulation instability. In fact, provided that the eigenvalue is purely imaginary, thereby enforcing that the modulations of the wavefunction are parallel to a dimensional axis, the frequency of that modulation is defined as $\kappa = 2\sqrt{1 + \lambda^2}$. To form a rogue wave solution of order m, it follows that the zero-frequency limit $\lambda_j \to i$ $(\kappa_j \to 0)$ must be applied to all components, with the highly unintuitive proviso that no two κ_j variables can ever be equal. Following this procedure, the resulting solution should appear as a finite number of spatiotemporally localized peaks.

4.3   ROGUE WAVE TRIPLET

For the continuous case, the NLSE supports second-order rogue wave triplet solutions, where three constituent peaks are located equidistantly on a circle [23]. This solution features two arbitrary parameters, β and γ, of which the radius of the circle is a function. It is a particularly important structure, as it has led to the discovery of circular rogue wave clusters [26], cascades [27] and the full hierarchy of fundamental rogue wave structures [24].

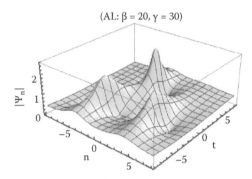

FIGURE 4.2: An AL discrete rogue wave triplet as given by Eqs. (4.7) and (4.8). Here $\beta = 20, \gamma = 30, h = 1$ and $q = 1/2$. As discussed in Sec. 4.2.1, the solution is only technically defined along the lines where n is an integer, but can be extended as a continuous function.

We now present, for the first time with arbitrary grid-spacing h, the corresponding solution for the AL equation. Previous study [28] only considered the basic "fused" ($\beta = 0$ and $\gamma = 0$) form and took $h = 1$. Here we find a more involved family, incorporating arbitrary parameters β and γ along with the dependency on h. It can be written in the form

$$\psi_n(t) = q \left[\frac{4}{D_n} \left(1 + h^2 q^2 \right) \left(G_n + \frac{iH_n}{\sqrt{1 + h^2 q^2}} \right) + 1 \right] e^{iq^2 t}, \qquad (4.7)$$

with

$$
\begin{aligned}
G_n(t) &= 3[3 - 6(N^2 - 2q^2 h^2) + N^2(4q^2 h^2 - N^2) \\
&\quad -6(3 + 2q^2 h^2 + N^2)T^2 - 5T^4] - 6\beta N + 6\gamma T, \\
K_n(t) &= T[3[15 + 6(N^2 + 4q^2 h^2) + N^2(16q^2 h^2 - N^2) \\
&\quad -2(1 + N^2)T^2 - T^4] - 6\beta N] \\
&\quad +3\gamma(T^2 - N^2 - 1), \\
D_n(t) &= 9 + (27 + 24q^2 h^2 + 16q^4 h^4)N^2 + N^6 \\
&\quad +3(33 + 72q^2 h^2 - 6N^2 \\
&\quad +48q^4 h^4 - 16q^2 h^2 N^2 + N^4)T^2 \\
&\quad +3(9 + 8q^2 h^2 + N^2)T^4 \\
&\quad +T^6 + N^4(3 - 8q^2 h^2) \\
&\quad +\beta[\beta + 2N(3T^2 - N^2 + 3 + 4q^2 h^2)] \\
&\quad +\gamma[\gamma + 2T(3N^2 - T^2 - 9 - 12q^2 h^2)], \qquad (4.8)
\end{aligned}
$$

where $N = 2qhn$ and $T = 2q^2\sqrt{1 + q^2 h^2}t$. This solution directly satisfies the AL equation, and an example of it is shown in Fig. 4.2, with $\beta = 20, \gamma =$

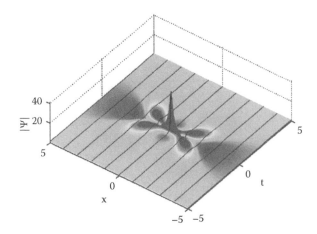

FIGURE 4.3: Fused second-order AL rogue wave produced numerically by the Darboux scheme. Here $x_1 = x_2 = 0$ ($\beta = 0$), $t_1 = t_2 = 0$ ($\gamma = 0$) and $h = 1$, so that $x = n$. Modulation frequencies $\kappa_1 = 1 \times 10^{-4}$ and $\kappa_2 = 1.189 \times 10^{-4}$ are sufficiently close to zero so as to bring the peak maximum to 40.97, in good agreement with theoretical maximum 41.

$30, h = 1$ and $q = 1/2$. While true that a triplet of arbitrary spacing is unlikely to align with the discrete grid, it is always possible to translate the whole structure analytically (by replacing n with $n - n_s$) so that the lattice experiences the maximum amplitude of at least one peak. In any case, the continuous wavefunction displayed is a natural extension of the discrete rogue wave, according to the $x = nh$ relation and the discussion in Sec. 4.2.1. Indeed, by setting $n = x/h$ and taking the limit $h \to 0$, the discrete triplet represented by Eq. (4.8) reduces to the continuous triplet in Eq. (4.6), as expected.

Although we strictly reserve the label "shifts" in this work for x_j and t_j, parameters used within the Darboux scheme, β and γ are directly related to the shifts [26] and control the location of the triplet peaks. Naturally, when $\beta = \gamma = 0$, all three are unshifted and fused together at the origin. They form one peak, with height relative to the background given by

$$\psi_0(0) - q = 4q(1 + h^2 q^2)(1 + 4h^2 q^2). \tag{4.9}$$

So, for $q = h = 1$, the central maximum amplitude is 41, as verified in Fig. 4.3 by a numerical implementation of the Darboux scheme.

4.4   DISCRETIZATION EFFECTS

Despite the apparent bijectivity between AL and NLSE rogue waves, discounting the intricacies involving continuous and discrete domains, it is clear that discretization in the AL manner has two primary effects. As Fig. 4.3 indicates, the first is that an increase in h dramatically boosts the height of a rogue wave

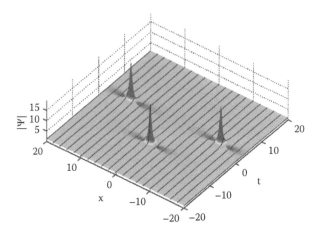

FIGURE 4.4: Fused second-order AL rogue wave produced numerically by the Darboux scheme. Here $x_1 = x_2 = 0$, $t_1 = (5\kappa)^2$, $t_2 = 2(5\kappa)^2$ and $h = 2$, so that $x = 2n$. Modulation frequencies are $\kappa_1 = \kappa$ and $\kappa_2 = \sqrt{2}\kappa$. Constant factor κ is numerically close to zero.

solution. It also does not do this with linear scaling, resulting in particularly steep and sharp amplitude bursts. Indeed, the secondary peaks around the central maximum of the second-order rogue wave are much smaller in relative terms than in the NLSE case.

This amplitude effect is again emphasized by the difference between AL Peregrine solitons for a system with $h = 2$, shown in Fig. 4.4, and for a system with $h = 1/2$, shown in Fig. 4.1. However, the former figure also provides clear evidence of the second consequence of discretization. An increase in h leads to the stretching of a rogue wave solution along the x (or n) axis. This distortion is important, because one of the curiosities regarding the NLSE rogue wave triplet is related to the ease with which the equation can be scaled so as to produce a perfectly circular arrangement of Peregrine solitons [22]. This geometric elegance was retained even when higher-order clusters were discovered [26]. Yet it appears that discretization forces triplet peaks to be confined to an ellipse, rather than a circle, with the major axis dependent on h. In contrast, there is no evidence that the minor axis of this ellipse changes length with discretization, which is not unexpected as the nodes are periodic solely along the x axis. It is nonetheless noteworthy that the new physics arising from the AL equation is able to reproduce rogue wave peak events at the same times as for the NLSE by only modifying amplitudes and spatial settings. This temporal alignment between continuous and discrete wavefunctions strongly supports the bijectivity of rogue wave solutions.

Further insight into how wavefunction behavior changes along the spatial axis between the discrete AL system and the continuous NLSE case can be derived from closer examination of the Darboux transformation scheme. Although rogue wave solutions of either type are strictly quasi-rational, with polynomial numerators and denominators, they are still limiting cases of breather solutions. In the continuous case, breathers are algebraic combinations of exponential functions with base e [25]. However, even though evolution along the t axis retains this form, recurrence relations result in a modified rate of growth along the x axis. This distinction between discrete and continuous behavior is mathematically elegant, particularly as the two are related by

$$\lim_{h \to 0} \left(\sqrt{1 - h^2\lambda^2} \pm ih\sqrt{1 + \lambda^2} \right)^n$$

$$= \lim_{h \to 0} (1 - h^2\lambda^2)^{\frac{n}{2}} \left(1 \pm \frac{ih\sqrt{1 + \lambda^2}}{\sqrt{1 - h^2\lambda^2}} \right)^n$$

$$= \lim_{h \to 0} \left(1 \pm \frac{i\kappa}{2} \frac{h}{\sqrt{1 - h^2\lambda^2}} \right)^n$$

$$= \lim_{n \to \infty} \left(1 \pm \frac{i\kappa}{2} \left(\frac{x}{n} + O\left[\left(\frac{x}{n} \right)^3 \right] \right) \right)^n$$

$$= e^{\pm \frac{i\kappa x}{2}}. \tag{4.10}$$

This proof requires the understanding that $n \to \infty$ for fixed x when $h \to 0$, as well as knowledge of the transformation $\lim_{n \to \infty} (1 + x/n)^n = e^x$.

Beyond the spatiotemporal domain, it is worth examining the effect of discretization on rogue wave spectra. Indeed, the NLSE Peregrine soliton is notable for being linked with an unusual triangular spectral shape [29]. Comparing this with the AL case then is not difficult, particularly considering that the ubiquitous fast Fourier transform (FFT) method is but a numerical implementation of the discrete Fourier transform (DFT). Certainly, AL rogue waves are well suited for such an investigation, where the FFT operates only along their periodic lattice. Given this, grid-spacing h provides a limit on the angular sampling frequency, $w_s = 2\pi/h$, and the effective "spectral density" is accordingly defined as h^2 times the squared modulus of a rogue wave DFT.

Applying this, we find that discrete first-order rogue waves all appear to maintain the same triangular spectral shape to a certain extent, as shown by Fig. 4.5. This is particularly true for low values of h, even though spectra with peak-lattice alignment appear to have upturned tails and spectra that are shifted by half a node (i.e., $n_s = 1/2$) have tails that rapidly fall off. However, as the sampling frequency decreases, all spectral components eventually start increasing noticeably in amplitude. Moreover, in the extreme case where h becomes comparable with the width of the spatiotemporal event, peak-lattice alignment begins to play a crucial role in spectral density. Nonetheless, this scenario is relatively uninteresting, as few nodes of the lattice pass through and register the rogue event. This is why, in Fig. 4.5, both $n_s = 0$ and $n_s = 1/2$ spectra for $h = 2$ approximate pure delta spikes with barely a triangular fea-

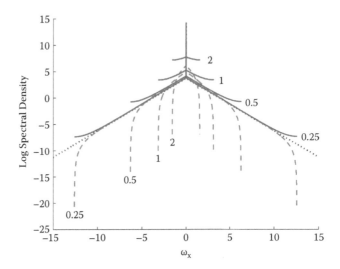

FIGURE 4.5: Spectra of first-order and fused second-order AL rogue waves. Solid lines represent unshifted structures, while dashed lines represent rogue waves shifted by half a node with respect to their lattice (i.e., $n_s = 1/2$ or $x_s = h/2$). Numbers next to spectral tails denote h values. All spectra are symmetric and have similar level DC component spikes at $\omega_x = 0$. Dotted black lines indicate asymptotic spectra for $h \to 0$. (a) Rogue waves of order 1. (b) Rogue waves of order 2.

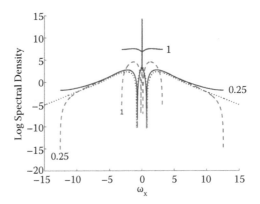

FIGURE 4.6: As for Figure 4.5 (a) Rogue waves of order 1. (b) Rogue waves of order 2.

ture. Wavefunction values are almost universally constant across the periodic grid in this extreme.

Discrete second-order rogue waves in fused form show similar behavior to the first-order case. The spectra have a triangular appearance at moderate frequencies, turn up or down in the far field depending on peak-lattice alignment, but possess a sharp dip on either side of the DC component. This similarity to the continuous NLSE case is shown in Fig. 4.5. Again, spectral features become difficult to classify for values of h that compare with the width of a second-order rogue wave, and this is simply due to poor resolution of the event across nodes of the lattice. However, provided that h is small enough, we can confirm that the spectra of continuous and discrete rogue waves are quite similar.

As we have presented a rogue wave triplet in this work, it is also of interest to examine the spectrum of a cross-sectional profile through two of the three peaks when separated. A possible arrangement with an aligned doublet is shown in Fig. 4.4, but we select triplets with smaller h values so as to increase the sampling frequency. Curiously, Fig. 4.7 shows that these spectra maintain the qualitative triangular appearance of a single AL Peregrine soliton. The only difference arises as periodic dips in spectral density, assuming the form of an interference pattern. Notably, increased values of h correspond to an increased frequency of spectral holes, but this tuning of the interference pattern is only an indirect result of discretization, thanks to the already established spreading effect along the spatial axis. The two "interfering" peaks move further apart as the discrete grid becomes more sparse. So, while these results admit that discretization and lattice spacing do have some effect on rogue wave spectra, the salient features of the continuous case remain present.

4.5 ABLOWITZ–LADIK ROGUE WAVE HIERARCHY

As mentioned, the Darboux transformation procedure can be used recursively and numerically to obtain higher-order discrete rogue waves while avoiding tedious algebra. Notably, the entire hierarchy of fundamental structures [24] is reproducible for the Ablowitz–Ladik discretization of the NLSE.

Rogue wave triplets, as an element of the circular cluster subset [26], are not complicated to produce in either the NLSE or AL case. If a rogue wave of order m is produced by taking the modulation frequency limit $\kappa \to 0$, then shifting an arbitrary component by a constant multiple of $\kappa^{2(m-1)}$ is sufficient to pull out the cluster structure, be it circular or elliptical in shape. In fact, this class of solution can be considered as a "default" shifted wavefunction as its existence conditions will always be met, even if parameters are poorly chosen so as not to produce the other fundamental structures [24].

In general, though, higher-order rogue waves in the fundamental hierarchy require precise parametric relations involving pre-limit component frequencies and shifts. The details for the NLSE are found elsewhere [24], but it is noteworthy that discretization does not affect the existence conditions for these structures in any way. For instance, Fig. 4.9 shows that a triangular cascade is still produced when there is a linear relation between shifts and squared

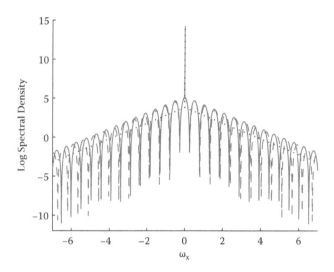

FIGURE 4.7: Spectral interference patterns for two aligned first-order AL rogue waves, as part of a numerically generated rogue wave triplet. Darboux scheme parameters are $x_1 = x_2 = 0$, $t_1 = (5\kappa)^2$ and $t_2 = 2(5\kappa)^2$. Modulation frequencies are $\kappa_1 = \kappa$ and $\kappa_2 = \sqrt{2}\kappa$, with constant κ numerically close to zero. The solid line represents the spectrum for $h = 1/4$. The dashed line represents the spectrum for $h = 1/16$. The triangular spectrum marked by a dotted black line is that of a first-order rogue wave, with $h = 1/4$, and serves as a visual aid.

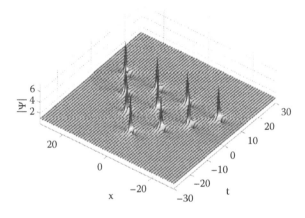

FIGURE 4.8: As for Figure 4.7.

pre-limit component frequencies, despite a lattice spacing of $h = 1$. Features of the circumellipse will require further analytic investigation to constrain, but

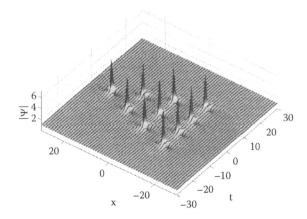

FIGURE 4.9: Fourth-order triangular cascades constructed via the Darboux scheme, with $h = 1$, so that $x = n$. The pre-limit component frequency ratio is $\kappa_1 : \kappa_2 : \kappa_3 : \kappa_4 = 1 : \sqrt{2} : \sqrt{3} : \sqrt{4}$. (a) Shifts are non-zero only in t, with $t_j = j(5\kappa)^2$. (b) Shifts are non-zero only in x, with $x_j = j(5\kappa)^2$.

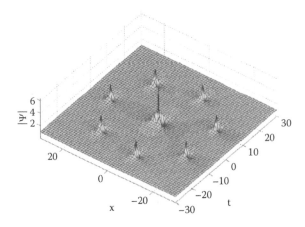

FIGURE 4.10: Fourth-order triangular cascades constructed via the Darboux scheme, with $h = 1$, so that $x = n$. Pre-limit component frequency ratio is $\kappa_1 : \kappa_2 : \kappa_3 : \kappa_4 = 1 : \sqrt{2} : \sqrt{3} : \sqrt{4}$. (a) Shifts are non-zero only in t, with $t_j = j(5\kappa)^2$. (b) Shifts are non-zero only in x, with $x_j = j(5\kappa)^2$.

Fig. 4.10 proves that a non-zero h does not affect the existence of a cascade even when component shifts are along the x (or n) axis.

Given that the methodology for producing the entire rogue wave hierarchy

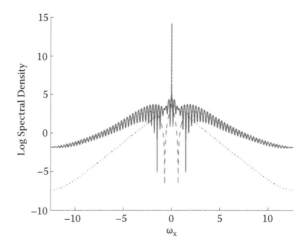

FIGURE 4.11: Spectral interference patterns for a first-order and second-order AL rogue wave in alignment, as part of a numerically generated fourth-order rogue wave heptagram. Darboux scheme parameters are $x_j = j(5\kappa)^6$, $t_j = 0$ and $h = 1/4$, so that $x = n/4$. Modulation frequencies are $\kappa_j = j^{1/6}\kappa$, with constant κ numerically close to zero. (a) Wavefunction in spatiotemporal domain. Grid does not represent lattice and is a visual aid. (b) Solid line represents spectrum for profile at $t = 0$. Serving as visual aids, the dotted black and dashed lines denote spectra for first-order and fused second-order rogue waves, respectively, with $h = 1/4$.

has been reconfirmed for the AL case, it is possible to explore the spectra of higher-order structures with standard application of DFTs, as mentioned in Sec. 4.4. One interesting scenario revolves around how the spectrum is resolved when its spatial profile cuts through both a first- and second-order peak, as an extension of the interference concept presented in Fig. 4.7. This situation is not possible as part of a fundamental structure until order 4, specifically within the heptagram presented by Fig. 4.11. With this structure, a Fourier transform of its $t = 0$ profile produces the unique image in Fig. 4.8. Notably, the spectrum can be treated as a compound, produced from those of an AL Peregrine soliton and a fused second-order rogue wave, as exemplified by Figs. 4.5 and 4.6, respectively. Indeed, the oscillating interference pattern appears to trace the maximum of both spectra, even appearing to produce Gibbs-like phenomena at the non-differentiable points that denote a switch-over. Furthermore, although not shown to avoid visual clutter, reducing the distance between the center of the heptagram and its outlying peak results in the expected spectral effect. Similar to Fig. 4.7, a decreased h increases the wavelength of the interference pattern and spreads out the spectral ripples.

4.6 CONCLUSION

In summary, our main results are as follows:

1. We have provided a closed-form expression for the Ablowitz–Ladik (AL) rogue wave triplet of arbitrary lattice-spacing h. This expression was notably derived from that of the continuous NLSE triplet solution by including additional polynomial terms involving h^2 and h^4.

2. Investigation of the triplet indicates that discretization appears to affect only the amplitude and spatial location of its peaks. Each of the three sufficiently separated events are amplified with nonlinear scaling so as to appear particularly steep and sharp. An increase in h also leads to an increase in inter-peak distance along the spatial axis, such that a triplet appears to be arranged in an ellipse, as opposed to a typical NLSE circle.

3. The Darboux transformation allows for the construction of solutions beyond the rogue wave triplet. The mathematics of the procedure become effectively identical to the scheme for the NLSE in the $h \to 0$ continuous limit. Accordingly, we find that there is a bijectivity of discrete and continuous rogue wave solutions, such that the entire fundamental hierarchy can be reproduced. Discretization does not appear to affect the existence conditions for any of the more complicated structures, with the effects of h remaining purely associated with amplitude and spatial setting.

4. Due to the suitability of the discrete Fourier transform (DFT) for Ablowitz–Ladik solutions, we have shown that spectra of discrete rogue waves retain all the regular qualitative features of continuous NLSE rogue waves, provided that grid-spacing h is sufficiently small to allow good resolution of the structure in the spatiotemporal domain. We have also presented the first examples of spectral interference patterns arising from rogue waves in temporal alignment. When the structures are of different order, the resulting DFT shape appears as a compound of the individual spectra. We also find that an increase in h separates the rogue waves and decreases the wavelength of the resulting spectral interference pattern.

ACKNOWLEDGMENTS

The authors acknowledge the support of the Australian Research Council (Discovery Project number DP140100265). N.A. and A.A. acknowledge support from the Volkswagen Stiftung.

REFERENCES

1. A. R. Osborne, *Nonlinear Ocean Waves And The Inverse Scattering Transform*. Elsevier, 2010.

2. H. Yuen and B. Lake, "Nonlinear dynamics of deep-water gravity waves," *Adv. Appl. Mech*, 22, 67, 1982.

3. G. P. Agrawal, *Nonlinear Fiber Optics (Optics and Photonics)*. Academic Press, 5th edition, 2012.

4. A. V. Gurevich, *Nonlinear Phenomena in the Ionosphere*. Springer, Berlin, 1978.

5. A. B. Zakharov and V. E. Shabat, "Exact theory of two-dimensional self-focusing and one-dimensional self-modulation of waves in nonlinear media," *Journal of Experimental and Theoretical Physics*, 34, 62, 1972.

6. N. Akhmediev and V. I. Korneev, "Modulation instability and periodic solutions of the nonlinear Schrödinger equation," *Teoreticheskaya i Matematicheskaya Fizika (USSR)*, 69, 189, 1986. English translation in *Theor. Math. Phys.*, 69, 1089 (1986).

7. D. J. Kedziora, A. Ankiewicz, and N. Akhmediev, "Rogue waves and solitons on a cnoidal background," *The European Physical Journal Special Topics*, 223, 43, 2014.

8. N. Akhmediev, A. Ankiewicz, and M. Taki, "Waves that appear from nowhere and disappear without a trace," *Phys. Lett. A*, 373, 675, 2009.

9. Y. Kartashov, B. A. Malomed, and L. Torner, "Solitons in nonlinear lattices," *Review of Modern Physics*, 83, 247, 2011.

10. J. C. Eilbeck and M. Johansson, "The discrete nonlinear Schrödinger equation-20 years on," in *Conference on Localization and Energy Transfer in Nonlinear Systems*, p. 44, 2003.

11. Q. Zhenyun, "A generalized Ablowitz–Ladik hierarchy, multi-Hamiltonian structure and Darboux transformation," *Journal of Mathematical Physics*, 49, 063505, 2008.

12. X. Wu, W. Rui, and X. Hong, "A new discrete integrable system derived from a generalized Ablowitz-Ladik hierarchy and its Darboux transformation," *Discrete Dynamics in Nature and Society*, 2012, 2012.

13. X. Liu and Y. Zeng, "On the Ablowitz–Ladik equations with self-consistent sources," *Journal of Physics A: Mathematical and Theoretical*, 40, 8765, 2007.

14. O. O. Vakhnenko and V. O. Vakhnenko, "Physically corrected Ablowitz-Ladik model and its application to the Peierls-Nabarro problem," *Physics Letters A*, 196, 307, 1995.

15. N. Akhmediev and A. Ankiewicz, "Modulation instability, Fermi-Pasta-Ulam recurrence, rogue waves, nonlinear phase shift, and exact solutions of the Ablowitz-Ladik equation," *Physical Review E*, 83, 046603, 2011.

16. A. Ankiewicz, N. Devine, M. Ünal, A. Chowdury, and N. Akhmediev, "Rogue waves and other solutions of single and coupled Ablowitz–Ladik and nonlinear Schrödinger equations," *Journal of Optics*, 15, 064008, 2013.

17. N. Akhmediev, A. Ankiewicz, and J. M. Soto-Crespo, "Rogue waves and rational solutions of the nonlinear Schrödinger equation," *Phys. Rev. E*, 80, 026601, 2009.

18. M. J. Ablowitz and J. F. Ladik, "Nonlinear difference scheme and inverse scattering," *Studies in Applied Mathematics*, 55, 213, 1976.

19. M. J. Ablowitz, B. Prinari, and A. D. Trubatch, *Discrete and Continuous Nonlinear Schrödinger Systems*, Cambridge University Press, 2004.
20. K. W. Chow, R. Conte, and N. Xu, "Analytic doubly periodic wave patterns for the integrable discrete nonlinear Schrödinger (Ablowitz–Ladik) model," *Physics Letters A*, 349, 422, 2006.
21. N. Akhmediev and A. Ankiewicz, *Solitons: Nonlinear Pulses and Beams.* Chapman & Hall, London, 1997.
22. A. Ankiewicz, N. Akhmediev, and F. Lederer, "Approach to first-order exact solutions of the Ablowitz-Ladik equation," *Physical Review E*, 83, 056602, 2011.
23. A. Ankiewicz, D. J. Kedziora, and N. Akhmediev, "Rogue wave triplets," *Physics Letters A*, 375, 2782, 2011.
24. D. J. Kedziora, A. Ankiewicz, N. Akhmediev, "Classifying the hierarchy of nonlinear-Schrödinger-equation rogue-wave solutions," *Phys. Rev. E*, 88, 013207, 2013.
25. D. J. Kedziora, A. Ankiewicz, N. Akhmediev, "Second-order nonlinear Schrödinger equation breather solutions in the degenerate and rogue wave limits," *Phys. Rev. E*, 85, 066601, Jun 2012.
26. D. J. Kedziora, A. Ankiewicz, N. Akhmediev, "Circular rogue wave clusters," *Phys. Rev. E*, 84, 056611, 2011.
27. D. J. Kedziora, A. Ankiewicz, N. Akhmediev, "Triangular rogue wave cascades," *Phys. Rev. E*, 86, 056602, 2012.
28. A. Ankiewicz, N. Akhmediev, J. M. Soto-Crespo, "Discrete rogue waves of the Ablowitz-Ladik and Hirota equations," *Physical Review E*, 82, 026602, 2010.
29. N. Akhmediev, A. Ankiewicz, and J. M. Soto-Crespo, J. M. Dudley, "Rogue wave early warning through spectral measurements?" *Physics Letters A*, 375, 541, 2011.

5 A theoretical study on modulational instability in relaxing saturable nonlinear optical media

Kuppuswamy Porsezian and *K. Nithyanandan*

5.1 INTRODUCTION

In classical optics, the intensity dependent process in the optical system seeds the emergence to the field of nonlinear optics [1], which deserves considerable attention both fundamental and application perspectives. The interaction between the linear dispersion due to the frequency dependent refractive index and the nonlinearity due to the intensity dependent refractive index admits interesting features which have aroused a great deal of interest for more than five decades. The stability of an input light beam under various physical situations is one of the most exciting issues of modern nonlinear optics. One such stability criterion leads to the phenomenon of modulational instability (MI). MI is a universal instability process which is identified as a characteristic of wave propagation in dispersive nonlinear systems, such as in fluid dynamics [2], nonlinear optics [3], and plasma physics [4]. In the context of nonlinear optics, MI can be explained as the continuous wave (cw) or quasi-cw wave propagating in the nonlinear dispersive media becomes inherently unstable under weak perturbation and evolves into a train of ultrashort pulses as a consequence of temporal modulation due to MI.

The history of MI dates back to mid 1960s, courtesy of Lighthill, who is responsible for the famous Lighthill criterion, which sets the conditions for the stability of the plane wave in a nonlinear medium [5]. Such an instability condition, later called MI, was observed first in hydrodynamics by Benjamin and Feir in 1967 [2]. In the same year, Ostrovskii predicted the possibility of MI in nonlinear optics [6], and it was later explained in detail by Hasegawa and Tappert in the context of optical fibers [7]. Ever since these pioneering works, MI has gained momentum and evolved as one of the most fascinating phenomena in the field of nonlinear optics. The continuing interest in MI for more than five decades in various disciplines of science and technology is a clear manifestation of the importance of MI in science and technology.

Until the present time MI has been addressed in two distinct directions based on its fundamental and applied interest. The two distinct perceptions are exactly opposite; one deals with catastrophic effects and the other presents the exuberance of its usefulness in various applications. The deleterious effects of MI are detrimental to long-haul optical fiber communication systems; the non-return-to-zero code in optical communication, the drastic enhancement of MI gain in Wavelength Division Multiplexing (WDM) systems, sets the limitation to the bandwidth window of communication systems, MI lasers; new frequency generation in optical systems is the ultimate concern for the realization of repeaterless long haul optical communication systems [8,9]. Quite a number of works have been devoted to addressing the above highlighted issues with an objective of reducing the disastrous effects caused by MI. Equally, another side of the literature, especially contemporary research activities, focused primarily on the constructive part of MI. For example, the generation of pulse trains at a high repetition rate is a useful technique to produce ultra-short pulses; frequency conversion and the generation of new frequencies can be effectively made useful in a multi-frequency source under the context of super-continuum generation, which has been recognized in modern days as a "white-light laser" [10–12]. In addition, MI has also find important applications in optical amplification of weak signals, material absorption and loss compensation [13, 14], dispersion management, all-optical switching [15], frequency comb for metrology, and so on [16,17]. Thus suitable manipulation with clinical tailoring can make MI a suitable contender for a wide class of applications.

After a detailed introduction with historical perspectives about the subject of this investigation, it is useful to provide an idea about the current status of the field and the motivation for choosing the selected problems. The dynamical equation governing MI and super continuum generation (SCG) is deeply connected with the nonlinear Schrödinger equation (NLSE), which admits the formation of envelope solitons or envelope solitary pulses due to a delicate balance between anomalous group velocity dispersion (GVD) the self-focusing Kerr nonlinearity [7, 18–21]. MI of this kind is represented as a conventional MI, which leads to a symmetrical pair of sidebands on either side of the pump wave frequency, and it is said to occur in the anomalous dispersion regime. However, MI is also observed in the normal dispersion regime with the aid of higher order dispersion (HOD); a comprehensive analysis of MI in the normal dispersion regime can be seen in the [22]. Moreover, one breakthrough in the context of MI is the observation of MI in normal GVD [23–25]. It is a fact that the propagation of optical beam in the normal GVD regime is not subject to the MI process, due to the lack of phase matching between the dispersion and nonlinear components of the system. But the nonlinear coupling between the two co-propagating beams due to cross-phase modulation (XPM) (i.e., the refractive index seen by one wave depends on the intensity of the co-propagating wave through the XPM coefficient) destabilizes, the steady state leading to frequency modulation even in the normal GVD regime [26–28]. The pioneering work of Agrawal et al. set the benchmark for the extensive work on two color light propagation in optical fiber systems.

We consider both the scalar and vector MI corresponding to a single beam and a co-propagating beam in the optical fibers. It is instructive that the above description of MI considered the nonlinear response of the medium to an optical beam instantaneous, but, strictly speaking, the response of the medium is no longer instantaneous, rather delayed in nature. It is worth noting that the usual assumption of instantaneous Kerr response fails for ultra-short pulses; thus the inclusion of delay in the nonlinear response is essential. On the other hand, for any material medium, there is an upper limit for the optically induced nonlinear refractive index, beyond which the higher order nonlinear susceptibilities inevitably are excited and after a certain power threshold the nonlinear response saturates [29, 30]. Thus indices relaxation and saturation are an integral part of high index nonlinear material in the ultra-short regime, and hence both effects have to be treated together [31].

Considering the importance of higher order dispersion (HOD), we investigated the impact of fourth order dispersion in the MI dynamics of a relaxing saturable nonlinear system, featured in Sec. 5.1 [32]. The influence of nonlinear relaxation in the MI dynamics of vector MI with higher order dispersion is discussed in Sec. 5.2 [33]. The interplay between nonlinear relaxation and coupling coefficient dispersion in a nonlinear directional coupler is discussed in Sec. 5.3 [34]. The influence of nonlinear saturation in the MI of a two core fiber with the effect of coupling coefficient dispersion is discussed in Sec. 5.4 [35]. Section 5.5 features the observation of novel two state behavior in the instability spectrum of saturable nonlinear media [36]. A novel semiconductor doped dispersion decreasing (SD-DDF) proposed in view of ultrashort pulse generation is presented in Sec. 5.6 [37]. Section 5.7 concludes with a brief summary of results.

5.2   SCALAR MI IN THE RELAXING SATURABLE NONLINEARITY (SNL) SYSTEM

5.2.1   THEORETICAL FRAMEWORK

The dynamics of an intense optical beam near the zero dispersion wavelength (ZDW) of saturable nonlinear media is governed by a modified nonlinear Schrödinger equation (MNLSE). The general form of the MNLSE for the slowly varying envelope $E(z,t)$ along the z-axis in a retarded reference time, centered at the frequency ω_0, is given by [32, 38]

$$\frac{\partial E}{\partial z} + \sum_{n=2}^{4} \beta_k \frac{i^{n-1}}{n!} \frac{\partial^n E}{\partial t^n} = \gamma \frac{\Gamma |E|^2}{1 + \Gamma |E|^2} E \qquad (5.1)$$

where z and t are the longitudinal coordinate and time in the moving reference frame, respectively. The dispersion coefficient β_k ($k = 1, 2, 3, 4$) is attributed to the Taylor expansion of the propagation constant around the center frequency

ω_0. We go only up to fourth order dispersion (FOD), since the dispersion coefficient beyond FOD is relatively insignificant due to its small magnitude.
Furthermore, $\gamma = \frac{n_2 \omega_0}{c A_{eff}}$ is the Kerr parameter, n_2 the nonlinear index coefficient and A_{eff} is the effective core area. The function in the right-hand side
of Eq. (5.1) accounts for the saturation of the nonlinear response.

5.2.2   LINEAR STABILITY ANALYSIS

The stability of steady-state solutions of Eq. (5.1) against small harmonic
perturbations can be studied by using linear stability analysis [18]. Steady-
state solutions, or continuous wave (CW) solutions, for our purposes are

$$E_s = E_0 \exp[-i\tilde{f}(\Gamma|E_0|^2)z], \quad N_s = \tilde{f}(\Gamma,|E_0|^2). \tag{5.2}$$

In order to analyze the stability of our steady-state solutions against harmonic
perturbations, we let

$$E_p = (E_0 + a(z,t)) \exp[-i\tilde{f}(\Gamma|E_0|^2)z], \tag{5.3}$$
$$N_p = \tilde{f}(\Gamma,|E_0|^2) + n(z,t) \tag{5.4}$$

where $a(z,t)$ and $n(z,t)$ are small deviations from the stationary solutions of
the electric field and the nonlinear index of refraction, respectively. After some
mathematical manipulations, the dynamical equations for the perturbation
can be written as

$$i\frac{\partial a}{\partial z} = \frac{\beta_2}{2}\frac{\partial^2 a}{\partial t^2} + i\frac{\beta_3}{6}\frac{\partial^3 a}{\partial t^3} - \frac{\beta_4}{24}\frac{\partial^4 a}{\partial t^4} + \gamma\, n E_0, \tag{5.5}$$

$$i\frac{\partial n}{\partial t} = \frac{1}{\tau}[-n + \Gamma \tilde{f}'(\Gamma,|E_0|^2)\, E_0(a + a^*)]. \tag{5.6}$$

We now assume the following ansatz for the perturbation with the frequency
detuning from the pump:

$$a(z,t) = a_1 \exp[-i(Kz - \Omega t)] + a_2 \exp[i(Kz - \Omega t)] \tag{5.7}$$

where a_1 and a_2 are the perturbation amplitudes corresponding to the anti-
Stokes and Stokes sideband s, respectively, and Ω and K are the frequency
and the wavenumber of the perturbation. The compatibility condition for two
equations in a_1 and a_2 will lead to the dispersion relation

$$K = \beta_3 \frac{\Omega^3}{6} \pm \left[\left(\beta_2 \frac{\Omega^2}{2} + \beta_4 \frac{\Omega^4}{24} \right) \left(\beta_2 \frac{\Omega^2}{2} + \beta_4 \frac{\Omega^4}{24} + 2\tilde{\gamma} E_0^2 \right) \right]^{1/2}. \tag{5.8}$$

5.2.3   RESULTS AND DISCUSSION

Using Eq. (5.8), we study MI in relaxing saturable nonlinearity with the effect
of fourth order dispersion (FOD) for the various combinations of dispersion
coefficients. The nonlinear response of the medium is a function of time, using

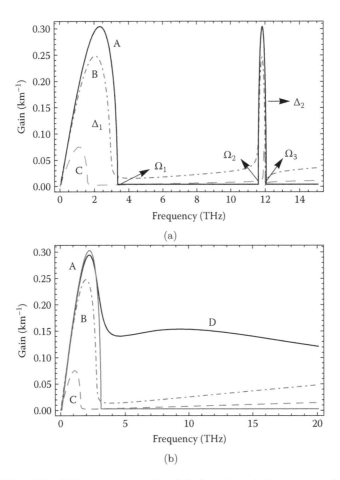

FIGURE 5.1: The MI gain spectra for (a) $\beta_2 < 0$ and $\beta_4 > 0$ and (b) $\beta_2 < 0$ and $\beta_4 < 0$.

the Debye relaxational model. Furthermore, the saturable behavior of the nonlinear response at high intensity is included in the relaxational model. The propagation of the input beam is near the ZDW of the fiber, so the effect of HOD, especially FOD, has been investigated through various possible combinations of group dispersion coefficients.

The relaxing nonlinear system is characterized by a finite delay time; as a result, the nonlinear response of the medium becomes complex. This leads to the parametric and Raman bands, corresponding to the real and imaginary parts of the complex nonlinearity, respectively. The parametric MI regime is certainly determined by the signs of the group dispersion coefficient. The conventional MI band can be observed for the case of a negative second-order dispersion (SOD) coefficient irrespective of the sign of the FOD, as shown in Fig. 5.1a and b. The sign of the FOD coefficient is crucial in determining the

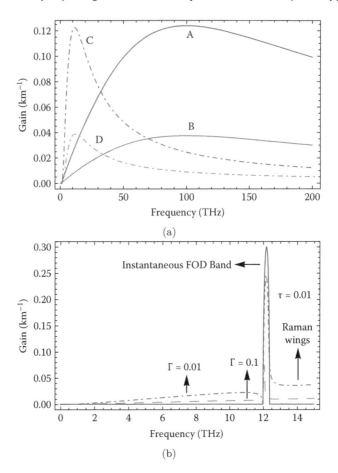

FIGURE 5.2: The MI gain spectra for (a) $\beta_2 > 0$ and $\beta_4 > 0$ and (b) $\beta_2 > 0$ and $\beta_4 < 0$.

dynamics of the MI; for instance, the nonconventional MI band is observed for a positive value of FOD, as in Fig. 5.1a. For negative values of FOD, the MI is made possible irrespective of the sign of SOD. The basic concept behind all the above behavior is determined by the so-called phase matching condition between the dispersion and the nonlinearity. Among the various cases, the combination of $\beta_2 > 0$ and $\beta_4 > 0$ is interesting since the much needed phase matching condition for MI is not possible as both dispersion coefficients are on the positive side. However, an instability band is still observed and is purely due to the delayed response of the nonlinearity, as shown in Fig. 5.2. Due to the complex nature of the nonlinearity, the ranges of unstable harmonic frequencies run down to infinite, which is identified as the main signature of a relaxing nonlinear system.

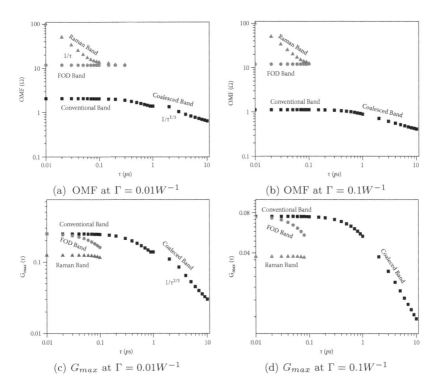

FIGURE 5.3: Plot of OMF and G_{max} as a function of delay time for different saturation parameters Γ.

We have analyzed the MI in two nonlinear response regimes, namely, the fast and slow responses, and the behavior of the system is found to be dramatically different. From our entire discussion, the role of saturation is identified to be unique, and inherently suppresses MI by depleting the nonlinear contribution of the medium. In the regime of the fast response, instability bands are well defined and clearly localized at characteristic detuning frequencies. For instance, a system in the anomalous dispersion regime with a positive FOD coefficient possesses three instability bands $(2 + 1)$, two parametric MI bands (conventional and FOD bands) and the Raman instability band. For a negative FOD coefficient, only two instability bands are observed $(1+1)$, i.e., a conventional band and the Raman band, as shown in the Fig. 5.3. For the case of a normal group dispersion regime in the fast response limit, when the FOD coefficient takes a negative value, only two instability bands are observed at different detuning frequencies $(1 + 1 \rightarrow$ FOD band + Raman band), as is evident from Fig. 5.4. When both dispersion coefficients take positive values, then only the band due to the Raman instability is observed. The Raman band is found to be a highly sensitive function of τ; with an increase in the delay time the optimum frequency (OMF) decreases by a factor of $1/\tau$. The

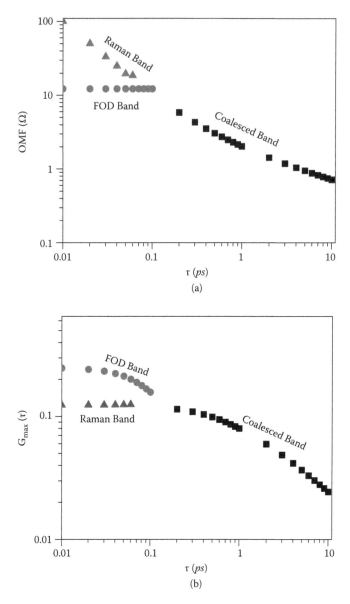

FIGURE 5.4: Plot of (a) OMF and (b) G_{max} as a function of delay time for $\beta_2 > 0$ and $\beta_4 < 0$.

maximum gain of each band in the fast response regime is independent of τ. The dynamics of MI in the slow response regime is unique irrespective of the nature of the dispersion regime. This is attributed to the fact that an increase in delay suppresses the overall MI, and thereby favors the Raman instability process over the parametric MI process. At a particular value of τ, instability bands merge together, resulting in a single coalesced band. A further increase in τ downshifts the coalesced band toward the center frequency by a factor of $1/\tau^{1/3}$. The maximum gain of the coalesced band decreases by $1/\tau^{2/3}$.

To conclude, our detailed theoretical analysis reveals that, depending upon the nature of the nonlinear response, either fast or slow, the behavior of MI qualitatively differs. This will pave the way for the design of experimental systems with the desired nonlinear response time based on its utility [32].

5.3 VECTOR MI IN A RELAXING SYSTEM WITH THE EFFECT OF WALK-OFF AND HIGHER-ORDER DISPERSION

5.3.1 THEORETICAL FRAMEWORK

The co-propagation of two optical waves of different frequencies and the same polarization in a single-mode optical fiber is given by the coupled nonlinear Schrödinger equation (CNLSE). The general form of the CNLSE for the slowly varying envelope $E_j(z,t)$ corresponding to the co-propagation of optical beams along the z-axis with a group velocity v_{gj} in a retarded time frame $t = (T - z/v_g)$ is given by [23, 33, 39]

$$\frac{\partial E_j}{\partial z} + \frac{1}{V_{gj}}\frac{\partial E_j}{\partial t} + i\frac{\beta_{2j}}{2}\frac{\partial^2 E_j}{\partial t^2} - \frac{\beta_{3j}}{6}\frac{\partial^3 E_j}{\partial t^3} + i\frac{\beta_{4j}}{24}\frac{\partial^4 E_j}{\partial t^4} = i\gamma_j N_j E_j, \quad (5.9\text{a})$$

$$\frac{\partial N_j}{\partial t} = \frac{1}{\tau}(-N_j + |E_j|^2 + 2|E_{3-j}|^2). \quad (5.9\text{b})$$

5.3.2 RESULTS AND DISCUSSION

Using standard linear stability analysis as described earlier, the dispersion relation is given by

$$[(k - \frac{1}{V_{g1}}\Omega - \frac{1}{6}\beta_{31}\Omega^3)^2 - h_1][(k - \frac{1}{V_{g2}}\Omega - \frac{1}{6}\beta_{32}\Omega^3)^2 - h_2] = C_{XPM}. \quad (5.10)$$

$$h_j = (\frac{1}{2}\beta_{2j}\Omega^2 - \frac{1}{24}\beta_{4j}\Omega^4)[\frac{1}{2}\beta_{2j}\Omega^2 - \frac{1}{24}\beta_{4j}\Omega^4 + 2\tilde{\gamma}_j|E_j^0|^2] \quad (5.11)$$

$$C_{XPM} = 16\tilde{\gamma}_1\tilde{\gamma}_2|E_1^0|^2|E_2^0|^2(\frac{1}{2}\beta_{2j}\Omega^2 - \frac{1}{24}\beta_{4j}\Omega^4)(\frac{1}{2}\beta_{2j}\Omega^2 - \frac{1}{24}\beta_{4j}\Omega^4). \quad (5.12)$$

Using Eq. (5.10), we discuss the MI scenario in the case of co-propagation of two optical beams with the effect of walk-off and higher order dispersion

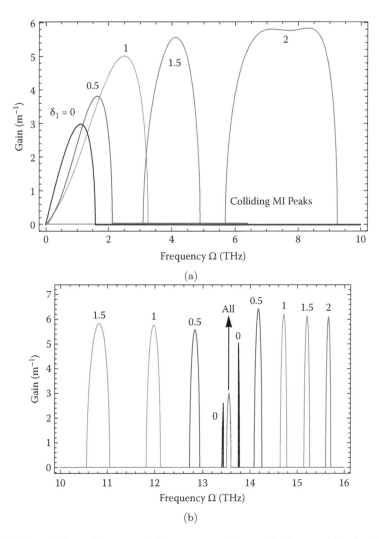

FIGURE 5.5: MI in the normal dispersion regime with the combined effect of GVM and FOD. (a) First spectral windows. (b) Second spectral window.

in the case of relaxing saturable nonlinear media. In the normal group velocity dispersion (GVD) regime, the HOD and walk-off effect brings new characteristic spectral bands at definite frequency windows. The walk-off effect consists of both the group velocity mismatch (GVM) and third-order dispersion (TOD) difference, and it is observed that both play an identical role in MI dynamics. In general, an increase in the walk-off (δ) shifts the MI band toward the higher frequency side and also leads to new spectral bands. Four discrete spectral bands are observed, which have been divided into two spectral windows, as shown in the Fig. 5.5. The first spectral window consists

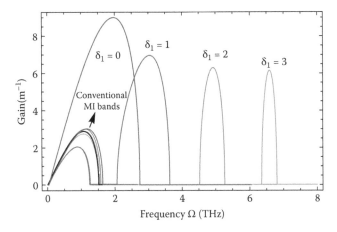

FIGURE 5.6: The MI spectra $G(\Omega)$ with the effect of FOD and GVM for various values of δ_1.

of a single band near the center frequency and the spectral window consists of three bands, one immobile primary band and two mobile secondary bands on either side of the primary band, respectively. An increase in (δ) shifts the mobile band away from the primary center band. In the anomalous GVD regime, there exist two bands, one near and the other away from the center frequency, as shown in the Fig. 5.6. This is attributed to the fact that for the anomalous GVD regime, FOD is relatively suppressed due to the dominance of SOD. There is no observation of a secondary spectral band, and the role of the walk-off effect is found to be the same as that of the normal GVD case.

Later we discuss exclusively the influence of relaxation in the nonlinear response for both dispersion regimes. The role of delay is addressed in two typical cases, namely, fast response and slow response. The cumulative effect of HOD, walk-off and delay leads to interesting behavior as follows: (i) any finite relaxation extends the range of unstable frequencies and there exist two unstable modes called the Raman modes. (ii) In the normal GVD regime, there exist four discrete unstable MI bands for the case of instantaneous response. Inclusion of delay (τ) makes the MI curve run through the local instability bands and thereby extends the range of unstable frequencies literally down to infinite frequency. For the case of fast response, as shown in the Fig. 5.7a, the incorporation of delay hardly has any impact on the first band, whereas the rest of the MI bands undergo a relative fall in gain. The second unstable mode possesses the larger gain factor and a hump corresponding to the FOD band. The slow response reduces the MI gain, thereby suppressing the overall MI, as portrayed in Fig. 5.7b. (iii) In the case of the anomalous GVD regime, there exist two discrete MI bands. The delay, as discussed earlier, extends the

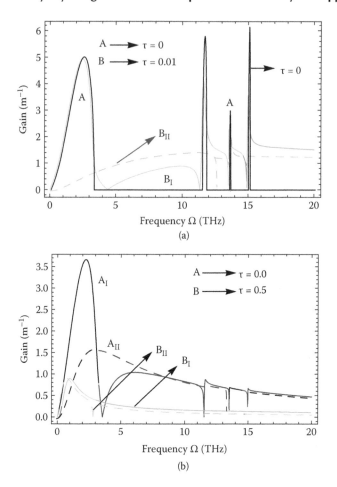

FIGURE 5.7: The MI gain spectra under the combined action of FOD and the walk-off effect ($\delta_1 = 1$ psm^{-1} and $\delta_2 = 4 \times 10^{-4}$ ps^{3}m^{-1}). (a) Fast response. (b) Slow response.

unstable frequencies and connects the two MI bands. For the fast response, as shown in the Fig. 5.8a, the gain of the first band remains unaffected, whereas the second band suffers a slight fall in gain. In the case of slow response, the delay extends the unstable frequencies but affects MI by depleting the overall MI gain, as evident from Fig. 5.8b.

Overall, the relaxation in the nonlinear response offers infinite unstable frequencies. The walk-off effect and HOD, on the other hand, bring new characteristic local MI bands. Thus the combined action of both walk-off and HOD effects is responsible for the generation of MI spectrum, where the MI curve appears to be sailing over the local instability bands. Thus, the combination

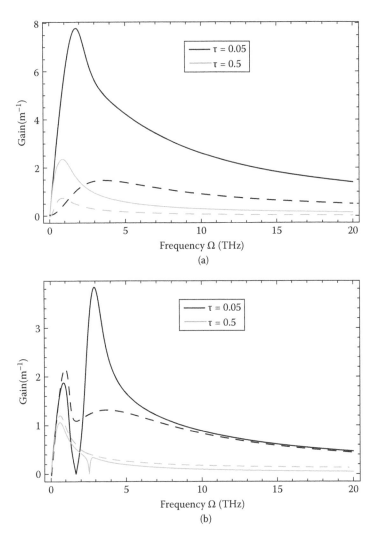

FIGURE 5.8: The MI gain spectra under the combined action of FOD and the walk-off effect ($\delta_1 = 1\mathrm{psm}^{-1}$ and $\delta_2 = 4 \times 10^{-4}\mathrm{ps}^3\mathrm{m}^{-1}$. (a) Fast response. (b) Slow response.

of HOD and the relaxation effect in coupled systems leads to a rich variety of information [33], which requires experimental realization. We believe that the outcome of the results can set the benchmark for experiments pertaining to the investigation of various nonlinear effects in the relaxing nonlinear system.

5.4   MI IN A TWO-CORE NONLINEAR DIRECTIONAL COUPLER WITH RELAXING NONLINEARITY

5.4.1   THEORETICAL FRAMEWORK

The equation governing the dynamics of pulse propagation in a non-instantaneous dual core fiber with a finite nonlinear response time (τ) can be written as [34]

$$i\frac{\partial E_j}{\partial z} - \frac{1}{2}\beta_2\frac{\partial^2 E_j}{\partial t^2} + \kappa\, E_{3-j} + i\kappa_1\frac{\partial E_{3-j}}{\partial t} + \gamma N_j E_j, \tag{5.13a}$$

$$\frac{\partial N_j}{\partial t} = \frac{1}{\tau}(-N_j + |E_j|^2). \tag{5.13b}$$

The parameter $N_j = N_j(z,t)$ is equivalent to $|E_j^2|$ of the original NLSE. The dynamics of N depends on the finite response time of the medium (τ) and the local intensity of the field. κ is the coupling coefficient and $\kappa_1 = d\kappa/d\omega$ is the coupling coefficient dispersion corresponding to the wavelength dependence of the coupling coefficient and is equivalent to the intermodal dispersion.

5.4.2   RESULTS AND DISCUSSION

Using standard linear stability analysis as described in Sec. 5.2.2, the dispersion relation corresponding to MI is given by

$$\left[\left(K + \frac{\kappa_1\Omega}{\sqrt{2}}\right)^2 - h_1\right]\left[\left(K + \frac{\kappa_1\Omega}{\sqrt{2}}\right)^2 - h_2\right] = h, \tag{5.14}$$

$$h_1 = \frac{1}{2}\left[\tilde{\gamma}^2 E_0^4 - 4\kappa^2 + (\kappa_1^2 - 2\sqrt{2}\beta_2\kappa)\Omega^2 + \frac{\beta_2^2\Omega^4}{2}\right], \tag{5.15a}$$

$$h_1 = \frac{1}{2}\left[\tilde{\gamma}^2 E_0^4 - 4\kappa^2 + (\kappa_1^2 + 2\sqrt{2}\beta_2\kappa)\Omega^2 + \frac{\beta_2^2\Omega^4}{2}\right], \tag{5.15b}$$

$$h = \frac{1}{4}(\tilde{\gamma}^2 E_0^4 - 4\kappa^2)^2 - (\tilde{\gamma}^2 E_0^4 - 4\kappa^2)(\beta_2\tilde{\gamma}E_0^2 + 2\kappa_1^2)\Omega^2 - \tag{5.15c}$$

$$(5\beta_2^2\kappa^2 + \kappa_1^4 - \beta_2^2\tilde{\gamma}^2 E_0^4)\Omega^4 + \frac{1}{2}\beta_2^2\kappa_1^2\Omega^6, \tag{5.15d}$$

$$\tilde{\gamma} = \gamma/(1 + i\Omega\tau). \tag{5.15e}$$

Using Eq. (5.14), we study MI in the two core nonlinear directional coupler with the effect of relaxing nonlinearity and coupling coefficient dispersion. First, symmetric/antisymmetric CW is considered, and it is found that coupling coefficient dispersion (CCD) does not have any significant effect apart from shifting the range of unstable frequencies. However, the inclusion of delay in the nonlinear response leads to an additional instability band at higher frequencies, which is recognized as the Raman band, due the delayed nonlinear response. Following that we have extended our study to the case of the asymmetric CW state, where both delay and CCD have a huge impact on the MI spectrum. From the dispersion relation corresponding to the asymmetric case Eq. (5.14), it is observed for any finite value of delay, the equation becomes fourth order polynomial with complex coefficients. Therefore, there exist two unstable modes in addition to the parametric instability band; this is in contrast to the single parametric instability band observed in the instantaneous Kerr case. The unstable modes are represented as Raman bands due to the delayed nonlinear response.

To give a comprehensive picture of the problem, we studied the impact of various system parameters on the MI spectrum, in both anomalous and normal dispersion regimes. In the anomalous dispersion regime, any increase in power monotonously increases the gain of the instability bands (one parametric band + two Raman bands), as shown in Fig. 5.9.

To emphasize the role of delay (τ) in the MI spectrum, MI was studied for a range of delay times. Two typical delay regimes are considered based on the time scale, namely, the fast and slow response regimes. In fast response, the parametric MI band dominates, of which the gain is twice that of the Raman band. As τ increases at a particular frequency, both the parametric and Raman bands collide, leading to a single coalesced band, as shown in Fig. 5.10. With further increase in τ, the gain of the coalesced band decreases by a factor of $1/\tau^{2/3}$, as shown in Fig. 5.10. Thus, one can infer that the slow response leads to the overall supersession of MI. On the other hand, the OMF at fast response remains constant for the parametric band, and the Raman band decreases by a factor of $1/\tau$. At the slow response regime, the optimum modulation frequency (OMF) of the coalesced band decreases at the rate of $1/\tau^{1/3}$, as shown in the Fig. 5.11. The role of CCD in the MI spectrum is interesting, since CCD leads to the emergence of new spectral bands near the center frequency, and the Raman band at the far detuning frequency does not vary significantly. There exists a critical CCD (measured as κ_{1cr}), where the dynamics of MI dramatically differs. To emphasize the role of critical CCD, κ_{1cr} as a function of system parameters is shown in Fig. 5.12.

For better illustration, the normal dispersion case has also been considered for discussion, and it is observed that MI can still be achieved even without the aid of the coupling coefficient. The instability band at the zero coupling coefficient is known to be the self-phase matched Raman band. For any finite value of κ, one can notice an instability band near the center frequency. The variation of gain with power and coupling coefficient is observed to be the same as in the anomalous dispersion regime. CCD, on the other hand, leads to new spectral bands, and at threshold value ($\kappa 1cr$), the MI spectrum behaves qualitatively in a different manner.

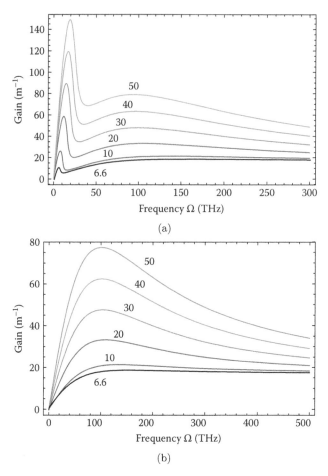

FIGURE 5.9: (a) Plot showing the variation of MI gain as a function of power. (b) Variation of the gain of the Raman as a function of power.

Overall, the study based on the incorporation of relaxation in the nonlinear response offers a rich variety of information such as (i) existence of infinite unstable frequencies, (ii) possibilities of an instability band at the normal dispersion regime even at the zero coupling coefficient, (iii) observation of Raman band at the dispersionless limit, (iv) enhancement of the gain of the Raman band at higher values of coupling coefficient. Also, based on the nature of the medium (or dopant), the nonlinear response can vary from picoseconds to nanoseconds, and hence the present result showing the dynamics of MI in the two core fibers for a wide class of relaxation times can be very informative and can set the benchmark for future prospects. Furthermore, the study of CCD in the two core fiber now attracts a great deal of interest and new results

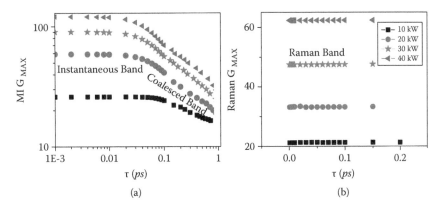

FIGURE 5.10: Variation of G_{max} of MI and Raman bands as a function of τ for different values of power with $\kappa = 10\,\mathrm{m}^{-1}$.

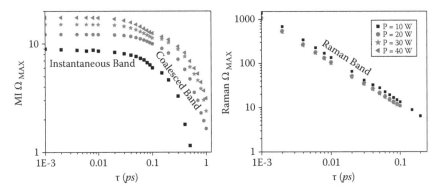

FIGURE 5.11: Variation of OMG of MI and Raman bands as a function of τ for different values of power.

have been achieved. Thus the present chapter with the aforementioned results featuring the interplay between CCD and the relaxation of nonlinearity is interesting and we hope that our theoretical results will provide guidelines in the designing of novel two core fiber systems for various applications [34].

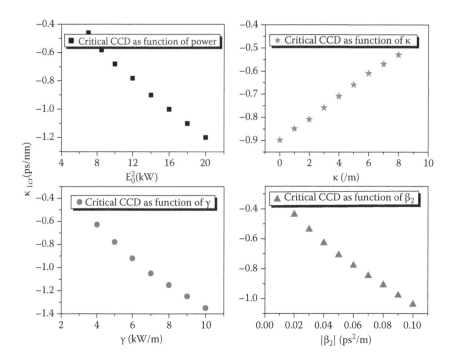

FIGURE 5.12: Plot showing the variation of critical CCD ($\kappa_1$cr) with system parameters as depicted in the inset of the figure.

5.5   MI IN A TWO-CORE FIBER WITH THE EFFECTS OF SATURABLE NONLINEARITY AND COUPLING COEFFICIENT DISPERSION

5.5.1   THEORETICAL FRAMEWORK

The propagation of a high intensity optical beam in the single-mode twin core liquid filled photonic crystal fiber in the limit of slowly varying envelope approximation is given by the pair of linearly coupled nonlinear Schrödinger equations (CNLSE). The CNLSE describing the evolution of the slowly varying envelopes (U_j, $j = 1, 2$) with the effect of SNL is given by the equation [35, 40–42]

$$i\frac{\partial U_j}{\partial z} - \frac{1}{2}\beta_2\frac{\partial^2 U_j}{\partial t^2} + \kappa U_{3-j} + i\kappa_1\frac{\partial U_{3-j}}{\partial t} + \gamma\frac{f(\Gamma|U_j|^2)}{\Gamma}U_j = 0. \qquad (5.16)$$

5.5.2   RESULTS AND DISCUSSION

Following the conventional linear stability analysis as described earlier in Sec. 5.2.2, the MI dispersion relation with the effect of SNL and coupling coefficient dispersion is given by

$$\left[\left(K + \frac{\kappa_1 \Omega}{\sqrt{2}}\right)^2 - h_1\right]\left[\left(K + \frac{\kappa_1 \Omega}{\sqrt{2}}\right)^2 - h_2\right] = h, \qquad (5.17)$$

$$h_1 = \frac{1}{2}\left[\tilde{\gamma}^2 P^2 - 4\kappa^2 + (\kappa_1^2 - 2\sqrt{2}\beta_2\kappa)\Omega^2 + \frac{\beta_2^2\Omega^4}{2}\right], \qquad (5.18a)$$

$$h_1 = \frac{1}{2}\left[\tilde{\gamma}^2 P^2 - 4\kappa^2 + (\kappa_1^2 + 2\sqrt{2}\beta_2\kappa)\Omega^2 + \frac{\beta_2^2\Omega^4}{2}\right], \qquad (5.18b)$$

$$h = \frac{1}{4}(\tilde{\gamma}^2 P^2 - 4\kappa^2)^2 - (\tilde{\gamma}^2 P^2 - 4\kappa^2)(\beta_2\tilde{\gamma}P + 2\kappa_1^2)\Omega^2$$

$$- (5\beta_2^2\kappa^2 + \kappa_1^4 - \beta_2^2\tilde{\gamma}^2 P^2)\Omega^4 + \frac{1}{2}\beta_2^2\kappa_1^2\Omega^6, \qquad (5.18c)$$

$$\tilde{\gamma} = \gamma\exp^{-\Gamma P}. \qquad (5.18d)$$

Using the above equation, we investigated MI in twin core liquid filled photonic crystal fiber with the effects of saturation of the nonlinear response and coupling coefficient dispersion. First, the symmetric/antisymmetric case is considered and it is observed that CCD does not dramatically modify the spectrum and SNL behaves in a perceptible manner such that the gain and the unstable region decrease. Hence, our spotlight is on the asymmetric case. First we study the impact of various system parameters such as power (P) and coupling coefficient κ in the instability spectrum as shown in Figures 5.13 and 5.14. It is observed that with an increase in power (coupling coefficient) the gain of the instability band increases (decreases) monotonously. For better insight, we have divided the operating power regime into Type-I and Type-II, corresponding to input power below and above the saturation power, as shown in Fig. 5.15. To illustrate, the conventional Kerr type nonlinear response has also been considered along with the SNL case. It is observed that both the Kerr and SNL cases behave similarly in the Type-I regime, but a slight decrease in the instability gain and frequency is observed for SNL, owing to the inherent depletion in the effective nonlinearity. On the other side of the picture, the Type-II regime behaves in the opposite way such that the increase in power above the saturation power leads to the suppression of MI by decreasing the gain and the unstable frequency window. To account for these interesting observations we plotted various system parameters such as nonlinearity, nonlinear factor and power threshold for sustaining a continuous wave (cw) as a function of power. It can be inferred from Fig. 5.16 that SNL and Kerr behave qualitatively in different ways under the influence of power. For instance, nonlinearity and P_{min} are found to be independent of input power in the Kerr case, whereas both are functions of power in the SNL case. Also, the product of effective nonlinearity and power is identified to be a crucial factor in the observed effects and we called it the nonlinear factor. It is evident from Fig. 5.16b that the nonlinear factor increases monotonously

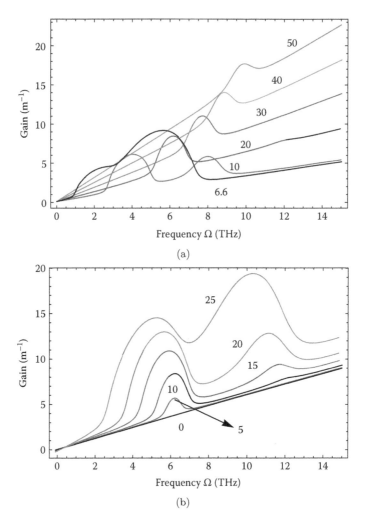

FIGURE 5.13: (a) Plot showing the variation of MI gain as a function of power (b) Variation of the gain of the Raman as a function of power.

for the Kerr case and behaves in a unique manner in the saturable nonlinear media (SNL) system. It is observed that the nonlinear factor increases with an increase in power in the Type-I regime and reaches a maximum at $P = P_s$ and further increases in power takes it to the Type-II regime, where the nonlinear factor progressively decreases. There exist two powers corresponding to the same nonlinear factor and the behavior of the system at the two respective powers is indeed same, that is, the physical quantities like the MI gain and the unstable frequency window are preserved.

Following a detailed interpretation of the role of SNL in the MI spectrum, the subsequent section was dedicated to investigating the interplay between

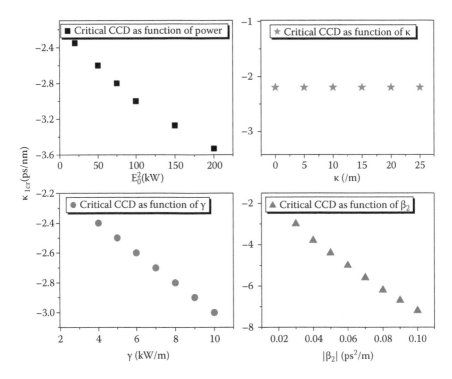

FIGURE 5.14: Variation of critical CCD (κ_{1cr}) with variation in system parameters such as $E_0^2\,\mathrm{kW}$, $\kappa\,\mathrm{m}^{-1}$, $\gamma\,\mathrm{kW\,m}^{-1}$ and $\beta\,\mathrm{ps}^2\,\mathrm{m}^{-1}$.

CCD and SNL in the anomalous dispersion regime. CCD leads to the emergence of a new spectral band; the number of bands, gain and the unstable window are all characteristics of the value of the CCD. The saturation, on the other hand, inherently suppresses MI by depleting the effective nonlinear coefficient. Basically, we consider two power schemes, one near and the other far from P_{min}. Along similar lines as [40] for Kerr case, the instability spectrum of the SNL case also differs at the two power schemes, as shown in Fig. 5.15. The interesting observation is the existence of the critical CCD at which the system evolves in a different manner, as shown in Figure 5.17. For better insight, the critical CCD is plotted as a function of different system parameters such as power, coupling coefficient, nonlinear coefficient and dispersion coefficient . A linear variation is observed for all the cases (refer to Fig. 5.18). An interesting feature which we speculate is unique in the system is the observation of system invariance at two different powers. Irrespective of the power scheme, the system remains invariant at two powers for the same nonlinear factor and the two powers are identified to be in two regimes, Type-I and Type-II, respectively. Further, we extended the study to the case of the

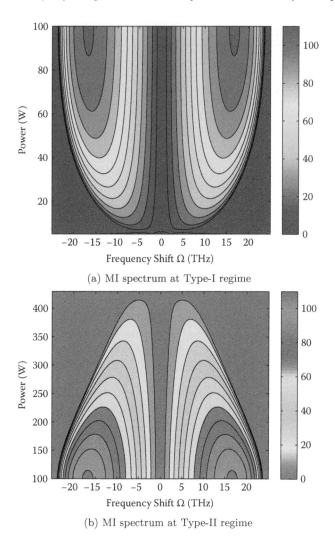

(a) MI spectrum at Type-I regime

(b) MI spectrum at Type-II regime

FIGURE 5.15: Plot showing the variation of MI gain spectra of SNL as a function of input power at $\Gamma = 0.01\text{W}^{-1}$.

normal dispersion regime, where it is obvious that MI generally is not feasible. However the presence of the non-zero coupling coefficient ensures the required phase matching for MI. The behavior of MI in the normal dispersion regime differs a great deal from the anomalous dispersion case, such that two instability bands are observed against the single band in the anomalous dispersion regime. However as in the anomalous case the gain of the SNL system is less than the conventional counterpart. As power increases the P_{min}, two instability bands are observed and with an increase in power two bands drift

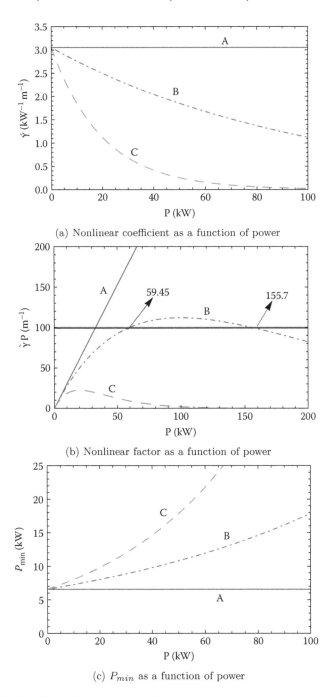

(a) Nonlinear coefficient as a function of power

(b) Nonlinear factor as a function of power

(c) P_{min} as a function of power

FIGURE 5.16: Plot showing the variation of system parameters as a function of power.

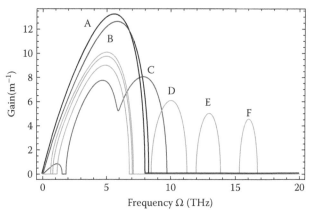

(a) The MI spectra of Kerr at low power with the effect of coupling coefficient dispersion (κ_1)

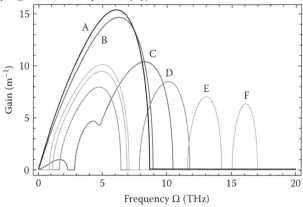

(b) The MI spectra of SNL at low power with the effect of coupling coefficient dispersion (κ_1)

FIGURE 5.17: Plot showing the instability spectra at P = 8 kW for different coupling coefficient dispersions $\kappa_1(A = 0; B = -0.2; C = -0.45; D = -0.6; E = -0.8; F = -1)\,\mathrm{ps\,m^{-1}}$.

away from the center frequency; at higher power only a single instability band survives. The coupling coefficient, on the other hand, rather behaves similarly to the power, such that any non-zero value of κ leads to two instability bands. It is worth noting that the instability band vanishes for κ equal to zero, owing to the lack of phase matching for MI. The interplay between CCD and SNL is observed to be nearly same as that of the anomalous dispersion case, except the change in the definition of the critical CCD. Similar to the case of the anomalous dispersion regime, the system invariance is observed at two different powers at two regimes.

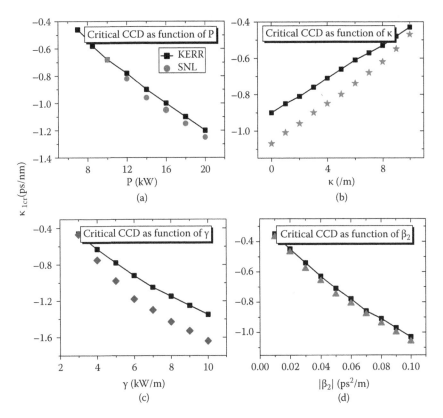

FIGURE 5.18: Illustration of the variation of critical CCD (κ_{1cr}) with system parameters in the anomalous dispersion regime. Parameters are $\beta_2 = -0.02 \text{ ps}^2 \text{ m}^{-1}$, $\gamma = 3.05 \text{ kW}^{-1} \text{ m}^{-1}$, $P = 10 \text{ kW}$, $\kappa = 10 \text{ m}^{-1}$.

Overall, the study based on the incorporation of saturation of the nonlinear response is interesting and offers a rich variety of information as follows: (i) the observation of the unique role of pump powers such that the increasing power either increases or decreases the MI gain relative to the saturation power, (ii) the existence of two different powers for the same value of a nonlinear factor and (iii) the observation of the system invariance at two different powers. The above results are interesting and can pave the way for the design and development of twin core liquid filled photonic crystal fibers based couplers, since one can maneuver the different characteristics of the fibers by merely tailoring the input power relative to the saturation power [35].

5.6 TWO-STATE BEHAVIOR IN THE INSTABILITY SPECTRUM OF A SATURABLE NONLINEAR SYSTEM

5.6.1 THEORETICAL FRAMEWORK

The propagation of an optical beam in the SNL is given by the modified nonlinear Schrödinger equation (MNLSE). The MNLSE of the slowly varying envelope $E(z,t)$ along the z-axis in a retarded reference time centered at frequency ω_0 is given by [36, 38, 44]

$$\frac{\partial E}{\partial z} + \sum_{n=2}^{4} \beta_n \frac{i^{n-1}}{n!} \frac{\partial^n E}{\partial t^n} = i \frac{\gamma |E|^2}{1 + \Gamma |E|^2} E, \qquad (5.19)$$

where t and z are the time and longitudinal coordinates in the moving reference frame, respectively. γ is the Kerr coefficient and β_n is the dispersion coefficient resulting from the Taylor expansion of the propagation constant around the center frequency ω_0.

$$K = -\beta_3 \frac{\Omega^3}{6} \pm \sqrt{(\tilde{\gamma} P_0 + \beta_2 \frac{\Omega^2}{2} + \beta_4 \frac{\Omega^4}{24})^2 - \tilde{\gamma}^2 P_0^2}; \qquad \tilde{\gamma} = \frac{\gamma}{(1 + \Gamma P_0)^2}. \qquad (5.20)$$

Using Eq. (5.20), the MI spectra for some representative cases of power is shown in Fig. 5.19a. The choice of parameter is made in such a way that the saturation power is made equal to 1 W and the input power is chosen below (curve A) and above (curve C) the saturation power. From Fig. 5.19a, unlike the case of the conventional Kerr medium, here the increase in power does not lead to the monotonous increase in the MI gain; rather it follows a unique behavior which is identified as the signature of the SNL system [44]. It is evident from the figure that MI gain is least for curve A $(P = 0.1\text{W})$ and maximum for curve B $(P = 1W)$. Curve C falls in between the two curves in spite of registering the highest power of all; this confirms that MI gain is not proportional to the input power.

To give insight into the influence of power in MI dynamics, critical modulation frequency (CMF) as a function of power is shown in Fig. 5.19b. For better understanding, we consider the conventional Kerr case alongside SNL, as shown in Fig. 5.19b. It is obvious from Fig. 5.19b, that the conventional Kerr type nonlinear response leads to a monotonous increase of CMF for all values of input power, whereas CMF in SNL is no longer proportional to the power. To exclusively analyze the role of power in the MI dynamics of the SNL system, we plotted the MI gain as a function of Γ in 3D, as shown in Fig. 5.20. Figure 5.20 shows that increasing Γ leads to a monotonous decrease in the MI gain. This is attributed to the fact that increasing Γ depletes the nonlinear response of the medium and thereby inherently suppresses MI, which is in agreement with references [44].

Unlike the Kerr case, here the SNL system exhibits two state behavior, which is identified as an interesting scenario in the context of MI. By two

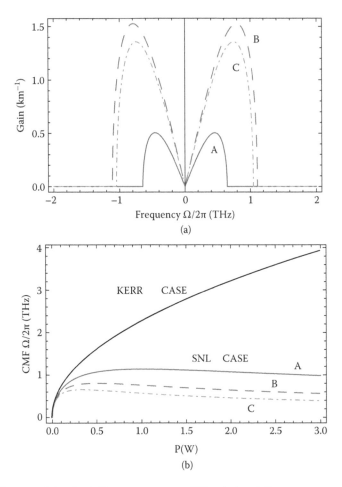

FIGURE 5.19: (a) The MI gain spectra $G(\Omega, z)$ for different powers $P(A = 0.1, B = 1, C = 2)$W. (b) Plot of CMF as a function of power in unsaturated nonlinearity (USN) and SNL for different saturation parameters $\Gamma(A = 1, B = 2, C = 3)$W^{-1}.

state behavior we mean the dual play of input pump power with reference to the saturation power. In other words, increasing power increases the CMF proportionally until the input power reaches the saturation threshold (P_s) called as the UP phase $(P_0 < P_s)$ and reaches a maximum at $P_0 = P_s$. A further increase in power beyond P_s leads to a decrease in the CMF or G_{max}, known as the DOWN phase $(P_0 > P_s)$, as portrayed in Fig. 5.21. Thus it is observed that an increase in power either increases or decreases CMF and G_{max} relative to P_s, which we speculate is a unique feature of the SNL system.

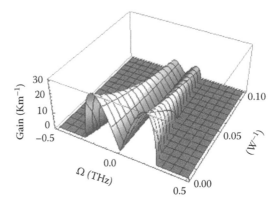

FIGURE 5.20: The variation of MI gain as a function of Γ for power $P = 10$ W.

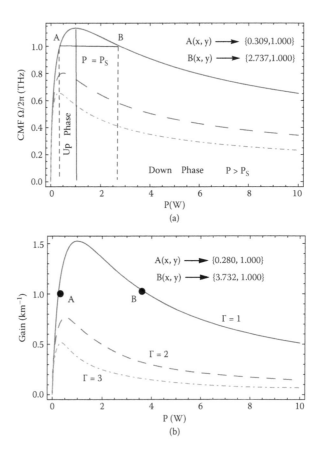

FIGURE 5.21: Plot of CMF and G_{max} as a function of power for different saturation parameter.

The nonlinear parameter is given by the expression ($\Gamma = 1/P_s$); therefore, any decreases in P_s reduces the power difference between P and P_s. This reduction in the power difference narrows the UP phase of CMF ($P_0 < P_s$) and extends the DOWN phase ($P_0 > P_s$); i.e., curve B and C. An arbitrary line meets the CMF curve at two different powers (A and B) for the same CMF, as shown in Fig. 5.21. It is evident from Fig. 5.21 that there exists a unique value of CMF (1 THz) at two different powers, one on the UP phase ($A \rightarrow 0.309$ W) and other on the DOWN phase ($B \rightarrow 2.737$ W).

The similar nature of the curve can also be observed in Fig. 5.21 for G_{max}, where the same value of G_{max} (1 km^{-1}) exists at ($A \rightarrow 0.280$ W) and ($B \rightarrow 3.732$ W). Also, the CMF and G_{max} are found to be maximum at $P = P_s$, which is identified as the active regime. Thus operating the system around the saturation power (active regime) leads to the maximum enhancement of MI and subsequent nonlinear effects [36].

5.7 MI IN A SEMICONDUCTOR DOPED DISPERSION DECREASING FIBER

5.7.1 THEORETICAL FRAMEWORK

The propagation of an optical beam in a semiconductor doped fiber (SDF) with SNL for the slowly varying envelope $E(z,t)$ of pulse propagation along the z-axis in a retarded reference time, centered at frequency ω_0, is given by [37, 43, 44]

$$\frac{\partial E}{\partial \xi} + i\frac{\beta_2 \exp(\alpha z)}{2}\frac{\partial^2 E}{\partial \tau^2} - \frac{\beta_3 \exp(\alpha z)}{6}\frac{\partial^3 E}{\partial \tau^3} = \frac{i\gamma|E|^2}{1+\tilde{\Gamma}|E|^2}E. \tag{5.21}$$

5.7.2 RESULTS AND DISCUSSION

Using standard linear stability analysis described earlier in the chapter, the explicit expression for the MI gain in a semiconductor doped dispersion decreasing fiber (SD-DDF) is given by

$$G(\Omega) = \frac{1}{2}|\beta_2| \exp(-\mu z) \exp(\alpha z)\Omega[\Omega_C^2 - \Omega^2]^{1/2} \tag{5.22}$$

where

$$\Omega_C = \left[\frac{4\gamma P_0 \exp(\mu - \alpha)z}{|\beta_2|(1 + \tilde{\Gamma} P_0)^2}\right]^{1/2}. \tag{5.23}$$

Using the above equation, MI analysis is performed under various combinations of fiber systems and the influence of SNL and DDF. Figure 5.22 summarizes our results for the various types of fiber systems considered, in the context. Curve b shows the broad spectral shift in DDF due to the depletion in dispersion, as a consequence of an increase in CMF. Curve c is the case of SDF, where an increase in Γ diminishes the nonlinear contribution and

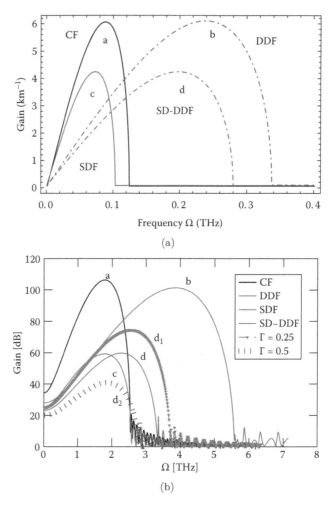

FIGURE 5.22: Plot illustrates the MI dynamics of the proposed SD-DDF (a) analytically and (b) numerically.

reduces the CMF and thereby suppresses both the MI gain and bandwidth considerably. Curve d is the case of SD-DDF, where the two physical effects of an opposite nature interact and result in a rich variety of information. For numerical appreciation and for better understanding, we extracted the required data from Fig. 5.22 and tabulated it in Table 5.1.

For further insight, the observed results are interpreted in the soliton context, since MI is considered to be a soliton precursor. MI or a soliton is due to the constructive interaction of the nonlinear effects and the dispersion effects. To put things in perspective and to emphasize the soliton in SD-DDF, we numerically solved the MNLSE using the split-step Fourier method and

TABLE 5.1

Summary of SD-DDF

S.No.	TYPE	L_{NL}(km)	Ω_C(THz)	Ω_{max}(THz)	G_{max}(km^{-1})	BF	$\Delta\omega$(THz)
1	CF	0.3333	0.1233	0.5477	6	1	0.2466
2	DDF	0.3333	0.3353	0.5477	6	2.718	0.6706
3	SNL	0.4800	0.1027	0.4564	4.166	0.8333	0.2054
4	SNL-DDF	0.4800	0.2794	0.4564	4.166	2.2652	0.5588

plotted the results in Fig. 5.23. We consider DDF alongside SD-DDF to explicitly analyze the cutting edge of the proposed SD-DDF over conventional DDF. Figure 5.23 illustrates that for the case of DDF the soliton is observed to be compressed and amplified in the temporal domain in comparison to the input profile. On the other hand, the soliton observed in SDF is found to be broader in width. These two distinct observations are due to the fact that the decreasing dispersion $\beta(z)$ in DDF compresses the soliton due to the phenomena known as adiabatic compression (refer to [45] and references therein); on the other hand, the depleting nonlinearity due to the doping of highly nonlinear semiconducting material broadens the temporal width of the soliton pulse [44, 46]. Interestingly, the temporal width of the SD-DDF falls between the two systems, owing to the fact that decreasing dispersion due to DDF compresses the temporal width; on the other side, saturation diminishes the nonlinearity and thereby broadens the same. Thus the two opposite physical effects perform action-reaction and thereby enable one to design fibers with the desired temporal profile (can compress or broaden the width, amplify or suppress the intensity) by simply maneuvering the operating power.

In the frequency domain, as far as the bandwidth of MI is concerned, depletion of the dispersion enhances and broadens (DDF), and saturation counteracts (SDF) and reduces the bandwidth. The relative dominance is given by the ratio of $\left[\frac{\hat{\gamma}}{\beta_z}\right]$. If γ depletes rapidly, then the bandwidth falls considerably, and vice versa. Thus the window between SD and DDF can enable one to realize fibers with the desired bandwidth profile.

The proposed SD-DDF finds immense interest over the conventional DDF, since the dispersion growth rate is fixed in DDF during fabrication, and there is no room for the end user to tailor the desired parameters, thereby setting the ultimate constraint and limiting the usage of such fibers for a wide range of applications. However, SD-DDF could keep the pace of future endeavors, realizing the required bandwidth profile by suitably maneuvering the value of Γ by changing the power [37]. This is a relatively simple means of achieving the desired bandwidth profile instead of going for a different fiber for a specific

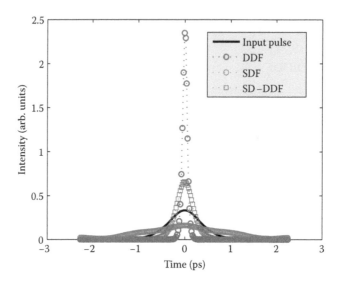

FIGURE 5.23: Evolution of a soliton in different fiber systems.

requirement. Thus the SD-DDF over DDF is useful and it would be the obvious choice for future prospects of ultrashort pulse generation using MI.

5.8   SUMMARY AND CONCLUSION

A theoretical investigation of modulational instability in the propagation dynamics of ultrashort pulses through optical fibers is presented. In the first part of the chapter a detailed MI analysis is performed with an emphasize on the role of relaxation and saturation of the nonlinear response. Particularly, the impact of higher order dispersion (HOD) effects is analyzed and it is observed that among the higher order dispersion effects fourth order dispersion is found to be crucial and significantly effects MI dynamics. One critical observation is the emergence of unconventional MI sidebands and the promotion of MI bands in the normal dispersion regime. In the case of cross-phase modulation, the combined effect of walk-off and fourth order dispersion leads to new spectral sidebands on either side of the pump frequency. The relaxation is broadly classified as slow and fast response based on the time scale. In the fast response regime, parametric MI bands are more pronounced and in the slow response regime parametric bands inevitably are suppressed and Raman bands due to the delayed nonlinear response are prominent. The saturable nonlinearity suppresses the MI by depleting the effective nonlinearity. The interplay between coupling coefficient dispersion and relaxation is interesting since the CCD brings new spectral bands near the center frequency and the relaxation leads to two unstable modes at higher detuning frequencies. MI is observed even in the normal dispersion regime with the aid of the delay nonlinear response. The influence of nonlinear saturation on MI dynamics of two core fibers is studied in detail and it is observed that there exist two power regimes, where the system resembles identical, which is identified to be the signature of a saturable nonlinear system. The coupling coefficient leads to new spectral bands, similar to the case of intermodal dispersion. Finally, a novel semiconductor doped dispersion decreasing fiber is proposed and it s cutting edge over the existing fiber system is emphasized in view of the ultra high repetition rate of soliton pulse generation using MI.

To wrap up, our detailed theoretical analysis reveals many interesting results in the context of MI, particularly in the regime of nonlinear saturation and relaxation. To the best of our knowledge, there are not many reports from the experiential point of view, so we believe our theoretical results will help in stimulating many new experiments in the context of MI in relaxing saturable nonlinear systems.

ACKNOWLEDGMENTS

KP thanks DST, IFCPAR, NBHM, FCT and CSIR Government of India for financial support through major projects. K. Nithyanandan thanks CSIR Government of India for providing financial support by awarding a Senior Research Fellowship (SRF).

REFERENCES

1. J. A. Armstrong, N. Bloembergen, J. Ducuing, P. S. Pershan, "Interactions between light waves in a nonlinear dielectric," *Physical Review*, 127, 1918, 1962.
2. T. B. Benjamin, J. E. J. Fier, *Fluid Mechn.*, 27, 417, 1967.
3. V. I. Karpman, "Self-modulation of nonlinear plane waves in dispersive media," *JETP Lett.*, 6, 277, 1967.
4. T. Taniuti, H. Washimi, "Self-trapping and instability of hydromagnetic waves along the magnetic field in a cold plasma," *Physical Review Letters*, 21, 209, 1968.
5. M. J. Lighthill, "Contributions to the theory of waves in non-linear dispersive systems," *IMA Journal of Applied Mathematics*, 1, 269, 1965.
6. L. A. Ostrovskii, "Propagation of wave packets and space-time self-focusing in a nonlinear medium," *Zh. Eksp. Teor. Fiz.*, 51, 1189, 1966.
7. Akira Hasegawa, F. Tappert, "Generation of a train of soliton pulses by induced modulational instability in optical fibers," *Appl. Phys. Lett.*, 23, 142, 1973.
8. Sergei K. Turitsyn, Alexander M. Rubenchik, Michail P. Fedoruk, "On the theory of the modulation instability in optical fiber amplifiers," *Opt. Lett.*, 35, 2684, 2010.
9. Alexander M. Rubenchik, Sergey K. Turitsyn, Michail P. Fedoruk, "Modulation instability in high power laser amplifiers," *Opt. Express*, 18, 1380, 2010.
10. John M. Dudley, Goëry Genty, Stéphane Coen, "Supercontinuum generation in photonic crystal fiber," *Rev. Mod. Phys.*, 78, 1135, 2006.

11. R. Vasantha Jayakantha Raja, K. Porsezian, K. Nithyanandan, "Modulational-instability-induced supercontinuum generation with saturable nonlinear response," *Phys. Rev. A*, 82, 013825, 2010.

12. K. Hammani, B. Wetzel, B. Kibler, J. Fatome, C. Finot, G. Millot, N. Akhmediev, J. M. Dudley, "Spectral dynamics of modulation instability described using Akhmediev breather theory," *Opt. Lett.*, 36, 2140, 2011.

13. M. N. Zambo Abou'ou, P. Tchofo Dinda, C. M. Ngabireng, B. Kibler, F. Smektala, "Impact of the material absorption on the modulational instability spectra of wave propagation in high-index glass fibers," *J. Opt. Soc. Am. B*, 28, 1518, 2011.

14. Ajit Kumar, A. Labruyere, P. Tchofo Dinda, "Modulational instability in fiber systems with periodic loss compensation and dispersion management," *Opt. Commun.*, 219, 221, 2003.

15. S. Trillo, S. Wabnitz, G. I. Stegeman, E. M. Wright, "Parametric amplification and modulational instabilities in dispersive nonlinear directional couplers with relaxing nonlinearity," *J. Opt. Soc. Am. B*, 6, 889, 1989.

16. Thibaut Sylvestre, Stéphane Coen, Philippe Emplit, Marc Haelterman, "Self-induced modulational instability laser revisited: Normal dispersion and dark-pulse train generation," *Opt. Lett.*, 27, 482, 2002.

17. Stéphane Coen, Marc Haelterman, "Continuous-wave ultrahigh-repetition-rate pulse-train generation through modulational instability in a passive fiber cavity," *Opt. Lett.*, 26, 39, 2001.

18. G. P. Agrawal, *"Nonlinear Fiber Optics,"* Academic Press, 2012.

19. Akira Hasegawa, "Generation of a train of soliton pulses by induced modulational instability in optical fibers," *Opt. Lett.*, 9, 288, 1984.

20. K. Tai, A. Hasegawa, A. Tomita "Observation of modulational instability in optical fibers," 56, 135, 1986.

21. M. J. Potasek, "Modulation instability in an extended nonlinear Schrödinger equation," *Opt. Lett.*, 12, 921, 1987.

22. John D. Harvey, Rainer Leonhardt, Stéphane Coen, Gordon K. L. Wong, Jonathan C. Knight, William J. Wadsworth, Philip St. J. Russell, "Scalar modulation instability in the normal dispersion regime by use of a photonic crystal fiber," *Opt. Lett.*, 28, 2225, 2003.

23. Govind P. Agrawal "Modulation instability induced by cross-phase modulation," *Phys. Rev. Lett.*, 59, 880, 1987.

24. Govind P. Agrawal, P. L. Baldeck, R. R. Alfano, "Modulation instability induced by cross-phase modulation in optical fibers," *Phys. Rev. A*, 39, 3406, 1989.

25. M. Yu, C. J. McKinstrie, Govind P. Agrawal, "Instability due to cross-phase modulation in the normal-dispersion regime," *Phys. Rev. E*, 48, 2178, 1993.

26. J. E. Rothenberg, "Modulation instability for normal dispersion," *Phys. Rev. A*, 42, 1990.

27. Takuo Tanemura, Kazuro Kikuchi, "Unified analysis of modulational instability induced by cross-phase modulation in optical fibers," *J. Opt. Soc. Am. B*, 20, 2502, 2003.

28. P. D. Drummond, T. A. B. Kennedy, J. M. Dudley, R. Leonhardt, J. D. Harvey, "Cross-phase modulational instability in high-birefringence fibers," *Opt. Commun.*, 78, 137, 1990.

29. U. Langbein, F. Lederer, T. Peschel, H. E. Ponath, "Nonlinear guided waves in saturable nonlinear media," *Opt. Lett.*, 10, 571, 1985.

30. Jean-Louis Coutaz, Martin Kull, "Saturation of the nonlinear index of refraction in semiconductor-doped glass," *J. Opt. Soc. Am. B*, 8, 95, 1991.

31. G. L. da Silva, Iram Gleria, M. L. Lyra, A. S. B. Sombra, "Modulational instability in lossless fibers with saturable delayed nonlinear response," *J. Opt. Soc. Am. B*, 26, 183, 2009.

32. K. Porsezian, K. Nithyanandan, R. Vasantha Jayakantha Raja, P. K. Shukla, "Modulational instability at the proximity of zero dispersion wavelength in the relaxing saturable nonlinear system," *J. Opt. Soc. Am. B*, 29, 2803, 2012.

33. K. Nithyanandan, R. V. J. Raja, K. Porsezian, B. Kalithasan, "Modulational instability with higher-order dispersion and walk-off in Kerr media with cross-phase modulation," *Physical Review A*, 86, 2012.

34. "Interplay between relaxation of nonlinear response and coupling coefficient dispersion in the instability spectra of dual core optical fiber," *Optics Communications*, 303, 46 2013.

35. K. Nithyanandan, R. V. J. Raja, K. Porsezian, "Modulational instability in a twin-core fiber with the effect of saturable nonlinear response and coupling coefficient dispersion," *Physical Review A*, 87, 043805 2013.

36. K. Nithyanandan, K. Porsezian, "Observation of two state behavior in the instability spectra of saturable nonlinear media," *European Physical Journal: Special Topics*, 222, 821, 2013.

37. K. Nithyanandan, R. V. J. Raja, K. Porsezian, "Theoretical investigation of modulational instability in semiconductor doped dispersion decreasing fiber and its cutting edge over the existing fiber systems," *J. Opt. Soc. Am. B*, 30, 178, 2013.

38. P. T. Dinda, K. Porsezian, "Impact of fourth-order dispersion in the modulational instability spectra of wave propagation in glass fibers with saturable nonlinearity," *J. Opt. Soc. Am. B*, 27, 1143, 2010.

39. G. Millot, P. Tchofo Dinda, E. Seve, S. Wabnitz, "Modulational instability and stimulated Raman scattering in normally dispersive highly birefringent fibers," *Optical Fiber Technology*, 7, 170, 2001.

40. J. H. Li, K. S. Chiang, K. W. Chow "Modulation instabilities in two-core optical fibers," *J. Opt. Soc. Am. B*, 28, 1693, 2011.

41. G. I. Stegeman, C. T. Seaton, C. N. Ironside, T. Cullen, A. C. Walker, "Effects of saturation and loss on nonlinear directional couplers," *Applied Physics Letters*, 50, 1035, 1987.

42. Y. Chen, "Twin core nonlinear couplers with saturable nonlinearity," *Electronics Letters*, 26, 1374, 1990.

43. W. C. Xu, S. M. Zhang, W. C. Chen, A. P. Luo, S. H. Liu, "Modulation instability of femtosecond pulses in dispersion-decreasing fibers," *Opt. Commun.*, 199, 355, 2001.

44. J. M. Hickmann, S. B. Cavalcanti, N. M. Borges, E. A. Gouveia, A. S. Gouveia-Neto, "Modulational instability in semiconductor-doped glass fibers with saturable nonlinearity," *Opt. Lett.*, 18, 182, 1993.

45. P. K. A. Wai, Wen-hua Cao, "Ultrashort soliton generation through higher-order soliton compression in a nonlinear optical loop mirror constructed from dispersion-decreasing fiber," *J. Opt. Soc. Am. B*, 20, 1346, 2003.

46. S. Gatz, J. Herrmann, "Soliton propagation in materials with saturable non-linearity," *J. Opt. Soc. Am. B*, 8, 2296, 1991.

6 Modulational instabilities in a system of four coupled, nonlinear Schrödinger equations with the effect of a coupling coefficient

H. Tagwo, S. Abdoulkary, A. Mohamadou, C. G. Latchio Tiofack, and *T. C. Kofane*

6.1 INTRODUCTION

Self-focusing and self-guiding light beams organized in the solitonic structure have received much attention in recent years. Spatial and spatiotemporal solitons can find applications in all-optical switching and integrated logic, since they are self-guided in bulk media. Extensive research has been carried out in the field of pulse propagation in optical fibers [1–4]. On par with pulse propagation, continuous wave propagation in optical fibers has also demanded special attention [4]. A continuous wave with a cubic nonlinearity in an anomalous dispersion regime is known to develop instability with respect to small modulations in amplitude or in phase in the presence of noise or any other weak perturbation, called modulational instability (MI) [2, 4, 5]. Generally, the perturbation has its origin in quantum noise or a frequency shifted signal wave [4]. The MI phenomenon was discovered in fluids [6], in nonlinear optics [7] and in plasmas [8]. MI of a light wave in an optical fiber was suggested by Hasegawa [9] as a means to generate a far infrared light source, and since then it has attracted extensive attention for both its fundamental and applied interests [4, 5, 10]. In the optics community, MI is relevant to many topics, including Bragg grating s [11, 12], cross-phase modulations [13, 14], four-wave mixing, novel materials [15, 16], parametric oscillators [17], polarization and birefringence [18, 19], saturable nonlinearity [20], spatial instability [21], supercontinuum generation [22], and temporal solitons in fibers [23, 24].

As regards applications, MI provides a natural means of generating ultrashort pulses at ultrahigh repetition rates and is thus potentially useful for the development of high speed optical communication systems in the future; hence it has been exploited a great deal in many theoretical and experimental

133

studies for the realization of laser sources adapted to ultrahigh bit-rate optical transmissions [25, 26]. The MI phenomenon is accompanied by sideband evolution at a frequency separation from the carrier which is proportional to the square root of the optical pump power [27]. This represents the simplest case of MI in an anomalous dispersion medium with a simple Kerr nonlinearity. When two or more optical waves copropagate through a birefringent optical fiber, they interact with each other through the fiber nonlinearity in such a way that the effective refractive index of a wave depends not only on the intensity of that wave but also on the intensity of other copropagating waves, a phenomenon known as cross-phase modulation (XPM) [1–3, 5].

MI is governed by the nonlinear Schrödinger (NLS) equation, which inherently admits the formation of solitary pulses or envelope solitons as a result of the delicate balance between anomalous group velocity dispersion (GVD) and self-focusing Kerr nonlinearity [28–30]. The periodic pulse trains that emerge as the result of the temporal modulation due to MI are in fact identified as the train of the ideal soliton pulses. The propagation of vectorial pulses in birefringence fibers could be modeled using two NLS equations coupled with a cross-phase modulation (XPM) term [31]. An extension of this model uses four NLS equations.

Systems of coupled nonlinear Schrödinger (CNLS) equations have been demonstrated to be relevant in many scientific applications, especially in hydrodynamics and optics [1, 32]. For a single NLS equation, MI will arise if the signs of dispersive and nonlinear terms are the same (the anomalous dispersion regime). Conversely, this instability is absent if the signs are different (for a normal dispersion regime). In fluid mechanics, growing/decaying modes NLS have been employed to model giant/rogue waves [33]. For CNLS, the dynamics is subtle and more intriguing. The motivation comes from an experimental investigation where four channels for light propagation are permitted. Two of the four waveguides are in the normal dispersion regime while the other two are in the anomalous regime. Two pump waves are installed together in either the anomalous or normal dispersion regime. Therefore, generally speaking, both of the coupling coefficient effects and self-phase modulation (SPM) terms should be taken into consideration in the study of XPM-induced MI in the four CNLS.

The evolution of the optical beam in two twin-core fiber could be given by four linear CNLS equations. The periodic power transfer between the two cores of the fiber is governed by the linear coupling coefficient [34, 35]. However, in the ultrashort pulse regime, the so-called coupling coefficient dispersion (CCD) arising as a result of the wavelength-dependent coupling coefficient is found to be crucial [5], especially in long fibers, and can cause severe pulse distortion, which eventually leads to pulse break-up [37]. The effect of CCD and its relative influence on various physical mechanisms were studied in detail both theoretically and also through experiments by Chiang and his group through various analysis [38–42]. Considering the importance of CCD, Li et al. extended the results of [43, 44] with the inclusion of CCD [45]. The authors have reported that CCD does not affect the symmetric or asymmetric CW state but dramatically changes the MI of the asymmetric CW state. Nithyanandan et al. [46] predicted a critical CCD, where the system of twin-core fiber with the effect of saturable nonlinear response evolves dramatically in a different manner. However, the study of MI in a system of four coupled NLS equations is not novel [47, 48], but what differentiates our present problem from the rest is the investigation of MI in four coupled NLS equations with the effect of the interplay with CCD, which has not been discussed yet to our knowledge and hence deserves attention. The remainder of the chapter is arranged as follows. In Sec. 6.2, a system of four CNLS is presented. In Sec. 6.3, we study the linear stability analysis of this model equation. The results of MI gain are investigated in Sec. 6.4. Typical outcomes of the nonlinear development of MI are reported in Sec. 6.5 and Sec. 6.6 concludes the chapter.

6.2 MODEL

The propagation of pulse in a system of four coupled nonlinear Schrödinger equations with the effect of coupling coefficient may be described by the following NLS equations:

$$i\frac{\partial A_1}{\partial z} + \lambda\frac{\partial^2 A_1}{\partial t^2} + [A_1 A_1^* + \sigma(A_2 A_2^* + A_3 A_3^* + A_4 A_4^*)]A_1 + R(A_2 + A_3 + A_4) +$$

$$iR_1\frac{\partial}{\partial t}(A_2 + A_3 + A_4) = 0,$$

$$i\frac{\partial A_2}{\partial z} + \frac{\partial^2 A_2}{\partial t^2} + [A_2 A_2^* + \sigma(A_1 A_1^* + A_3 A_3^* + A_4 A_4^*)]A_2 + R(A_1 + A_3 + A_4) +$$

$$iR_1\frac{\partial}{\partial t}(A_1 + A_3 + A_4) = 0,$$

$$i\frac{\partial A_3}{\partial z} - \frac{\partial^2 A_3}{\partial t^2} + [A_3 A_3^* + \sigma(A_1 A_1^* + A_2 A_2^* + A_4 A_4^*)]A_3 + R(A_1 + A_2 + A_4) +$$

$$iR_1\frac{\partial}{\partial t}(A_1 + A_2 + A_4) = 0,$$

$$i\frac{\partial A_4}{\partial z} - \lambda\frac{\partial^2 A_4}{\partial t^2} + [A_4 A_4^* + \sigma(A_1 A_1^* + A_2 A_2^* + A_3 A_3^*)]A_4 + R(A_1 + A_2 + A_3) +$$

$$iR_1\frac{\partial}{\partial t}(A_1 + A_2 + A_3) = 0 \quad (6.1)$$

where $A_n (n = 1, 2, 3, 4)$ are envelopes of the axial electric fields, z is the distance and t is the retarded time. The dispersion coefficients are ranked as $(+\lambda, +1, -1, -\lambda)$, SPM coefficients are normalized to one, σ is the coefficient of XPM, R is the linear coupling coefficient which is responsible for the periodic power transfer between the four wave guides. R_1 represents the coupling-coefficient dispersion (CCD) [46]. Previous studies of MI in a system of four coupled, nonlinear Schrödinger equations [47] do not account for the effects of the R_1 terms, and our goal is to perform such an investigation.

6.3 LINEAR STABILITY ANALYSIS

In order to investigate the evolution of weak perturbations along the propagation distance, we carry out a linear stability analysis. First, we look for a CW solution of the system described by Eq. (6.1), given by:

$$A_n = \sqrt{P_0}\exp(i\phi_{NL}z), \quad n = 1, 2, 3, 4 \tag{6.2}$$

where P_0 is the input pump power and ϕ_{NL} is the nonlinear phase shift given by

$$\phi_{NL} = P_0 + 3\sigma P_0 + 3R. \tag{6.3}$$

The linear stability of the steady state can be examined by introducing the perturbed fields of the following form:

$$A_1 = (\sqrt{P_0} + u_1)\exp(i\phi_{NL}z),$$

$$A_2 = (\sqrt{P_0} + v_1)\exp(i\phi_{NL}z),$$

$$A_3 = (\sqrt{P_0} + u_2)\exp(i\phi_{NL}z),$$

$$A_4 = (\sqrt{P_0} + v_2)\exp(i\phi_{NL}z) \tag{6.4}$$

where $u_1, v_1, u_2, v_2 \ll \sqrt{P_0}$. Thus, if the perturbed fields grow exponentially, the steady state becomes unstable. By substituting Eq. (6.4) into Eq. (6.1) and collecting terms in u_1, v_1, u_2, v_2, we obtain four linearized equations:

$$i\frac{\partial u_1}{\partial z} + \lambda\frac{\partial^2 u_1}{\partial t^2} - 3Ru_1 + P_0(u_1 + u_1^*) + \sigma P_0(v_1 + v_1^*) + \sigma P_0(u_2 + u_2^*) + \sigma P_0(v_2 + v_2^*) +$$

$$R(v_1 + u_2 + v_2) + iR_1(\frac{\partial v_1}{\partial t} + \frac{\partial u_2}{\partial t} + \frac{\partial v_2}{\partial t}) = 0,$$

$$i\frac{\partial v_1}{\partial z} + \frac{\partial^2 v_1}{\partial t^2} - 3Rv_1 + P_0(v_1 + v_1^*) + \sigma P_0(u_1 + u_1^*) + \sigma P_0(u_2 + u_2^*) + \sigma P_0(v_2 + v_2^*) +$$

$$R(u_1 + u_2 + v_2) + iR_1(\frac{\partial u_1}{\partial t} + \frac{\partial u_2}{\partial t} + \frac{\partial v_2}{\partial t}) = 0,$$

$$i\frac{\partial u_2}{\partial z} - \frac{\partial^2 u_2}{\partial t^2} - 3Ru_2 + P_0(u_2 + u_2^*) + \sigma P_0(u_1 + u_1^*) + \sigma P_0(v_1 + v_1^*) + \sigma P_0(v_2 + v_2^*) +$$

$$R(u_1 + v_1 + v_2) + iR_1(\frac{\partial u_1}{\partial t} + \frac{\partial v_1}{\partial t} + \frac{\partial v_2}{\partial t}) = 0,$$

$$i\frac{\partial v_2}{\partial z} - \lambda\frac{\partial^2 v_2}{\partial t^2} - 3Rv_2 + P_0(v_2 + v_2^*) + \sigma P_0(u_1 + u_1^*) + \sigma P_0(v_1 + v_1^*) + \sigma P_0(u_2 + u_2^*) +$$

$$R(u_1 + v_1 + u_2) + iR_1(\frac{\partial u_1}{\partial t} + \frac{\partial v_1}{\partial t} + \frac{\partial u_2}{\partial t}) = 0 \tag{6.5}$$

where $*$ denotes complex conjugate. We now search for sidebands in the following ansatz with frequency detuning from the pump Ω:

$$u_1 = a_1 \exp(i\Omega t) + b_1 \exp(-i\Omega t),$$

$$v_1 = a_2 \exp(i\Omega t) + b_2 \exp(-i\Omega t),$$

$$u_2 = a_3 \exp(i\Omega t) + b_3 \exp(-i\Omega t),$$

$$v_2 = a_4 \exp(i\Omega t) + b_4 \exp(-i\Omega t). \tag{6.6}$$

Inserting Eq. (6.6) into Eq. (6.5), one obtains a set of eight linear ordinary differential equations

$$\frac{d[Y]}{dz} = i[M][Y], \tag{6.7}$$

where $[M]$ is an 8×8 matrix with $[Y] = [a_1, b_1^*, a_2, b_2^*, a_3, b_3^*, a_4, b_4^*]^T$. The elements of $[M]$ are given in the appendix. The above 8×8 matrix is referred to as the stability matrix which is used to study the stability of the system under consideration. MI occurs only when at least one of the eigenvalues of the stability 8×8 matrix possesses a nonzero and negative imaginary part, which results in an exponential growth of the amplitude with perturbation. By requiring the determinant of the associated matrix in Eq. (6.7) to vanish, we can find the solvability condition for these equations, which amount to a polynomial of degree eight. The MI phenomenon is measured by a power gain given by $G = 2|Im(K)|$ where K is the eigenvalue of M that possesses the largest imaginary part [13].

6.4   MODULATIONAL INSTABILITY GAIN

We study the effect of CCD on the MI gain. Several qualitatively different situations emerge depending on the values of CCD. Figure 6.1 shows the MI gain in the absence of CCD. As one can see in this figure, MI can develop. Figure 6.1(a) displays the MI gain when the XPM effect is less than the SPM for $\sigma = 2/3$. Only one sideband has been obtained. Figure 6.1 (b) depicts the MI gain when the SPM is less than the XPM for $\sigma = 2$. In this case, we see that the magnitude of the MI gain as well as the bandwidth increase [36].

Let us take into account the effect of linear coupling through the system. Figure 6.2 describes such a case. Figures 6.2(a), 6.2(c), 6.2(e) and 6.2(g) correspond to the case where the XPM effect is less than the SPM ($\sigma = 2/3$), from which one sees that the magnitude of the MI gain increases when the coupling coefficient increases. An interesting case occurs in Fig. 6.2(e) for $R = 0.8$: a nonconventional band is obtained. On the other hand, when the SPM is less than the XPM ($\sigma = 2$) one obtains Figs. 6.2(b), 6.2(d), 6.2(f) and 6.2(h). The magnitude of the MI gain increases with the increase of the linear coefficient. In this case, the nonconventional band appears for $R = 1.0$.

Now we take into account the CCD effect through the system. Let us begin with the case where the XPM effect is less than the SPM, that is, Figs. 6.3(a), 6.4(a), 6.4(c) and 6.4(e). The magnitude of the maximum gain decreases with the increase of the coefficient.

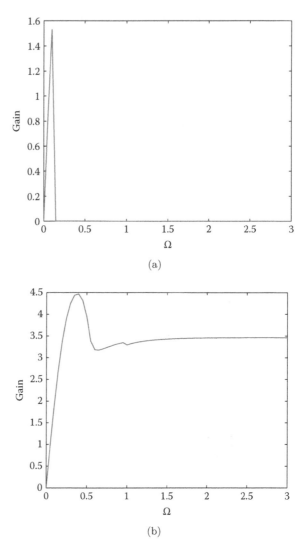

FIGURE 6.1: The MI gain spectra for different parameters. Two of the waveguides propagate in the normal dispersion regime, while the other experiences anomalous GVD. (a) With XPM < SPM ($\sigma = 2/3$) (b) with XPM > SPM ($\sigma = 2$). ($\lambda = 0.8, R = 0, R_1 = 0, P_0 = 1$).

The width of the sideband becomes narrow as the CCD coefficient increases. We also see that the nonconventional band disappears with the increase of the CCD coefficient. When the SPM effect is less than the XPM, we also see that the magnitude of the MI gain decreases with the increase of the CCD coefficient (see Figs. 6.3(b), 6.4(b), 6.4(d) and 6.4(f)). In this case, nonconventional bands are obtained for the values of the CCD used. Such sidebands have the

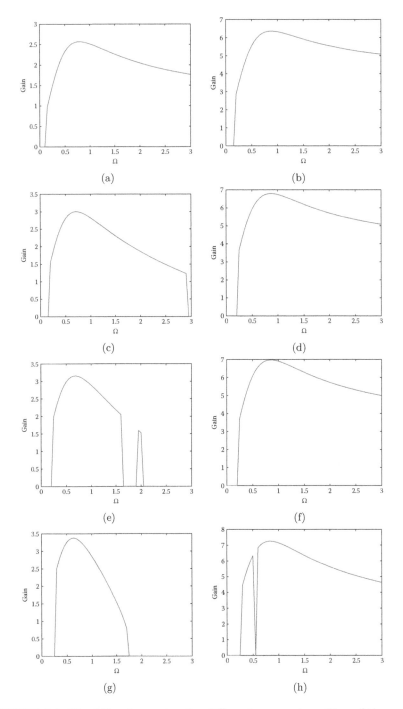

FIGURE 6.2: The MI gain spectra for different parameters. Two of the waveguides propagate in the normal dispersion regime, while the other experiences anomalous GVD. (a) With XPM < SPM ($\sigma = 2/3$); (b) with XPM > SPM ($\sigma = 2$). ($\lambda = 0.2, R_1 = -1, P_0 = 1$). For (a), (b), $R = 0.5$; (c), (d), $R = 0.7$; (e), (f), $R = 0.8$; (g), (h), $R = 1$.

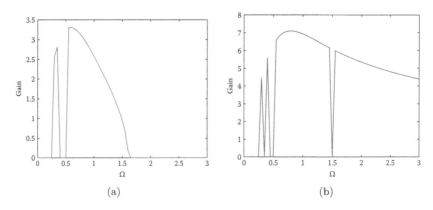

(a)   (b)

FIGURE 6.3: The MI gain spectra for different parameters. (a) With XPM $<$ SPM ($\sigma = 2/3$); (b) with XPM $>$ SPM ($\sigma = 2$). ($\lambda = 0.2, R = 1, P_0 = 1$) for (a) and (b).

following general features: The MI frequency region is narrow. The peak MI frequencies are only slightly dependent on the CCD coefficient, but are very sensitive to the XPM or SPM coefficient involved in the MI process. This special MI process leads to the coexistence of two pairs of sidebands of conventional and nonconventional natures, respectively, in the same system. In fact, the nonconventional MI has a substantial advantage in applications in the field of optical telecommunications over ordinary instability, offering more possibilities of obtaining a large MI bandwidth.

6.5   PROPAGATION OF WAVES THROUGH THE SYSTEM

Direct numerical simulations were run in order to understand the dynamics of CW states under MI in the nonlinear regime. We solved Eq. (6.1) using a split-step Fourier method applying the fast Fourier transform [49]. The onset of the modulation should occur at the dominant MI frequency, i.e., the MI frequency that has the maximum gain. Wave propagation analysis can serve as a verification of the MI analysis. The CW state was initially perturbed as

$$A_1 = A_2 = A_3 = A_4 = \sqrt{P_0}(1 + 0.003 \times \cos(\Omega t)), \qquad (6.8)$$

where $P_0 = 1$ is the power of the CW and $\Omega = 0.3$ is the angular frequency of the weak sinusoidal modulation imposed on the CW. Figure 6.5 shows the evolution of the CW solution given by Eq. (6.8). In Figs. 6.5(a) and (c), we have plotted the case where XPM $<$ SPM ($\sigma = 2/3$) and in Figs. 6.5(b) and (d) the case where XPM $>$ SPM ($\sigma = 2$).

Figures 6.5(a), (b), (c), (d) show the evolution of the initial solution through the system. The CCD have been taken into account, and we observe that the

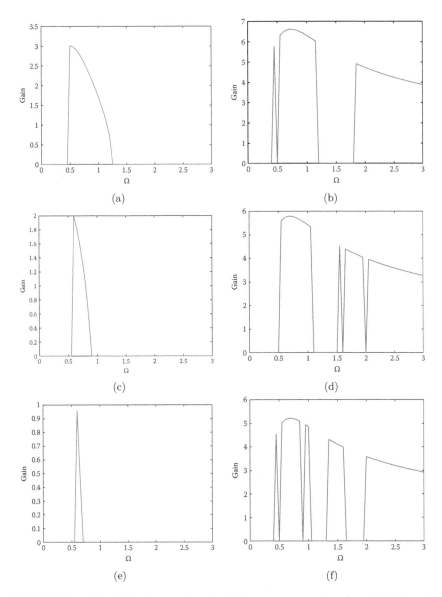

FIGURE 6.4: The MI gain spectra for different parameters ($\lambda = 0.2, P_0 = 1$). $R_1 = -0.9$, for (a) and (b); $R_1 = -0.7$ for (c) and (d); $R_1 = -0.5$ for (e) and $R_1 = -0.4$. for (f).

pulses propagate stably without any distortion when XPM < SPM ($\sigma = 2/3$). However, when XPM > SPM ($\sigma = 2$), the pulses are distorted at the end of propagation. The present result, especially the formation of the stable periodic

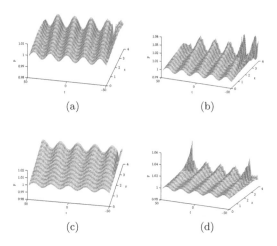

(a)   (b)

(c)   (d)

FIGURE 6.5: Evolution of MI from a CW input in the presence of CCD. Two of the waveguides propagate in the normal dispersion regime, while the other experiences anomalous GVD. (a) With XPM < SPM ($\sigma = 2/3$); (b) with XPM > SPM ($\sigma = 2$). ($\lambda = 0.8, R_1 = 0, R = 0, P_0 = 1$) for (a), (b); $\lambda = 0.2, R = 1, R_1 = -0.4, P_0 = 1 for (c), (d)$.

array of localized pulses, may find straightforward application in nonlinear optics.

6.6 CONCLUSION

The full treatment of MI of continuous waves for a system of four coupled, nonlinear Schrödinger equations has been studied. In contrast to previous works, the coupling coefficient effect has been considered. The results indicate that SPM, XPM and CCD terms will alter the gain spectra of MI considerably in terms of the shape and spectral regions of MI. Moreover, the nonlinear evolution of MI with the presence of a coupling coefficient has been studied numerically using the split-step Fourier method. The most notable change occurs when XPM > SPM ($\sigma = 2$). We have observed here that when SPM < XPM ($\sigma = 2/3$), MI leads to a pattern in the form of a periodic pulse array. However, when XPM > SPM ($\sigma = 2$), MI leads to a pattern in the form of a periodic pulse array which is distorted at the end of the propagation. As regards applications, nonconventional MI processes have a substantial advantage over ordinary MI process, that is, the possibility to obtain large MI frequencies without having to make use of large pump powers. This advantage should soften the conditions of generation of ultrashort pulses at ultrahigh repetition rates and is thus potentially useful for the development of high-speed optical communication systems.

ACKNOWLEDGMENT

AM acknowledges financial support from the Abdus Salam International Center for Theoretical Physics (ICTP) Trieste-Italy through the Associate Programme.

APPENDIX

Matrix elements of Eq. (6.7) are given by

$m_{11} = -m_{22} = -\lambda\Omega^2 - 3R + P_0,$

$m_{12} = -m_{21} = m_{34} = -m_{43} = m_{56} = -m_{65} = m_{78} = -m_{87} = P_0,$

$m_{13} = m_{15} = m_{17} = m_{31} = m_{35} = m_{37} = m_{51} = m_{53} = m_{57} = m_{71} = m_{73} = m_{75} = \sigma P_0 + R - R_1\Omega,$

$m_{14} = m_{16} = m_{18} = -m_{23} = -m_{25} = -m_{27} = m_{32} = m_{36} = m_{38} = -m_{41} = -m_{45} = -m_{47} = m_{52} = m_{54} = m_{58} = -m_{61} = -m_{63} = -m_{67} = m_{72} = m_{74} = m_{76} = -m_{81} = -m_{83} = -m_{85} = \sigma P_0,$

$m_{33} = -m_{44} = -\Omega^2 - 3R + P_0,$

$m_{24} = m_{26} = m_{28} = m_{42} = m_{46} = m_{48} = m_{62} = m_{64} = m_{68} = m_{82} = m_{84} = m_{86} = -(\sigma P_0 + R + R_1\Omega),$

$m_{55} = -m_{66} = \Omega^2 - 3R + P_0,$

$m_{77} = -m_{88} = \lambda\Omega^2 - 3R + P_0.$

REFERENCES

1. Y. S. Kivshar and G. P. Agrawal, *Optical Solitons*, Academic Press, 2003.
2. A. Hasegawa, *Optical Solitons—Theoretical and Experimental Challenges* 613, eds., K. Porsezian and V. C. Kuriakose, Berlin: Springer, 2003.
3. N. N. Akhmediev and A. Ankiewicz, *Solitons-Nonlinear Pulses and Beams*, Chapman and Hall, 1997.
4. G. P Agrawal, *Applications of Nonlinear Fiber Optics*, San Diego, CA: Academic, 2001.
5. G. Millot, S. Pitois, J. M. Dudley, and M. Haelterman, *Optical Solitons Theoretical and Experimental Challenges* vol 613, ed K Porsezian and V C Kuriakose (Berlin: Springer,2003).
6. L. A. Ostrovskii, *Sov. Phys.-JETP* 24, 797, 1966.
7. V. I. Karpman, *JETP Lett.* 6, 277, 1967.
8. A. Hasegawa, *Phys. Rev. Lett.* 24, 1165, 1970.
9. A. Hasegawa and W. F. Brinkman, *IEEE J. Quantum Electron.* 16, 694, 1980.
10. S. Pitois, G. Millot, P. Grelu and M. Haelterman, *Phys. Rev. A* 60, 994, 1999.
11. C. Martijn de Sterke, *J. Opt. Soc. Am. B* 15, 2660, 1998.
12. A. Parini, G. Bellanca, S. Trillo, M. Conforti, A. Locatelli, C. De Angelis, *J. Opt. Soc. Am B* 24, 2229, 2007.
13. G. P. Agrawal, *Phys. Rev. Lett.* 59, 880, 1987.
14. T. Tanemura and K. Kikuchi, *J. Opt. Soc. Am. B* 20, 2502, 2003.
15. N. C. Panoiu, X. Chen, and R. M. Osgoof, *Opt. Lett.* 31, 3609, 2006.
16. X. Dai, Y. Xiang, S. Wen, and D. Fan, *J. Opt. Soc. Am. B* 26, 564, 2009.
17. C. R. Phillips and M. M. Feger, *J. Opt. Soc. Am. B* 27, 2687, 2010.

18. G. Millot, S. Pitois, and P. Tchofo Dinda, *J. Opt. Soc. Am. B* 19, 454, 2002.
19. H. S. Chiu and K. W. Chow, *Phys. Rev. A* 79, 065803, 2009.
20. P. Tchofo Dinda and K. Porsezian, *J. Opt. Soc. Am. B* 27, 1143, 2010.
21. E. A. Ultanir, D. N. Chritodoulides, and G. I. Stegeman, *J. Opt. Soc. Am. B* 23, 341, 2006.
22. J. M. Dudley, G. Genty, F. Dias, B. Kibler, and N. Akhmediev, *Opt. Express* 17, 21497, 2009.
23. Y. S. Kivshar and G. P. Agrawal, *Optical Solitons: From Fibers to Photonic Crystals*, Academic, 2003.
24. T. Tanemura, Y. Ozeki, and K. Kikuchi, *Phys. Rev. Lett.* 93, 163902, 2004.
25. M. Yu, C. J. McKinistric, and G. P. Agrawal, *J. Opt. Soc. Am. B* 15, 607, 1998.
26. M. J. Steel, T. P. White, and C. M. de Sterke, *Opt. Lett.* 26, 488, 2001.
27. E. Seve, P. T. Dinda, G. Millot, M. Remoissenet, J. M. Bilbault, and M. Haelterman, *Phys. Rev. A* 54, 3519, 1996.
28. G. Agrawal, *Nonlinear Fiber Optics*, 4th ed., Academic Press, San Diego, 2007.
29. A. Hasegawa and F. Tappert, *Appl. Phys. Lett.* 23, 142, 1973.
30. K. Tai, A. Hasegawa, and A. Tomita, *Phys. Rev. Lett.* 56, 135, 1986.
31. N. Akhmediev, V. Korneev, and N. Mitskevich, *Sov. Phys. Radiophysics*, 34, 73, 1991.
32. C. C. Mei, *The Applied Dynamics of Ocean Waves*, World Scientific, Singapore, 1989.
33. A. R. Osborne, M. Onorato, and M. Serio, *Phys. Lett. A* 275, 386, 2000.
34. K. Saitoh, Y. Sato, and M. Koshiba, *Opt. Express* 11, 3188, 2003.
35. R. Ganapathy, B. A. Malomed, and K. Porsezian, *J. Opt. B: Quantum Semiclass. Opt.* 6, S436, 2004.
36. Q. Li, Y. Xie, Y. Zhu, and S. Qian, *Opt. Commun.* 281, 2811, 2008.
37. K. S. Chiang, Y. T. Chow, D. J. Richardson, D. Taverner, L. Dong, L. Reekie, and K. M. Lo, *Opt. Commun.* 143, 189, 1997.
38. A. W. Snyder, *J. Opt. Soc. Am.* 62, 1267, 1972.
39. K. Chiang, *J. Opt. Soc. Am. B* 14, 1437, 1997.
40. Z. Wang, T. Taru, T. A. Birks, J. C. Knight, Y. Liu, and J. Du, *Opt. Express* 15, 4795, 2007.
41. M. Liu, D. Li, and Z. Liao, *JETP Lett.* 95, 10, 2012.
42. P. Peterka, P. Honzatko, J. Kanka, V. Matejec, and I. Kasik, *Proc. SPIE* 5036, 376, 2003.
43. S. Trillo, S. Wabnitz, G. I. Stegeman, and E. M. Wright, *J. Opt. Soc. Am. B* 6, 889, 1989.
44. R. Tasgal and B. Malomed, *Phys. Scr.* 60, 418, 1999.
45. J. H. Li, K. S. Chiang, and K. W. Chow, *J. Opt. Soc. Am. B* 28, 1693, 2011.
46. K. Nithyanandan, R. V. J. Raja, and K. Porsezian, *Phys. Rev. A* 87, 043805, 2013.
47. K. W. Chow, K. K. Wong, and K. Lam, *Phys. Lett. A* 372, 4596, 2008.
48. R. Ganapathy, B. A. Malomed, and K. Porsezian, *Phys. Lett. A* 354, 366, 2006.
49. Y. Kodama, A. Maruta, and A. Hasegawa, *Quantum and Semiclassical Optics* 6, 463, 1994.

7 Hidden symmetry reductions and the Ablowitz–Kaup–Newell– Segur hierarchies for nonautonomous solitons

V. N. Serkin, *A. Hasegawa*, and *T. L. Belyaeva*

7.1 INTRODUCTION

Nonlinear science is considered to be the most important frontier for understanding nature [1, 2]. The interpenetration of the main ideas and methods being used in different fields of science and technology has become one of the decisive factors in the progress of science as a whole. Among the most spectacular examples of such an interchange of ideas and theoretical methods for analysis of various physical phenomena is the problem of soliton formation.

Solitons—self-localized robust and long-lived nonlinear solitary waves that do not disperse and preserve their identity during propagation and after a collision—arise in any physical system possessing both nonlinearity and dispersion, diffraction or diffusion (in time and/or space). Being a product of the high-speed computer revolution of the 20th century, today the soliton paradigm provides a remarkable example in which an abstract mathematical concept has produced a large impact on the real world of high technologies. The soliton determines the spirit of nonlinear science, overcoming traditional boundaries between different fields, and during the last decades of its development, the soliton paradigm has brought together mathematicians and physicists from different areas, in particular, from atomic, low-temperature and condensed matter physics, pico and femtosecond nonlinear optics and lasers, fluid mechanics and fundamental general and particle physics.

The canonical soliton concept was developed for nonlinear and dispersive systems that have been autonomous (described by autonomous constant-coefficient nonlinear equations); namely, time (and/or space) has only played the role of the independent variable and has not appeared explicitly in the nonlinear evolution equation [6–19]. Nonautonomous physical systems subjected to some form of external input are of great interest because they are much more common and realistic, and they are not as idealized as canonical autonomous ones. Such situations could include repeated stress testing of a soliton in nonuniform media with space- and/or time-dependent density gradients. These situations are typical for experiments with temporal/spatial optical solitons, soliton lasers and ultrafast soliton switches and logic gates. The formation of matter-wave solitons in Bose–Einstein condensates (BECs) by magnetically tuning the interatomic interaction near a Feshbach resonance provides a good example of a nonautonomous nonlinear system [20, 21].

In general terms, the nonautonomous soliton concept was introduced for nonlinear and dispersive systems governed by nonautonomous (variable-coefficient) nonlinear models with external potentials varying in time and space. Soliton solutions for the nonautonomous nonlinear Schrödinger equation models with varying dispersion, nonlinearity, gain or absorption, and, in more general cases, subjected to external potentials varying in time and space, substantially extend the concept of classical solitons and generalize it to the plethora of nonautonomous solitons that interact elastically and generally move with varying amplitudes, speeds and spectra. It was shown that nonautonomous solitons exist only under certain conditions, and the parameter functions describing varying dispersion, nonlinearity, gain or absorption and external inhomogeneous potentials cannot be chosen independently: they satisfy the so-called soliton adaptation laws both to the external potentials and to the dispersion and nonlinearity variations.

It should be specially emphasized that our primary purposes in this chapter are conceptual and pedagogical. So far, in contrast to canonical solitons, very little of detail is known about the complete integrability of nonautonomous nonlinear models. As we go, we will find that nonautonomous solitons are in some ways different from canonical solitons in autonomous systems, and in other ways alike. The lesson here is that, when working with the nonautonomous soliton concept, one often discovers novel interesting matters "during one's visits to nonautonomous soliton islands in the sea of nonlinear waves" [22].

In order to motivate our pedagogical treatment of nonautonomous solitons, it cannot be too highly stressed that the celebrated soliton paradigm clearly demonstrates the existence of two famous principles on which all mathematical theories must be based: "Physical laws should have mathematical beauty" developed by Dirac [23], and "Nature is simple in its essence" introduced by Yukawa [24].

The importance and beauty of the soliton paradigm cannot be overrated, and we have every reason to characterize soliton research by Dirac's words: "As time goes on, it becomes increasingly evident that the rules which the mathematician finds interesting are the same as those which Nature has chosen" [25]. On the another hand, concerning the beauty and complexity of soliton studies, and in particular, presented in this book, the Yukawa concept, may be restated in a different way: "The nature is simple in its essence, but yet is great" [26].

We believe that our chapter will be pedagogically helpful, so that the interested reader can directly enjoy the Dirac law of mathematical beauty applied to the nonautonomous soliton concept. We demonstrate the simplicity of this concept in its essence. Genuinely, the "roots" of nonautonomous solitons are deeplying in the so-called "hidden" symmetries of the famous quantum mechanical equations, and there exist many fruitful analogies with coherent and squeezed states of a quantum mechanical harmonic oscillator. And what is more, the "tree" of the nonautonomous soliton concept is cultivated on the basis of generalized nonisospectral Ablowitz–Kaup–Newell–Segur (AKNS) hierarchies [27] in the framework of the inverse scattering transform (IST) method [28–35] with varying spectral parameter introduced for the first time by Chen and Liu [36] to the nonlinear Schrödinger equation model, Hirota and Satsuma to the Korteweg–de Vries (KdV) equation in nonuniform media [37], and Calogero and Degasperis [38] to the KdV boomeron model.

7.2 HUSIMI–TANIUTI AND TALANOV TRANSFORMATIONS IN QUANTUM MECHANICS AND THE SOLITON THEORY

"Coming events cast their shadows before them." We start with this famous English proverb to emphasize once again that the history of science provides many cases where outstanding scientists were far ahead of their time.

The remarkable harmony of the linear Schrödinger equation (SE) lies in the fact that there exists group symmetry, often called the "hidden" or nontrivial Schrödinger group symmetry, under which this equation is invariant [39–41].

For the first time, this symmetry was considered by Husimi [42, 43] on the example of the forced harmonic oscillator model (known today in quantum mechanics as the Husimi parametric oscillator).

$$i\frac{\partial \Psi}{\partial t} + \frac{1}{2}\frac{\partial^2 \Psi}{\partial x^2} - \frac{\omega^2(t)}{2}x^2\Psi - f(t)x\Psi = 0. \tag{7.1}$$

Husimi demonstrated that this equation can be satisfied by a function of the form

$$\Psi(x,t) \to \exp\left\{\frac{i}{2}\left[a(t)x^2 + 2b(t)x + c(t)\right]\right\} \tag{7.2}$$

so that there are enough degrees of freedom to be able to find solutions for three coefficients $a(t), b(t)$ and $c(t)$. In doing that, Husimi used the reduction introduced by Taniuti to relate the free ($f(t) = 0$) and parametrically driven ($f(t) \neq 0$) oscillators. Taniuti proposed an elegant method to study the forced harmonic oscillator (7.1) in a commoving frame:

$$\xi = x - \alpha(t) \tag{7.3}$$

and discovered that two cases of free and forced oscillators are intimately connected. In general, once a solution of the free case (Eq. (7.1)) with $f(t) = 0$) has been obtained, the corresponding solution under arbitrary external force $f(t)$ can immediately be constructed by means of the following simple transformation:

$$\Psi(x,t) = \Upsilon \left[x - \alpha(t), t \right] \exp \left\{ i \frac{d\alpha}{dt} \left[x - \alpha(t) \right] + i \int\limits_0^t L(t')dt' \right\}, \qquad (7.4)$$

where L is the classical Lagrangian

$$L(t) = \frac{1}{2} \left(\frac{d\alpha}{dt} \right)^2 - \frac{1}{2} \omega^2 \alpha^2(t) + f(t)\alpha(t) \qquad (7.5)$$

and a new function $\Upsilon(\xi,t)$ represents the solution of the free Schrödinger harmonic oscillator

$$i \frac{\partial \Upsilon}{\partial t} + \frac{1}{2} \frac{\partial^2 \Upsilon}{\partial \xi^2} - \frac{1}{2} \omega^2 \xi^2 \Upsilon = 0. \qquad (7.6)$$

The interested reader can find the first detailed presentation of the Taniuti transformation (7.4) in [43].

The most surprising discovery was made by Talanov in 1970 [44]: he revealed the existence of this symmetry transformation for the nonlinear parabolic equation in the self-focusing theory. This equation, known today as the nonlinear Schrödinger equation (NLSE), is written here in its canonical form

$$i \frac{\partial \Psi_{d=2}}{\partial z} + \frac{1}{2} \left(\frac{\partial^2}{\partial x^2} + \frac{\partial^2}{\partial y^2} \right) \Psi_{d=2} + |\Psi_{d=2}|^2 \Psi_{d=2} = 0. \qquad (7.7)$$

In the context of nonlinear optics, the Talanov transformation (also called "the lens transformation" in different applications from nonlinear optics and plasma to BEC [45], can be written in the form

$$(x,y) \Rightarrow \left[\frac{x}{L(z)}, \frac{y}{L(z)} \right], \qquad (7.8)$$

$$\zeta = \int\limits_0^z L^{-2}(z')dz' = \frac{z}{1 + z/F_0}, \qquad (7.9)$$

$$\Psi_{d=2}(x,y,z) = \frac{1}{L(z)} \exp \left\{ \frac{i}{2} \frac{1}{L(z)} \frac{dL}{dz} (x^2 + y^2) \right\} \Phi \left(\frac{x}{L(z)}, \frac{y}{L(z)}, \zeta \right), \qquad (7.10)$$

$$L(z) = 1 + \frac{z}{F_0}. \qquad (7.11)$$

It should be emphasized that for the linear Schrödinger equation models, Talanov symmetry transformation exists in any dimension, $d = 1, 2, 3$, but in the general case of the d-dimensional NLSE, the invariance under the Talanov symmetry transformation exists if and only if $d = 2$ (two-dimensional laser beams) [46]. The term lens transformation originates from the fact that

Eqs. (7.8–7.11) can be represented as a thin lens transformation with a focal distance F_0 located at $z = 0$:

$$\frac{1}{\zeta} = \frac{1}{z} + \frac{1}{F_0}. \tag{7.12}$$

That is why the general kind of symmetry transformation Eqs. (7.3–7.5), (7.8–7.11), can be named for its discoverers the Husimi–Taniuti and Talanov (HT&T) transformation in the theory of linear and nonlinear waves.

It is useful to introduce the so-called "hidden" symmetry parameter $s(t)$ directly into the HT&T transformation so that it takes the most mathematically beautiful form

$$
\begin{aligned}
\Psi_d(x,t) \;=\;& \exp\left\{\frac{i}{2}\frac{1}{s(t)}\frac{ds}{dt}\left[x_d - \alpha(t)\right]^2 + i\frac{d\alpha}{dt}\left[x_d - \frac{\alpha(t)}{2}\right]\right\} \\
& \times \exp\left\{-\frac{i}{2}\int_0^t f(t')\alpha(t')dt' - \frac{d}{2}\log s(t)\right\} \\
& \times \Phi\left\{\xi_d = \frac{(x_d - \alpha(t))}{s(t)}, \eta = \int_0^t s^{-2}(t')dt'\right\}. \tag{7.13}
\end{aligned}
$$

In order to demonstrate the remarkable mathematical beauty and the exceptional ease of the HT&T transformation, as the first step, let us learn how to apply this transformation to attack canonical linear problems, and only then consider more complicated features of nonlinear and nonautonomous models.

Since we have enough degrees of freedom, we will seek possible conversions of the linear (and nonlinear) Schrödinger equation into itself, or different so-called conditional symmetry reductions to other equations that are easier to solve.

7.2.1 DARWIN WAVE PACKET AND CHEN AND LIU ACCELERATING SOLITON

As a first example, consider the existing remarkable and deep "connection in time" between Darwin's solution [47] in 1928 for the linear SE with gravitational potential and the Chen and Liu solution [36] in 1976 for the NLSE soliton.

Substituting the HT&T symmetry transformation (7.13) into the linear SE with gravitational time-dependent linear potential,

$$i\frac{\partial\Psi}{\partial t} + \frac{1}{2}\frac{\partial^2\Psi}{\partial x^2} - f(t)x\Psi = 0, \tag{7.14}$$

in the simplest case of the hidden symmetry condition $s(t) = 1$, we obtain

$$i\frac{\partial \Phi}{\partial \eta} + \frac{1}{2}\frac{\partial^2 \Phi}{\partial \xi^2} - \left(\frac{d^2\alpha}{dt^2} + f(t)\right)\xi\Phi - \frac{1}{2}\alpha\left(\frac{d^2\alpha}{dt^2} + f(t)\right)\Phi = 0. \qquad (7.15)$$

As follows from this equation, the symmetry reduction

$$\frac{d^2\alpha}{dt^2} + f(t) = 0 \qquad (7.16)$$

connects the forced SE (7.14) with the free SE

$$i\frac{\partial \Phi}{\partial \eta} + \frac{1}{2}\frac{\partial^2 \Phi}{\partial \xi^2} = 0, \qquad (7.17)$$

so that once an arbitrary solution of the free equation (7.17) has been obtained, the corresponding solution of the SE with a gravitational potential (7.14) can immediately be constructed by means of the following simple transformation:

$$\Psi(x,t) = \exp\left\{i\frac{d\alpha}{dt}\left(x - \frac{\alpha(t)}{2}\right) + \frac{i}{2}\int_0^t \frac{d^2\alpha}{dt'^2}\alpha(t')dt'\right\}\Phi\left\{\xi = x - \alpha(t), \eta = t\right\}$$

$$(7.18)$$

with $\alpha(t)$ given from Eq. (7.16). In particular, if $f(t) = f_0 = const$, and initial conditions $\alpha(t = 0) = 0$, and $d\alpha/dt(t = 0) = 0$, according to the relation

$$\xi = x - \alpha(t) = x + \frac{1}{2}f_0 t^2, \qquad (7.19)$$

we arrive at the symmetry reduction

$$\Psi(x,t) = \exp\left\{-if_0 t\left(x + \frac{1}{6}f_0 t^2\right)\right\}\Phi\left\{\xi = x + \frac{1}{2}f_0 t^2, \eta = t\right\}. \qquad (7.20)$$

In order to show the details of this symmetry reduction, let us consider the simplest case where the solution of the linear SE (7.17) $\Phi\left\{\xi = x + \frac{1}{2}f_0 t^2, t\right\}$ represents the spreading in a free space Gaussian wave packet. The corresponding solution of Eq. (7.14) with gravitational potential shows the same spreading behavior, but it accelerates according to Newton's law, given by Eq. (7.19). This solution is known in quantum mechanics as the Darwin wave packet [47].

We need to stress that the cubic nonlinearity appearing in the NLSE does not affect the considered symmetry transformation Eq. (7.18). This fact is known as the Tappert transformation [36] of the NLSE with a linear x-dependent external potential:

$$i\frac{\partial \Psi}{\partial t} + \frac{1}{2}\frac{\partial^2 \Psi}{\partial x^2} - f(t)x\Psi + |\Psi|^2\Psi = 0.$$

Thus we see that there exists a remarkable "connection in time" between the Darwin quantum-mechanical spreading wave packet in external gravitational potential and the NLSE nonspreading soliton solution in the linear x-dependent potential [36]. And this "connection in time" is much deeper than it seems at first sight. Really, if we substitute the *sech*-type envelope of the Schrödinger soliton to the NLSE, we obtain exactly a linear SE with a $sech^2(x)$ trapping potential. Obviously, we can associate the soliton propagation factor $\varkappa_0^2 t/2$ with its own eigenvalue energy level $E_0 = \varkappa_0^2/2$ representing the well-known exact eigenvalue in the quantum-mechanical $sech^2(x)$ trapping potential. As the self-localized wave object, the soliton, by virtue of Galilean symmetry (given, for example, by the HT&T transformation (7.13) in the simplest case $f(t) = 0$ and $s(t) = 1$), is characterized by its own solitonic analog of the de Broglie wavelength, sending us in search of a solitonic analog of the Ramsauer–Townsend effect [48, 49]. On the other hand, the extended particle-like object, the soliton, because of the nonlinear self-interaction, becomes a bound state in its own self-induced trapping potential and, as a consequence, acquires a negative self-interaction (binding) energy. The soliton binding energy provides the shape and structural stability of solitons and, similar to nuclear binding energy, can be considered as the degree of how strongly the quasi-particles that make up the soliton are bound together, for example, photons in optical spatial or temporal solitons, or Bose atoms trapped in matter–wave solitons in a BEC. This form of soliton energy (that is yet hidden from us) is sending us in search of the nonlinear soliton analogs of the Gamow tunneling effect and the Geiger–Nuttall law for solitons [50, 51]. Investigation of soliton tunneling phenomena and soliton energy accumulation and ejection from potential traps is the most important research field for understanding hidden soliton symmetries and the wave-particle duality of solitons [52–57] and references therein].

7.2.2   AIRY AND HERMITE ACCELERATING WAVE PACKETS IN FREE SPACE

As a second example, let us consider the famous "unexpected" result of quantum mechanics obtained by Berry and Ballazs in 1979 [58]: the accelerating Airy function solution for the free Schrödinger equation

$$i\frac{\partial \Psi}{\partial t} + \frac{1}{2}\frac{\partial^2 \Psi}{\partial x^2} = 0. \tag{7.21}$$

Substituting (7.13) into (7.21),

$$i\frac{\partial \Phi}{\partial \eta} + \frac{1}{2}\frac{\partial^2 \Phi}{\partial \xi^2} - \frac{d^2\alpha}{dt^2}\xi\Phi - \frac{1}{2}\alpha\left(\frac{d^2\alpha}{dt^2} - f(t)\right)\Phi = 0 \tag{7.22}$$

under the simplest hidden symmetry condition $s(t) = 1$, we easily obtain the canonical Airy equation for the function $\Phi \{\xi = (x - \alpha(t)), \eta = t\}$

$$i\frac{\partial \Phi}{\partial \eta} + \frac{1}{2}\frac{\partial^2 \Phi}{\partial \xi^2} - f(\eta)\xi\Phi = 0 \tag{7.23}$$

with the condition

$$\frac{d^2\alpha}{d\eta^2} - f(\eta) = 0. \tag{7.24}$$

In the simplest case $f(\eta) = f_0 = const$, and with the initial conditions $d\alpha/dt(t = 0) = 0$ and $\alpha(t = 0) = 0$, we obtain $\alpha(t) = f_0 t^2/2$. Substituting it into the symmetry transformation given by (7.13), we directly confirm the Berry and Ballazs solution for the free SE:

$$\Psi(x, t) = Ai\left(x - \frac{1}{2}f_0 t^2\right) \exp\left\{i\left(f_0 tx - \frac{f_0^2 t^3}{3}\right)\right\}. \tag{7.25}$$

It is evident from (7.13) that the Berry and Ballazs solution is not unique, and there exists a more general (spreading) solution which takes into account the hidden symmetry parameter $s(t)$.

As the next example, let us show that the easiest method to obtain a canonical spreading wave packet in free space can be based on the hidden symmetry of the linear SE given by Eq. (7.6).

Substitution of (7.13) into (7.21) transforms the linear Schrödinger equation in free space to the canonical SE with a harmonic oscillator potential

$$i\frac{\partial\Phi}{\partial\eta} + \frac{1}{2}\frac{\partial^2\Phi}{\partial\xi^2} - \frac{1}{2}\omega_0^2\xi^2\Phi = 0, \tag{7.26}$$

if and only if the following two conditions on the parameters α and s are satisfied:

$$s^3\frac{d^2s}{dt^2} = \omega_0^2, \tag{7.27}$$

$$\frac{d^2\alpha}{dt^2} = 0. \tag{7.28}$$

The solutions of Eqs. (7.27 and 7.28) read

$$s^2(t) = s_0^2 + \frac{\omega_0^2 t^2}{s_0^2} \tag{7.29}$$

$$\alpha(t) = \alpha_0 + v_0 t. \tag{7.30}$$

The most interesting example represents the situation when solutions of Eq. (7.26) are given by the eigenfunctions of the harmonic oscillator potential, Hermite polynomials: $\Phi_n(x - \alpha_0 - v_0 t) = H_n(x - \alpha_0 - v_0 t)$. In this case, according to (7.29) and (7.30),

$$|\Psi(x, t)|^2 = \frac{1}{\sqrt{1 + \omega_0^2 t^2}}|H_0(x - \alpha_0 - v_0 t)|^2 \tag{7.31}$$

where $H_0(x - \alpha_0 - v_0 t)$ represents the Gaussian wave function for the ground state of the harmonic oscillator (7.26), and the hidden symmetry parameter is chosen to be $s_0^2 = 1$ at $t = 0$.

It is evident from (7.13) that the dispersion spreading of a Gaussian wave packet in free space is not unique, and there exist much more general (spreading) solutions given by high-order Hermite polynomials $H_n(x - a_0 - v_0 t)$. Surprisingly in quantum mechanics, these solutions have been forgotten for a long time despite the fact that these solutions are well known in laser physics as the high-order Hermite–Gaussian modes [59]. Substituting Eqs. (7.29) and (7.30) into (7.13), we obtain the exact solutions by the simplest way practically without any calculations and without using the propagator method of quantum mechanics. Accelerating finite energy Airy beams and self-accelerating optical beams in highly nonlocal nonlinear media have been discovered recently [60–62].

7.2.3 COHERENT STATES, SQUEEZED STATES, AND SQUEEZIONS

Now let us consider the most important "precursor" of the nonautonomous soliton concept to demonstrate further an important deep "connection in time": the so-called coherent and squeezed states of the quantum mechanical linear harmonic oscillator.

The remarkable harmony of the linear Schrödinger harmonic oscillator consists of the existence of the group symmetry given by Eq. (7.13), often called the "hidden" or nontrivial Schrödinger group symmetry, under which this equation is invariant. Thus, according to the symmetry transformation Eq. (7.13) which we called the HT&T transformation, there exists the following one-to-one correspondence between the harmonic oscillators in coordinates x, t

$$i\frac{\partial \Psi}{\partial t} = -\frac{1}{2}\frac{\partial^2 \Psi}{\partial x^2} + \frac{\omega_0^2}{2}x^2 \Psi \tag{7.32}$$

and coordinates $\xi = ([x - \alpha(t)]/s(t), \quad \eta = \int_0^t s^{-2}(t')dt'$

$$i\frac{\partial \Phi}{\partial \eta} = -\frac{1}{2}\frac{\partial^2 \Phi}{\partial \xi^2} + \frac{\omega_0^2}{2}\xi^2 \Phi, \tag{7.33}$$

so that

$$\frac{d^2 s}{dt^2} + \omega_0^2 s = \frac{\omega_0^2}{s^3} \tag{7.34}$$

$$\frac{d^2 \alpha}{dt^2} + \omega_0^2 \alpha = 0. \tag{7.35}$$

The system (7.34)–(7.35) is known in nonlinear oscillation theory as the famous Ermakov equation system with solutions given by Ermakov in 1880 [63]:

$$\begin{aligned} s^2(t) &= s_0^2 \cos^2(\omega_0 t) + s_0^{-2} \sin^2(\omega_0 t) \tag{7.36} \\ &= \frac{1}{2}\left[(s_0^2 + s_0^{-2}) + (s_0^2 - s_0^{-2})\cos(2\omega_0 t)\right], \end{aligned}$$

$$\alpha(t) = x_0 \cos(\omega_0 t) + p_0 \sin(\omega_0 t). \tag{7.37}$$

There are two cases of physical meaning: the first one is given by the hidden symmetry parameter $s^2(t) = 1$ and is known as the Schrödinger coherent state [64]; the second one is given by arbitrary $s^2(t) \neq 1$, known in quantum mechanics as the Kennard squeezed state [65].

In particular, substituting the harmonic oscillator ground state E_0 into the symmetry transformation (7.13), we can obtain the textbook solution of Eq. (7.32) for the coherent $s^2(t) = 1$ and squeezed states $s^2(t) \neq 1$:

$$\psi(x,t) = \frac{1}{\sqrt{s(t)}\sqrt{\pi}} \exp\left(-\frac{1}{2}\left(\frac{x - x_0 \cos t}{s(t)}\right)^2 + i\theta(x,t) - iE_0 t\right) \tag{7.38}$$

$$\theta(x,t) = -\frac{x^2(a_0^4 - 1)\sin 2t}{4a_0^2 s^2(t)} + \frac{x_0^2 \sin 2t}{4a_0^2 s^2(t)} - \frac{x x_0 \sin t}{a_0^2 s^2(t)} \tag{7.39}$$

$$a_0^2 s^2(t) = a_0^4 \cos^2 t + \sin^2 t. \tag{7.40}$$

It should be especially stressed that according, to the symmetry transformation Eq. (7.13), we obtain all solutions practically without any calculations and without using a complicated quantum mechanical method based on the so-called Mehler kernel, Green's function or propagator of the Hamiltonian for the quantum harmonic oscillator [65, 66]. It is evident from (7.13) that both the coherent and squeezed states are not unique, and there exist much more general coherent and squeezed states given by high-order Hermite eigenfunctions $H_n [x - x_0 \cos(\omega_0 t) - p_0 \sin(\omega_0 t)]$ with energy eigenvalues E_n.

The importance of the coherent and squeezed states in modern physics cannot be overstated. When quantum mechanics was just few a months old, in 1926, E. Schrödinger in his paper titled "The continuous transition from micro- to macro-mechanics" [64], introduced the concept of *oscillating wave group as the representation of a particle in wave mechanics.* Inspirited by the de Broglie wave-particle duality, Schrödinger recognized that if the initial state of a quantum oscillator represents a ground-state wave function displaced from the origin, then this state "*may represent a 'particle', which is executing the 'motion' expected from the usual mechanics ... Our wave group always remains compact, and does not spread out into larger regions as time goes on.*" The oscillator potential prevents the position probability density from spreading, and this property explains the name "coherent state" given by Glauber for a quantum description of coherent light sources with the advent of the laser and quantum optics ages [67, 68].

Historically, the discovery of squeezed states of a harmonic oscillator dates back to 1927 by E. Kennard [65]. It was in this research that the Greens function or propagator of the Hamiltonian with quantum harmonic oscillator potential was discovered. And it was in this research that the more general class of oscillating wave packets was discovered for arbitrary initial widths and localizations of wave functions not restricted simply by the shifted ground state, but, for example, wider or sharper of it. The Kennard states (known today as squeezed states) are really remarkable for several reasons. The wave packet remains Gaussian at all times, while its width and amplitude oscillate and the wave function can be more highly localized in position space than the coherent state. This feature explains the name "squeezed state" given by Hollenhorst to underscore the increasing sensitivity of gravitational antenna in this state [69]. Over the years, there have been many significant contributions to the development of coherent and squeezed states theory and experiments (see, for example, [70–75] and references therein).

The coherent state is not the only one in which the shape of the wave packet remains constant during the time evolution, and the squeezed state is not the only one in which localization in position and momentum space periodically oscillates. The even and odd coherent and squeezed states (known also as the "Schrödinger cat male and female states") have been discovered and investigated [76]. All initially translated high-energy Hermite–Gaussian eigenstates of a linear harmonic oscillator can form the higher-order coherent and squeezed states [69, 77].

It should be especially emphasized that the less known fact is that both even and odd coherent and squeezed states can be considered as "elastically interacting" wave packets, even though they are not orthogonal [69]. In some sense, their dynamics is remarkably similar to NLSE soliton interaction dynamics.

The fact that the de Broglie wavelength of the squeezed state is being "chirped" (referred to here as the "chirped" de Broglie wave "within the particle") has escaped the attention of researchers up to this point. We stress that the chirping of the de Broglie wave is a key physical condition for the existence of squeezed states in a linear harmonic oscillator. It is precisely this wave mechanics effect that opens a remarkable deep "connection in time" and that opens the way to the fundamental extension of the squeezed states concept to different nonlinear physical systems, which may be described by the NLSE model with harmonic oscillator potential.

The most surprising thing is that so mathematically beautiful "hidden" symmetry reductions (7.13) existing for linear SE models can be applied for nonautonomous NLSE with arbitrary nonlinearity of the form $F(|\Psi|^2)$:

$$i\frac{\partial \Psi}{\partial t} + \frac{1}{2}D(t)\frac{\partial^2 \Psi}{\partial x^2} - \frac{\omega^2(t)}{2}x^2\Psi - f(t)x\Psi + R(t)F(|\Psi|^2)\Psi = 0, \qquad (7.41)$$

as well as for the generalized nonautonomous Schrödinger equation model:

$$i\frac{\partial \Psi_d}{\partial t} + \frac{1}{2}D(t)\triangle_d\Psi_d - \frac{\omega^2(t)}{2}r_d^2\Psi + \sigma R(t)|\Psi_d|^2\Psi_d = i\Gamma(t)\Psi_d, \qquad (7.42)$$

which in the one-dimensional case $d = 1$ is written here in the most general form:

$$i\frac{\partial \Psi}{\partial t} + \frac{D(t)}{2}\frac{\partial^2 \Psi}{\partial x^2} + \sigma R(t)|\Psi|^2\Psi - \frac{\omega^2(t)}{2}x^2\Psi = i\Gamma(t)\Psi. \qquad (7.43)$$

Inserting

$$\Psi(x,t) = Q\left(x, \tau = \int^t D(t')dt'\right) \exp \int^t \Gamma(t')dt' \qquad (7.44)$$

$$\Omega^2(t) = \omega^2(t)/D(t) \qquad (7.45)$$

$$\widetilde{R(t)} = \frac{R(t)}{D(t)} \exp 2 \int^t \Gamma(t')dt'$$

into Eq.(7.42), we obtain

$$i\frac{\partial Q_d}{\partial \tau} + \triangle_d Q_d - \frac{\Omega^2}{2}r_d^2\Psi + \tilde{R}|Q_d|^2 Q_d = 0. \qquad (7.46)$$

It is convenient to demonstrate that Eq. (7.46) allows many physically important hidden HT&T symmetry reductions.

$$Q_d(x,\tau) = \exp\left\{\frac{i}{2}\frac{1}{s(\tau)}\frac{ds}{d\tau}(r_d - \alpha(\tau))^2 + i\frac{d\alpha}{d\tau}\left(r_d - \frac{\alpha(\tau)}{2}\right)\right\}$$

$$\times \exp\left\{-\frac{i}{2}\int_0^\tau f(t')\alpha(t')dt' - \frac{d}{2}\log s(\tau)\right\}$$

$$\times \Phi_d\left\{\xi_d = \frac{(r_d - \alpha(\tau))}{s(\tau)}, \eta = \int_0^\tau s^{-2}(t')dt'\right\}. \qquad (7.47)$$

One of the more studied HT&T symmetry reductions can be obtained after substitution of Eq. (7.47) into (7.46):

$$i\frac{\partial \Phi_d}{\partial \eta} + \frac{1}{2}\triangle(\xi_d)\Phi_d - \frac{1}{2}\xi_d^2 s^3\left[\frac{d^2 s}{d\tau^2} + \Omega^2(t)s\right]\Phi_d$$

$$+\sigma s^{2-d}(\tau)\widetilde{R(t)}|\Phi_d|^2\Phi_d = 0, \qquad (7.48)$$

$$\frac{d^2\alpha}{d\tau^2} + \Omega^2(t)\alpha = 0. \qquad (7.49)$$

As follows from this equation system, in the special case $\Omega^2(t) = 0$; $\widetilde{R(t)} = 1$, and $\Gamma(t) = 0$, the exact invariance of Eq. (7.46) under the transformation (7.47) exists if and only if $d = 2$, i.e., for cylindrically symmetric two-dimensional laser beam geometry.

The hidden symmetry parameter $s(t)$ in this case is determined by the simple condition

$$\frac{d^2 s}{dt^2} = 0,$$

which (in terms of the axes geometry for laser beam propagation, $t \rightarrow z$) immediately leads to the remarkable Talanov transformation (see Eqs. (7.8)–(7.12)

$$\frac{d^2 s}{dz^2} = 0,$$
$$s(z) = az + b.$$

That is why, as we need to underline especially, the general hidden symmetry transformation (Eqs. (7.13) and (7.47)) can be named for its discoverers, the Husimi–Taniuti and Talanov (HT&T) transformation in the theory of nonlinear waves.

In conclusion, it is straightforward to verify that in the framework of the unidimensional nonautonomous model (7.43), the conditional ("hidden parameter") symmetry relates Eq. (7.43) to the canonical autonomous model only under the condition that dispersion $D(t)$, nonlinearity $R(t)$ and confining harmonic potential satisfy the exact integrability scenario

$$\Omega^2(t)D(t) - 2D(t)\frac{d}{dt}\left(\frac{\Gamma(t)}{D(t)}\right) + 4\Gamma^2(t) \tag{7.50}$$
$$= \frac{W(R,D)}{R(t)D(t)}\left(4\Gamma(t) + \frac{d}{dt}\ln R(t)\right) - \frac{d}{dt}\left(\frac{W(R,D)}{R(t)D(t)}\right),$$

where $W(R,D) = R\left(D\right)_t - D\left(R\right)_t$ is the Wronskian of the functions $R(t)$ and $D(t)$.

It is evident that the conditional squeezing symmetry (7.47) of Eq. (7.43) is not unique, and there exist much more general squeezing-like symmetries. For example, if we apply the HT&T symmetry transformation to the quasi-soliton concept introduced by Hasegawa and Kumar [78–81], we can immediately conclude that there exists one more remarkable deep "connection in time": an analogy between the Hasegawa–Kumar quasi-soliton in parabolic trapping potential and the Schrödinger coherent states one hand and an analogy between self-compressing Hasegawa–Kumar quasi-solitons and the Kennard squeezed states from on other hand. The last transformation is described by Eq. (7.47) with the phase profiles given by formulas (7.39) and (7.40). This fact and the long-living practically elastic characters of their interactions in parabolic traps allow us to use the term "squeezions" to emphasize their deep "connection in time" with NLSE solitons and squeezed states of the linear quantum mechanical harmonic oscillator [82].

7.3 LAX OPERATOR METHOD AND EXACT INTEGRABILITY OF NONAUTONOMOUS NONLINEAR AND DISPERSIVE MODELS WITH EXTERNAL POTENTIALS

The classification of dynamic systems into autonomous and nonautonomous ones is commonly used in science to characterize different physical situations in which, respectively, an external time-dependent driving force is present or absent. The mathematical treatment of a nonautonomous system of equations is much more complicated than that of a traditional autonomous one. As a typical illustration, we may mention both a simple pendulum whose length changes with time and a parametrically driven nonlinear Duffing oscillator [83].

Let us consider the nonisospectral modification of the AKNS hierarchy, which allows us to construct the main evolution equations by allowing the time varying spectral parameter $\Lambda(T)$. We start from the (2×2) linear eigenvalue problem

$$\Psi_x = \widehat{\mathcal{L}}\Psi(x,t), \qquad \Psi_t = \widehat{\mathcal{M}}\Psi(x,t), \qquad (7.51)$$

where $\Psi(x,t) = \{\psi_1, \psi_2\}^T$ is a 2-component complex function and $\widehat{\mathcal{L}}$ and $\widehat{\mathcal{M}}$ are complex-valued (2×2) matrices

$$\widehat{\mathcal{L}}(\Lambda; S, T) = \widehat{\mathcal{L}}\left\{\Lambda(T), u\left[S(x,t), T\right]; \frac{\partial u}{\partial S}S_x; \frac{\partial^2 u}{\partial S^2}S_x^2; \dots\right\}$$

$$\widehat{\mathcal{M}}(\Lambda; S, T) = \widehat{\mathcal{M}}\left\{\Lambda(T), u\left[S(x,t), T\right]; \frac{\partial u}{\partial S}S_x; \frac{\partial^2 u}{\partial S^2}S_x^2; \dots\right\},$$

dependent on the generalized coordinates $S = S(x,t)$ and $T(t) = t$, where the function $u\left[S(x,t), T\right]$ and its derivatives denote the scattering potentials $q(S,T)$, $r(S,T)$ and their derivatives, correspondingly.

The nonlinear integrable equations arise as the compatibility condition of the system of the linear matrix differential equations

$$\frac{\partial \widehat{\mathcal{L}}}{\partial T} + \frac{\partial \widehat{\mathcal{L}}}{\partial S}S_t - \frac{\partial \widehat{\mathcal{M}}}{\partial S}S_x + \left[\widehat{\mathcal{L}}, \widehat{\mathcal{M}}\right] = 0 \qquad (7.52)$$

Here $F(T)$ and $\varphi(S,T)$ are real unknown functions, γ is an arbitrary constant and $\sigma = \pm 1$. The desired elements of the $\widehat{\mathcal{M}}$ matrix are expanded in the powers of the spectral parameter $\Lambda(T)$: $\widehat{\mathcal{M}} = \sum_{k=0}^{k=3} G_k \Lambda^k$, as well as a derivative of the time-varying spectral parameter

$$d\Lambda/dT \equiv \Lambda_T = \lambda_0(T) + \lambda_1(T)\Lambda(T), \qquad (7.53)$$

where $\lambda_0(T)$ and $\lambda_1(T)$ are the time-dependent functions.

Solving the system (7.52)–(7.53), we find both the AKNS matrix elements A, B, C:

$$A = -i\lambda_0 S/S_x + a_0 - \frac{1}{4}a_3\sigma F^{2\gamma}(qr\varphi_S S_x + iqr_S S_x - irq_S S_x)$$

$$+ \frac{1}{2}a_2\sigma F^{2\gamma}qr + \Lambda\left(-i\lambda_1 S/S_x + \frac{1}{2}a_3\sigma F^{2\gamma}qr + a_1\right) + a_2\Lambda^2 + a_3\Lambda^3,$$

$$B = \sqrt{\sigma} F^\gamma \exp[i\varphi S/2]\{-\frac{i}{4}a_3 S_x^2 \left(q_{SS} + \frac{i}{2}q\varphi_{SS} - \frac{1}{4}q\varphi_S^2 + iq_S\varphi_S \right)$$

$$-\frac{i}{4}a_2 q\varphi_S S_x - \frac{1}{2}a_2 q_S S_x + iq \left(-i\lambda_1 S/S_x + \frac{1}{2}a_3\sigma F^{2\gamma} qr + a_1 \right)$$

$$+\Lambda \left(-\frac{i}{4}a_3 q\varphi_S S_x - \frac{1}{2}a_3 q_S S_x + ia_2 q \right) + ia_3\Lambda^2 q\},$$

$$C = \sqrt{\sigma} F^\gamma \exp[-i\varphi S/2]\{-\frac{i}{4}a_3 S_x^2 \left(r_{SS} - \frac{i}{2}r\varphi_{SS} - \frac{1}{4}r\varphi_S^2 - ir_S\varphi_S \right)$$

$$-\frac{i}{4}a_2 r\varphi_S S_x + \frac{1}{2}a_2 r_S S_x + ir \left(-i\lambda_1 S/S_x + \frac{1}{2}a_3\sigma F^{2\gamma} qr + a_1 \right)$$

$$+\Lambda \left(-\frac{i}{4}a_3 r\varphi_S S_x + \frac{1}{2}a_3 r_S S_x + ia_2 r \right) + ia_3\Lambda^2 r\},$$

and evolution equations for the scattering potentials $q(S,T)$:

$$iq_T = \frac{1}{4}a_3 q_{SSS} S_x^3 + \frac{3i}{8}a_3 q_{SS}\varphi_S S_x^3 - \frac{3i}{4}a_3\sigma F^{2\gamma} q^2 r\varphi_S S_x$$

$$-\frac{3}{2}a_3\sigma F^{2\gamma} qr q_S S_x - \frac{i}{2}a_2 q_{SS} S_x^2 + ia_2\sigma F^{2\gamma} q^2 r$$

$$(7.54)$$

$$+iq_S \left(-S_t + \lambda_1 S + ia_1 S_x - \frac{i}{2}a_2\varphi_S S_x^2 + \frac{3}{8}a_3\varphi_{SS} S_x^3 + \frac{3i}{16}a_3\varphi_S^2 S_x^3 \right)$$

$$+q \left(i\lambda_1 - i\gamma\frac{F_T}{F} + \frac{1}{4}a_2\varphi_{SS} S_x^2 - \frac{3}{16}a_3\varphi_S\varphi_{SS} S_x^3 \right)$$

$$+q \left[2\lambda_0 S/S_x + 2ia_0 + \frac{1}{2}(\varphi_T + \varphi_S S_t) - \frac{1}{2}\lambda_1 S\varphi_S - \frac{i}{2}a_1\varphi_S S_x \right]$$

$$+q \left(\frac{i}{8}a_2\varphi_S^2 S_x^2 - \frac{i}{32}a_3\varphi_S^3 S_x^3 + \frac{i}{8}a_3\varphi_{SSS} S_x^3 \right)$$

and $r(S,T)$:

$$ir_T = \frac{1}{4}a_3 r_{SSS} S_x^3 - \frac{3i}{8}a_3 r_{SS}\varphi_S S_x^3 + \frac{3i}{4}a_3\sigma F^{2\gamma} r^2 q\varphi_S S_x$$

$$-\frac{3}{2}a_3\sigma F^{2\gamma} rq r_S S_x + \frac{i}{2}a_2 r_{SS} S_x^2 - ia_2\sigma F^{2\gamma} r^2 q \qquad (7.55)$$

$$+ir_S \left(-S_t + \lambda_1 S + ia_1 S_x - \frac{i}{2}a_2\varphi_S S_x^2 - \frac{3}{8}a_3\varphi_{SS} S_x^3 + \frac{3i}{16}a_3\varphi_S^2 S_x^3 \right)$$

$$+r \left(i\lambda_1 - i\gamma\frac{F_T}{F} + \frac{1}{4}a_2\varphi_{SS} S_x^2 - \frac{3}{16}a_3\varphi_S\varphi_{SS} S_x^3 \right)$$

$$+r\left[-2\lambda_0 S/S_x - 2ia_0 - \frac{1}{2}(\varphi_T + \varphi_S S_t) + \frac{1}{2}\lambda_1 S\varphi_S + \frac{i}{2}a_1\varphi_S S_x\right]$$

$$+r\left(-\frac{i}{8}a_2\varphi_S^2 S_x^2 + \frac{i}{32}a_3\varphi_S^3 S_x^3 - \frac{i}{8}a_3\varphi_{SSS}S_x^3\right),$$

where the arbitrary time-dependent functions $a_0(T)$, $a_1(T)$, $a_2(T)$, and $a_3(T)$ were introduced within corresponding integrations.

Accepted in the IST reduction procedure, $r = -q^*$ reduces Eqs. (7.54) and (7.55) to the same form if the conditions

$$a_0 = -a_0^*, \quad a_1 = -a_1^*, \quad a_2 = -a_2^*, \quad a_3 = -a_3^*, \tag{7.56}$$

$$\lambda_0 = \lambda_0^*, \quad \lambda_1 = \lambda_1^*, \quad F = F^* \tag{7.57}$$

are fulfilled.

Notice that the nonlinear evolution equations that arise in the approach of a variable spectral parameter contain, as a rule, an explicit dependence on the coordinates. Our general nonisospectral approach makes it possible to construct not only the well-known equations, but also a number of new integrable equations (nonautonomous Hirota, KdV, modified KdV, NLSE, and so on) by extending the Zakharov–Shabat (ZS) and AKNS formalism.

7.4 NONAUTONOMOUS NONLINEAR EVOLUTION EQUATIONS

7.4.1 NONAUTONOMOUS HIROTA EQUATION

In accordance with conditions (7.57), the imaginary functions $a_0(T)$, $a_1(T)$, $a_2(T)$ and $a_3(T)$ can be defined in the following way:

$$a_0(T) = i\gamma_0(T), \quad a_1(T) = iV(T), \quad a_2(T) = -iD_2(T), \tag{7.58}$$

$$a_3(T) = -4iD_3(T) \tag{7.59}$$

where $D_3(T), D_2(T), V(T)$ and $\gamma_0(T)$ are arbitrary real functions.

Then Eqs. (7.54) and (7.55) read

$$iq_T = -iD_3 q_{SSS} S_x^3 - 6i\sigma F^{2\gamma} D_3 |q|^2 q_S S_x + \frac{3}{2} q_S D_3 \varphi_{SS} S_x^3$$

$$+\frac{1}{2}(3D_3\varphi_S S_x - D_2) q_{SS} S_x^2 + \sigma F^{2\gamma}(3D_3\varphi_S S_x - D_2)|q|^2 q$$

$$-iq_S \widetilde{V} + iq\Gamma + Uq, \tag{7.60}$$

where

$$\widetilde{V} = -\frac{3}{4}D_3\varphi_S^2 S_x^3 + \frac{1}{2}D_2\varphi_S S_x^2 + S_t - \lambda_1 S + V S_x, \tag{7.61}$$

$$\Gamma = \lambda_1 - \gamma\frac{F_T}{F} + \frac{1}{4}(3D_3\varphi_S S_x - D_2)\varphi_{SS}S_x^2, \tag{7.62}$$

$$U = 2\lambda_0 S/S_x - 2\gamma_0 + \frac{1}{2}\left(\varphi_T + \varphi_S S_t\right) - \frac{1}{2}\lambda_1 S\varphi_S + \frac{1}{2}V\varphi_S S_x$$

$$+\frac{1}{2}D_3\varphi_{SSS}S_x^3 - \frac{1}{8}\left(D_3\varphi_S S_x - D_2\right)\varphi_S^2 S_x^2. \tag{7.63}$$

Equation (7.60) can be written in the independent variables (x, t), taking into account that $q_T = q_t - q_S S_t$ and $q_S = q_x/S_x$.

$$iq_t = -iD_3 q_{xxx} - 6i\sigma F^{2\gamma} D_3 |q|^2 q_x + \frac{3}{2}D_3\varphi_{xx}q_x$$

$$+\frac{1}{2}\left(3D_3\varphi_x - D_2\right)q_{xx} + \sigma F^{2\gamma}\left(3D_3\varphi_x - D_2\right)|q|^2 q \tag{7.64}$$

$$-iVq_x + i\Gamma q + Uq,$$

where

$$V = -\frac{3}{4}D_3\varphi_S^2 S_x^2 + \frac{1}{2}D_2\varphi_S S_x - \lambda_1 S/S_x + V. \tag{7.65}$$

Now we transform Eq. (7.64) into a more convenient form which does not contain terms with Q_x, and assuming $V = 0$, $\varphi_{SS} = 0$. Applying commonly accepted in the IST method [27] reduction $V = -ia_1 = 0$, we get

$$\lambda_1 S = \left(-\frac{3}{4}D_3\varphi_S S_x + \frac{1}{2}D_2\right)\varphi_S S_x^2.$$

In the case $S(x, t) = P(t)x$, $\varphi(x, t) = C(t)x$, and $\lambda_1 = 0$, the real potential $U(x, t)$ and the gain (or absorption) coefficient $\Gamma(t)$ become

$$U(x, t) = 2\lambda_0 S/S_x - 2\gamma_0 + \frac{1}{2}\left(\varphi_T + \varphi_S S_t\right) - \frac{1}{8}\left(D_3\varphi_x - D_2\right)\varphi_S^2 S_x^2$$

$$= \left[2\lambda_0(t) + \frac{1}{2}C_t(t)\right]x - \frac{1}{8}\left(D_3 C - D_2\right)C^2 - 2\gamma_0(t) \tag{7.66}$$

$$\Gamma(t) = -\gamma\frac{F_t}{F} = -\frac{1}{2}\frac{(R_3/D_3)_t}{R_3/D_3} = \frac{1}{2}\frac{W(R_3, D_3)}{R_3 D_3}, \tag{7.67}$$

where the Wronskian $W(R_3, D_3) = R_3 D_{3t} - D_3 R_{3t}$ of the functions $R_3(t)$ and $D_3(t)$ is introduced.

The resulting equation

$$iq_t = -iD_3(t)q_{xxx} - 6i\sigma R_3(t)|q|^2 q_x$$

$$-\frac{1}{2}\widetilde{D}_2(t)q_{xx} - \sigma\widetilde{R}_2(t)|q|^2 q + i\Gamma(t)q + U(x, t)q \tag{7.68}$$

represents the nonautonomous Hirota equation with time-dependent nonlinearity, dispersion, gain or absorption, and linear x-dependent potential, and can be considered a nonisospectral generalization of the famous Hirota equation [84]. In Eqs. (7.66)–(7.68), we introduced notations $R_3(T) = F^{2\gamma}D_3(T)$,

$\widetilde{D}_2(t) = -(3D_3C - D_2)$, $\widetilde{R}_2(t) = -F^{2\gamma}(3D_3C - D_2)$. The D_3, R_3, $\widetilde{D}_2$ and $\widetilde{R}_2$ coefficients are represented by positively defined functions and are connected by the condition

$$\frac{R_3}{D_3} = \frac{\widetilde{R}_2}{\widetilde{D}_2} \tag{7.69}$$

(for $\sigma = -1$, γ is assumed to be a semi-entire number).

We note that external potential (7.66)

$$U(x,t) = \left[2\lambda_0(t) + \frac{1}{2}C_t(t)\right]x + \frac{1}{8}\widetilde{D}_2C^2 + \frac{1}{4}D_3C^3 - 2\gamma_0(t) \tag{7.70}$$

is determined by coefficients D_3 and $\widetilde{D}_2$, phase shift $\varphi(x,t) = C(t)x$ and the functions $\lambda_0(t)$, $\gamma_0(t)$.

The autonomous model (7.68) ($R_3, D_3, \widetilde{R}_2, \widetilde{D}_2 = const$) introduced by Kodama and Hasegawa [85, 86] is one of the most intensively studied models of nonlinear femtosecond optics.

The isospectral version of Eq. (7.68) with a constant spectral parameter Λ, $U(x,t) = 0$ and $C(t) = 0$

$$iq_t + iD_3(t)q_{xxx} + 6iR_3(t)|q|^2 q_x + \frac{1}{2}D_2(t)q_{xx} + R_2(t)|q|^2 q - i\Gamma(t)q$$

$$\equiv iq_t + iD_3(t)q_{xxx} + 6iR_3(t)\left[\frac{\partial(|q|^2 q)}{\partial x} - q\frac{\partial(|q|^2)}{\partial x}\right]$$

$$+\frac{1}{2}D_2(t)q_{xx} + R_2(t)|q|^2 q - i\Gamma(t)q = 0 \tag{7.71}$$

was investigated by Porsezian et al. [87] in the context of femtosecond soliton management. Porsezian et al. constructed the single and two soliton solutions of Eq. (7.71) by using the auto-Bäcklund transformation. The exact combined (bright and dark) solitary wave solutions of the higher-order nonautonomous NLSE (7.71) were found by Yang et al. [88] by means of the AKNS formalism. Recently, special solutions of the generalized variable-coefficient equation (7.71) were investigated by [89–91].

7.4.2 SOLITONS OF THE NONAUTONOMOUS KORTEWEG–DE VRIES EQUATION

Equations (7.54) and (7.55) under the reduction $a_1(T) = iV(T)$, $a_2(T) = 0$, $a_3(T) = -4iD_3(T)$, and $r = 1$, as well as $\varphi = 0$, become

$$iq_T = -iD_3q_{SSS}S_x^3 + 6i\sigma D_3F^{2\gamma}qq_S S_x + iq_S(-S_t + \lambda_1 S - VS_x)$$

$$+iq\left(\lambda_1 - \gamma\frac{F_t}{F}\right) + q(2\lambda_0 S/S_x + 2ia_0), \tag{7.72}$$

$$0 = i\left(\lambda_1 - \gamma\frac{F_t}{F}\right) - 2\lambda_0 S/S_x - 2ia_0. \tag{7.73}$$

Thus, to fulfill compatibility conditions for Eqs. (7.72) and (7.73), we have to assume that $\lambda_0 = 0$ and $\lambda_1 = \gamma F_t/F + 2a_0$.

We rewrite Eq. (7.72) in the independent variables (x, t) and consider the simplest option to choose a function $S(x, t) = x$:

$$q_t + D_3(t)q_{xxx} - 6\sigma R_3(t)qq_x + q_x \left[V(t) - \lambda_1(t)x\right] + \gamma_0 q = 0, \qquad (7.74)$$

where $R_3(t) = D_3(t)F^2(t)$ and $\gamma_0(t) = -4a_0(t)$:

Equation (7.74) represents the generalized KdV equation with variable coefficients, gain or loss, and nonuniformity terms. It is a straightforward calculation to show that Eq. (7.74) has a nonautonomous soliton solution

$$q(x, t) = -\frac{1}{2}\sigma\Lambda_0^2\frac{D_{03}}{R_{03}}\exp\left(-\int_0^t \gamma_0 dt'\right)\operatorname{sech}^2\xi, \qquad (7.75)$$

where

$$\xi(x, t) = 1/2\left[\Lambda(t)x + \varphi(t)\right],$$

$$\varphi(t) = -\int_0^t \left(D_3\Lambda^3 + V\Lambda\right) dt',$$

and

$$\Lambda(t) = \Lambda_0 \left[\frac{R_3(t)}{D_3(t)}\frac{D_{03}}{R_{03}}\right]^{1/2}\exp\left(-\frac{1}{2}\int_0^t \gamma_0(t')dt'\right), \qquad (7.76)$$

is a time-dependent spectral parameter with a derivative

$$d\Lambda/dt = \lambda_1(t)\,\Lambda(t) = \frac{1}{2}\left[\frac{W(D_3, R_3)}{D_3 R_3} - \gamma_0(t)\right]\Lambda(t).$$

D_{03} and R_{03} are the initial values of dispersion and nonlinearity $D_{03} = D_3(t = 0)$ and $R_{03} = R_3(t = 0)$ and W is the Wronskian of dispersion $D_3(t)$ and nonlinearity $R_3(t)$.

Equation (7.74) is a generalization of the KdV equation

$$u_t + \gamma u + (C_0 - \alpha x)\,u_x + 6uu_x + u_{xxx} = 0, \qquad (7.77)$$

describing damping solitons without radiation moving in nonuniform media with a relaxation effect studied by Hirota and Satsuma (1976). The solution of Eq. (7.77) arises as a special case of the solution (7.75) when $D_3 = R_3 = D_{03} = R_{03} = 1$, $\lambda_1(t) = -\alpha = const$, $\sigma = -1$, $\gamma_0 = \gamma$. Hirota and Satsuma were the first to study a nonisospectral problem for the KdV equation inspired by the work of Chen and Liu [36], who studied for the first time the nonisospectral problem for the NLSE in a linear inhomogeneous media. The condition $\alpha = \gamma/2$ found by Hirota and Satsuma shows that the exact soliton solution supported by the Lax pair is realized only by rigid balancing between the loss (gain) and the linear nonuniformity of the media.

Our more general solution (7.75) describes a nonautonomous KdV soliton, which exists even when $\gamma_0(t) = 0$, because the time-dependent spectral parameter $\Lambda(t)$ (7.76) is supported by the linearly independent varying in time dispersion and nonlinearity. In the limiting case of a isospectral parameter $\Lambda(t) = \Lambda_0 = 2k = const$, the solution (7.75) is transformed to the standard KdV soliton

$$q(x,t) = 2\sigma k^2 \mathrm{sech}^2 \left(kx - 4k^3 t \right).$$

Different kinds of soliton-like solutions for nonisospectral KdV-type equations have been intensively studied in the last four decades [92, 94–96, 98, 99, 101, 102, 106].

7.4.3 NONAUTONOMOUS MODIFIED KORTEWEG–DE VRIES EQUATION

Equations (7.54) and (7.55) under the reduction $r = -q^*$ and compatibility conditions

$$
\begin{aligned}
a_0 &= -a_0^*, \quad a_1 = -a_1^*, \ a_3 = -a_3^* \\
\lambda_0 &= \lambda_0^*, \quad \lambda_1 = \lambda_1^*, \quad F = F^*
\end{aligned}
$$

are reduced to the same form

$$
\begin{aligned}
iq_T &= -iD_3 q_{SSS} S_x^3 - 6i\sigma R_3 |q|^2 q_S S_x + \frac{3}{2} D_3 q_S \varphi_{SS} S_x^3 \\
&\quad + \frac{3}{2} D_3 q_{SS} \varphi_S S_x^3 + \sigma 3 R_3 |q|^2 q \varphi_S S_x - iq_S \widetilde{V} + iq\Gamma + Uq, \quad (7.78)
\end{aligned}
$$

where we used notations

$$
\begin{aligned}
a_0(T) &= i\gamma_0(T), a_1(T) = iV(T), \ a_2(T) = 0, \\
a_3(T) &= -4iD_3(T), \ R_3 = F^{2\gamma} D_3,
\end{aligned}
$$

$$\widetilde{V} = -\frac{3}{4} D_3 \varphi_S^2 S_x^3 + S_t - \lambda_1 S + V S_x,$$

$$\Gamma = \lambda_1 - \gamma \frac{F_T}{F} + \frac{3}{4} D_3 \varphi_S \varphi_{SS} S_x^3,$$

$$U = 2\lambda_0 S/S_x - 2\gamma_0 + \frac{1}{2} (\varphi_T + \varphi_S S_t) - \frac{1}{2}\lambda_1 S \varphi_S + \frac{1}{2} V \varphi_S S_x$$

$$+ \frac{1}{2} D_3 \varphi_{SSS} S_x^3 - \frac{1}{8} D_3 \varphi_S^3 S_x^3.$$

Let us study a more convenient and simple case when $\varphi_s = 0$ ($\varphi = \theta(t)$) and $S = x$. Then Eq. (7.78) gets a form of the variable-coefficient complex modified Korteweg–de Vries equation (vc-cmKdV) with a linear potential, variable gain or loss, and nonuniformity terms

$$q_t + D_3(t)q_{xxx} + 6\sigma R_3(t)\,|q|^2\,q_x = q_x\left[\lambda_1(t)x - V(t)\right]$$

$$+q\left(\lambda_1(t) - \frac{1}{2}\frac{W(R_3, D_3)}{R_3 D_3}\right) + iq\left(2\gamma_0 - \frac{1}{2}\theta_t - 2\lambda_0 x\right). \tag{7.79}$$

Equation (7.79) presents an example of a nonisospectral problem for the mKdV with time-dependent spectral parameter $\Lambda(t)$. In spite of the fact that exact solutions of modified Korteweg–de Vries equation (vc-cmKdV) equations are not very well studied, there exist in the literature attempts to find solutions of some special kinds of this equation [105–110].

7.5   GENERALIZED NLSE AND NONAUTONOMOUS SOLITONS

Let us study a special case of the reduction procedure for Eq. (7.55) when $a_3 = 0$.

$$A = -i\lambda_0 S/S_x + a_0(T) - \frac{1}{2}a_2(T)\sigma F^{2\gamma}\,|q|^2 - i\lambda_1 S/S_x\Lambda + a_1(T)\Lambda + a_2(T)\Lambda^2,$$

$$\begin{aligned}
B &= \sqrt{\sigma}F^\gamma \exp\left(i\varphi/2\right)\left\{-\frac{i}{4}a_2(T)q\varphi_S S_x - \frac{1}{2}a_2(T)q_S S_x\right\} + \\
&\quad i\left\{q\left[-i\lambda_1 S/S_x + a_1(T) + \Lambda a_2(T)\right]\right\},
\end{aligned}$$

$$\begin{aligned}
C &= \sqrt{\sigma}F^\gamma \exp\left(-i\varphi/2\right)\left\{\frac{i}{4}a_2(T)q^*\varphi_S S_x - \frac{1}{2}a_2(T)q_S^* S_x\right\} \\
&\quad -i\left\{q^*\left[-i\lambda_1 x + a_1(T) + \Lambda a_2(T)\right]\right\}. \tag{7.80}
\end{aligned}$$

Taking into account conditions (7.57), Eqs. (7.54) and (7.55) are transformed into the same equation.

$$iq_T = -\frac{1}{2}D_2 q_{SS}S_x^2 - \sigma R_2\,|q|^2\,q - i\widetilde{\mathcal{V}}q_S + i\Gamma q + Uq, \tag{7.81}$$

where

$$\widetilde{\mathcal{V}}(S,T) = \frac{1}{2}D_2 S_x^2\varphi_S + VS_x + S_t - \lambda_1 S, \tag{7.82}$$

$$U(S,T) = \frac{1}{8}D_2 S_x^2\varphi_S^2 - 2\gamma_0 + \frac{1}{2}\left(\varphi_T + \varphi_S S_t + VS_x\varphi_S\right) + 2\lambda_0 S/S_x - \frac{1}{2}\lambda_1\varphi_S S, \tag{7.83}$$

$$\Gamma = \left(-\gamma\frac{F_T}{F} - \frac{1}{4}D_2 S_x^2\varphi_{SS} + \lambda_1\right) = \left(\frac{1}{2}\frac{W(R_2, D_2)}{R_2 D_2} - \frac{1}{4}D_2 S_x^2\varphi_{SS} + \lambda_1\right). \tag{7.84}$$

Equation (7.81) can be written in more convenient form in the independent variables (x, t)

$$iq_t + \frac{1}{2}D_2 q_{xx} + \sigma R_2 |q|^2 q - Uq = i\Gamma q \qquad (7.85)$$

using the reduction $V = -ia_1 = 0$ and a condition $\widetilde{V} = 0$, which defines the parameter λ_1 from (7.82)

$$\lambda_1 = \frac{1}{2}D_2 S_x^2 \varphi_S / S. \qquad (7.86)$$

Now we rewrite a potential $U(S, T)$ from Eq. (7.83):

$$U(S, T) = -2\gamma_0(T) + 2\lambda_0(T)S/S_x + \frac{1}{2}(\varphi_T + \varphi_S S_t) - \frac{1}{8}D_2 S_x^2 \varphi_S^2 \qquad (7.87)$$

and the gain or absorption coefficient (7.84)

$$\Gamma(T) = \frac{1}{2}\frac{W(R_2, D_2)}{R_2 D_2} - \frac{1}{4}D_2 S_x^2 \varphi_{SS} + \frac{1}{2}D_2 S_x^2 \varphi_S / S. \qquad (7.88)$$

Let us consider some special choice of variables to specify the solutions of (7.81). First of all, we assume that variables are factorized in the phase profile: $\varphi(S, T) = C(T)S^\alpha$. The first term in the real potential (7.87) represents some additional time-dependent phase of the solution $q(x, t)$ of Eq. (7.85) and, without loss of generality, we use below $\gamma_0 = 0$. Now, taking into account the three last terms in (7.87), we write

$$U(S, T) = 2\lambda_0 S/S_x + \frac{1}{2}C_T S^\alpha + 1/2\alpha C S^{\alpha-1} S_t - \frac{1}{8}D_2 C^2 S_x^2 \alpha^2 S^{2\alpha-2}. \qquad (7.89)$$

The gain or absorption coefficient (7.88) becomes

$$\Gamma(T) = \frac{1}{2}\frac{W(R_2, D_2)}{R_2 D_2} + \frac{\alpha}{4}(3 - \alpha)D_2 S_x^2 C S^{\alpha-2}$$

and the parameter λ_1 is

$$\lambda_1 = \frac{1}{2}D_2 S_x^2 C\alpha S^{\alpha-2}. \qquad (7.90)$$

If we assume that the functions $\Gamma(T)$ and $\lambda_1(T)$ depend only on T and do not depend on S, we conclude that $\alpha = 0$ or $\alpha = 2$. Let us consider the simplest option to choose the variable $S(x, t)$ when the variables (x, t) are factorized: $S(x, t) = P(t)x$.

In the first case of $\alpha = 0$, Eq. (7.85)

$$iq_t + \frac{1}{2}D_2(t)q_{xx} + \sigma R_2(t)|q|^2 q - 2\lambda_0(t)xq = i\frac{1}{2}\frac{W(R_2, D_2)}{R_2 D_2}q \qquad (7.91)$$

with a spectral parameter $\Lambda(t)$ dependent on time belongs to the class of the generalized NLSE with varying coefficients and a linear spatial potential. The Lax pair for Eq. (7.91) directly follows from (7.81) and was found in [111] and [112]. The solutions of Eq. (7.91) without time and space phase modulation (chirp) are a generalization of the solutions found by Chen and Liu [36] of the NLSE with the linear spatial potential and constant λ_0, describing the Alfen wave propagation in plasma.

Now let us discuss the second case of $\alpha = 2$ and $\varphi(S,T) = C(T)S^2$, when Eq. (7.85) has solutions with chirp. Corresponding real spatial-temporal potential (7.89)

$$U(x,t) = 2\lambda_0 x + \frac{1}{2}\left(\Theta_t - D_2\Theta^2\right)x^2 \equiv 2\lambda_0(t)x + \frac{1}{2}\Omega^2(t)x^2, \qquad (7.92)$$

the phase modulation

$$\varphi(x,t) = \Theta(t)x^2, \qquad (7.93)$$

the gain (or absorption) coefficient

$$\Gamma(t) = \frac{1}{2}\left(\frac{W(R_2,D_2)}{R_2 D_2} + D_2 P^2 C\right) = \frac{1}{2}\left(\frac{W(R_2,D_2)}{R_2 D_2} + D_2\Theta\right) \qquad (7.94)$$

and the derivative of the spectral parameter λ_1

$$\lambda_1(t) = D_2 P^2 C = D_2(t)\Theta(t) \qquad (7.95)$$

are found to be dependent on the self-induced soliton phase shift $\Theta(t)$. Notice that the definition

$$\Omega^2(t) \equiv \Theta_t - D_2\Theta^2 \qquad (7.96)$$

was introduced in Eq. (7.92).

Now we can rewrite the generalized NLSE (7.85) with time-dependent nonlinearity, dispersion and gain or absorption in the form of the nonautonomous NLSE with linear and parabolic potentials

$$iq_t + \frac{1}{2}D_2(t)q_{xx} + \sigma R_2(t)\left|q\right|^2 q - 2\lambda_0(t)x - \frac{1}{2}\Omega^2(t)x^2 q = i\Gamma(t)q. \qquad (7.97)$$

Historically, the study of soliton propagation through density gradients began with the pioneering work of Tappert and Zabusky [113]. Chen and Liu [36] substantially extended the concept of classical solitons to the accelerated motion of a soliton in a linearly inhomogeneous plasma. The most interesting aspects of nonisospectral IST problems and dynamics of the NLSE solitons in nonuniform media have been investigated in [114–123]. Experimental observation of nonautonomous solitons was reported for the first time by Murphy [124] and then by [125].

7.6   SOLITON ADAPTATION LAW TO EXTERNAL POTENTIALS

As we have seen, solitary waves in nonautonomous nonlinear and dispersive systems can propagate in the form of nonautonomous solitons (see [22, 87, 103, 127–140, 142–146, 148–151, 153–158, 160, 162–167] and references therein).

Nonautonomous solitons interact elastically and generally move with varying amplitudes, speeds and spectra adapted both to the external potentials and to the dispersion and nonlinearity variations. In general, soliton solutions for nonautonomous nonlinear equations with varying dispersion, nonlinearity, gain or absorption, and, in general, subjected to varying in time external potentials, exist only under certain conditions, and the varying functions satisfy the soliton adaptation laws both to the external potentials and to the dispersion and nonlinearity changes [22, 126, 133]. From the physical point of view, the adaptation means that solitons conserve their functional form and do not emit dispersive waves during their interactions with external potentials and with each other.

The soliton adaptation laws are known today as the Serkin–Hasegawa theorems (SH theorems) [186–188]. Serkin and Hasegawa obtained their SH theorems by using symmetry reduction methods when the initial nonautonomous NLSE can be transformed to the canonical autonomous NLSE under specific conditions found in [22, 126].

It is straightforward to verify, substituting the phase profile $\Theta(t)$ from Eq. (7.96) into Eq. (7.94), that the frequency of the harmonic potential $\Omega(t)$ is related to the dispersion $D_2(t)$, nonlinearity $R_2(t)$ and gain or absorption coefficient $\Gamma(t)$ by the following conditions:

$$
\begin{aligned}
\Omega^2(t)D_2(t) &= D_2(t)\frac{d}{dt}\left(\frac{\Gamma(t)}{D_2(t)}\right) - \Gamma^2(t) \\
&\quad -\frac{d}{dt}\left(\frac{W(R_2,D_2)}{R_2 D_2}\right) + \left(2\Gamma(t) + \frac{d}{dt}\ln R_2(t)\right)\frac{W(R_2,D_2)}{R_2 D_2} \\
&= D_2(t)\frac{d}{dt}\left(\frac{\Gamma(t)}{D_2(t)}\right) - \Gamma^2(t) + \left(2\Gamma(t) + \frac{d}{dt}\ln R_2(t)\right)\frac{d}{dt}\ln\frac{D_2(t)}{R_2(t)} - \frac{d^2}{dt^2}\ln\frac{D_2(t)}{R_2(t)}.
\end{aligned}
$$

After the substitutions

$$
q(x,t) = Q(x,t)\exp\left[\int_0^t \Gamma(\tau)d\tau\right], \quad R(t) = R_2(t)\exp\left[2\int_0^t \Gamma(\tau)d\tau\right],
$$

$$
D(t) = D_2(t), \quad (7.98)
$$

Eq. (7.97) is transformed into the generalized NLSE without gain or loss terms:

$$
i\frac{\partial Q}{\partial t} + \frac{1}{2}D(t)\frac{\partial^2 Q}{\partial x^2} + \left[\sigma R(t)\,|Q|^2 - 2\lambda_0(t)x - \frac{1}{2}\Omega^2(t)x^2\right]Q = 0. \quad (7.99)
$$

Finally, the Lax equation (7.52) with matrices (7.53) and (7.54) provides the nonautonomous model (7.99) under the condition that the dispersion $D(t)$, nonlinearity $R(t)$, and harmonic potential satisfy the exact integrability conditions

$$\Omega^2(t)D(t) = \frac{W(R,D)}{RD}\frac{d}{dt}\ln R(t) - \frac{d}{dt}\left(\frac{W(R,D)}{RD}\right) \tag{7.100}$$

$$= \frac{d}{dt}\ln D(t)\frac{d}{dt}\ln R(t) - \frac{d^2}{dt^2}\ln D(t) - R(t)\frac{d^2}{dt^2}\frac{1}{R(t)}. \tag{7.101}$$

The self-induced soliton phase shift is given by

$$\Theta(t) = -\frac{W\left[(R(t),D(t)\right]}{D^2(t)R(t)} \tag{7.102}$$

and the time-dependent spectral parameter is represented by

$$\Lambda(t) = \kappa(t) + i\eta(t) = \frac{D_0 R(t)}{R_0 D(t)}\left[\Lambda(0) + \frac{R_0}{D_0}\int_0^t \frac{\lambda_0(\tau)D(\tau)}{R(\tau)}d\tau\right], \tag{7.103}$$

where the main parameters, time-invariant eigenvalue $\Lambda(0) = \kappa_0 + i\eta_0$, $D_0 = D(t=0)$, $R_0 = R(t=0)$, are defined by the initial conditions.

Thus Eq. (7.99) represents the law of soliton adaptation to external potentials, which comprises the essence of the Serkin–Hasegawa theorem. The basic property of classical solitons to interact elastically holds true, but the novel features of nonautonomous solitons arise. Namely, both amplitudes and speeds of the solitons, and consequently, their spectra, during propagation and after the interaction are no longer the same as those prior to the interaction. All nonautonomous solitons generally move with varying amplitudes $\eta(t)$ and speeds $\kappa(t)$ adapted both to the external potentials and to the dispersion $D(t)$ and nonlinearity $R(t)$ variations.

7.7 BRIGHT AND DARK NLSE NONAUTONOMOUS SOLITONS

Having obtained the elements of the Lax pair matrices (7.52)–(7.54), we can apply the auto-Bäcklund transformation [189] and the recurrent relation [190]

$$Q_n(x,t) = -Q_{n-1}(x,t) - \frac{4\eta_n \widetilde{\Gamma}_{n-1}(x,t)}{1 + \left|\widetilde{\Gamma}_{n-1}(x,t)\right|^2} \times \sqrt{\frac{D(t)}{R(t)}}\exp[-i\Theta x^2/2], \tag{7.104}$$

which connects the $(n-1)$ and n soliton solutions by means of the so-called pseudo-potential $\widetilde{\Gamma}_{n-1}(x,t) = \psi_1(x,t)/\psi_2(x,t)$ for the $(n-1)$ soliton scattering functions $\Psi(x,t) = (\psi_1\psi_2)^T$ and write the general solutions of Eq. (6.3) for bright $(\sigma = +1)$ and dark $(\sigma = -1)$ nonautonomous solitons:

$$Q_1^+(x,t\mid\sigma=+1) = 2\eta_1(t)\sqrt{\frac{D(t)}{R(t)}}\mathrm{sech}[\xi_1(x,t)]\times\exp\left\{-i\left(\frac{\Theta(t)}{2}x^2+\chi_1(x,t)\right)\right\};$$
$$\tag{7.105}$$

$$\xi_1(x,t) = 2\eta_1(t)x + 4\int_0^t D(\tau)\eta_1(\tau)\kappa_1(\tau)d\tau, \tag{7.106}$$

$$\chi_1(x,t) = 2\kappa_1(t)x + 2\int_0^t D(\tau)\left[\kappa_1^2(\tau) - \eta_1^2(\tau)\right]d\tau; \tag{7.107}$$

$$Q_1^-(x,t \mid \sigma = -1) = 2\eta_1(t)\sqrt{\frac{D(t)}{R(t)}}\left[\sqrt{(1-a^2)} + ia\tanh\zeta(x,t)\right]$$
$$\times \exp\left\{-i\left(\frac{\Theta(t)}{2}x^2 + \phi(x,t)\right)\right\}, \tag{7.108}$$

$$\zeta(x,t) = 2a\eta_1(t)x + 4a\int_0^t D(\tau)\eta_1(\tau)\kappa_1(\tau)d\tau, \tag{7.109}$$

$$\phi(x,t) = 2\left[\kappa_1(t) - \eta_1(t)\sqrt{(1-a^2)}\right]x$$
$$+ 2\int_0^t D(\tau)\left[\kappa_1^2 + \eta_1^2\left(3-a^2\right) - 2\kappa_1\eta_1\sqrt{(1-a^2)}\right]d\tau. \tag{7.110}$$

Equation (7.108) for dark (gray) solitons contains an additional parameter, $0 \le a \le 1$, which designates the depth of modulation (the blackness of the gray soliton) and its velocity against the background. When $a = 1$, the depth of modulation of the gray soliton reaches the maximum and it becomes black. For optical applications, Eq. (7.108) can be easily transformed into the Hasegawa and Tappert form for nonautonomous dark solitons [9] under the condition $\kappa_0 = \eta_0\sqrt{(1-a^2)}$, which corresponds to the special choice of the retarded frame associated with the group velocity of the soliton.

$$Q_1^-(x,t \mid \sigma = -1) = 2\eta_1(t)\sqrt{\frac{D(t)}{R(t)}}\left[\sqrt{(1-a^2)} + ia\tanh\varpi(x,t)\right]$$
$$\times \exp\left\{-i\left(\frac{\Theta(t)}{2}x^2 + \vartheta(x,t)\right)\right\}, \tag{7.111}$$

$$\varpi(x,t) = 2a\eta_1(t)x + 4a\int_0^t D(\tau)\eta_1(\tau)\left[\eta_1(\tau)\sqrt{(1-a^2)} + \mathcal{K}(\tau)\right]d\tau,$$

$$\vartheta(x,t) = 2\mathcal{K}(t)x + 2\int_0^t D(\tau)\left[\mathcal{K}^2(\tau) + 2\eta_1^2(\tau)\right]d\tau, \tag{7.112}$$

$$\mathcal{K}(t) = \frac{R(t)}{D(t)} \int_0^t \lambda_0(\tau) \frac{D(\tau)}{R(\tau)} d\tau. \tag{7.113}$$

The exact solutions (7.105)–(7.113) are realized only when the nonlinearity, dispersion and confining harmonic potential are connected by Eq. (7.101), and both $D(t) \neq 0$ and $R(t) \neq 0$ for all times by definition.

A two-soliton $Q_2(x,t)$ solution for $\sigma = +1$ is generated by the recurrent expression Eq. (7.1) using the one-soliton solution (7.105):

$$Q_2(x,t) = 4\sqrt{\frac{D(t)}{R(t)} \frac{N\,(x,t)}{D\,(x,t)}} \exp\left[-\frac{i}{2}\Theta(t)x^2\right], \tag{7.114}$$

where the numerator N (x,t) is given by

$$
\begin{aligned}
N \;=\; & \cosh \xi_2 \exp\left(-i\chi_1\right) \\
& \times \left[(\kappa_2 - \kappa_1)^2 + 2i\eta_2(\kappa_2 - \kappa_1)\tanh \xi_2 + \eta_1^2 - \eta_2^2\right] + \eta_2 \cosh \xi_1 \exp\left(-i\chi_2\right) \\
& \times \left[(\kappa_2 - \kappa_1)^2 - 2i\eta_1(\kappa_2 - \kappa_1)\tanh \xi_1 - \eta_1^2 + \eta_2^2\right], \tag{7.115}
\end{aligned}
$$

and the denominator D (x,t) is represented by

$$
\begin{aligned}
D \;=\; & \cosh(\xi_1 + \xi_2)\left[(\kappa_2 - \kappa_1)^2 + (\eta_2 - \eta_1)^2\right] \\
& + \cosh(\xi_1 - \xi_2)\left[(\kappa_2 - \kappa_1)^2 + (\eta_2 + \eta_1)^2\right] - 4\eta_1\eta_2 \cos\left(\chi_2 - \chi_1\right), \tag{7.116}
\end{aligned}
$$

which is dependent on the arguments

$$\xi_i(x,t) = 2\eta_i(t)x + 4\int_0^t D(\tau)\eta_i(\tau)\kappa_i(\tau)d\tau, \tag{7.117}$$

and phase shifts

$$\chi_i(x,t) = 2\kappa_i(t)x + 2\int_0^t D(\tau)\left[\kappa_i^2(\tau) - \eta_i^2(\tau)\right]d\tau \tag{7.118}$$

related to the amplitudes

$$\eta_i(t) = \frac{D_0 R(t)}{R_0 D(t)}\eta_{0i} \tag{7.119}$$

and velocities

$$\kappa_i(t) = \frac{D_0 R(t)}{R_0 D(t)}\left[\kappa_{0i} + \frac{R_0}{D_0}\int_0^t \frac{\lambda_0(\tau)D(\tau)}{R(\tau)}d\tau\right] \tag{7.120}$$

of the nonautonomous solitons, where κ_{0i} and η_{0i} correspond to the initial velocity and amplitude of the i-th soliton ($i = 1, 2$).

Equations (7.114)–(7.120) describe the temporal and spatial dynamics of two bound solitons in nonautonomous systems with external linear and oscillator potentials. Obviously, these soliton solutions reduce to classical soliton solutions in the limit of autonomous nonlinear and dispersive systems given by conditions $R(t) = D(t) = 1$, and $\lambda_0(t) = \Omega(t) \equiv 0$ for canonical NLSE without external potentials.

7.8   COLORED NONAUTONOMOUS SOLITONS

The generalized nonlinear Schrödinger equations (7.97) and (7.99) and their solutions can be rewritten in the form accepted in nonlinear optics by the substitution $x \to \tau$ (or $x \to X$); $t \to z$ and $Q^+(x,t) \to \varphi^+ [z, \tau(or\ X)]$ for bright solitons, and $[Q^-(x,t)]^* \to \varphi^- [z, \tau(or X)]$ for dark solitons, where the asterisk denotes the complex conjugate, z is the normalized distance, and τ is the retarded time for temporal solitons, while X is the transverse coordinate for spatial solitons,

$$i\varphi_z + \frac{1}{2}D_2(z)\varphi_{\tau\tau} + \sigma R_2(z) |\varphi|^2 \varphi - 2\lambda_0(z)\tau - \frac{1}{2}\Omega^2(z)\tau^2\varphi = i\Gamma(z)\varphi. \quad (7.121)$$

An important special case of Eq. (7.121) arises under the conditions $\Omega^2(z) = 0$ (we can get $\Gamma(z) = 0$ without loss of the generality, taking into account Eq. (7.98)):

$$i\frac{\partial\psi}{\partial z} + \frac{\sigma}{2}D(z)\frac{\partial^2\psi}{\partial\tau^2} + R(z) |\psi|^2 \psi - 2\sigma\lambda_0(z)\tau\psi = 0. \quad (7.122)$$

The generalized NLSE, (7.122), possesses the exact soliton solutions with time and space phase modulation (chirp) and corresponds to the case of $\alpha = 2$ in Section 7.5, Eq. (7.97), with a spectral parameter (7.103).

The self-induced soliton phase shift $\Theta(z)$, dispersion $D(z)$, and nonlinearity $R(z)$ are related by the law of soliton adaptation to external linear potential:

$$D(z)/D_0 = R(z)/R_0 \exp\left\{-\frac{\Theta_0 D_0}{R_0}\int_0^z R(z')dz'\right\}. \quad (7.123)$$

The soliton solutions of Eq. (7.122), nonautonomous solitons with nontrivial self-induced phase shifts (7.102) and varying amplitudes, speeds and spectra, are given in quadratures by Eqs. (7.105) and (7.120) under the condition $\Omega^2(z) = 0$.

The nonautonomous exactly integrable NLSE model given by Eqs. (7.122, 7.123) is a generalization of the Chen and Liu model [36] with linear potential $\lambda_0(z) \equiv \alpha_0 = const$ and $D(z) = D_0 = R(z) = R_0 = 1$, $\sigma = +1$, $\Theta_0 = 0$. The accelerated solitons predicted by Chen and Liu in plasma were discovered in nonlinear fiber optics only a decade later as the self-induced Raman effect (also called the soliton self-frequency shift) [191,192]. In 1987, the concept of colored solitons related to chirped optical solitons with moving spectra was introduced by Serkin [193]. Now this concept can be extended to nonautonomous systems.

Let us show that Raman colored optical solitons can be approximated by Eq. (7.122). The self-induced Raman effect is described by an additional term, $-\sigma_R \psi \partial \left(| \psi |^2 \right) / \partial \tau$, in the NLSE, where σ_R originates from the frequency-dependent Raman gain [7,86,194]. Assuming that soliton amplitude does not significantly vary during self-scattering, we can estimate this term using the exact solution $| \psi |^2 = \eta^2 \mathrm{sech}^2(\eta\tau)$,

$$\sigma_R \frac{\partial | \psi |^2}{\partial \tau} \approx -2\sigma_R \eta^4 \tau = 2\alpha_0 \tau,$$

and calculate an acceleration of the soliton $dV/dz = \sigma_R \eta^4 / 2$, where $V = \kappa/2$. The result of soliton perturbation theory [7,8,13] gives $dV/dz = 8\sigma_R \eta^4 / 15$. This fact explains the remarkable stability of colored Raman solitons that is guaranteed by the property of the exact integrability of the Chen and Liu model.

A more general model (7.121) and (7.123) and its exact soliton solutions open the possibility of designing an effective soliton compressor, for example, by drawing a fiber with $R(z) = 1$ and $D(z) = \exp(-c_0 z)$, where $c_0 = \Theta_0 D_0$. It seems very attractive to use the results of the nonautonomous soliton concept in ultrashort photonic applications and soliton laser design.

As we have mentioned before, another interesting feature of novel solitons, which we call colored nonautonomous solitons, is associated with the nontrivial dynamics of their spectra. The frequency spectrum of the chirped nonautonomous optical soliton moves in the frequency domain. In particular, if dispersion and nonlinearity evolve in unison $D(z) = R(z)$ or $D = R = 1$, the solitons propagate with identical spectra, but with totally different time-space behavior [133].

Let us study Eq. (7.122) with linear potential that, for simplicity, does not depend on time: $\lambda_0(z) = \alpha_0 = const$. We assume constant nonlinearity $R = R_0$, but dispersion $D(z)$, which varies exponentially along the propagation distance

$$D(z) = D_0 \exp\left(-c_0 z\right), \qquad \Theta(z) = \Theta_0 \exp\left(c_0 z\right). \qquad (7.124)$$

The one and two bright soliton solutions read

$$\psi_1(z,\tau) = 2\eta_{01}\sqrt{D_0 \exp\left(c_0 z\right)}\mathrm{sech}\left[\xi_1(z,\tau)\right]$$

$$\times \exp\left[-\frac{i}{2}\Theta_0 \exp\left(c_0 z\right)\tau^2 - i\chi_1(z,\tau)\right], \qquad (7.125)$$

$$\psi_2(z,\tau) = 4\sqrt{D_0 \exp\left(-c_0 z\right)}\frac{N(z,\tau)}{D(z,\tau)}\exp\left[-\frac{i}{2}\Theta_0 \exp\left(c_0 z\right)T^2\right], \qquad (7.126)$$

where the nominator $N(z, \tau)$ and denominator $D(z, \tau)$ are given by Eqs. (7.115) and (7.116) and

$$\xi_i(z, \tau) = 2\eta_{0i}\tau \exp(c_0 z) + 4D_0 \eta_{0i} \tag{7.127}$$

$$\times \left\{ \frac{\kappa_{0i}}{c_0} [\exp(c_0 z) - 1] + \frac{\alpha_0}{c_0} \left[\frac{\exp(c_0 z) - 1}{c_0} - z \right] \right\},$$

$$\chi_i(z, \tau) = 2\kappa_{0i}\tau \exp(c_0 z) + 2D_0 \left(\kappa_{0i}^2 - \eta_{0i}^2 \right) \frac{\exp(2c_0 z) - 1}{2c_0}$$

$$+ 2\tau \frac{\alpha_0}{c_0} [\exp(c_0 z) - 1] + 4D_0 \kappa_{0i} \frac{\alpha_0}{c_0} \left[\frac{\exp(c_0 z) - 1}{c_0} - z \right]$$

$$+ 2D_0 \left(\frac{\alpha_0}{c_0} \right)^2 \left[\frac{\exp(c_0 z) - \exp(-c_0 z)}{c_0} - 2z \right]. \tag{7.128}$$

The initial velocity and amplitude of the i-th soliton ($i = 1, 2$) are denoted by κ_{0i} and η_{0i}.

Nonautonomous colored solitons described by Eqs. (7.125)–(7.128) not only are accelerated and reflected from the linear potential, but also are compressed due to amplitude amplification. The same dynamics is observed for dark nonautonomous colored solitons.

The Chen and Liu model [36] appears as a limit case of Eqs.(7.125)–(7.128) when $c_0 \to \infty$ (that means $D(Z) = D_0 = $ constant). The solitons with arguments and phase shifts

$$\xi(z, \tau) = 2\eta_0 \left(\tau + 2\kappa_0 z + \alpha_0 z^2 - \tau_0 \right),$$

$$\chi(z, \tau) = 2\kappa_0 \tau + 2\alpha_0 \tau z + 2 \left(\kappa_0^2 - \eta_0^2 \right) z + 2\kappa_0 \alpha_0 z^2 + \frac{2}{3} \alpha_0^2 z^3$$

represent particle-like solutions, which may be accelerated and reflected from the linear potential.

Now let us show that the law of soliton adaptation to external potentials (7.101) allows us to stabilize the soliton even without a trapping potential. Indeed, let us rewrite Eqs. (7.125–7.128) defining $\lambda_0(z) = \alpha_0 = 0$. To make formulae clearer, we choose all eigenvalues to be purely imaginary, which means that the initial velocities of the solitons are equal to zero. The one- and two-soliton solutions become

$$\psi_1(z, \tau) = 2\eta_{01} \sqrt{D_0 \exp(c_0 z)} \, \text{sech} \left[2\eta_{01}\tau \exp(c_0 z) \right] \tag{7.129}$$

$$\times \exp \left[-\frac{i}{2}\Theta_0 \exp(c_0 z) \tau^2 + i2D_0 \eta_{01}^2 \frac{\exp(2c_0 z) - 1}{2c_0} \right],$$

$$\psi_2(z, \tau) = 4\sqrt{D_0 \exp(-c_0 z)} \frac{N(z, \tau)}{D(z, \tau)} \exp \left[-\frac{i}{2}\Theta_0 \exp(c_0 z) \tau^2 \right], \tag{7.130}$$

where

$$
\begin{aligned}
N &= \left(\eta_{01}^2 - \eta_{02}^2\right) \exp\left(c_0 z\right) \left[\eta_{01} \cosh \xi_2 \exp\left(-i\chi_1\right)\right. \\
&\quad \left. - \eta_{02} \cosh \xi_1 \exp\left(-i\chi_2\right)\right], \quad (7.131) \\
D &= \cosh(\xi_1 + \xi_2)\left(\eta_{01} - \eta_{02}\right)^2 + \cosh(\xi_1 - \xi_2)\left(\eta_{01} + \eta_{02}\right)^2 \\
&\quad - 4\eta_{01}\eta_{02}\cos\left(\chi_2 - \chi_1\right), \quad (7.132)
\end{aligned}
$$

and

$$
\xi_i(z, \tau) = 2\eta_{0i}\tau \exp\left(c_0 z\right), \quad (7.133)
$$

$$
\chi_i(z, \tau) = -2D_0\eta_{0i}^2 \frac{\exp\left(2c_0 z\right) - 1}{2c_0} + \chi_{i0}. \quad (7.134)
$$

This result indicates the possibility for the optimal compression of chirped colored nonautonomous solitons.

For the particular case of $\eta_{10} = 1/2$, $\eta_{20} = 3/2$, Eqs. (7.130)–(7.134) are transformed to a nonautonomous analog of the Satsuma–Yajima breather [195], a so-called "agitated breather."

$$
\begin{aligned}
\psi_2(z, \tau) &= 4\sqrt{D_0} \exp\left(-c_0 z\right) \exp\left[-\frac{i}{2}\Theta_0 \exp\left(c_0 z\right)\tau^2\right] \quad (7.135) \\
&\quad \times \exp\left[\frac{i}{4c_0}D_0\left[\exp\left(2c_0 z\right) - 1\right] + \chi_{10}\right] \\
&\quad \times \frac{\cosh 3X - 3\cosh X \exp\left\{i2D_0\left[\exp\left(2c_0 z\right) - 1\right]/c_0 + i\Delta\varphi\right\}}{\cosh 4X + 4\cosh 2X - 3\cos\left\{2D_0\left[\exp\left(2c_0 z\right) - 1\right]/c_0 + \Delta\varphi\right\}},
\end{aligned}
$$

where $X = \tau\exp(c_0 z)$, $\Delta\varphi = \chi_{20} - \chi_{10}$.

The solution (7.136) is reduced to the Satsuma–Yajima breather in the $c_0 = 0$ limit and $D(z) = D_0 = 1$.

$$
\psi_2(z, \tau) = 4\frac{\cosh 3\tau + 3\cosh\tau \exp\left(4iz\right)}{\cosh 4\tau + 4\cosh 2\tau + 3\cos 4z} \exp\left(\frac{iz}{2}\right).
$$

At $z = 0$ it takes the simple form $\psi_2(z = 0, \tau) = 2sech(\tau)$. An interesting property of this solution is that its form oscillates with the soliton period $T_{sol} = \pi/2$.

In the more general case of varying dispersion, $D(z) = D_0 \exp\left(-c_0 z\right)$, used in Eq. (7.135), the soliton period depends on time. The Satsuma–Yajima breather solution can be obtained from the general solution if and only if the soliton phase shift is chosen properly, precisely when $\Delta\varphi = \pi$. Then the intensity profiles $|\psi_2(z, \tau)|^2$ of the wave build up a complex landscape of peaks and valleys and reach their peaks at the points of the maximum. In nonautonomous systems, decreasing group velocity dispersion (8.4) (or increasing nonlinearity) stimulates the nonautonomous analog of the Satsuma–Yajima breather to accelerate its period of "breathing" and to increase its peak amplitudes of "breathing"; that is why we call this effect "agitated breather."

7.9   CONCLUSION

In this chapter, we have tried to demonstrate the concept of nonautonomous solitons in its essence and simplicity based on the "hidden" symmetries of the famous quantum mechanical equations and the generalized Lax pair operator method. This approach opens the possibility to study in detail the nonlinear dynamics of solitons in nonautonomous nonlinear and dispersive physical systems. All the cases considered have similar qualitative features for nonautonomous solitons: their properties and dynamics have been completely determined by the "hidden" symmetry parameters. Symmetry reduction represents today one of the most powerful methods of finding previously unknown soliton-like solutions. However, we need to emphasize that these methods do not provide the complete integrability of the models under discussion. Really, as we have already demonstrated in the representative example for the two–three-dimensional NLSE model (see Eq. (7.48), under the condition $s^{2-d} = 1$, there exists exact HT&T symmetry for the two-dimensional case, if and only if $d = 2$, but, as is well known, in two dimensions the NLSE model is not completely integrable by the IST method. What is more, its soliton-like spatial solutions (beams) are not stable under modulation instability, collapse, and filament formation. However, the conditional "hidden-parameter" symmetry does exist! And what is more, this kind of symmetry allows us to obtain different solutions, for example, analogical to coherent and squeezed states. That is why we need to conclude that Eq. (7.48) clearly demonstrates that to find completely integrable models we need to reduce the dimensionality of the nonautonomous models from $d = 2, 3$ to $d = 1$, and, on this basis, we can construct completely integrable nonautonomous models.

The problem of integrability of nonlinear dynamic systems, primarily of Hamiltonian systems with an infinite number of degrees of freedom, is one of the most important in modern science [196]. How can we determine whether a given nonlinear evolution equation is integrable or not? The ingenious method to answer this question was discovered by Gardner et al. [28]. Following this work, Lax [29] formulated a general principle for associating nonlinear evolution equations with linear operators so that the eigenvalues of the linear operator are integrals of the nonlinear equation. Ablowitz et al. [27] have found that many nonlinear evolution equations can be solved by IST. Thus, the IST method provides a general research strategy to understand the integrability of the known equations, to test the integrability of new ones, and to discover in this manner novel completely integrable nonlinear and nonautonomous models. The way thereby opened for the search and discovery of nonlinear evolution equations solvable by IST techniques, a problem that is currently under intense study.

We would like to conclude by saying that the concept of adaptation is of primary importance in nature, and nonautonomous solitons that interact elastically and generally move with varying amplitudes, speeds, and spectra adapted both to the external potentials and to the dispersion and nonlinearity changes can be fundamental objects of nonlinear science [42,60,81,97,104,146, 151,161,166,168,169,171,174–179,179,180,182–185].

REFERENCES

1. J. A. Krumhansl, "Unity in the science of physics," *Physics Today*, 44, 33, 1991.
2. P. L. Christiansen, M. P. Sorensen, A. C. Scott, "Nonlinear science at the dawn of the 21st century," *Lecture Notes in Physics*, Springer, Berlin.
3. A. Hasegawa, F. Tappert, "Transmission of stationary nonlinear optical pulses in dispersive dielectric fibers. I. Anomalous dispersion," *Appl. Phys. Lett.*, 23, 142, 1973.
4. A. Hasegawa, F. Tappert, "Transmission of stationary nonlinear optical pulses in dispersive dielectric fibers. II. Normal dispersion," *Appl. Phys. Lett.* 23, 171, 1973.
5. L. F. Mollenauer, R. H. Stolen, J. P. Gordon, "Experimental observation of picosecond pulse narrowing and solitons in optical fibers," *Phys. Rev. Lett.* 45, 1095, 1980.
6. A. Hasegawa, *Optical Solitons in Fibers*, Springer-Verlag, 1989.
7. E. M. Dianov, P. V. Mamyshev, A. M. Prokhorov, V. N. Serkin, *Nonlinear Effects in Optical Fibers*, Harwood Academic Publ., New York, 1989.
8. J. R. Taylor, *Optical solitons - theory and experiment*, Cambridge Univ, 1992.
9. A. Hasegawa, Y. Kodama, *Solitons in Optical Communications*, Oxford University Press, 1995.
10. N. N. Akhmediev, A. Ankiewicz, *Solitons. Nonlinear pulses and beams*, Chapman & Hall, 1997.
11. A. I. Maimistov, A. M. Basharov, *Nonlinear Optics Waves*, Kluwer Academic Publishers, Dordrecht, 1999.
12. A. Hasegawa, *Massive WDM and TDM Soliton Transmission Systems*, Kluwer Academic Publishers, Boston, 2000.
13. G. P. Agrawal, *Nonlinear Fiber Optics*, 3rd ed., Academic Press, 2001.
14. A. Hasegawa, M. Matsumoto, *Optical Solitons in Fibers*, 3rd Edition, Springer-Verlag, Berlin, 2003.
15. A. Biswas, S. Konar, "Theory of dispersion-managed optical solitons," *Progress In Electromagnetics Research*, 50, 83, 2005.
16. L. F. Mollenauer, J. P. Gordon, *Solitons in Optical Fibers*, Academic Press, Boston, 2006.
17. N. N. Akhmediev, A. Ankiewicz, *Dissipative Solitons: From Optics to Biology and Medicine*, Springer-Verlag, 2008.
18. J. M. Dudley, J. R. Taylor, *Supercontinuum Generation in Optical Fibers*, Cambridge University Press, 2010.
19. D. E. Pelinovsky, *Localization in Periodic Potentials: From Schrödinger Operators to the Gross-Pitaevskii Equation, Cambridge University Press*, 2011.
20. V. N. Serkin, A. Hasegawa, T. L. Belyaeva, "Nonautonomous matter-wave solitons near the Feshbach resonance", *Phys. Rev. A*, 81, 023610, 2010.
21. Z. Y. Sun, Y. T. Gao, X. Yu, Y. Liu. "Ampli.cation of nonautonomous solitons in the Bose-Einstein condensates and nonlinear optics", *Europhys. Lett.*, 93, 40004, 2011.

22. V. N. Serkin, A. Hasegawa, "Novel soliton solutions of the nonlinear Schrödinger equation model",. *Phys. Rev. Lett.,* 85, 4502, 2000.

23. P. A. M. Dirac, A self-chosen inscription written on the wall of Prof. D. D. Ivanenko office 4-59 October 3, 1956 and saved today at the Physical Department of Moscow State University.

24. H. Yukawa, A self-chosen inscription written on the wall of Prof. D. D. Ivanenko office 4-59 July 28, 1959 and saved today at the Physical Department of Moscow State University.

25. A. Pais, Playing with equations, the Dirac Way. In *Paul Adrien Maurice Dirac: Reminiscences about a Great Physicist,* eds. B. N. Kursunoglu and E. P. Wigner, Cambridge Univ. Press, Cambridge, 1990.

26. A. Hasegawa, "Maintenance of Order and the Entropy Law-Examples in Life, Earth Environment and Fusion Device," *Technonet* 564, 10, Osaka University, *Japan, 2014.*

27. M. J. Ablowitz, D. J. Kaup, A. C. Newell, H. Segur, "Nonlinear evolution equations of physical significance," *Phys. Rev. Lett.,* 3 1, 125, 1973.

28. C. S. Gardner, J. M. Greene, M. D. Kruskal, R. M. Miura, "Method for solving the Korteweg-deVries equation", *Phys. Rev. Lett.,* 19, 1095, 1967.

29. P. D. Lax, "Integrals of nonlinear equations of evolution and solitary waves", *Commun. Pure and Applied Math.,* 21, 467, 1968.

30. V. E. Zakharov, A. B. Shabat, "Exact theory of two-dimensional selffocusing and one-dimensional self-modulation of waves in nonlinear media," *JETP,* 34, 62, 1972.

31. V. E. Zakharov, A. B. Shabat, "Interaction between solitons in a stable medium," *JETP,* 37, 823, 1973.

32. V. E. Zakharov, The inverse scattering method. In R. Bullough and P. J. Caudrey (eds), *Solitons.* Berlin, Springer-Verlag, 1980.

33. F. Calogero, A. Degasperis, *Spectral transform and solitons: tools to solve and investigate nonlinear evolution equations,* Amsterdam, Elsevier Science Ltd, 1982.

34. M. Ablowitz, H. Segur, *Solitons and the inverse scattering transform,* SIAM, 1981.

35. M. J. Ablowitz, P. A. Clarkson, *Solitons, nonlinear evolution equations and inverse scattering,* Cambridge, 1991.

36. H. H. Chen, C. S. Liu, "Solitons in nonuniform media," *Phys. Rev. Lett.,* 37, 693, 1976.

37. R. Hirota, J. Satsuma, "N-soliton solutions of the K-dV equation with loss and nonuniformity terms," *J. Phys. Soc. Japan Lett.,* 41, 2141, 1976.

38. F. Calogero, A. Degasperis, "Coupled nonlinear evolution equations solvable via the inverse spectral transform, and solitons that come back: the boomeron", *Lett. Nuovo Cimento,* 16, 425, 1976.

39. U. Niederer, "The maximal kinematical invariance group of the free Schrödinger equations", *Helv. Phys. Acta ,* 45, 802, 1972.

40. U. Niederer, "The maximal kinematical invariance group of the harmonic Oscillator", *Helv. Phys. Acta,* 46, 191, 1973.

41. W. T. Fushchich, A. G. Nikitin, *Symmetries of equations of quantum mechanics,* Allerton Press Inc., New York, 1994.

42. K. Husimi, "Miscellanea in elementary quantum mechanics, I", *Prog. Theor. Phys.*, 9, 238, 1953.

43. K. Husimi, "Miscellanea in elementary quantum mechanics, II", *Prog. Theor. Phys.* , 9, 381, 1953.

44. V. I. Talanov, "Focusing of light in cubic media", *JETP Lett.* 11, 199, 1970.

45. C. Sulem, P. L. Sulem, *The Nonlinear Schrödinger Equation*, Springer-Verlag, 1999.

46. E. A. Kuznetsov, S. K. Turitsyn, "Talanov transformations in self-focusing problems and instability of stationary waveguides", *Phys. Lett. A*, 112, 273, 1985.

47. C. G. Darwin, "Free Motion in the Wave Mechanics", *Proc. Royal Soc. London*, 117, 258, 1928.

48. T. L. Belyaeva, V. N. Serkin, C. Hernandez-Tenorio, F. Garcia-Santibañez, "Enigmas of optical and matter-wave soliton nonlinear tunneling", *J. Mod. Optics*, 57, 1087, 2010.

49. T. L. Belyaeva, V. N. Serkin, "Wave-particle duality of solitons and solitonic analog of the Ramsauer–Townsend effect", *Eur. Phys. J. D*, 66, 153, 2012.

50. V. N. Serkin, A. Hasegawa, T. L. Belyaeva, "Geiger-Nuttall law for Schrödinger solitons", *J. Mod. Optics* , 60, 116, 2013.

51. V. N. Serkin, A. Hasegawa, T. L. Belyaeva, "Soliton self-induced sub-barrier transparency and the controllable -shooting out.effect", *J. Mod. Optics*, 60, 444, 2013.

52. V. N. Serkin, A. Hasegawa, Femtosecond soliton ampli.cation in nonlinear dispersive traps and soliton dispersion management. In *Optical Pulse and Beam Propagation II*, Yehuda B. Band, Editor, Proceedings of SPIE 3927, 302, 2000.

53. A. Barak, Or. Peleg, C. Stucchio, A. Soffer, M. Segev, "Observation of soliton tunneling phenomena and soliton ejection", *Phys. Rev. Lett.*, 100, 153901, 2008.

54. A. Barak, Or. Peleg, A. Soďer, M. Segev, "Multisoliton ejection from an amplifying potential trap", *Opt. Lett.*, 33, 1798, 2008.

55. Ch. P. Jisha, A. Alberucci, R. K. Lee, G. Assanto, "Optical solitons and wave-particle duality", *Opt. Lett.*, 36, 1848, 2011.

56. Ch. P. Jisha, A. Alberucci, R. K. Lee, G. Assanto, .Deflection and trapping of spatial solitons in linear photonic potentials", *Opt. Exp.*, 21, 18646, 2013.

57. Y. Linzon, R. Morandotti, M. Volatier, V. Aimez, R. Ares, S. Bar-Ad, "Nonlinear scattering and trapping by local photonic potentials", *Phys. Rev. Lett.*, 99, 133901, 2007.

58. M. V. Berry, N. L. Balazs, "Nonspreading wave packets", *Am. J. Phys.*, 47, 264, 1979.

59. G. D. Boyd and J. P. Gordon, "Confocal Multimode Resonator for Millimeter Through Optical Wavelength Masers", *The Bell System Tech. J.*, March, 489, 1961.

60. G. A. Siviloglou, D. N. Christodoulides, "Accelerating finite energy Airy Beams", *Opt. Lett.*, 32, 979, 2007.

61. R. Bekenstein, M. Segev, "Self-accelerating optical beams in highly nonlocal nonlinear media", *Opt. Express*, 19, 23706, 2011.

62. M. A. Bandres, I. Kaminer, M. Segev, D. N. Christodoulides, "Accelerating Optical Beams", *Optics & Photonics News* , 32, 2013.

63. V. P. Ermakov, "Second-order differential equations. Integrability conditions in closed form [in Russian]", *Universitetskie Izvestiya, Kiev* , 9, 1, 1880.

64. E. Schrödinger, "The continuous transition from micro- to macro-mechanics," *Die Naturwissenschaften* , 28, 664 (Eng. transl. in Collected Papers on Wave Mechanics (1928), Blackie and Son, London).

65. E. H. Kennard, "Zur Quantenmechanik einfacher Bewegungstypen", *Zeitschrift fur Physik*, 44, 326, 1927.

66. R. P. Feynman, A. R. Hibbs, *Quantum Mechanics and Path Integrals*, McGraw-Hill,1965.

67. R. J. Glauber, "Photon correlations", *Phys. Rev. Lett.*, 10, 84, 1963.

68. R. J. Glauber, "Coherent and incoherent states of the radiation", *eld,.Phys. Rev.*, 131, 2766, 1963.

69. J. N. Hollenhorst, "Quantum limits on resonant-mass gravitational-radiation Detectors", *Phys. Rev. D*, 19, 1669, 1979.

70. M. M. Nieto, L. M. Simmons, "Coherent states for general potentials", *Phys. Rev. Lett.*, 41, 207, 1978.

71. M. M. Nieto, D. R. Truax, "Squeezed states for general systems", *Phys. Rev. Lett.*, 71, 2843, 1993.

72. M. M. Nieto, D. R. Truax, "Arbitrary-order Hermite generating functions for obtaining arbitrary-order coherent and squeezed states", *Phys. Lett. A*, 208, 8, 1995.

73. M. M Nieto, "The discovery of squeezed states - in 1927", arXiv:quantph/9708012v1., 1997.

74. V. V. Dodonov, "Nonclassical states in quantum optics: a 'squeezed' review of the first 75 years, *J. Opt. B: Quantum Semiclass. Opt.*, 4, R1, 2002.

75. T. L. Belyaeva, C. A. Ramírez-Medina, V. N. Serkin, "Dinámica de estados comprimidos de Kennard y squeezions en el potencial armónico: A la memoria de Ehrenfest y Kennard por los 85 años de la publicación de sus artículos de 1927", *Internet. Electron J. Nanociencia. Moletrón.* 10, 1859, 2012.

76. V. I. Manko, "Introduction to quantum optics",. arXiv:quant-ph/9509018,1995.

77. M. E. Marhic, "Oscillating Hermite-Gaussian wave functions of the harmonic Oscillator", *Lett. Nuovo Cim.*, 22, 376, 1978.

78. Sh. Kumar, A. Hasegawa, "Quasi-soliton propagation in dispersion-managed optical fibers", *Opt. Lett.*, 22, 372, 1997.

79. A. Hasegawa, "Quasi-soliton for ultra-high speed communications", *Physica D*, 123, 267, 1998.

80. T. Hirooka, A. Hasegawa, "Chirped soliton interaction in strongly dispersion-managed wavelength-division-multiplexing systems",*Opt. Lett.*, 23, 10, 1998.

81. Yu. S. Kivshar, T. J. Alexander, S. K. Turitsyn, "Nonlinear modes of a macroscopic quantum oscillator", *Phys. Lett. A*, 278, 225, 2001.

82. V. Serkin, Hidden symmetries in nonlinear .ber optics: nonautonomous solitons and squeezions. Proceedings of "The Lasers, Applications, and Technologies (LAT2013)", June 18-22, Moscow, Russia, Fiber Optics III, LWI1, 1, 2013.

83. A. H. Nayfeh, B. Balachandran, *Applied Nonlinear Dynamics*, Wiley-VCH Verlag GmbH & Co. KGaA, Weinheim, 2004.

84. R. Hirota, "Exact envelope-soliton solutions of a nonlinear wave equation", *J. Math. Phys.*, 14, 805, 1973.

85. Y. Kodama, "Optical solitons in a monomode fiber", *J. Stat. Phys.*, 39, 597, 1985.

86. Y. Kodama, A. Hasegawa, "Nonlinear pulse propagation in a monomode dielectric guide", *IEEE J. Quantum Electron*, 23, 510, 1987.

87. K. Porsezian, A. Hasegawa, V. Serkin, T. Belyaeva, R. Ganapathy, "Dispersion and nonlinear management for femtosecond optical solitons", *Phys. Lett. A*, 361, 504, 2007.

88. R. Yang, L. Li, R. Hao, Z. Li, G. Zhou, "Combined solitary wave solutions for the inhomogeneous higher-order nonlinear Schrödinger equation", *Phys. Rev. E*, 71, 036616, 2005.

89. C. Q. Dai, J. F. Zhang, "New solitons for the Hirota equation and generalized higher-order nonlinear Schrödinger equation with variable coefficient", *J. Phys. A: Math and General*, 39, 723, 2006.

90. P. Wang, B. Tian,W. J. Liu, M. Li, K. Sun,"Soliton solutions for a generalized inhomogeneous variable-coefficient Hirota equation with symbolic computation", *Studies Applied Math.*, 125, 213, 2010.

91. Z. P. Liu, L. M. Ling, Y. R. Shi, Ch. Ye, L. Ch. Zhao, "Nonautonomous optical bright soliton under generalized Hirota equation frame", *Chaos, Solitons & Fractals*, 48, 38, 2013.

92. F. Calogero, A. Degasperis, "Extension of the Spectral transform Method for Solving Nonlinear Evolution Equation", *Lett. Nuovo Cimento*, 22, 131, 1978.

93. R. Grimshaw, "Slowly varying solitary waves. I. Korteweg-de Vries equation", *Proc. R. Soc. Lond.*, 368, 359, 1979.

94. N. Nirmala, M. J. Vedan, B. V. Baby, "AutoBäcklund transformation, Lax pairs, and Painlevé property of a variable coefficient Korteweg-de Vries equation I", *J. Math. Phys.*, 27, 2640, 1986.

95. N. Joshi, "Painlevé property of general variable-coefficient versions of the Korteweg- de Vries and nonlinear Schrödinger equations", *Phys. Lett. A*, 125, 456, 1987.

96. W. L. Chen, Y. K. Zheng, "Solutons of a nonisospectral and variable coefficient Korteweg-de Vries equation", *Lett. Math. Phys.*, 14, 293, 1987.

97. T. Chou, "Symmetries and a hierarchy of the general KdV equation", *J. Phys. A: Math. Gen.*, 20, 359, 1987.

98. T. Brugarino, "Painlevé property, autoBäcklund transformation, Lax pairs, and reduction to the standard form for the Kortewe-De Vries equation with nonuniformities", *J. Math. Phys.*, 30, 1013, 1989.

99. J. P. Gazeau, P. Winternitz, "Symmetries of variable coefficient Korteweg-de Vries equations", *J. Math. Phys.*, 33, 4087, 1992.

100. M. Vlieg-Hulstman, W. D. Halford, "Exact solutions to KdV Equations with variable coefficients and/or nonuniformities", *Computers Math. Applic.*, 29, 39, 1995.

101. M. Gürses, A. Karasu,"Variable coefficient third order Korteweg-de Vries type of equations", *J. Math. Phys.*, 36, 3485, 1995.

102. T. Ning, D. Chen, D. Zhang, "Soliton-like solutions for a nonisospectral KdV Hierarchy", *Chaos, Solitons and Fractals.*, 21, 395, 2004.

103. Q. Li, D. Zhang, D. J. Chen, "Solving the hierarchy of the nonisospectral KdV equation with self-consistent sources via the inverse scattering transform",. *Phys. A: Math. Theor.*, 41, 355209, 2008.

104. M. S. Abdel Latif, "Some exact solutions of KdV equation with variable Coefficients", *Commun Nonlinear Sci. Numer. Simulat.*, 16, 1783, 2011.

105. P. Wang, B. Tian, W.-J. Liu, Y. Jiang, Y.-Sh. Xue, "Interactions of breathers and solitons of a generalized variable-coefficient Korteweg-de Vries-modified Korteweg-de Vries equation with symbolic computation", *Eur. Phys. J. D*, 66, 233, 2012.

106. M. Russo, S. R. Choudhury, "Building generalized Lax integrable KdV and mKdV equations with spatiotemporally varying coefficients", *J. Phys.: Conference Series*, 482, 012038, 2014.

107. W. L. Chan, K. S. Li, "Non-propagating solitons of the non-isospectral and variable coefficient modified KdV equation", *Phys. A: Math. Gen.*, 27, 883, 1994.

108. A. Korkmaz, I. Dag, "Solitary wave simulations of complex modified Korteweg-de Vries equation using differential quadrature method",*Comput. Phys. Commun.*, 180, 1516, 2009.

109. Zh. Y. Sun, Y. T. Gao, Y. Liu, X. Yu, "Soliton management for a variable-coefficient modified Korteweg-de Vries equation", *Phys. Rev. E* , 84, 026606, 2011.

110. Y. Lin, Ch. Li, J. He, "Nonsingular position solutions of a variable-coefficient modified KdV equation", *Open J. Applied Sci.*, 3, 102, 2013.

111. V. N. Serkin, T. L. Belyaeva, "High-energy optical Schrödinger solitons", *JETP Lett.*, 74, 573, 2001.

112. V. N. Serkin, T. L. Belyaeva, "The Lax representation in the problem of soliton management", *Quant. Electron.*, 31, 1007, 2001.

113. F. D. Tappert, N. J. Zabusky, "Gradient-induced fission of solitons", *Phys. Rev. Lett.*, 27, 1774, 1971.

114. M. R. Gupta, J. Ray, "Extension of inverse scattering method to nonlinear evolution equation in nonuniform medium", *J. Math. Phys.*, 22, 2180, 1981.

115. J. J. E. Herrera, "Envelope solitons in inhomogeneous media", *J. Phys. A: Math. Gen.*, 17, 95, 1984.

116. R. Balakrishnan, "Soliton propagation in nonuniform media", *Phys. Rev. A*, 32, 1144, 1985.

117. S. P. Burtsev, V. E. Zakharov, A. V. Mikhailov, "Inverse scattering method with variable spectral parameter", *Theor. Math. Phys*, 70, 227, 1987.

118. L. Gagnon, P. Winternitz, "Lie symmetries of a generalized nonlinear Schrödinger equation: I. The symmetry group and its subgroups", *J. Phys. A: Math. Gen*, 21, 1493, 1988.

119. L. Gagnon, P. Winternitz, "Lie symmetries of a generalized nonlinear Schrödinger equation: II. Exact solutions", *J. Phys. A: Math. Gen.*, 22, 469, 1989.

120. L. Gagnon P. Winternitz "Symmetry classes of variable coefficient nonlinear Schrödinger equations", *J. Phys. A: Math. Gen*, 26, 7061, 1993.

121. B. S. Azimov, M. M. Sagatov, A. P. Sukhorukov, "Self-similar self-compression pulses in media with a modified gain", *Sov. J. Quantum Elec-tron.* 21, 785, 1991.

122. V. Y. Khasilev, Optimal control of all-optical communication soliton systems, *Proc. SPIE*, 2919, 177, 1996.

123. J. D. Moores, "Nonlinear compression of chirped solitary waves with and without phase modulation", *Opt. Lett.*, 21, 555, 1996.

124. T. E. Murphy, "10-GHz 1.3-ps Pulse Generation Using Chirped Soliton Compression in a Raman Gain Medium", *IEEE Photon. Tech.. Lett*, 14 (10), 2002.

125. A. A. Sysoliatin, A. I. Konyukhov, L. A. Melnikov, Dynamics of Optical Pulses Propagating in Fibers with Variable Dispersion, In *Numerical Simulations of Physical and Engineering Processes*, ed. Jan Awrejcewicz, InTech, Rijeka, Croatia, 2011.

126. V. N. Serkin, A. Hasegawa, "Soliton management in the nonlinear Schrödinger equation model with varying dispersion, nonlinearity, and gain", *JETP Lett.*, 72, 89, 2000.

127. V. N. Serkin, A. Hasegawa, "Exactly integrable nonlinear Schr ödinger equation models with varying dispersion, nonlinearity and gain: application for soliton dispersion", *IEEE J. Select. Topics Quant. Electron.*, 8, 418, 2002.

128. T. C. Hernandez, V. E. Villargan, V. N. Serkin, G. M. Aguero, T. L. Belyaeva, M. R. Pena, L. L. Morales, "Dynamics of solitons in the model of nonlinear Schrödinger equation with an external harmonic potential: 1. Bright solitons", *Quant. Electron.*, 35, 778, 2005.

129. C. H. Tenorio, E. Villagran-Vargas, V. N. Serkin, M. Agüero-Granados, T. L. Belyaeva, R. Pena-Moreno, L. Morales-Lara, "Dynamics of solitons in the model of nonlinear Schrödinger equation with an external harmonic potential: 2. dark solitons", *Quant. Electron.*, 35, 929, 2005.

130. R. Atre, P. K. Panigrahi, G. S. Agarwal, "Class of solitary wave solutions of the one-dimensional Gross-Pitaevskii equation", *Phys. Rev. E*, 73, 056611, 2006.

131. S. A. Ponomarenko, G. P. Agrawal, "Do solitonlike self-similar waves exist in nonlinear optical media?", *Phys. Rev. Lett* ., 97, 013901, 2006.

132. C. Hernandez-Tenorio, T. L. Belyaeva, V. N. Serkin, "Parametric resonance for solitons in the nonlinear Schrödinger equation model with time-dependent harmonic oscillator potential", *Physica B: Condensed Matter*, 398, 460, 2007.

133. V. N. Serkin, A. Hasegawa, T. L. Belyaeva, "Nonautonomous solitons in external potentials", *Phys. Rev. Lett.*, 98, 074102, 2007.

134. S. Chen, Y. H. Yang, L. Yi, P. Lu, D. S. Guo, "Phase fluctuations of linearly chirped solitons in a noisy optical fiber channel with varying dispersion, nonlinearity, and gain", *Phys. Rev. E*, 75, 036617, 2007.

135. J. Belmonte-Beitia, V. M. Perez-Garc¬a, V. Vekslerchik, V. V. Konotop, "Localized nonlinear waves in systems with time- and space-modulated nonlinearities", *Phys. Rev. Lett*, 100, 164102, 2008.

136. R. Hao, G. Zhou, "Exact multi-soliton solutions in nonlinear optical systems", *Opt. Communic.*, 281, 4474, 2008.

137. W. J. Liu, B. Tian, H. Q. Zhang, "Types of solutions of the variable-coefficient nonlinear Schrödinger equation with symbolic computation", *Phys. Rev. E*, 78, 066613, 2008.

138. H. J. Shin, "Darboux invariants of integrable equations with variable spectral Parameters", *J. Phys. A: Math. and Theor.*, 41, 285201, 2008.

139. L. Wu, J. F. Zhang, L. Li, C. Finot, K. Porsezian, "Similariton interactions in nonlinear graded-index waveguide amplifiers", *Phys. Rev. A*, 78, 053807, 2008.

140. K. H. Han, H. J. Shin, "Nonautonomous integrable nonlinear Schrödinger equations with generalized external potentials", *J. Phys. A: Math. Theor.*, 42, 335202, 2009.

141. U. Al Khawaja, "Soliton localization in Bose.Einstein condensates with time-dependent harmonic potential and scattering length", *J. Phys. A: Math. Theor*., 42, 265206, 2009.

142. D. Zhao, X. G. He, H. G. Luo, "Transformation from the nonautonomous to standard NLS equations", *Eur. Phys. J. D*, 53, 213, 2009.

143. A. T. Avelar, D. Bazeia, W.B. Cardoso, "Solitons with cubic and quintic nonlinearities modulated in space and time", *Phys. Rev. E*, 79, 025602, 2009.

144. H. Luo, D. Zhao, X. He, "Exactly controllable transmission of nonautonomous optical solitons", *Phys. Rev. A*, 79, 063802, 2009.

145. K. Porsezian, R. Ganapathy, A. Hasegawa, V. N. Serkin, "Nonautonomous soliton dispersion management", *IEEE J. Quant. Electron*, 45, 1577, 2009.

146. X. G. He, D. Zhao, L. Li, H. G. Luo, "Engineering integrable nonautonomous nonlinear Schrödinger equations", *Phys. Rev. E*, 79, 056610, 2009.

147. S. Rajendran, P. Muruganandam, M. Lakshmanan, "Bright and dark solitons in a quasi-1D Bose-Einstein condensates modelled by 1D Gross-Pitaevskii equation with time-dependent parameters", *Physica D*, 239, 366, 2010.

148. Zh. Yan, "Nonautonomous rogons in the inhomogeneous nonlinear Schrödinger equation with variable coefficients", *Phys. Lett. A*, 374, 672, 2010.

149. Y. Zh. Yang, L.-Ch. Zhao, T. Zhang, "Snakelike nonautonomous solitons in a graded-index grating waveguide", *Phys. Rev. A*, 81, 043826, 2010.

150. Y. Zh. Yang, L.-Ch. Zhao, T. Zhang, "Bright chirp-free and chirped nonautonomous solitons under dispersion and nonlinearity management", *J. Opt. Soc. Amer. B*, 28, 236, 2011.

151. L.-C. Zhao, Z.-Y. Yang, L.-M..Ling, J. Liu, "Precisely controllable bright nonautonomous solitons in Bose-Einstein condensate", *Phys. Lett. A*, 375, 1839, 2011.

152. D. Zhao, Yu-J. Zhang, W. W. Lou, H. G. Luo, "AKNS hierarchy, Darboux transformation and conservation laws of the 1D nonautonomous nonlinear Schrödinger equations", *J. Math. Phys.*, 52, 043502, 2011.

153. Zh. Y. Sun, Y. T. Gao, X. Yu, Y. Liu, "Amplification of nonautonomous solitons in the Bose-Einstein condensates and nonlinear optics", *Eur. Phys. Lett.*, 93, 40004, 2011.

154. C. Q. Dai, Y. J. Xu, "Nonautonomous spatiotemporal solitons in the harmonic external potential", *Acta Physica Polonica B*, 43, 367, 2012.

155. C. Y. Liu, C. Q. Dai "Nonautonomous solitons in the (3+1)-dimensional inhomogeneous cubic-quintic nonlinear medium", *Commun. Theor. Phys.*, 57, 568, 2012.

156. M. S. Mani Rajan, A. Mahalingam, A. Uthayakumar, K. Porsezian, "Nonlinear tunneling of nonautonomous optical solitons in combined nonlinear Schrödinger and Maxwell-Bloch systems", *J. Optics*, 14, 105204, 2012.

157. R. Gupta, Sh. Loomba, Ch. N. Kumar, "Class of nonlinearity control parameter for bright solitons of non-autonomous NLSE with trapping potential", *IEEE J. Quant. Electron.*, 48, 847, 2012.

158. S. K. Suslov, "On integrability of nonautonomous nonlinear Schrödinger equations", *Proc. Amer. Math. Soc.*, 140, 3067, 2012.

159. C. Q. Dai, C. L. Zheng, H. P. Zhu, "Controllable rogue waves in the nonautonomous nonlinear system with a linear potential", *Eur. Phys. J. D*, 66, 2012.

160. T. Kanna, R. B. Mareeswaran, F. Tsitoura, H. E. Nistazakis, D. J. Frantzeskakis,"Nonautonomous bright-dark solitons and Rabi oscillations in multi-component Bose-Einstein condensates", *J. Phys. A* , 46, 475201, 2013.

161. Zh. Yan, Ch. Dai, "Optical rogue waves in the generalized inhomogeneous higher-order nonlinear Schrödinger equation with modulating coefficients", *J. Opt.*, 15, 064012, 2013.

162. J. R. He, H. M. Li "Nonautonomous solitary-wave solutions of the generalized nonautonomous cubic quintic nonlinear Schrödinger equation with time- and space-modulated coefficients", *Chin. Phys. B* , 22, 040310, 2013.

163. Ch. Liu, Zh.-Y. Yang, W.-Li Yang, R.-H. Yue, "Nonautonomous dark solitons and rogue waves in a graded-index grating waveguide",*Commun. Theor. Phys.*, 59, 311, 2013.

164. Li-Ch. Zhao, "Dynamics of nonautonomous rogue waves in Bose-Einstein Condensate",*Annals Phys.*, 329, 73, 2013.

165. T. Kanna, R. B. Mareeswaran, K. Sakkaravarthi, "Nonautonomous bright matter wave solitons in spinor Bose-Einstein condensates", *Phys. Lett. A*, 378, 158, 2014.

166. C. Q. Dai, Y. Y. Wang, "Nonautonomous solitons in parity-time symmetric Potentials", *Opt. Commun.*, 315, 303, 2014.

167. Yu. J. Shen, Yi. T. Gao, D. W. Zuo, Y. H. Sun, Y. J. Feng, L. Xue, "Nonautonomous matter waves in a spin-1 Bose-Einstein condensate", *Phys. Rev.E*, 89, 062915, 2014.

168. Q. Y. Li, Z. D Li, Sh. X. Wang, W. W. Song, F. F. Fu, "Nonautonomous solitons of Bose-Einstein condensation in a linear potential with an arbitrary time-dependence", *Opt. Commun.*, 282, 1676, 2009.

169. J. R. He, H. M. Li, "Nonautonomous bright matter-wave solitons and soliton collisions in Fourier-synthesized optical lattices", *Opt. Commun.* 284, 3084, 2011.

170. Li-Ch. Zhao, Y. Zh. Yang, Li-M. Ling, J. Liu, "Precisely controllable bright nonautonomous solitons in Bose-Einstein condensate", *Phys. Lett. A*, 375, 1839, 2011.

171. Y. Y. Wang, J. S. He, Y-Sh. Li, "Soliton and rogue wave solution of the new nonautonomous nonlinear Schrödinger equation", *Commun. Theor. Phys*, 56, 995, 2011.

172. Zh.-Y. Yang, L.-Ch. Zhao, T. Zhang, "Dynamics of a nonautonomous soliton in a generalized nonlinear Schrödinger equation", *Phys. Rev. E* , 83, 066602, 2011.

173. J. He, J. Zhang, M. Zhang, C. -Q., Dai, "Analytical nonautonomous soliton solutions for the cubic-quintic nonlinear Schrödinger equation with distributed coefficients", *Opt. Commun.*, 285, 755, 2012.

174. H. M. Li, L. Ge, J. R. He, "Nonautonomous bright solitons and soliton collisions in a nonlinear medium with an external potential", *Chinese Phys. B*, 21, 050512, 2012.

175. Z. Li, W. Hai, Y. Deng, Q. Xie, "A biperiodically driven matter-wave nonautonomous deformed soliton", *J. Phys. A-Math. Theor.*, 45, 435003, 2012.

176. J. Y. Zhou, H. M. Li, J. R. He "Analytical periodic wave and soliton solutions of the generalized nonautonomous nonlinear Schrödinger equation in harmonic and optical lattice potentials", *Commun. Theor. Phys*, 58, 393, 2012.

177. X. Lu, M. Peng, "Nonautonomous motion study on accelerated and decelerated solitons for the variable-coefficient Lenells-Fokas model", *Chaos* , 23, 013122, 2013.

178. W.-D. Xie, F. Ye, W. He, "Nonautonomous dark solitons in Bose-Einstein Condensate", *Mod. Phys. Lett. B*, 27, 1350229, 2013.

179. W. L. Chen, Ch. Q. Dai, L. H. Zhao, "Nonautonomous superposed Akhmediev breather in water waves", *Computers & Fluids*, 92, 1, 2014.

180. P. R. Gordoa, A. Pickering, J. A. D Wattis, "Nonisospectral scattering problems and similarity reductions", *Applied Math. Comput.* 237, 77, 2014.

181. Y. X. Chen, Ch. Q. Dai, X. G. Wang, "Two-dimensional nonautonomous solitons in parity-time symmetric optical media", *Opt. Commun.*, 324, 10, 2014.

182. Zh. G. Liu, X. X. Ma, "Dynamics of nonautonomous matter breather wave in a time-dependent harmonic trap with nonlinear interaction management", *Mod. Phys. Lett. B*, 28, 1450026, 2014.

183. L. Wang, X. Q. Feng, Li-Ch. Zhao, "Dynamics and trajectory of nonautonomous rogue wave in a graded-index planar waveguide with oscillating refractive index", *Opt. Commun.*, 329, 135, 2014.

184. H. J. Jiang, J. J. Xiang, Ch.-Q. Dai, Y.-Y. Wang, "Nonautonomous bright soliton solutions on continuous wave and cnoidal wave backgrounds in blood vessels", *Nonlin. Dynamics* , 75, 201, 2014.

185. Q. Y. Li, Sh. J. Wang, Z. D. Li "Nonautonomous dark soliton solutions in two-component Bose-Einstein condensates with a linear time-dependent potential", *Chinese Phys. B*, 23, 060310, 2014.

186. T. Belyaeva, V. Serkin, M. Agüero, C. Hernandez-Tenorio, L. Kovachev, "Hidden features of the soliton adaptation law to external potentials", *Laser Phys.*, 21, 258, 2011.

187. T. L. Belyaeva, V. N. Serkin, Nonautonomous Solitons: Applications from Nonlinear Optics to BEC and Hydrodynamics. In *Hydrodynamics – Advanced Topics,* ed. Harry Edmar Schulz, InTech, Rijeka, Croatia, 2011.

188. M. S. Mani Rajan, J. Hakkim, A. Mahalingam, A. Uthayakumar "Dispersion management and cascade compression of femtosecond nonautonomous soliton in birefringent", *ber,.Eur. Phys. J. D*, 67, 150, 2013.

189. H. H. Chen, "General derivation of Bäcklund transformations from inverse scattering problems", *Phys. Rev. Lett.*, 33, 925, 1974.

190. V. N. Serkin, A. Hasegawa, T. L. Belyaeva, "Solitary waves in nonautonomous nonlinear and dispersive systems: nonautonomous solitons", *J. Modern Optics*, 57, 1456, 2010.

191. E. M. Dianov, A. Ya. Karasik, P. V. Mamyshev, A. M. Prokhorov, V. N. Serkin, M. F. Stel.makh, A. A. Fomichev, "Stimulated-Raman conversion of multisoliton pulses in quartz optical fibers", *Pis.ma Zh. Eksp. Teor. Fiz*, 41, 242, 1985. Translation: *JETP Lett.* 41, 294, 1985.

192. F. M. Mitschke, L. F. Mollenauer, "Discovery of the soliton self-frequency Shift",*Opt. Lett.*, 11, 659, 1986.

193. V. N. Serkin, "Colored envelope solitons in optical fibers", *Sov. Technic. Phys. Letters*, 13, 320, 1987.
194. J. P. Gordon, "Theory of the soliton self-frequency shift", *Opt. Lett.*, 11, 662, 1986.
195. J. Satsuma, N. Yajima, "Initial value problems of one-dimensional self-modulation of nonlinear waves in dispersive media", *Prog. Theor. Phys. Suppl.*, 55, 284, 1974.
196. V. E. Zakharov, Ed., *What Is Integrability?* Springer Series in Nonlinear Dynamics. Berlin, Springer-Verlag, 1991.

8 Hot solitons, cold solitons, and hybrid solitons in fiber optic waveguides

P. Tchofo Dinda, E. Tchomgo Felenou, and *C. M. Ngabireng*

8.1 INTRODUCTION

After its appearance in the literature in the 1960s, the concept of the soliton has quickly spread in most of the branches of physics and then extended beyond this discipline to reach areas as diverse as fluids, plasmas, condensed matter, biology and geology, thus becoming one of the most multidisciplinary concepts of the modern sciences [1]. The experimental demonstration of the reality of this concept stimulated the emergence of a stream of scientific topics having fundamental or applied interests. The soliton is a physical object that results from an energy localization effect having a relatively long lifetime. This object is capable not only of moving as a whole entity, but may also execute internal vibrations. A wide variety of solitons has already been demonstrated in the literature, ranging from the conventional soliton [2] to the Peregrine soliton [3], through the Akhmediev breathers [4], guiding-center solitons [5], dispersion-managed solitons [6], to mention a few. Such solitary waves, as mathematical objects, correspond to idealized representations of the real world, where the wave propagates through a perfect physical medium without defects or perturbations. Obviously, real physical systems are always more or less perturbed, and the presence of small perturbations in a soliton system may generate several major effects, such as an alteration of the soliton profile as compared to its exact profile [7], or the creation of internal dynamics within the soliton [8]. In certain situations, the soliton may generate radiation waves, i.e., wave packets of low amplitude which follow or sometimes precede the soliton [8]. A widespread idea well beyond the area of optics considers the radiation produced by a perturbed soliton as a restructuring process during which the soliton is endeavoring to get rid of the perturbation. According to this idea, the radiation would express the natural tendency of a perturbed soliton to always evolve toward its non-perturbed state. In the same vein, it is also very common to assimilate a slightly perturbed soliton and its non-perturbed counterpart to the same collective entity. In this work, we present dynamical behaviors that go against these ideas. We establish the existence of non-conventional behaviors in which the slightly perturbed soliton seems to have lost the memory of its fundamental (non-perturbed) state as it evolves and stabilizes in a state different from this fundamental state. More generally, we present results that suggest that in a slightly perturbed environment, the dynamical behavior of a soliton ramifies into several families, which we call *hot solitons* (or *hyperthermic solitons*), *cold solitons* (or *hypothermic solitons*) and *ideal soliton* (*isothermic soliton*), which correspond, respectively, to pulses that stabilize with an energy level higher than, lower than, and equal to the energy of the non-perturbed soliton. On the borders of the domains of existence of those families of solitons, we have identified *hybrid solitons*, corresponding to solitons that cool during propagation, due to a significant loss of energy by radiation. This cooling process is accompanied by a change of state, from hyperthermia to hypothermia, or from isothermal to hypothermia.

Bearing in mind that the soliton refers to a propagation phenomenon having a relatively long lifetime, to illustrate the ideas just mentioned, it is preferable to use a physical medium in which the soliton is sufficiently robust, so that its dynamical state can be clearly identified, without any ambiguity. In this context, the guiding-center soliton and the dispersion-managed soliton are well-known examples of robust solitons that can propagate over thousands of kilometers [5,6]. Such solitons propagate in fiber-optic waveguides (FOWGs), which are necessarily made up of the repetition of the same basic structure, called *amplification span*. Within each span, the pulse executes an internal dynamic, before going back (at the end of the span) to a profile identical or close to the one it had at the beginning of the span. If one disregards the internal dynamic within each span (*fast dynamic*), and if one considers only the pulse profile at the end of each span (*slow dynamic*), then the pulse will display a stationary behavior similar (at least in appearance) to that of a conventional soliton (which propagates without change of profile) [2]. Here, the terminology *stationary pulse* (SP) refers to a pulse that moves by executing a periodic deformation of profile, whose periodicity corresponds exactly to that of that waveguide. However, at this juncture, it is worth noting that there is a major qualitative difference between the ideal system, where the (conventional) soliton propagates with a perfectly smooth profile (of Sech shape) [2], and the periodically structured waveguides, where the SP profile is never perfectly smooth. Indeed, the periodic structure of the FOWG acts on the SP like a perturbation which restructures its profile and makes it somewhat rough at the level of the pulse's wings. Consequently, the exact profiles of SPs for most FOWGs, denoted hereafter as A_S, are virtually impossible to synthesize by means of the currently available light sources. In practice, for better stability of pulse propagation, one endeavors to make it so that the input pulse, say $A(z = 0, t)$, is as close as possible to the SP, $A_S(0, t)$. Here, z and t denote the distance and time, respectively. It is crucial to realize that the input pulse, $A(0, t)$, which is different but very close to $A_S(0, t)$, is nevertheless felt by the waveguide as a perturbation of the SP, with a perturbation field $q(0, t)$ given by

$$q = A - A_S, \tag{8.1}$$

with $|q(0, t)| \ll |A_S(0, t)|$.

Our analysis of pulse behavior is fundamentally based on the assumption that the injection of the pulse $A(0, t)$ in the system is felt by the waveguide as an event equivalent to a collision between the SP $A_S(0, t)$ and the perturbation $q(0, t)$, and that during this collision, the fields $A_S(0, t)$ and $q(0, t)$ interfere. The input power in the waveguide can then decompose in the following way:

$$
\begin{aligned}
P(0, t) &= (A_S(0, t) + q(0, t)) \left(A_S^*(0, t) + q^*(0, t)\right) \\
&= P_S(0, t) + P_q(0, t) + P_{IF}(0, t),
\end{aligned} \tag{8.2}
$$

where $P_S(0, t) = |A_S(0, t)|^2$ represents the input power of the SP, and $P_q(0, t) = |q(0, t)|^2$ the input power of the perturbation field. The power of the interference field associated with this collision is given by

$$P_{IF}(0, t) = q^*(0, t) A_S(0, t) + q(0, t) A_S^*(0, t). \tag{8.3}$$

Here, the quantity $P_{IF}(0, t)$ is in fact a *pseudo power* in the sense that it can be positive, negative, or zero. Consequently, the energy of the interference field, which is defined by

$$E_{IF}(0, t) = \int_{-\infty}^{+\infty} P_{IF}(0, t) dt, \tag{8.4}$$

is in fact a *pseudo energy* that can be positive, negative, or zero. So, at the waveguide input, the pulse energy can decompose in the following way:

$$E_0 = E_S + E_q + E_{IF}, \tag{8.5}$$

where $E_S \equiv \int_{-\infty}^{+\infty} |A_S(0, t)|^2 dt$ is the energy of the SP, and $E_q \equiv \int_{-\infty}^{+\infty} |q(0, t)|^2 dt$ the energy of the perturbation field just before the collision process. In the present study, we have identified the interference field energy, E_{IF}, as one of the major tools of prediction of the dynamical behavior of optical pulses. More important, by carefully exploiting the three major energy quantities associated with the input pulse, (E_S, E_q, E_{IF}), we discovered that perturbed solitons can be classified into six major families of perturbed solitons, which are schematically represented in Fig. 8.1. In other words, we assert that whatever be the type of the perturbation field affecting the soliton, its dynamical behavior will necessarily fall in one of the six scenarios of dynamic behaviors displayed in Fig. 8.1, to which we have assigned the qualifiers *hyperthermic soliton, isothermic soliton, hypothermic soliton*, and *hybrid soliton*, defined in detail below.

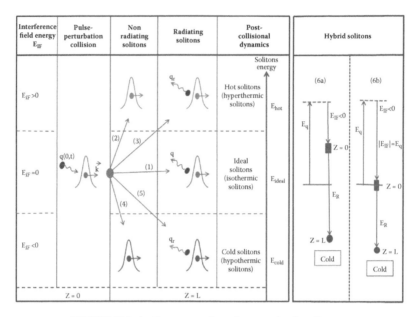

FIGURE 8.1: Cartography of perturbed solitons.

8.2   ISOTHERMIC SOLITONS

The family of *isothermic solitons* refers to pulses for which the interference field energy is zero ($E_{IF} = 0$). In this case, the pulse expels entirely the perturbation field (by radiation) and recovers the energy that it had before the collision with the perturbation. We have also assigned to this kind of pulse the qualifier *ideal soliton* because of its ability to expel a perturbation field.

8.3   HYPERTHERMIC SOLITONS

When the energy of the interference field is not zero, the perturbed pulse becomes incapable of expelling the totality of the perturbation, but it remains able to execute a highly stable propagation with an excess of energy as compared to that of the SP. This family of solitons consists of two subclasses which differ in the ability of the pulse to generate radiation: The *non-radiating hyperthermic solitons* are indicated by the arrow labeled (2) in Fig. 8.1, whereas the *radiating hyperthermic solitons* are indicated by the arrow (3). Thus the hyperthermic solitons behave as objects that have been heated during their initial interaction with the perturbation. Consequently, for the sake of simplicity, we call them also *hot solitons*.

8.4 HYPOTHERMIC SOLITONS

Certain pulses can execute a highly stable propagation with a deficit of energy as compared to that of the SP. This family of solitons consists also of two subclasses, namely, *non-radiating hypothermic solitons*, which are indicated by the arrow labeled (4) in Fig. 8.1, whereas *radiating hypothermic solitons* are indicated by the arrow (5) in Fig. 8.1. Clearly, hypothermic solitons behave as objects that have cooled during their initial interaction with the perturbation. Consequently, we can call them *cold solitons*.

8.5 HYBRID SOLITONS

A soliton which ends its propagation in the state of hypothermia may also have evolved in one of the two configurations below:

(i) The case $|E_{IF}| < E_q$, represented by configuration (6a) in Fig. 8.1, corresponds to a very interesting special case, where the pulse begins its propagation in a state of hyperthermia before switching to a state of hypothermia because of a loss of energy by radiation. We refer to this kind of pulse as a *hybrid soliton* of type I. The terminology *hybrid soliton* refers to the fact that this pulse executes two types of behavior during the same propagation.

(ii) Another special situation arises when $|E_{IF}| = E_q$ [configuration (6b) in Fig. 8.1]. In this case, the pulse preserves its energy during the initial interaction, and begins its propagation with an energy exactly equal to that of the SP. Then the loss of energy by radiation causes the switching of the pulse to a state of hypothermia.

To demonstrate quantitatively the existence of the families of perturbed solitons displayed in Fig. 8.1, we have carried out numerical simulations of pulse propagation over several thousands of kilometers, based on the generalized non-linear Schrödinger equation (GNLSE) which follows [5,6]:

$$A_z = -i\frac{\beta(z)}{2}A_{tt} + i\gamma |A|^2 A - \frac{\alpha}{2}A$$
$$+ \left(\sqrt{G} - 1\right) \times \sum_{n=1}^{N} \delta\left(z - nZ_A\right)A, \qquad (8.6)$$

where A refers to the electric field of the pulse, β, γ and α designate the dispersion, non-linearity, and linear-attenuation coefficient s, respectively. The parameter $G = \exp\left(\alpha Z_A\right)$ refers to the gain of each amplifier. Equation (8.6) may describe in a realistic way the pulse propagation within a FOWG where the amplification span consists of only one section of fiber with constant dispersion (which is required for a guiding-center soliton) [5], as well as within a FOWG where the amplification span is made up of a concatenation of sections of fibers with alternately positive and negative dispersion coefficients (which is required for a dispersion-managed soliton) [6]. For our numerical simulations, we have focused on these two types of waveguides. This choice is dictated only by a concern for simplicity and does not restrict at all the generality of the concepts underlying our cartography of perturbed solitons. Here, it is worth noting that, in the literature, there exists no exact analytical expression for the SP profiles for the two FOWGs under consideration. The profiles of those SPs are accessible only numerically, by means of specialized techniques. By applying the procedure of [9], we have obtained the results depicted in Fig. 8.2 (a) for a dispersion-managed system with the following typical parameters: dispersion $\beta_\pm = \pm 2.5$ ps^2/km, losses $\alpha^\pm = 0.22$ dB/km, effective core area $A_{\text{eff}}^\pm = 55\,\mu\text{m}^2$, and $\gamma^\pm = 2 \times 10^{-3}$W^{-1}m^{-1}. The fiber lengths of the dispersion map, $L_\pm$ ($L_+ = 439.015$ m, $L_- = 438.018$ m), have been obtained by following the procedure of [10]. Figure 8.2 (a) illustrates the roughness of the spectral profile of the SP in the waveguide. In particular, one can easily identify the Kelly sidebands (labeled 1, 2, and 3), which are induced by the periodic structure of the waveguide and the resulting internal dynamics. Figure 8.2 (b) shows the perturbation field $q(0,t)$ corresponding to the Gaussian pulse $A(0,t)$ closest to the SP. Figures 8.2 (c) and (d) show the evolution of the perturbation field corresponding to the propagation of the Gaussian pulse over 6000 kilometers. Figure 8.2 (c) shows two waves of radiation (indicated by the small horizontal arrows) moving away from the center of the pulse rest frame. At the center of this frame, one can clearly distinguish a trapped field (corresponding to the non-radiating part of the perturbation field), indicated by the small vertical arrow. Figure 8.2 (d) illustrates the spectral profile of the perturbation field. To illustrate quantitatively all the new families of perturbed solitons displayed in Fig. 8.1, it is necessary to specify a profile of perturbation field taking into account the wide variety of perturbations that can distort the pulses. To this end, we have chosen a perturbation field of the following form: $q(0,t) = \xi(0,t)A_S(0,t)$, where we make use of $A_S(0,t)$ to restrict the action of the perturbation to a temporal window surrounding the pulse. The field $\xi(0,t)$ is arbitrarily chosen to be of the following general form: $\xi(0,t) = \frac{\varepsilon}{2}\left[\eta \sin(\Omega_0 t + \phi) + \delta\right]\exp\left[i\left(\phi_0 + \frac{a}{2}t^2\right)\right]$, where the physical meaning of the parameters of $\xi(0,t)$ is obvious. The results of our numerical simulations are visible in Fig. 8.3, which shows that an SP that comes into contact with a perturbation on entering the waveguide, transforms into one of the solitons listed in our general cartography of Fig. 8.1. Depending on the nature of the perturbation field, the pulse may transform into a hot soliton [Fig. 8.3 (a)], cold soliton [Fig. 8.3 (b)], ideal soliton [Fig. 8.3 (c)], or a hybrid soliton [Fig. 8.3 (d)]. In Fig. 8.3, the non-radiating solitons are easily recognizable by their energy curves, which are horizontal dotted lines. The radiating solitons are also easily recognizable by their curves endowed with a *stair* which corresponds to the loss of energy resulting from radiation. The presence of this *stair* demonstrates clearly that the process of radiation (by a light pulse) does not take place in a continual way, but rather in the form of the emission of an energy packet, which occurs in the beginning of propagation and in a strictly time-limited way.

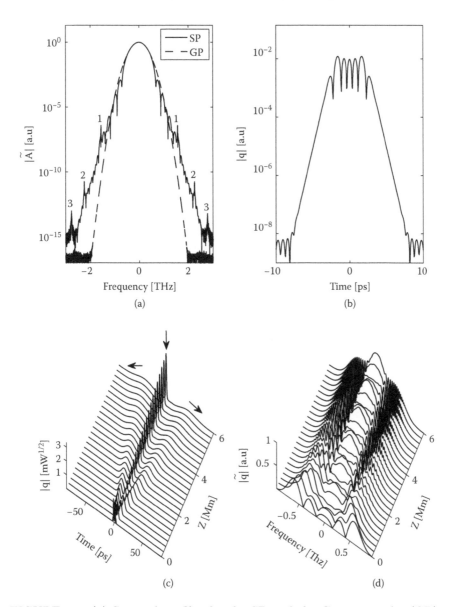

FIGURE 8.2: (a) Spectral profiles for the SP and the Gaussian pulse (GP). (b) : Temporal profile of the perturbation field q corresponding to the GP closest to the SP. (c) and (d) Evolution of the temporal and spectral profiles of q.

On the other hand, we have also carried out numerical simulations using the waveguide of a guiding-center soliton, with the same system parameters

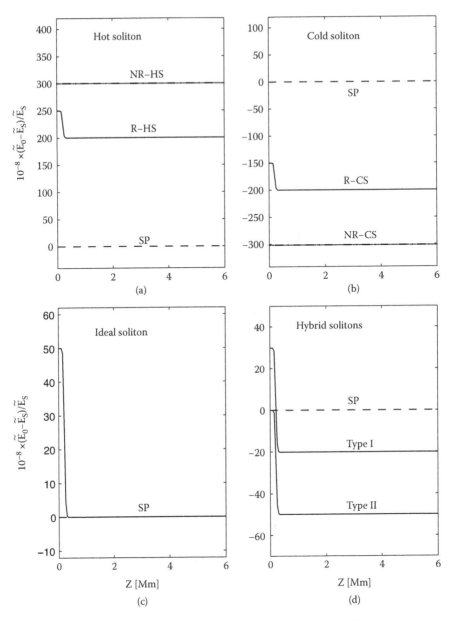

FIGURE 8.3: Evolution of the energies of perturbed solitons, $\tilde{E}_0$, for $a = \phi_0 = 0$, $\varepsilon = 2 \times 10^{-3}$. $\tilde{E}_S$ is the energy of the SP. HS *hot soliton*. CS, *cold soliton*. (a) $\eta = \phi = 0$, $\delta = 15 \times 10^{-4}$ for the NR-HS (non-radiating HS). $\eta = 1$, $\phi = \pi/10$, $\delta = 10^{-3}$ for the R-HS (radiating HS). (b) $\eta = \phi = 0$, $\delta = -15.1 \times 10^{-4}$ for the NR-CS. $\eta = 1$, $\phi = 11\pi/10$, $\delta = -10^{-3}$ for the R-CS. (c) $\eta = 1$, $\phi = \delta = 0$ for the ideal soliton. (d) $\eta = 1$, $\phi = 11\pi/10$, $\delta = -10^{-4}$ for the soliton of type I, and $\eta = 1$, $\phi = 0$, $\delta = -25 \times 10^{-5}$ for the type II.

as those of [5], and qualitatively, we have found the same results as those of Figs. 8.2 and 8.3. The general cartography of the perturbed solitons displayed in Fig. 8.1 is thus remarkably confirmed by the numerical simulations of pulse propagation in the realistic conditions of a perturbed environment. The behavior of the hybrid soliton of type I in Fig. 8.3 (d) is even more outstanding. Indeed, this soliton begins its propagation in a state of hyperthermia, then undergoes a loss of energy (by radiation) and gets closer to its stationary state. While we expect that it stabilizes in this state, this soliton passes through this state and stabilizes rather in a state of lower energy. Finally, the present study gives a new perspective on the dynamics of solitons. In particular, we show that, contrary to a widespread idea, a slightly perturbed soliton does not evolve necessarily in a way that tends to bring it back toward its fundamental (non-perturbed) state. Everything happens as if in a slightly perturbed environment, the fundamental state is no more easily accessible, and it into several perturbed states having their own identities (and energies), displayed in Fig. 8.1. The perturbed soliton accesses more easily perturbed states, without losing its stability. Without being too speculative, we think that the general cartography of perturbed solitons which we have drawn up in this study, possesses a general character that should extend to other types of solitons that have emerged in the literature.

ACKNOWLEDGMENT

E. T. Felenou acknowledges the SCAC (Cameroun) for financial support.

REFERENCES

1. Michel Remoissenet. *Waves Called Solitons: Concepts and Experiments.* 3rd ed., Springer-Verlag, Berlin, Heidelberg, 1993.
2. A. Hasegawa and F. Tappert. "Transmission of stationary nonlinear optical pulses in dispersive dielectric fibers. I. Anomalous dispersion." *Appl. Phys. Lett.*, 23, 142, 1973.
3. B. Kibler, J. Fatome, C. Finot, G. Millot, F. Dias, G. Genty, N. Akhmediev, and J. M. Dudley. "The Peregrine soliton in nonlinear fiber optics." *Nature Physics*, 6, 790, 2010.
4. M. Erkintalo, G. Genty, B. Wetzel, and J. M. Dudley. "Akhmediev breather evolution in optical fiber for realistic initial conditions." *Phys. Lett. A*, 375, 2029, 2011.
5. A. Hasegawa and Y. Kodama. "Guiding-center soliton in optical fibers." *Opt. Lett.*, 15, 1443, 1990.
6. S. K. Turitsyn, V. K. Mezentsev, and E. G. Shapiro. "Dispersion-managed solitons and optimization of the dispersion management." *Opt. Fiber Techn.*, 4, 384, 1998.
7. J. P. Gordon. "Dispersive pertubations of solitons of non linear Schrödinger equation." *J. Opt. Soc. Am. B*, 9, 91, 1992.
8. C. M. Ngabireng and P. Tchofo Dinda. "Radiating and non-radiating dispersion-managed solitons." *Opt. Lett.*, 30, 595, 2005.

9. J. H. B. Nijhof, N. J. Doran, W. Forysiak, and F. M. Knox. "Stable soliton-like propagation in dispersion managed systems with net anomalous, zero and normal dispersion." *Electron Lett.*, 33, 1726, 1997.

10. K. Nakkeeran, A. B. Moubissi, P. Tchofo Dinda, and S. Wabnitz. "Analytical method for designing dispersion managed fiber systems." *Opt. Lett.*, 26, 1544, 2001.

9 Optical solitary modes pumped by localized gain

Boris A. Malomed

9.1 INTRODUCTION AND MODELS

Spatial dissipative solitons (SDSs) are self-trapped beams of light [1–3] or plasmonic waves [4–12] propagating in planar or bulk waveguides. They result from the balance between diffraction and self-focusing (SF) nonlinearity, which is maintained simultaneously with the balance between the material loss and compensating gain. Due to their basic nature, SDSs are modes of profound significance to nonlinear photonics (optics and plasmonics), as concerns the fundamental studies and potential applications alike. In particular, a straightforward possibility is to use each sufficiently narrow SDS beam as a signal carrier in all-optical data-processing schemes. This application, as well as other settings in which the solitons occur, stresses the importance of the stabilization of the SDSs modes, and of development of enabling techniques for the generation and steering of such planar and bulk beams.

In terms of the theoretical description, basic models of SDS dynamics make use of complex Ginzburg–Landau equations (CGLEs). The prototypical one is the CGLE with cubic nonlinearity, which includes conservative paraxial-diffraction and Kerr terms, nonlinear (cubic) loss with coefficient $\epsilon > 0$, which represents two-photon absorption in the medium, and spatially uniform linear gain, with strength $\gamma > 0$, aiming to compensate the loss [1,2]:

$$\frac{\partial u}{\partial z} = \frac{i}{2}\nabla_\perp^2 u - (\epsilon - i\beta)|u|^2 u + \gamma u. \tag{9.1}$$

Here u is the complex amplitude of the electromagnetic wave in the spatial domain, z is the propagation distance, the paraxial-diffraction operator $\nabla_\perp^2$ acts on transverse coordinates (x, y) in the case of propagation in the bulk, or on the single coordinate, x, in the planar waveguide. Accordingly, Eq. (9.1) is considered as a two- or one-dimensional (2D or 1D) equation in those two cases. The equation is normalized so that the diffraction coefficient is 1, while β is the Kerr coefficient, $\beta > 0$ and $\beta < 0$ corresponding to the SF and self-defocusing (SDF) signs of nonlinearity, respectively.

A more general version of the CGLE may include an imaginary part of the diffraction coefficient [13–15], which is essential, in particular, for the use of the CGLE as a model of traveling-wave convection [16, 17]. However, in

199

optical models that coefficient, which would represent diffusivity of photons, is usually absent.

A well-known fact is that the 1D version of Eq. (9.1) gives rise to an exact solution in the form of an exact *chirped* SDS, which is often called a *Pereira–Stenflo soliton* [18, 19]:

$$u(x, z) = Ae^{ikz} [\text{sech}(\kappa x)]^{1+i\mu}, \qquad (9.2)$$

$$A^2 = 3\gamma/(2\epsilon), \quad \kappa^2 = \gamma/\mu, \quad k = (\gamma/2)(\mu^{-1} - \mu), \qquad (9.3)$$

where the chirp coefficient is

$$\mu = \sqrt{(3\beta/2\epsilon)^2 + 2 - 3\beta/(2\epsilon)}. \qquad (9.4)$$

This exact solution is subject to an obvious instability, due to the action of the uniform linear gain on the zero background far from the soliton's core. Therefore, an important problem is the design of physically relevant models which may produce stable SDSs.

One possibility is to achieve full stabilization of the solitons in systems of linearly coupled CGLEs modeling dual-core waveguides, with the linear gain and loss acting in different cores [10, 20–26]. This includes, inter alia, a $\mathcal{PT}$-symmetric version of the system that features the exact balance between spatially separated gain and loss [27, 28]. The simplest example of such a stabilization mechanism is offered by the following coupled CGLE system [22]:

$$\frac{\partial u}{\partial z} = \frac{i}{2}\nabla_\perp^2 u - (\epsilon - i\beta)|u|^2 u + \gamma u + i\lambda v, \qquad (9.5)$$

$$\frac{\partial v}{\partial z} = (iq - \Gamma)v + i\lambda u, \qquad (9.6)$$

where λ is the linear-coupling coefficient, $v(x, z)$ and $\Gamma > 0$ are the electromagnetic wave amplitude and the linear loss rate in the stabilizing dissipative core, and q is a possible wavenumber mismatch between the cores. In the case of $q = 0$, the zero background is stable in the framework of Eqs. (9.5) and (9.6) under the condition

$$\gamma < \Gamma < \lambda^2/\gamma. \qquad (9.7)$$

The same ansatz (9.2) which produced the Pereira–Stenflo soliton for the uncoupled CGLE yields an exact solution of the coupled system (9.5), (9.6):

$$\{u(x, z), v(x, z)\} = \{A, B\} e^{ikz} [\text{sech}(\kappa x)]^{1+i\mu}, \qquad (9.8)$$

with chirp μ given by the same expression (9.4) as above, and

$$B = i\lambda [\Gamma + i(k - q)]^{-1} A. \qquad (9.9)$$

A stable soliton is obtained if a pair of distinct solutions is found, compatible with the condition of the stability for the zero background [which is Eq. (9.7) in the case of $q = 0$], instead of the single solution in the case of Eq. (9.3). Then, the soliton with the larger amplitude is stable, coexisting, as an *attractor*, with the stable zero solution, while the additional soliton with a smaller amplitude plays the role of an unstable *separatrix* which delineates the boundary between attraction basins of the two coexisting stable solutions [22].

In the case of $q = 0$, the aforementioned condition of the existence of two solutions reduces to

$$\gamma\Gamma\left(1 - \mu^2\right) > 4\mu^2\left[(\lambda^2 - \gamma\Gamma) + 2\Gamma\left(\Gamma - \gamma\right)\right]. \tag{9.10}$$

In particular, it follows from Eqs. (9.10) and (9.4) that a related necessary condition, $\mu < 1$, implies $\epsilon < 3\beta$, i.e., the Kerr coefficient, β, must feature the SF sign, and the cubic-loss coefficient, ϵ, must be sufficiently small in comparison with β. If inequality $\mu < 1$ holds, and zero background is close to its stability boundaries, i.e., $0 < \Gamma - \gamma \ll \gamma$ and $0 < \lambda^2 - \gamma\Gamma \ll \gamma^2$ [see Eq. (9.7)], the parameters of the stable soliton with the larger amplitude are

$$\kappa_{\mathrm{st}} \approx \sqrt{\frac{\gamma}{\mu}\frac{1 - \mu^2}{1 + \mu^2}}, A_{\mathrm{st}} \approx \sqrt{\frac{3\gamma}{2\epsilon}\frac{1 - \mu^2}{1 + \mu^2}}, B_{\mathrm{st}} \approx \frac{2\mu}{(1 - i\mu)^2}A, \tag{9.11}$$

while propagation constant k is given, in the first approximation, by Eq. (9.3). In the same case, the unstable separatrix soliton has a small amplitude and large width, while its chirp keeps the above value (9.4):

$$\kappa_{\mathrm{sep}}^2 \approx \frac{2\left(\Gamma - \gamma\right)}{\gamma\left(1 - \mu^2\right)}\sqrt{\lambda^2 - \gamma\Gamma}, \; k_{\mathrm{sep}} \approx \sqrt{\lambda^2 - \gamma\Gamma}, \tag{9.12}$$

$$A_{\mathrm{sep}}^2 \approx \frac{3\mu}{2\epsilon}\kappa_{\mathrm{sep}}^2, \; B_{\mathrm{sep}} \approx iA_{\mathrm{sep}}. \tag{9.13}$$

Getting back to models based on the single CGLE, stable solitons can also be generated by the equation with cubic gain "sandwiched" between linear and quintic loss terms, which corresponds to the following generalization of Eq. (9.1):

$$\frac{\partial u}{\partial z} = \frac{i}{2}\nabla_{\perp}^2 u + (\epsilon_3 + i\beta_3)\left|u\right|^2 u - (\epsilon_5 + i\beta_5)\left|u\right|^2 u - \Gamma u, \tag{9.14}$$

with $\epsilon_3 > 0$, $\epsilon_5 > 0$, $\Gamma > 0$, and $\beta_5 \geq 0$. The linear loss, represented by coefficient Γ, provides for the stability of the zero solution to Eq. (9.14). The cubic-quintic (CQ) CGLE was first proposed, in a phenomenological form, by Petviashvili and Sergeev [29]. Later, it was demonstrated that the CQ model may be realized in optics as a combination of linear amplification and saturable absorption [36–38]. Stable dissipative solitons supported by this model were investigated in detail by means of numerical and analytical methods [30–35].

The subject of the present mini-review is the development of another method for creating stable localized modes which makes use of linear gain applied at a "hot spot" (HS), i.e., a localized amplifying region embedded into a bulk lossy waveguide. The experimental technique which allows one to create localized gain by means of strongly inhomogeneous distributions of dopants implanted into the lossy waveguide, which produce the gain if pumped by an external source of light, is well known [39]. Another possibility is even more feasible and versatile: the dopant density may be uniform, while the external pump beam is focused on the location where the HS should be created.

Supporting dissipative solitons by localized gain was first proposed not in the framework of CGLEs, but for a gap soliton pinned to an HS in a lossy Bragg grating (BG) [40]. In terms of the spatial-domain dynamics, the respective model is based on the system of coupled-mode equations (CMEs) for counterpropagating waves, $u(x, z)$ and $u(x, z)$, coupled by the Bragg reflection:

$$
iu_z + iu_x + v + \left(|u|^2 + 2|v|^2\right)u =
$$
$$
-i\gamma u + i\left(\Gamma_1 + i\Gamma_2\right)\delta(x)u, \tag{9.15}
$$
$$
iv_z - iv_x + u + \left(|v|^2 + 2|u|^2\right)v =
$$
$$
-i\gamma v + i\left(\Gamma_1 + i\Gamma_2\right)\delta(x)v, \tag{9.16}
$$

where the tilt of the light beam and the reflection coefficients are normalized to be 1, the nonlinear terms account for the self- and cross-phase modulation induced by the Kerr effect, $\gamma > 0$ is the linear-loss parameter, $\Gamma_1 > 0$ represents the local gain applied at the HS [$x = 0$, $\delta(x)$ being the Dirac delta function], and the imaginary part of the gain coefficient, $\Gamma_2 \geq 0$, accounts for a possible attractive potential induced by the HS (it approximates a local increase of the refractive index around the HS).

As mentioned in [40] too, and for the first time investigated in detail in [41], the HS embedded into the usual planar waveguide is described by the following modification of Eq. (9.1):

$$
\frac{\partial u}{\partial z} = \frac{i}{2}\nabla_\perp^2 u - (\epsilon - i\beta)|u|^2 u - \gamma u + (\Gamma_1 + i\Gamma_2)\delta(x)u, \tag{9.17}
$$

where, as well as in Eqs. (9.15) and (9.16), $\Gamma_1 > 0$ is assumed, and the negative sign in front of $\gamma \geq 0$ represents the linear loss in the bulk waveguide. Another HS model, based on the 1D CGLE with CQ nonlinearity, was introduced in [42]:

$$
\frac{\partial u}{\partial z} = \frac{i}{2}\frac{\partial^2 u}{\partial x^2} + i|u|^2 u - i\beta_5 |u|^2 u - \gamma u + \Gamma e^{-x^2/w^2}|u|^2 u, \tag{9.18}
$$

where $\beta_5 > 0$ represents the quintic self-defocusing term, $\gamma > 0$ and $\Gamma > 0$ are, as above, strengths of the bulk losses and localized *cubic* gain, and w is the width of the HS (an approximation corresponding to $w \to 0$, with the HS in the form of the delta function, may be applied here, too). While solitons in uniform media, supported by cubic gain, are always unstable against blowup in the absence of quintic loss [43], the analysis reported in [42] demonstrates that, quite counterintuitively, *stable* dissipative localized modes in the uniform lossy medium may be supported by *unsaturated* localized cubic gain in the model based on Eq. (9.18).

In addition to the "direct" linear gain assumed in the above-mentioned models, losses in photonic media may be compensated by parametric amplification, which, unlike the direct gain, is sensitive to the phase of the signal [44, 45]. This mechanism can be used for the creation of a HS, if the parametric gain is applied in a narrow segment of the waveguide. As proposed in [46], the respective 1D model is based on the following equation [cf. Eqs. (9.17) and (9.18)]:

$$\frac{\partial u}{\partial z} = \frac{i}{2}\frac{\partial^2 u}{\partial x^2} + i|u|^2 u - (\epsilon - i\beta)|u|^2 u - (\gamma - iq)u + \Gamma e^{-x^2/w^2} u^*, \quad (9.19)$$

where u^* is the complex conjugate field, q is a real phase-mismatch parameter, and, as well as in Eq. (9.18), the HS may be approximated by the delta function in the limit of $w \to 0$.

Models combining localized gain and uniformly distributed Kerr nonlinearity and linear loss have been recently developed in various directions. In particular, 1D models with two or multiple HSs [47–51] and periodic amplifying structures [52,53], as well as extended patterns [54,55], have been studied, chiefly by means of numerical methods. Numerical analysis has also made it possible to study 2D settings, in which, most notably, stable localized vortices are supported by the gain confined to an annular-shaped area [56–60]. The parametric amplification applied at a ring may support stable vortices, too, provided that the pump 2D beam itself has an inner vortical structure [60].

Another ramification of the topic is the development of symmetric combinations of "hot" and "cold" spots, which offer a realization of the concept of $\mathcal{PT}$-symmetric systems in optical media, which were proposed and built as settings integrating the balanced spatially separated gain and loss with a spatially symmetric profile of the local refractive index [61]– [64]. The study of solitons in nonlinear $\mathcal{PT}$-symmetric settings has drawn a great deal of attention [27, 28, 65–68]. In particular, it is possible to consider the 1D model in the form of a symmetric pair of hot and cold spots described by two delta functions embedded into a bulk conservative medium with cubic nonlinearity [69]. A limit case of this setting, which admits exact analytical solutions for $\mathcal{PT}$-symmetric solitons, corresponds to a $\mathcal{PT}$ *dipole*, which is represented by the derivative of the delta function in the following 1D equation [70,99]:

$$i\frac{\partial u}{\partial z} = -\frac{1}{2}\frac{\partial^2 u}{\partial x^2} - \sigma|u|^2 u - \varepsilon_0 u\delta(x) + i\gamma u\delta'(x). \quad (9.20)$$

Here $\sigma = +1$ and -1 correspond to the SF and SDF bulk nonlinearity, respectively, $\varepsilon_0 \geq 0$ is the strength of the attractive potential, which is a natural conservative component of the $\mathcal{PT}$ *dipole*, and γ is the strength of the $\mathcal{PT}$ dipole.

It is also natural to consider discrete photonic settings (lattices), which appear, in the form of discrete CGLEs, as models of arrayed optical [72–77] or plasmonics [78, 79] waveguides. In this context, lattice counterparts of the HSs amount to a single [80] or several [81] amplified site(s) embedded into a 1D or 2D [82] lossy array. Being interested in tightly localized discrete states, one can additionally simplify the model by assuming that the nonlinearity is carried only by the active cores, which gives rise to the following version of the discrete CGLE, written here in the general 2D form [82]:

$$\frac{du_{m,n}}{dz} = \frac{i}{2} \left(u_{m-1,n} + u_{m+1,n} + u_{m,n-1} + u_{m,n+1} - 4u_{m,n} \right)$$
$$-\gamma u_{m,n} + \left[(\Gamma_1 + i\Gamma_2) + (iB - E)|u_{m,n}|^2 \right] \delta_{m,0}\delta_{n,0} u_{m,n}, \qquad (9.21)$$

where $m, n = 0, \pm 1, \pm 2, \ldots$ are discrete coordinates on the lattice, $\delta_{m,0}$ and $\delta_{n,0}$ are the Kronecker symbols, and the coefficient of the linear coupling between adjacent cores is scaled to be 1. As above, $\gamma > 0$ is the linear loss in the bulk lattice, $\Gamma_1 > 0$ and $\Gamma_2 \geq 0$ represent the linear gain and linear potential applied at the HS site ($m = n = 0$), while B and E account for the SF ($B > 0$) or SDF ($B < 0$) Kerr nonlinearity and nonlinear loss ($E > 0$) or gain ($E < 0$) acting at the HS (the unsaturated cubic gain may be a meaningful feature in this setting [82]).

The $\mathcal{PT}$ symmetry can be introduced too in the framework of the lattice system. In particular, a discrete counterpart of the 1D continuous model (9.20) with the $\mathcal{PT}$-symmetric dipole was recently elaborated in [83]:

$$i\frac{du_n}{dz} = -\left(C_{n,n-1}u_{n-1} + C_{n+1,n}u_{n+1} \right) - g_n|u_n|^2 u_n + i\kappa_n u_n, \qquad (9.22)$$

where the $\mathcal{PT}$ dimer (discrete dipole) embedded into the Hamiltonian lattice is represented by $\kappa_n = +\kappa$ at $n = 0$, $-\kappa$ at $n = 1$, and 0 at $n \neq 0, 1$. A counterpart of the delta-functional attractive potential in Eq. (9.20) corresponds to a local defect in the inter-site couplings: $C_{1,0} = C_d$, $C_{n,n-1} = C_0 \neq C_d$ at $n \neq 1$. Finally, the nonlinearity is assumed to be carried solely by the dimer embedded into the lattice: $g_n = g$ at $n = 0, 1$, and 0 at $n \neq 0, 1$ [cf. Eq. (9.21)]. Equation (9.22) admits exact analytical solutions for all $\mathcal{PT}$-symmetric and antisymmetric discrete solitons pinned to the dimer.

In addition to the HS, one can naturally define a "warm spot" (WS) in the 2D CGLE with CQ nonlinearity, where the coefficient of the linear loss is given a spatial profile with a minimum at the WS ($\mathbf{r} = 0$) [84]. The equation may be taken as the 2D version of Eq. (9.14) with

$$\Gamma(r) = \Gamma_0 + \Gamma_2 r^2, \qquad (9.23)$$

where r is the radial coordinate, coefficients Γ_0 and Γ_2 being positive. This seemingly simple model gives rise to a great variety of stable 2D modes pinned to the WS. Depending on the values of the parameters in Eqs. (9.14) and

(9.23), these may be simple vortices, rotating elliptic, eccentric, and slanted vortices, spinning crescents, etc. [84].

Finally, the use of the spatial modulation of loss coefficients opens another way for the stabilization of the SDS: as shown in [85], the solitons may be readily made stable if the *spatially uniform* linear gain is combined with the local strength of the cubic loss, $\epsilon(r)$, growing from the center to the periphery at any rate faster than r^D, where r is the distance from the center and D the spatial dimension [11]. This setting is described by the following modification of Eq. (9.1):

$$\frac{\partial u}{\partial z} = \frac{i}{2}\nabla_\perp^2 u - [\epsilon(r) - i\beta]\,|u|^2 u + \gamma u, \qquad (9.24)$$

where, as above, γ and $\epsilon(r)$ are positive, so that $\lim_{r\to\infty}\left(r^D/\epsilon(r)\right) = 0$, for $D = 1$ or 2.

This article aims to present a survey of basic results obtained for SDSs pinned to HSs in the class of models outlined above. The stress is on the most fundamental results that can be obtained in an analytical or semi-analytical form, in combination with the related numerical findings, thus providing a deep insight into the dynamics of the underlying photonic systems. In fact, the possibility of obtaining many essential results in an analytical form is a certain asset of these models. First, in Sec. 9.2 findings are summarized for the most fundamental 1D model based on Eq. (9.17), which is followed, in Sec. 9.3, by consideration of the $\mathcal{PT}$-symmetric system Eq. (9.20). The BG model Eq. (9.15), Eq. (9.16) is the subject of Sec. 9.4, and Sec. 9.5 deals with the 1D version of the discrete model Eq. (9.21). The article is concluded by Sec. 9.6.

9.2   DISSIPATIVE SOLITONS PINNED TO HOT SPOTS IN THE ORDINARY WAVEGUIDE

The presentation in this section is focused on basic model (9.17) and its extension for two HSs, chiefly following works [41] and [47], for settings with a single and double HS, respectively. Both analytical and numerical results are presented which highlight the most fundamental properties of SDS supported by the tightly localized gain embedded into lossy optical media.

9.2.1   ANALYTICAL CONSIDERATIONS

9.2.2   EXACT RESULTS

Stationary solutions to Eq. (9.17) are looked for as $u(x,t) = e^{ikz}U(x)$, where complex function $U(x)$ satisfies an ordinary differential equation,

$$(\gamma + ik)\,U = \frac{i}{2}\frac{d^2 U}{dx^2} - (\epsilon - i\beta)\,|U|^2 U, \qquad (9.25)$$

at $x \neq 0$, supplemented by the boundary condition (b.c.) at $x = 0$, which is generated by the integration of Eq. (9.17) in an infinitesimal vicinity of $x = 0$,

$$\lim_{x \to +0} \frac{d}{dx} U(x) = (i\Gamma_1 - \Gamma_2) U(x = 0), \tag{9.26}$$

assuming even stationary solutions, $U(-x) = U(x)$. As seen from the expression for A^2 in Eq. (9.3) [recall that Eqs. (9.1) and (9.17) have opposite signs in front of γ], Eq. (9.25) with $\gamma > 0$ and $\epsilon > 0$ cannot be solved by a sech ansatz similar to that in Eq. (9.2). As an alternative, sech can be replaced by $1/\sinh$:

$$U(x) = A\left[\sinh\left(\kappa\left(|x| + \xi\right)\right)\right]^{-(1+i\mu)}, \tag{9.27}$$

where $\xi > 0$ prevents the singularity. This ansatz yields an exact *codimension-one* solution to Eq. (9.25) with b.c. (9.26), which is valid under a special constraint imposed on coefficients of the system:

$$\Gamma_1/\Gamma_2 - 2\Gamma_2/\Gamma_1 = 3\beta/\epsilon. \tag{9.28}$$

The parameters of this solution are

$$A^2 = \frac{3\gamma}{2\epsilon}, \quad \kappa^2 = \gamma\frac{\Gamma_2}{\Gamma_1}, \quad \mu = -\frac{\Gamma_1}{\Gamma_2}, \quad k = \frac{\gamma}{2}\left(\frac{\Gamma_2}{\Gamma_1} - \frac{\Gamma_1}{\Gamma_2}\right), \tag{9.29}$$

$$\xi = \frac{1}{2}\sqrt{\frac{\Gamma_1}{\gamma\Gamma_2}} \ln\left(\frac{\sqrt{\Gamma_1\Gamma_2} + \sqrt{\gamma}}{\sqrt{\Gamma_1\Gamma_2} - \sqrt{\gamma}}\right). \tag{9.30}$$

The squared amplitude of the solution is

$$|U(x = 0)|^2 = (3/2\epsilon)\left(\Gamma_1\Gamma_2 - \gamma\right). \tag{9.31}$$

The main characteristic of the localized beam is its total power,

$$P = \int_{-\infty}^{+\infty} |u(x)|^2 \, dx. \tag{9.32}$$

For the solution given by Eqs. (9.27)–(9.30),

$$P = (3/\epsilon)\sqrt{\Gamma_1}\left(\sqrt{\Gamma_1} - \sqrt{\gamma/\Gamma_2}\right). \tag{9.33}$$

Obviously, solution (9.27) exists if it yields $|U(x = 0)|^2 > 0$ and $P > 0$, i.e.,

$$\Gamma_1 > (\Gamma_1)_{\text{thr}} \equiv \gamma/\Gamma_2. \tag{9.34}$$

The meaning of threshold condition (9.34) is that, to support the stable pinned soliton, the local gain (Γ_1) must be sufficiently large in comparison with the background loss, γ. It is relevant to mention that, according to Eq. (9.28), the exact solution given by Eqs. (9.27), (9.29), and (9.30) emerges at threshold (9.34) in the SDF medium, with $\beta < 0$, provided that $\gamma < \sqrt{2}\Gamma_2^2$. In the opposite case, $\gamma > \sqrt{2}\Gamma_2^2$, the threshold is realized in the SF medium, with $\beta > 0$.

9.2.3 EXACT RESULTS FOR $\gamma = 0$ (NO LINEAR BACKGROUND LOSS)

The above analytical solution admits a nontrivial limit for $\gamma \to 0$, which implies that the local gain compensates only the nonlinear loss, accounted for by term $\sim \epsilon$ in Eq. (9.17). In this limit, the pinned state is weakly localized, instead of the exponentially localized one (9.27):

$$U_{\gamma=0}(x) = \sqrt{\frac{3}{2\epsilon}} \frac{\sqrt{\Gamma_1/\Gamma_2}}{\left(|x| + \Gamma_2^{-1}\right)^{1+i\mu}} \, , \quad k = 0, \tag{9.35}$$

with μ given by expression (9.29) (an overall phase shift is dropped here). Note that the existence of solution (9.35) does not require any threshold condition, unlike Eq. (9.34). This weakly localized state is a physically meaningful one, as its total power (9.32) converges, $P(\gamma = 0) = 3\Gamma_1/\epsilon$. Note that this power does not depend on the local-potential strength, Γ_2, unlike the generic expression (9.33).

9.2.4 PERTURBATIVE RESULTS FOR THE SELF-DEFOCUSING MEDIUM

In the limit case when the loss and gain vanish, $\gamma = \epsilon = \Gamma_1 = 0$, solution (9.27) goes over into an exact one in the SDF medium (with $\beta < 0$) pinned by the attractive potential:

$$U(x) = \frac{\sqrt{2k/|\beta|}}{\sinh\left(\sqrt{2k}\left(|x| + \xi_0\right)\right)}, \tag{9.36}$$

$$\xi_0 = \frac{1}{2\sqrt{2k}} \ln\left(\frac{\Gamma_2 + \sqrt{2k}}{\Gamma_2 - \sqrt{2k}}\right), \tag{9.37}$$

$$|U(x = 0)|^2 = |\beta|^{-1}\left(\Gamma_2 - 2k\right), \tag{9.38}$$

which in interval $0 < k < (1/2)\Gamma_2^2$ of the propagation constant. The total power of this solution is

$$P_0 = (2/|\beta|)\left(\Gamma_2 - \sqrt{2k}\right). \tag{9.39}$$

In the limit of $k = 0$, when the amplitude given by Eq. (9.38) and the power given by Eq. (9.39) attain their maxima, the solution degenerates from an exponentially localized into a weakly localized one [cf. Eq. (9.35)],

$$U_{k=0}(x) = \frac{1}{\sqrt{|\beta|}\left(|x| + \Gamma_2^{-1}\right)}, \tag{9.40}$$

whose total power converges, $P_0(k = 0) = 2\Gamma_2/|\beta|$, as per Eq. (9.39). The exact solutions given by Eqs. (9.36)–(9.40), which are generic ones in the conservative model [no spacial constraint, such as Eq. (9.28), is required], may be used to construct an approximate solution to the full system of Eqs. (9.25) and (9.26), assuming that the gain and loss parameters, Γ_1, γ, and ϵ, are all small. To this end, one can use the balance equation for the total power:

$$\frac{dP}{dz} = -2\gamma P - 2\epsilon \int_{-\infty}^{+\infty} |u(x)|^4 \, dx + 2\Gamma_1 |u(x=0)|^2 = 0. \qquad (9.41)$$

The substitution, in the zero-order approximation, of solution (9.36), (9.37) into Eq. (9.41) yields the gain strength which is required to compensate the linear and nonlinear losses in the solution with propagation constant k:

$$\Gamma_1 = \frac{2\gamma}{\Gamma_2 + \sqrt{2k}} + \frac{2\epsilon}{3|\beta|} \frac{\left(\Gamma_2 - \sqrt{2k}\right)\left(\Gamma_2 + 2\sqrt{2k}\right)}{\Gamma_2 + \sqrt{2k}}. \qquad (9.42)$$

As follows from Eq. (9.42), with the decrease of k from the largest possible value, $(1/2)\,\Gamma_2^2$, to 0, the necessary gain increases from the minimum, which exactly coincides with the threshold given by Eq. (9.34), to the largest value at which the perturbative treatment admits the existence of the stationary pinned mode,

$$(\Gamma_1)_{\max} = 2\gamma/\Gamma_2 + 2\epsilon\Gamma_2/\left(3|\beta|\right). \qquad (9.43)$$

The respective total power grows from 0 to the above-mentioned maximum, $2\Gamma_2/|\beta|$. It is expected that, at Γ_1 exceeding the limit value (9.43) admitted by the stationary mode, the solution becomes nonstationary, with the pinned mode emitting radiation waves, which makes the effective loss larger and thus balances $\Gamma_1 - (\Gamma_1)_{\max}$. However, this issue was not studied in detail. The perturbative result clearly suggests that, in the lossy SDF medium with local gain, the pinned modes exist not only under the special condition given by Eq. (9.28), at which they are available in the exact form, but as fully generic solutions too. Furthermore, the increase of power with gain strength implies that the modes are, most plausibly, stable ones.

9.2.5   PERTURBATIVE RESULTS FOR THE SELF-FOCUSING MEDIUM

In the case of $\beta > 0$, which corresponds to the SF sign of cubic nonlinearity, a commonly known exact solution for the pinned mode in the absence of the loss and gain, $\gamma = \epsilon = \Gamma_1 = 0$, is

$$U(x) \;=\; \sqrt{2k/|\beta|}\,\text{sech}\left(\sqrt{2k}\,(|x| + \xi_0)\right), \qquad (9.44)$$

$$\xi_0 \;=\; \frac{1}{2\sqrt{2k}} \ln\left(\frac{\sqrt{2k} + \Gamma_2}{\sqrt{2k} - \Gamma_2}\right), \qquad (9.45)$$

$$|U(x=0)|^2 \;=\; |\beta|^{-1}\left(2k - \Gamma_2^2\right), \qquad (9.46)$$

with the total power

$$P_0 = (2/\beta) \left(\sqrt{2k} - \Gamma_2 \right), \qquad (9.47)$$

which exists for propagation constant $k > (1/2) \Gamma_2^2$, [cf. Eqs. (9.36)–(9.39)]. In this case, the power-balance condition (9.41) yields a result which is essentially different from its counterpart, Eq.(9.42):

$$\Gamma_1 = \frac{2\gamma}{\sqrt{2k} + \Gamma_2} + \frac{2\epsilon}{3\beta} \frac{\left(\sqrt{2k} - \Gamma_2 \right) \left(2\sqrt{2k} + \Gamma_2 \right)}{\sqrt{2k} + \Gamma_2}. \qquad (9.48)$$

Straightforward consideration of Eq. (9.48) reveals the difference of the situation from that considered above for the SDF medium: if the strength of the nonlinear loss is relatively small,

$$\epsilon < \epsilon_{\mathrm{cr}} = \left(\beta/2\Gamma_2^2 \right) \gamma, \qquad (9.49)$$

the growth of power given by Eq. (9.47) from zero at $\sqrt{2k} = \Gamma_2$ with the increase of k initially (at small values of P_0) requires not the increase of the gain strength from the threshold value given by Eq. (9.34) to $\Gamma_1 > (\Gamma_1)_{\mathrm{thr}}$, but, on the contrary, a *decrease* of Γ_1 to $\Gamma_1 < (\Gamma_1)_{\mathrm{thr}}$ [40]. Only at $\epsilon > \epsilon_{\mathrm{cr}}$ [see Eq. (9.49)] power grows with Γ_1 starting from $\Gamma_1 = (\Gamma_1)_{\mathrm{thr}}$.

9.2.6 STABILITY OF THE ZERO SOLUTION AND ITS RELATION TO THE EXISTENCE OF PINNED SOLITONS

It is possible to check the stability of the zero solution, which is an obvious prerequisite for the soliton's stability, as mentioned above. To this end, one should use the linearized version of Eq. (9.17),

$$\frac{\partial u_{\mathrm{lin}}}{\partial z} = \frac{i}{2} \frac{\partial^2 u_{\mathrm{lin}}}{\partial x^2} - \gamma u_{\mathrm{lin}} + (\Gamma_1 + i\Gamma_2) \, \delta(x) u_{\mathrm{lin}}. \qquad (9.50)$$

The critical role is played by localized eigenmodes of Eq. (9.50),

$$u_{\mathrm{lin}}(x, t) = u_0 e^{\Lambda z} e^{il|x| - \lambda|x|}, \qquad (9.51)$$

where u_0 is an arbitrary amplitude, localization parameter λ must be positive, l is a wavenumber, and Λ is a complex instability growth rate. Straightforward analysis yields [41]

$$\lambda - il = \Gamma_2 - i\Gamma_1, \quad \Lambda = (i/2) \left(\Gamma_2 - i\Gamma_1 \right)^2 - \gamma. \qquad (9.52)$$

It follows from Eq. (9.52) that the stability condition for the zero solution, $\mathrm{Re}\,(\Lambda) < 0$, amounts to inequality $\gamma > \Gamma_1\Gamma_2$, which is exactly *opposite* to Eq. (9.34). In fact, Eq. (9.31) demonstrates that the exact pinned-soliton solution given by Eqs. (9.27)–(9.30) emerges, via the standard *forward* (alias *supercritical*) *pitchfork bifurcation* [86], precisely at the point where the zero solution loses its stability to the local perturbation. The previous analysis of Eq. (9.42) demonstrates that the same happens with the perturbative solution (9.36), (9.37) in the SDF model. On the contrary, analysis of Eq. (9.48) has revealed above that the same transition happens to the perturbative solution (9.44), (9.45) in the SF medium only at $\epsilon > \epsilon_{\rm cr}$ [see Eq. (9.49)], while at $\epsilon > \epsilon_{\rm cr}$ the pitchfork bifurcation is of the *backward* (alias *subcritical* [86]) type, featuring the power which originally grows with the *decrease* of the gain strength. Accordingly, the pinned modes emerging from the subcritical bifurcation are unstable. However, the contribution of the nonlinear loss ($\epsilon > 0$) eventually leads to the turn of the solution branch forward and its stabilization at the turning point. For very small ϵ, the turning point determined by Eq. (9.48) is located at $\Gamma_1 \approx 4\sqrt{2\gamma\epsilon/(3\beta)}$, $P_0 \approx \sqrt{6\gamma/(\beta\epsilon)}$. Finally, note that the zero solution is never destabilized, and the stable pinned soliton does not emerge in the absence of local attractive potential, i.e., at $\Gamma_2 \leq 0$.

9.2.7 NUMERICAL RESULTS

9.2.8 SELF-TRAPPING AND STABILITY OF PINNED SOLITONS

Numerical analysis of the model based on Eq. (9.17) was performed with the delta-function replaced by its Gaussian approximation, with the finite width σ.

$$\tilde{\delta}(x) = \left(\sqrt{\pi}\sigma\right)^{-1} \exp\left(-x^2/\sigma^2\right), \tag{9.53}$$

The shape of a typical analytical solution (9.27)–(9.31) for the pinned soliton, and a set of approximations to it provided by the use of approximation (9.53), are displayed in Fig. 9.1. All these pinned states are stable, as was checked by simulations of their perturbed evolution in the framework of Eq. (9.17) with $\delta(x)$ replaced by $\tilde{\delta}(x)$. The minimum (threshold) value of the local-gain strength, Γ_1, which is necessary for the existence of stable pinned solitons, is an important characteristic of the setting [see Eq. (9.34)]. Figure 9.2 displays the dependence of $(\Gamma_1)_{\rm thr}$ on the strength of the background loss, characterized by $\sqrt{\gamma}$, for $\beta = 0$ and three fixed values of the local-potential's strength, $\Gamma_2 = +1, 0, -1$ (in fact, the solutions corresponding to $\Gamma_2 = -1$ are unstable, as they are repelled by the HS). In addition, $(\Gamma_1)_{\rm thr}$ is also shown as a function of $\sqrt{\gamma}$ under constraint (9.28), which amounts to $\Gamma_2 = \Gamma_1/\sqrt{2}$, in the case of $\beta = 0$. The corresponding analytical prediction, as given by Eq. (9.34), is virtually identical to its numerical counterpart, despite the difference of approximation (9.53) from the ideal delta function.

Figure 9.2 corroborates that, as said above, the analytical solutions represent only a particular case of the family of generic dissipative solitons that can be found in the numerical form. In particular, the solution produces narrow and tall stable pinned solitons at large values of Γ_1. All the pinned solitons, including weakly localized ones predicted by analytical solution (9.35) for $\gamma = 0$, are stable at $\Gamma_1 > (\Gamma_1)_{\rm thr}$ and $\Gamma_2 > 0$. The situation is essentially different for

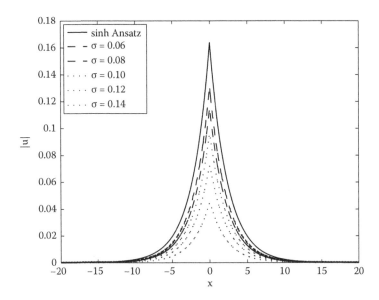

FIGURE 9.1: The exact solution for the pinned soliton given by Eqs. (9.27)–(9.31), and a set of approximations generated by the regularized delta function defined as per Eq. (9.53). All the profiles represent stable solutions. Other parameters are $\beta = 0$, $\gamma = 0.25$, $\Gamma_1 = 0.6155$, and $\Gamma_2 = \Gamma_1/\sqrt{2}$, as per Eq. (9.28).

large σ in Eq. (9.53), i.e., when the local gain is supplied in a broad region. In that case, simulations do not demonstrate self-trapping into stationary solitons; instead, a generic outcome is the formation of stable *breathers* featuring regular intrinsic oscillations, the breather's width being on the same order of magnitude as σ (see a typical example in Fig. 9.3). It seems plausible that, with the increase of σ, the static pinned soliton is destabilized via the Hopf bifurcation [86] which gives rise to the stable breather.

9.2.9 MODEL WITH A DOUBLE HOT SPOT

The extension of Eq. (9.17) for two mutually symmetric HSs separated by distance $2L$ was introduced in [47]:

$$\frac{\partial u}{\partial z} = \frac{i}{2}\frac{\partial^2 u}{\partial x^2} - \gamma u - (\epsilon - i\beta)|u|^2 u$$
$$+ (\Gamma_1 + i\Gamma_2)\left[\delta\left(x + L\right) + \delta\left(x - L\right)\right]u. \quad (9.54)$$

Numerical analysis has demonstrated that stationary symmetric solutions of this equation (they are not available in an analytical form) are stable (see a typical example in Fig. 9.4), while all antisymmetric states are unstable, spontaneously transforming into their symmetric counterparts. As shown above in

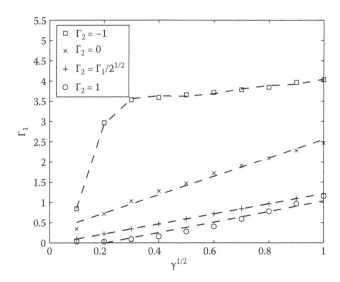

FIGURE 9.2: Chains of symbols show the minimum value of local gain, Γ_1, which is necessary for the creation of stationary pinned solitons in the framework of Eq. (9.17), as a function of the background loss (γ), with $\beta = 0$, $\varepsilon = 1$, and $\delta(x)$ approximated as per Eq. (9.53) with $\sigma = 0.1$. The strength of the local potential, Γ_2, is fixed, as indicated in the box. For $\Gamma_2 = \pm 1$ and 0, the lines are guides for the eye, while the straight line for the case of $\Gamma_2 = \Gamma_1/\sqrt{2}$, which corresponds to Eq. (9.28) with $\beta = 0$, is the analytical prediction given by Eq. (9.34).

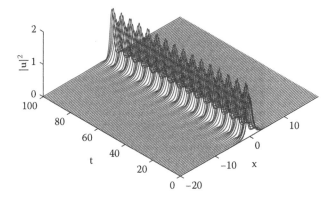

FIGURE 9.3: A typical example of a robust breather produced by simulations of Eq. (9.17) with $\delta(x)$ replaced by approximation $\tilde{\delta}(x)$ as per Eq. (9.53) with $\sigma = 2$, the other parameters being $\beta = \epsilon = \Gamma_2 = 1$, $\Gamma_1 = 4$, and $\gamma = 0.1$.

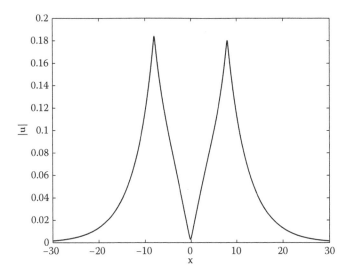

FIGURE 9.4: A stable symmetric mode generated by Eq. (9.54) with the delta functions approximated as per Eq. (9.53) with $\sigma = 0.1$. Other parameters are $\beta = 0, \epsilon = 1, \gamma = 0.057, \Gamma_1 = 0.334, \Gamma_2 = 0.236$, and $L = 8$.

Fig. 9.3, Eq. (9.17) with the single HS described by expression (9.53), where σ is large enough, supports breathers instead of stationary pinned modes. Simulations of Eq. (9.54) demonstrate that a pair of such broad HSs support unsynchronized breathers pinned by each HS if the distance between them is large enough, hence the breathers virtually do not interact. A completely different effect is displayed in Fig. 9.5 for two broad HSs which are set closer to each other: the interaction between the breathers pinned to each HS transforms them into a *stationary* stable symmetric mode. It is relevant to stress that the transformation of the breather pinned by the isolated HS into a stationary pinned state does not occur for the same parameters. This outcome is a generic result of the interaction between the breathers, provided that the distance between them is not too large.

9.2.10   RELATED MODELS

In addition to the HSs embedded into the medium with uniform nonlinearity, more specific models, in which nonlinearity is also concentrated at the HSs, were introduced in [47] and [51]. In particular, the system with the double HS

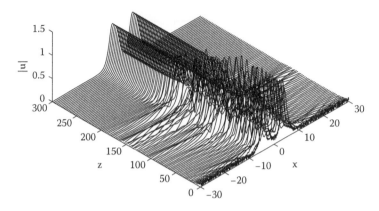

FIGURE 9.5: Spontaneous transformation of a pair of breathers pinned to two hot spots, described by Eq. (9.54), into a stationary symmetric mode. Parameters are $\beta = 0$, $\epsilon = 1$, $\gamma = 1$, $\Gamma_1 = 4$, $\Gamma_2 = 1$, and $L = 4$.

is described by the following equation:

$$\frac{\partial u}{\partial z} = \frac{i}{2}\frac{\partial^2 u}{\partial x^2} - \gamma u + \left[(\Gamma_1 + i\Gamma_2) - (E - iB)\,|u|^2\right] \times$$
$$[\delta\,(x + L) + \delta\,(x - L)]\,u, \qquad (9.55)$$

where B and E are coefficients of the localized Kerr nonlinearity and cubic loss, respectively. These settings may be realized if the nonlinear properties of the waveguides are dominated by narrow doped segments. Although model (9.55) seems somewhat artificial, its advantage is a possibility of finding both symmetric and antisymmetric pinned modes in an exact analytical form. These include both fundamental modes, with exactly two local power peaks, tacked to the HSs, and higher-order states, which feature additional peaks between the HSs. An essential finding pertains to the stability of such states, for which the sign of cubic nonlinearity plays a crucial role: in the SF case [$B > 0$ in Eq. (9.55)], only the fundamental symmetric and antisymmetric modes, with two local peaks tacked to the HSs, may be stable. In this case, all the higher-order multipeak modes, being unstable, evolve into the fundamental ones. In the case of the SDF cubic nonlinearity [$B < 0$ in Eq. (9.55)], the HS pair gives rise to *multistability*, with up to eight coexisting stable multi-peak patterns, both symmetric and antisymmetric ones. The system without the Kerr term ($B = 0$), the nonlinearity in Eq. (9.55) being represented solely by the local cubic loss ($\sim E$), is similar to one with self-focusing or defocusing nonlinearity, if the linear potential of the HS is, respectively, attractive or repulsive, i.e., $\Gamma_2 > 0$ or $\Gamma_2 < 0$ (note that a set of two local repulsive potentials may stably trap solitons in an effective *cavity* between them [40]). An additional noteworthy feature of the former setting is the coexistence of stable fundamental modes with robust breathers.

9.3 SOLITONS PINNED TO THE $\mathcal{PT}$-SYMMETRIC DIPOLE

9.3.1 ANALYTICAL RESULTS

The nonlinear Schrödinger equation (NLSE) in the form of Eq. (9.20) is a unique example of a $\mathcal{PT}$-symmetric system in which a full family of solitons can be found in an exact analytical form [70,99]. Indeed, looking for stationary solutions with real propagation constant k as $u(x,z) = e^{ikz}U(x)$, where $\mathcal{PT}$ symmetry is provided by condition $U^*(x) = U(-x)$, one can readily find, for the SF and SDF signs of nonlinearity, respectively,

$$U(x) = \sqrt{2k}\frac{\cos\theta + i\,\mathrm{sgn}(x)\sin\theta}{\cosh\left(\sqrt{2k}\,(|x| + \xi)\right)}, \quad \text{for } \sigma = +1, \tag{9.56}$$

$$U(x) = \sqrt{2k}\frac{\cos\theta + i\,\mathrm{sgn}(x)\sin\theta}{\sinh\left(\sqrt{2k}\,(|x| + \xi)\right)}, \quad \text{for } \sigma = -1, \tag{9.57}$$

with real constants θ and ξ [cf. Eq. (9.27)]. The form of this solution implies that $\mathrm{Im}\,(U(x=0)) = 0$, while jumps (Δ) of the imaginary part and the first derivative of the real part at $x = 0$ are determined by the b.c. produced by the integration of the δ- and δ'- functions in an infinitesimal vicinity of $x = 0$ [cf. Eq. (9.26)]:

$$\Delta\,\{\mathrm{Im}\,(U)\}\,|_{x=0} = 2\gamma\mathrm{Re}\,(U)\,|_{x=0}, \tag{9.58}$$

$$\Delta\left\{\left(\frac{d}{dx}\mathrm{Re}\,(U)\right)\right\}\,|_{x=0} = -2\varepsilon_0\,\mathrm{Re}\,(U)\,|_{x=0}. \tag{9.59}$$

The substitution of expressions (9.56) and (9.57) into these b.c. yields

$$\theta = \arctan(\gamma), \tag{9.60}$$

which does not depend on k and is the same for $\sigma = \pm 1$, and

$$\xi = \frac{1}{2\sqrt{2k}}\ln\left(\sigma\frac{\sqrt{2k} + \varepsilon_0}{\sqrt{2k} - \varepsilon_0}\right), \tag{9.61}$$

which does not depend on the $\mathcal{PT}$ coefficient, γ. The total power (9.32) of the localized mode is

$$P_\sigma = 2\sigma\left(\sqrt{2k} - \varepsilon_0\right). \tag{9.62}$$

As seen from Eq. (9.61), the solutions exist at

$$\begin{cases} \sqrt{2k} > \varepsilon_0 & \text{for } \sigma = +1, \\ \sqrt{2k} < \varepsilon_0 & \text{for } \sigma = -1. \end{cases} \tag{9.63}$$

As concerns stability of the solutions, it is relevant to mention that expression (9.62) with $\sigma = +1$ and -1 satisfy, respectively, the Vakhitov–Kolokolov (VK) [87,88] and "anti-VK" [89] criteria, i.e.,

$$dP_{+1}/dk > 0, \; dP_{-1}/dk < 0, \tag{9.64}$$

which are necessary conditions for the stability of localized modes supported, severally, by the SF and SDF nonlinearities, hence both solutions have a chance to be stable.

9.3.2 NUMERICAL FINDINGS

As above [see Eq. (9.53)], numerical analysis of the model needs to replace the exact δ-function by its finite-width regularization, $\tilde{\delta}(x)$. In the present context, the use of the Gaussian approximation is not convenient, as, replacing the exact solutions in the form of Eqs. (9.56) and (9.57) by their regularized counterparts, it is necessary, inter alia, to replace $\mathrm{sgn}(x) \equiv -1+2\int_{-\infty}^{x}\delta(x')dx'$ by a continuous function realized as $-1 + 2\int_{-\infty}^{x}\tilde{\delta}(x')dx'$, which would be a non-elementary function. Therefore, regularization was used in the form of the Lorentzian,

$$\delta(x) \to \frac{a}{\pi}\frac{1}{x^2+a^2}, \quad \delta'(x) \to -\frac{2a}{\pi}\frac{x}{\left(x^2+a^2\right)^2},$$

$$\mathrm{sgn}(x) \to \frac{2}{\pi}\arctan\left(\frac{x}{a}\right), \tag{9.65}$$

with $0 < a \ll k^{-1/2}$ [70].

The first result for SF nonlinearity, with $\sigma = +1$ (in this case, $\varepsilon_0 = +1$ is also fixed by scaling) is that, for given a in Eq. (9.65), there is a critical value, γ_{cr}, of the $\mathcal{PT}$ coefficient, such that at $\gamma < \gamma_{\mathrm{cr}}$ the numerical solution features a shape very close to that of the analytical solution (9.56), while at $\gamma > \gamma_{\mathrm{cr}}$ the single-peak shape of the solution transforms into a *double-peak* one, as shown in Fig. 9.6(a). In particular, $\gamma_{\mathrm{cr}}\,(a = 0.02) \approx 0.24$.

The difference between the single- and double-peak modes is that the former ones are completely stable, as was verified by simulations of Eq. (9.20) with regularization (9.65), while all the double-peak solutions are unstable. This correlation between the shape and (in)stability of the pinned modes is not surprising: the single- and double-peak structures imply that the pinned mode is feeling, respectively, effective attraction to or repulsion from the local defect. Accordingly, in the latter case the pinned soliton is unstable against spontaneous escape, transforming itself into an ordinary freely moving NLSE soliton, as shown in Fig. 9.7.

Figure 9.8 summarizes the findings in the plane of (a, γ) for a fixed propagation constant, $k = 3.0$. The region of the unstable double-peak solitons is a relatively narrow boundary layer between broad areas in which the stable single-peak solitons exist, as predicted by the analytical solution, or no solitons exist at all, at large values of γ. Note also that the stability area strongly expands to larger values of γ as the regularized profile (9.65) becomes smoother, with the increase of a. On the other hand, the stability region does not vanish even for very small a. For the same model but with SDF nonlinearity, $\sigma = -1$, the results are simpler: all the numerically found pinned solitons are close to

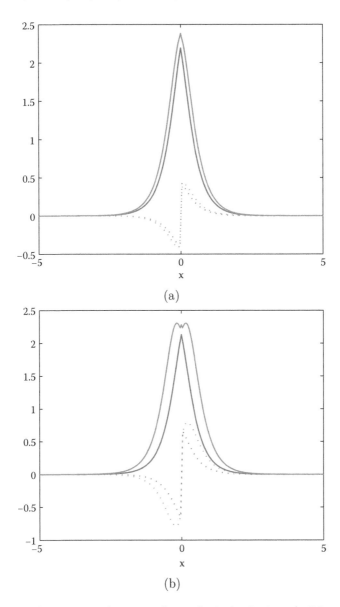

FIGURE 9.6: Comparison between the analytical solutions (solid and dotted curves show their real and imaginary parts, respectively), given by Eqs. (9.56), (9.60), and (9.61) with $\sigma = +1$ and $\varepsilon_0 = 1$, and their numerically found counterparts, obtained by means of regularization (9.65) with $a = 0.02$ (curves). The $\mathcal{PT}$ parameter is $\gamma = 0.20$ in (a) and 0.32 in (b). In both panels, the solutions are produced for propagation constant $k = 3$.

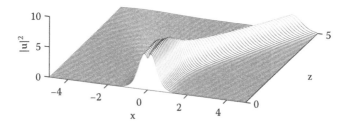

FIGURE 9.7: The unstable evolution (spontaneous escape) of the double-peak soliton whose stationary form is shown in Fig. 9.6(b).

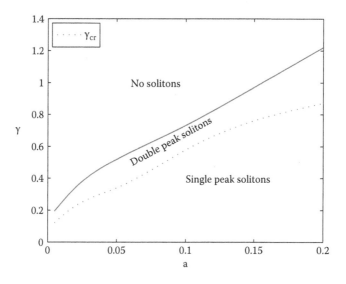

FIGURE 9.8: Regions of the existence of stable single-peak and unstable double-peak $\mathcal{PT}$-symmetric solitons, separated by $\gamma = \gamma_{\mathrm{cr}}(a)$, in the plane of the regularization scale, a, and the gain-loss parameter, γ, for fixed $k = 3.0$, in the system described by Eqs. (9.20) and (9.65) with $\sigma = +1$ (the SF non-linearity) and $\varepsilon_0 = 1$.

the analytical solution (9.57), featuring a single-peak form, and are completely stable.

A generalization of model (9.20), including a nonlinear part of the trapping potential, was studied in [70] too:

$$i\frac{\partial u}{\partial z} = -\frac{1}{2}\frac{\partial^2 u}{\partial x^2} - \sigma|u|^2 u - \left(\varepsilon_0 + \varepsilon_2|u|^2\right)u\delta\left(x\right) + i\gamma u\delta'\left(x\right). \qquad (9.66)$$

Equation (9.66) also admits an analytical solution for pinned solitons, which is rather cumbersome. It takes a simpler form in the case when the trapping potential at $x = 0$ is purely nonlinear, with $\varepsilon_0 = 0$, $\varepsilon_2 > 0$, Eq. (9.61) being replaced by

$$\xi = \frac{1}{2\sqrt{2k}}\ln\left[\frac{2\varepsilon_2\sqrt{2k}}{1+\gamma^2} + \sqrt{1 + \frac{8\varepsilon_2^2 k}{(1+\gamma^2)^2}}\right] \qquad (9.67)$$

[it is the same for both signs of the bulk nonlinearity, $\sigma = +1$ and -1, while the solution at $x \neq 0$ keeps the form of Eqs. (9.56) or (9.57), respectively], with total power

$$P_\sigma(k) = 2\left[\frac{1+\gamma^2}{2\varepsilon_2} + \sigma\sqrt{2k} - \sigma\sqrt{2k + \frac{(1+\gamma^2)^2}{4\varepsilon_2^2}}\right]. \qquad (9.68)$$

Note that expressions (9.67) and (9.68) depend on the $\mathcal{PT}$ strength γ, unlike their counterparts (9.61) and (9.62). Furthermore, solution (9.67) exists for all values of $k > 0$, unlike the one given by Eq. (9.61), whose existence region is limited by condition (9.63). Numerical analysis demonstrates that these solutions have a narrow stability area (at small γ) for the SF bulk medium, $\sigma = +1$, and are completely unstable for $\sigma = -1$.

9.4 GAP SOLITONS SUPPORTED BY A HOT SPOT IN THE BRAGG GRATING

As mentioned above, the first example of SDSs supported by an HS in lossy media was predicted in the framework of the CMEs (9.15) and (9.16) for the BG in a nonlinear waveguide [40]. It is relevant to outline this original result here. Unlike the basic models considered above, the CME system does not admit exact solutions for pinned solitons, because Eqs. (9.15) and (9.16) do not have analytical solutions in the bulk in the presence of the loss terms, $\gamma > 0$; therefore, analytical consideration (verified by numerical solutions) is only possible in the framework of the perturbation theory, which treats γ and Γ_1 as small parameters, while the strength of the local potential, Γ_2, does not need to be small. This example of the application of the perturbation theory is important, as it provided a paradigm for the analysis of other models, where exact solutions are not available either [42, 46].

9.4.1 ZERO-ORDER APPROXIMATION

A stationary solution to Eqs. (9.15) and (9.16) with $\gamma = \Gamma_1 = 0$ and $\Gamma_2 > 0$ is sought in the form which is common for quiescent BG solitons [90, 91]:

$$
\begin{aligned}
u(x, z) &= U(x) \exp(-iz \cos \theta), \\
v(x, z) &= -V^*(x) \exp(-iz \cos \theta),
\end{aligned}
\tag{9.69}
$$

where $*$ stands for the complex conjugate, θ is a parameter of the soliton family, and function $U(x)$ satisfies the equation

$$
\left[i \frac{d}{dx} U + \cos \theta + \Gamma_2 \delta(x) \right] U + 3|U|^2 U - U^* = 0.
\tag{9.70}
$$

The integration of Eq. (9.70) around $x = 0$ yields the respective b.c., $U(x = +0) = U(x = -0) \exp(i\Gamma_2)$ [cf. Eq. (9.26)]. As shown in [93], an exact soliton-like solution to Eq. (9.70), supplemented by the b.c., can be found, following the pattern of the exact solution for the ordinary gap solitons [90–92]:

$$
U(x) = \frac{1}{\sqrt{3}} \frac{\sin \theta}{\cosh \left[(|x| + \xi) \sin \theta - \frac{i}{2} \theta \right]},
\tag{9.71}
$$

where offset ξ [cf. Eqs. (9.27), (9.56), (9.57)], is determined by the relation

$$
\tanh(\xi \sin \theta) = \frac{\tan(\Gamma_2/2)}{\tan(\theta/2)}.
\tag{9.72}
$$

Accordingly, the soliton's squared amplitude (peak power) is

$$
|U(x = 0)|^2 = (2/3)(\cos \Gamma_2 - \cos \theta).
\tag{9.73}
$$

From Eq. (9.72) it follows that the solution exists not in the whole interval $0 < \theta < \pi$, where the ordinary gap solitons are found, but in a region determined by constraint $\tanh(\xi \sin \theta) < 1$, i.e.,

$$
\Gamma_2 < \theta < \pi.
\tag{9.74}
$$

In turn, Eq. (9.74) implies that the solutions exist only for $0 \leq \Gamma_2 < \pi$ ($\Gamma_2 < 0$ corresponds to the repulsive HS, hence the soliton pinned to it will be unstable).

9.4.2 FIRST-ORDER APPROXIMATION

In the case of $\gamma = \Gamma_1 = 0$, Eqs. (9.15) and (9.16) conserve the total power,

$$
P = \int_{-\infty}^{+\infty} \left[|u(x)|^2 + |v(x)|^2 \right]^2 dx.
\tag{9.75}
$$

In the presence of loss and gain, the exact evolution equation for the total power is

$$\frac{dP}{dz} = -2\gamma P + 2\Gamma_1 \left[|u(x)|^2 + |v(x)|^2 \right] |_{x=0} . \tag{9.76}$$

If γ and Γ_1 are treated as small perturbations, the balance condition for power, $dP/dz = 0$, should select a particular solution from the family of exact solutions (9.71) of the conservative model, which remains, to the first approximation, a stationary pinned soliton [cf. Eq. (9.41)].

The balance condition following from Eq. (9.76) demands $\gamma P = \Gamma_1 \left[|U(x = 0)|^2 + |V(x = 0)|^2 \right]$. Substituting, in the first approximation, the unperturbed solution (9.71) and (9.72) into this condition, it can be cast in the form of

$$\frac{\theta - \Gamma_2}{\cos \Gamma_2 - \cos \theta} = \frac{\Gamma_1}{\gamma} . \tag{9.77}$$

The pinned soliton selected by Eq. (9.77) appears, with the increase of the relative gain strength Γ_1/γ, as a result of a bifurcation. Inspection of Fig. 9.9, which displays $(\theta - \Gamma_2)$ vs. Γ_1/γ, as per Eq. (9.77), shows that the situation is qualitatively different for $\Gamma_2 < \pi/2$ and $\Gamma_2 > \pi/2$.

In the former case, a *tangent* (saddle-node) bifurcation [86] occurs at a minimum value $(\Gamma_1/\gamma)_{\min}$ of the relative gain, with two solutions existing at $\Gamma_1/\gamma > (\Gamma_1/\gamma)_{\min}$. Analysis of Eq. (9.77) demonstrates that, with the variation of Γ_2, the value $(\Gamma_1/\gamma)_{\min}$ attains an absolute minimum, $\Gamma_1/\gamma = 1$, at $\Gamma_2 = \pi/2$. With the increase of Γ_1/γ, the lower unstable solution branch that starts at the bifurcation point [see Fig. 9.9(a)] hits the limit point $\theta = \Gamma_2$ at $\Gamma_1/\gamma = 1/\sin \Gamma_2$, where it degenerates into the zero solution, according to Eq. (9.73). The upper branch generated, as a stable one, by the bifurcation in Fig. 9.9(a) continues until it attains the maximum possible value, $\theta = \pi$, which happens at

$$\frac{\Gamma_1}{\gamma} = \left(\frac{\Gamma_1}{\gamma} \right)_{\max} \equiv \frac{\pi - \Gamma_2}{1 + \cos \Gamma_2} . \tag{9.78}$$

In the course of its evolution, this branch may acquire an instability unrelated to the bifurcation (see below).

In the case of $\Gamma_2 > \pi/2$, the situation is different, as the saddle-node bifurcation is imaginary in this case, occurring in the unphysical region $\theta < \Gamma_2$ [see Fig. 9.9(b)]. The only physical branch of the solutions appears as a stable one at point $\Gamma_1/\gamma = 1/\sin \Gamma_2$, where it crosses the zero solution, making it unstable. However, as well as the above-mentioned branch, the present one may be subject to an instability of another type. This branch ceases to be a physical one at point (9.78). At the boundary between the two generic cases considered above, i.e., at $\Gamma_2 = \pi/2$, the bifurcation occurs exactly at $\theta = \pi/2$ [see Fig. 9.9(c)].

The situation is different too in the case of $\Gamma_2 = 0$ [see Fig. 1(d)], when the

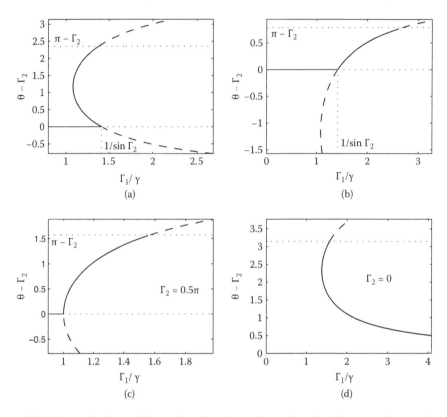

FIGURE 9.9: Analytically predicted solution branches for the pinned gap soliton in the BG with weak uniform loss and local gain. Shown is $\theta - \Gamma_2$ vs. the gain strength, Γ_1/γ. (a) $\Gamma_2 = \pi/4$; (b) $\Gamma_2 = 3\pi/4$; (c) $\Gamma_2 = \pi/2$; (d) $\Gamma_2 = 0$. In the last case, nontrivial solutions appear at point $\theta = 0.7442\pi$, $\Gamma_1/\gamma = 1.3801$, and at large values of Γ_1/γ the continuous curve asymptotically approaches the horizontal axis. In all the panels, the dashed lines show a formal continuation of the solutions in the unphysical regions, $\theta < \Gamma_2$, and $\theta > \pi$. In panels (a), (b), and (c), the trivial solution, $\theta - \Gamma_2 = 0$, is shown by the solid line where it is stable; in the case corresponding to panel (d), all the axis $\theta = 0$ corresponds to the stable trivial solution.

HS has no local-potential component, and Eq. (9.77) takes the form

$$\frac{\theta}{2\sin^2(\theta/2)} = \frac{\Gamma_1}{\gamma}. \tag{9.79}$$

In this case, the solution branches do not cross axis $\theta = 0$ in Fig. 9.9(d). The lower branch, which asymptotically approaches the $\theta = 0$ axis, is unstable, while the upper one might be stable within the framework of the present analysis. However, numerical results demonstrate that, in the case of $\Gamma_2 = 0$, the soliton is always unstable against displacement from $x = 0$.

9.4.3 STABILITY OF THE ZERO SOLUTION

As done above for the CGLE model with the HS [see Eqs. (9.50)–(9.52)], it is relevant to analyze the stability of the zero background, and the relation between the onset of the localized instability, driven by the HS, and the emergence of the stable pinned mode, in the framework of the CMEs. To this end, a solution to the linearized version of the CMEs is looked for as

$$\{u(x, z), v(x, z)\} = \{A_+, B_+\} e^{\Lambda z - Kx} \text{ at } x > 0, \tag{9.80}$$

$$\{u(x, z), v(x, z)\} = \{A_-, B_-\} e^{\Lambda z + Kx}, \text{ at } x < 0, \tag{9.81}$$

with $\mathrm{Re}\{K\} > 0$. A straightforward analysis makes it possible to eliminate constant K and find the instability growth rate [40]:

$$\mathrm{Re}\,\Lambda = -\gamma + (\sinh \Gamma_1)\,|\sin \Gamma_2|\,. \tag{9.82}$$

Thus the zero solution is subject to HS-induced instability, provided that the local gain is strong enough:

$$\sinh \Gamma_1 > \sinh((\Gamma_1)_{\mathrm{cr}}) \equiv \gamma/|\sin \Gamma_2|\,. \tag{9.83}$$

Note that instability is impossible in the absence of the local potential $\sim \Gamma_2$. The instability-onset condition (9.83) simplifies in the limit when both the loss and gain parameters are small (while Γ_2 is not necessarily small):

$$\Gamma_1 > (\Gamma_1)_{\mathrm{cr}} \approx \gamma/|\sin \Gamma_2|\,. \tag{9.84}$$

As seen in Fig. 9.9, the pinned mode emerges at critical point (9.84), for $\Gamma_2 \geq \pi/2$, while at $\Gamma_2 < \pi/2$ it emerges at $\Gamma_1 < (\Gamma_1)_{\mathrm{cr}}$.

9.4.4 NUMERICAL RESULTS

First, direct simulations of the conservative version of CMEs (9.15) and (9.16), with $\gamma = \Gamma_1 = 0$ (and $\Gamma_2 > 0$) and an appropriate approximation for the delta function, have demonstrated that there is a narrow stability interval in terms of parameter θ [see Eqs. (9.70)–(9.74)] close to (slightly larger than) $\theta_{\mathrm{stab}} \approx \pi/2$, such that the solitons with $\theta < \theta_{\mathrm{stab}}$ decay into radiation, while initially created solitons with $\theta > \theta_{\mathrm{stab}}$ spontaneously evolve, through emission of radiation waves, into a stable one with $\theta \approx \theta_{\mathrm{stab}}$ [40,93]. This value weakly depends on Γ_2. For instance, at $\Gamma_2 = 0.1$, the stability interval is limited to $0.49\pi < \theta < 0.52\pi$, while at a much larger value of local-potential strength, $\Gamma_2 = 1.1$, the interval is located at $0.51\pi < \theta < 0.55\pi$. In this connection, it is relevant to mention that in the usual BG model, based on Eqs. (9.15) and

(9.16) with $\Gamma_2 = 0$, the quiescent solitons are stable at $\theta < \theta_{cr}^{(0)} \approx 1.011 \cdot (\pi/2)$ [94, 95].

Direct simulations of the full CME system (9.15) and (9.16), which includes weak loss and local gain, which may be considered as perturbations, produce stable dissipative gap solitons with small but persistent internal oscillations, i.e., these are, strictly speaking, breathers, rather than stationary solitons [40]. An essential finding is that the average value of θ in such robust states may be essentially larger than the above-mentioned $\theta_{stab} \approx \pi/2$ selected by the conservative counterpart of the system (see particular examples in Table 9.1). With the increase of the gain strength Γ_1, the amplitude of the residual oscillations increases, and, eventually, the oscillatory state becomes chaotic, permanently emitting radiation waves, which increases the effective loss rate that must be compensated by the local gain.

Taking values of γ and Γ_1 small enough, it was checked if the numerically found solutions are close to those predicted by the perturbation theory in the form of Eqs. (9.71) and (9.72), with θ related to γ and Γ_1 as per Eq. (9.77). It has been found the quasi-stationary solitons (with the above-mentioned small-amplitude intrinsic oscillations) are indeed close to the analytical prediction. The corresponding values of θ were identified by means of the best fit to expression (9.71). Then, for the values of θ so found and given loss coefficient γ, the equilibrium gain strength Γ_1 was calculated as predicted by the analytical formula (9.77) (see a summary of the results in Table 9.1). It is seen in the table that the numerically found equilibrium values of Γ_1 exceed the analytically predicted counterparts by $\sim 10\% - 15\%$, which may be explained by the additional gain which is necessary to compensate the permanent power loss due to the emission of radiation.

Table 9.1. Values of the loss parameter γ at which quasi-stationary stable pinned solitons were found in simulations of the CME system, (9.15)–(9.16), by adjusting gain Γ_1, for fixed $\Gamma_2 = 0.5$. Values of the soliton parameter, θ, which provide for the best fit of the quasi-stationary solitons to the analytical solution (9.71) are displayed too. $(\Gamma_1)_{anal}$ is the gain coefficient predicted, for given γ, θ, and Γ_2, by the energy-balance equation (9.77), which does not take the radiation loss into account. The rightmost column shows the relative difference between the numerically found and analytically predicted

---

TABLE 9.1

Perturbation theory results

γ	θ	$(\Gamma_1)_{num}$	$(\Gamma_1)_{anal}$	$\frac{(\Gamma_1)_{num} - (\Gamma_1)_{anal}}{(\Gamma_1)_{anal}}$
0.000316	0.5π	0.000422	0.000386	0.0944
0.00316	0.595π	0.0042	0.00369	0.1373
0.01	0.608π	0.01333	0.01165	0.1442
0.1	0.826π	0.1327	0.121	0.0967

gain strength, which is explained by the necessity to compensate additional radiation loss.

It has also been checked that the presence of nonzero attractive potential with $\Gamma_2 > 0$ is necessary for the stability of the gap solitons pinned to the HS in the BG model. In addition to the analysis of stationary pinned modes, collisions between a gap soliton, freely moving in the weakly lossy medium, with the HS were also studied by means of direct simulations [40]. The collision splits the incident soliton into transmitted and trapped components.

9.5 DISCRETE SOLITONS PINNED TO THE HOT SPOT IN THE LOSSY LATTICE

The 1D version of the discrete model based on Eq. (9.21), i.e.,

$$\frac{du_n}{dz} = \frac{i}{2}\left(u_{n-1} + u_{n+1} + u_{n-1} - 2u_{m,n}\right)$$
$$-\gamma u_n + \left[(\Gamma_1 + i\Gamma_2) + (iB - E)|u_n|^2\right]\delta_{n,0}u_n, \tag{9.85}$$

makes it possible to gain insight into the structure of lattice solitons [12] supported by the "hot site," at which both the gain and nonlinearity are applied, as in that case an analytical solution is available [71]. It is relevant to stress that the present model admits stable localized states even in the case of unsaturated cubic gain, $E < 0$ (see details below).

9.5.1 ANALYTICAL RESULTS

The known *staggering transformation* [96,97], $u_m(t) \equiv (-1)^m e^{-2it}\tilde{u}_m^*$, where the asterisk stands for the complex conjugate, reverses the signs of Γ_2 and B in Eq. (9.85). Using this option, the signs are fixed by setting $\Gamma_2 > 0$ (the linear discrete potential is attractive), while $B = +1$ or $B = -1$ corresponds to the SF and SDF nonlinearity, respectively. Separately considered is the case of $B = 0$, when nonlinearity is represented solely by cubic dissipation localized at the HS.

Dissipative solitons with real propagation constant k are looked for by the substitution of $u_m = U_m e^{ikz}$ in Eq. (9.85). Outside of the HS site, $m = 0$, the linear lattice gives rise to the exact solution with real amplitude A,

$$U_m = A\exp(-\lambda|m|), \quad |m| \geq 1, \tag{9.86}$$

and complex $\lambda \equiv \lambda_1 + i\lambda_2$, localized modes corresponding to $\lambda_1 > 0$. The analysis demonstrates that the amplitude at the HS coincides with A, i.e., $U_0 = A$. Then, the remaining equations at $n = 0$ and $n = 1$ yield a system of four equations for four unknowns, A, λ_1, λ_2, and k:

$$-1 + \cosh\lambda_1 \cos\lambda_2 = k, \quad -\gamma - \sinh\lambda_1 \sin\lambda_2 = 0,$$
$$e^{-\lambda_1}\sin\lambda_2 - \gamma + \Gamma_1 - EA^2 = 0,$$
$$e^{-\lambda_1}\cos\lambda_2 - 1 + \Gamma_2 + BA^2 = k. \tag{9.87}$$

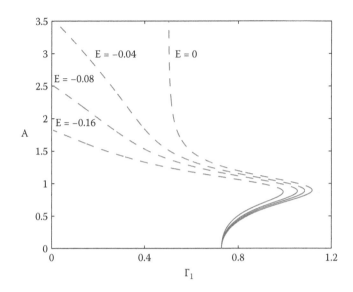

FIGURE 9.10: Amplitude A of the pinned 1D discrete soliton as a function of linear (Γ_1) and cubic ($E \leq 0$) gain. Other parameters in Eq. (9.85) are $\gamma = 0.5$, $\Gamma_2 = 0.8$, and $B = -1$ (SDF nonlinearity). Here and in Figs. 9.11 and 9.13 below, continuous and dashed curves denote stable and unstable solutions, respectively.

This system was solved numerically. The stability of the discrete SDSs was analyzed by the computation of eigenvalues for modes of small perturbations [97,98] and verified by means of simulations of the perturbed evolution. Examples of stable discrete solitons can be seen below in Fig. 9.12.

In the linear version of the model, with $B = E = 0$ in Eq. (9.85), amplitude A is arbitrary, dropping out from Eq. (9.87). In this case, Γ_1 may be considered as another unknown, determined by the balance between the background loss and localized gain, which implies structural instability of the stationary trapped modes in the linear model against small variations of Γ_1. In the presence of nonlinearity, the power balance is adjusted through the value of the amplitude at given Γ_1; therefore, solutions can be found in a range of values of Γ_1. Thus, families of pinned modes can be studied using linear gain Γ_1 and cubic gain/loss E as control parameters (in the underlying photonic lattice, their values may be adjusted by varying the intensity of the external pump).

9.5.2   NUMERICAL RESULTS

9.5.3   SELF-DEFOCUSING REGIME ($B = -1$)

The most interesting results were obtained for *unsaturated nonlinear gain*, i.e., $E < 0$ in Eq. (9.85). Figure 9.10 shows amplitude A of the stable (solid)

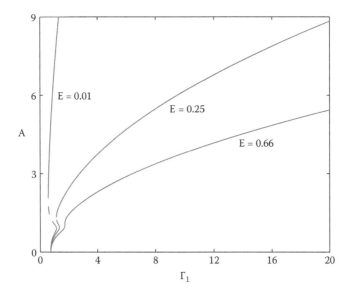

FIGURE 9.11: Solution branches for discrete solitons in the case of cubic loss ($E > 0$) and SDF nonlinearity ($B = -1$). The other parameters are same as those in Fig. 9.10.

and unstable (dashed) pinned modes as a function of linear gain $\Gamma_1 > 0$ and cubic gain. The existence of stable subfamilies in this case is a noteworthy finding. In particular, at $E = 0$ there exists a family of stable pinned modes in the region of $0.73 \leq \Gamma_1 \leq 1.11$, with amplitude ranging from $A = 0.08$ to $A = 0.89$. Outside this region, solutions decay to zero if the linear gain is too weak ($\Gamma_1 < 0.73$), or blow up at $\Gamma_1 > 1.11$. Figure 9.10 shows that the bifurcation of the zero solution $A = 0$ into the pinned mode takes place at a particular value $\Gamma_1 = 0.7286$, which is selected by the above-mentioned power-balance condition in the linear system.

The existence of stable pinned modes in the absence of gain saturation being a remarkable feature, the stability region is, naturally, much broader in the case of cubic loss, $E > 0$. Figure 9.11 shows the respective solution branches obtained with SDF nonlinearity. When the cubic loss is small, e.g., $E = 0.01$, there are two distinct families of stable modes, representing broad small-amplitude ($A \leq 0.89$) and narrow large-amplitude ($A \geq 2.11$) ones. These two stable families are linked by an unstable branch with the amplitudes in the interval of $0.89 < A < 2.11$. The two stable branches coexist in the interval of values of linear gain $0.73 \leq \Gamma_1 \leq 1.13$, where the system is *bistable*. Figure 9.12 shows an example of coexisting stable modes in the bistability region. In simulations, a localized input evolves into either of these two stable solutions, depending on the initial amplitude.

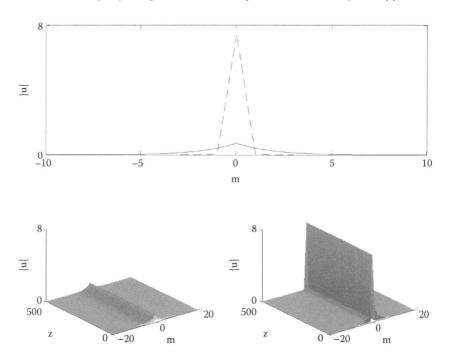

FIGURE 9.12: The coexistence of stable small- and large-amplitude pinned discrete modes (top), and the corresponding evolution of nonstationary solutions (bottom) at $E = 0.01$, in the bistability region. Inputs with amplitudes $A = 0.3$ and $A = 2$ evolve into the small-amplitude and large-amplitude stationary modes, respectively. The other parameters are $\gamma = 0.5$, $\Gamma_1 = 1$, $\Gamma_2 = 0.8$, and $B = -1$ (SDF nonlinearity).

While the pinned modes may be stable against small perturbations under the combined action of self-defocusing nonlinearity ($B = -1$) and unsaturated nonlinear gain ($E \leq 0$), one may expect fragility of these states against finite-amplitude perturbations. It was found indeed that sufficiently strong perturbations destroy those modes [71].

9.5.4 SELF-FOCUSING REGIME ($B = +1$)

In the case of SF nonlinearity, $B = +1$, the pinned-mode branches are shown in Fig. 9.13, all of them being unstable without cubic loss, i.e., at $E \leq 0$. For the present parameters, all the solutions originate, in the linear limit, from the power-balance point $\Gamma_1 = 0.73$. The localized modes remain stable even at very large values of Γ_1. Finally, in the case of $B = 0$, when nonlinearity is represented solely by cubic gain or loss, all the localized states are unstable under cubic gain, $E < 0$ and stable under cubic loss, $E > 0$.

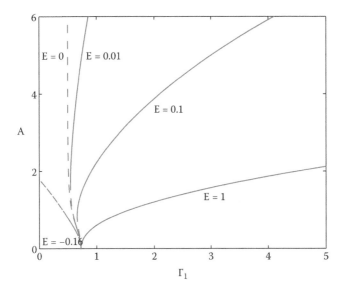

FIGURE 9.13: Amplitude A of the discrete pinned mode as a function of linear gain (Γ_1) and cubic loss (E) in the case of SF nonlinearity, $B = +1$. The other parameters are $\gamma = 0.5$ and $\Gamma_2 = 0.8$.

9.6   CONCLUSION

This article presents a compact review of theoretical results obtained in recently studied 1D and 2D models, which predict a generic method for supporting stable spatial solitons in dissipative optical media, based on the use of linear gain applied in narrow active segments (HSs, "hot spots") implanted into the lossy waveguide. In some cases, unsaturated cubic localized gain may also support stable spatial solitons, which is a counterintuitive finding. The presentation combines a review of the broad class of such models with a more detailed consideration of selected 1D models where exact or approximate analytical solutions for the dissipative solitons are available. Naturally, analytical solutions, in the combination with their numerical counterparts, provide a deeper understanding of the underlying physical models.

REFERENCES

1. N. N. Rosanov, *Spatial Hysteresis and Optical Patterns. Springer*, 2002.
2. M. Tlidi and P. Mandel, "Transverse dynamics in cavity nonlinear optics," *J. Opt. B*, 6, 60, 2004.
3. N. Akhmediev and A. Ankiewicz (Eds.), "Dissipative solitons," *Lect. Notes Phys.*, 661, *Springer*, 2005.

4. N. Lazarides and G. P. Tsironis, "Coupled nonlinear Schrödinger field equations for electromagnetic wave propagation in nonlinear left-handed materials," *Phys. Rev. E*, 71, 036614, 2005.

5. Y. M. Liu, G. Bartal, D. A. Genov, and X. Zhang, "Subwavelength discrete solitons in nonlinear metamaterials," *Phys. Rev. Lett.*, 99, 153901 2007.

6. E. Feigenbaum and M. Orenstein, "Plasmon-soliton," *Opt. Lett.* 32, 674, 2007.

7. I. R. Gabitov, A. O. Korotkevich, A. I. Maimistov, and J. B. Mcmahon, "Solitary waves in plasmonic Bragg gratings," *Appl. Phys. A*, 89, 277, 2007.

8. K. Y. Bliokh, Y. P. Bliokh, and A. Ferrando, "Resonant plasmon-soliton interaction," *Phys. Rev. A*, 79, 041803, 2009.

9. Y.-Y. Lin, R.-K. Lee, and Y. S. Kivshar, "Transverse instability of transverse-magnetic solitons and nonlinear surface plasmons," *Opt. Lett.*, 34, 2982, 2009.

10. A. Marini, D. V. Skryabin, and B. A. Malomed, "Stable spatial plasmon solitons in a dielectric-metal-dielectric geometry with gain and loss," *Opt. Exp.*, 19, 6616, 2011.

11. A. V. Gorbach and D. V. Skryabin, "Spatial solitons in periodic nanostructures," *Phys. Rev. A*, 79, 053812, 2009.

12. F. Ye, D. Mihalache, B. Hu, and N. C. Panoiu, "Subwavelength plasmonic lattice solitons in arrays of metallic nanowires," *Phys. Rev. Lett.*, 104, 10680, 2010.

13. M. C. Cross and P. C. Hohenberg, "Pattern formation outside of equilibrium," *Rev. Mod. Phys.*, 65, 851, 1993.

14. I. S. Aranson and L. Kramer, "The world of the complex Ginzburg-Landau equation," *Rev. Mod. Phys.*, 74, 99, 2002.

15. B. A. Malomed, in: *Encyclopedia of Nonlinear Science*, p. 157 (ed. by A. Scott; New York, Routledge, 2005).

16. P. Kolodner, J. A. Glazier, H. Williams, "Dispersive chaos in one-dimensional traveling-wave convection," *Phys. Rev. Lett.* 65, 1579, 1990.

17. S. K. Das, S. Puri, and M. C. Cross, "Nonequilibrium dynamics of the complex Ginzburg-Landau equation: Analytical results," *Phys. Rev. E*, 64, 046206, 2001.

18. L. M. Hocking and K. Stewartson, "Nonlinear response of a marginally unstable plane parallel flow to a 2-dimensional disturbance," *Proc. R. Soc. London Ser. A*, 326, 289, 1972.

19. N. R. Pereira, L. Stenflo, "Nonlinear Schrödinger equation including growth and damping," *Phys. Fluids*, 20, 1733, 1977.

20. B. A. Malomed and H. G. Winful, "Stable solitons in two-component active systems," *Phys. Rev. E*, 53, 5365, 1996.

21. J. Atai and B. A. Malomed, "Stability and interactions of solitons in two-component active systems," *Phys. Rev. E*, 54, 4371, 1996.

22. J. Atai and B. A. Malomed, "Exact stable pulses in asymmetric linearly coupled Ginzburg–Landau equations," *Phys. Lett. A*, 246, 412, 1998.

23. H. Sakaguchi and B. A. Malomed, "Breathing and randomly walking pulses in a semilinear Ginzburg-Landau system," *Physica D*, 147, 273, 2000.

24. B. A. Malomed, "Solitary pulses in linearly coupled Ginzburg-Landau equations," *Chaos*, 17, 037117, 2007.

25. P. V. Paulau, D. Gomila, P. Colet, N. A. Loiko, N. N. Rosanov, T. Ackemann, and W. J. Firth, "Vortex solitons in lasers with feedback," *Optics Express*, 18, 8859, 2010.

26. P. V. Paulau, D. Gomila, P. Colet, B. A. Malomed, and W. J. Firth, "From one- to two-dimensional solitons in the Ginzburg-Landau model of lasers with frequency-selective feedback," *Phys. Rev. E*, 84, 036213, 2011.

27. R. Driben and B. A. Malomed, "Stability of solitons in parity-time-symmetric couplers," *Opt. Lett.*, 36, 4323, 2011.

28. N. V. Alexeeva, I. V. Barashenkov, A. A. Sukhorukov, and Y. S. Kivshar, "Optical solitons in $\mathcal{PT}$-symmetric nonlinear couplers with gain and loss," *Phys. Rev. A*, 85, 063837, 2012.

29. V. I. Petviashvili and A. M. Sergeev, "Spiral solitons in active media with excitation thresholds," *Doklady AN SSSR (Sov. Phys. Doklady)*, 276, 1380, 1984.

30. B. A. Malomed, "Evolution of nonsoliton and "quasi-classical" wavetrains in nonlinear Schrödinger and Korteweg-de Vries equations with dissipative perturbations," *Physica D*, 29, 155, 1987.

31. W. van Saarloos, and P. Hohenberg, "Pulses and fronts in the complex Ginzburg-Landau equation near a subcritical bifurcation," *Phys. Rev. Lett.*, 64, 749, 1990.

32. V. Hakim, P. Jakobsen, and Y. Pomeau, "Fronts vs. solitary waves in nonequilibrium systems," *Europhys. Lett.*, 11, 19, 1990.

33. B. A. Malomed, and A. A. Nepomnyashchy, "Kinks and solitons in the generalized Ginzburg-Landau equation," *Phys. Rev. A*, 42, 6009, 1990.

34. P. Marcq, H. Chaté, and R. Conte, "Exact solutions of the one-dimensional quintic complex Ginzburg-Landau equation," *Physica D*, 73, 305, 1994.

35. T. Kapitula and B. Sandstede, "Instability mechanism for bright solitary-wave solutions to the cubic-quintic Ginzburg-Landau equation," *J. Opt. Soc. Am. B*, 15, 2757, 1998.

36. A. Komarov, H. Leblond, and F. Sanchez, "Quintic complex Ginzburg-Landau model for ring fiber lasers," *Phys. Rev. E*, 72, 025604, 2005.

37. W. Renninger, A. Chong, and E. Wise, "Dissipative solitons in normal-dispersion fiber lasers," *Phys. Rev. A*, 77, 023814, 2008.

38. E. Ding and J. N. Kutz, "Operating regimes, split-step modeling, and the Haus master mode-locking model," *J. Opt. Soc. Am. B*, 26, 2290, 2009.

39. J. Hukriede, D. Runde, and D. Kip, "Fabrication and application of holographic Bragg gratings in lithium niobate channel waveguides," *J. Phys. D*, 36, R1, 2003.

40. W. C. K. Mak, B. A. Malomed, and P. L. Chu, "Interaction of a soliton with a localized gain in a fiber Bragg grating," *Phys. Rev. E*, 67, 026608, 2003.

41. C. K. Lam, B. A. Malomed, K. W. Chow, and P. K. A. Wai, "Spatial solitons supported by localized gain in nonlinear optical waveguides," *Eur. Phys. J. Special Topics*, 173, 233, 2009.

42. O. V. Borovkova, V. E. Lobanov, and B. A. Malomed, "Stable nonlinear amplification of solitons without gain saturation," *EPL*, 97, 44003, 2012.

43. W. Schöpf and L. Kramer, "Small-amplitude periodic and chaotic solutions of the complex Ginzburg-Landau equation for a subcritical bifurcation," *Phys. Rev. Lett.*, 66, 2316, 2003.

44. D. K. Serkland, G. D. Bartolini, A. Agarwal, P. Kumar, and W. L. Kath, "Pulsed degenerate optical parametric oscillator based on a nonlinear-fiber Sagnac interferometer," *Opt. Lett.*, 23, 795, 1998.

45. M. A. Foster, A. C. Turner, J. E. Sharping, B. S. Schmidt, M. Lipson, A. L. Gaeta, "Broad-band optical parametric gain on a silicon photonic chip," *Nature*, 441, 960, 2006.

46. F. Ye, C. Huang, Y. V. Kartashov, and B. A. Malomed, "Solitons supported by localized parametric gain," *Opt. Lett.*, 38, 480, 2013.

47. C. H. Tsang, B. A. Malomed, C. K. Lam, and K. W. Chow, "Solitons pinned to hot spots," *Eur. Phys. J. D*, 59, 81, 2010.

48. D. A. Zezyulin, Y. V. Kartashov, and V. V. Konotop, "Solitons in a medium with linear dissipation and localized gain," *Opt. Lett.*, 36, 1200, 2011.

49. Y. V. Kartashov, V. V. Konotop, and V. V. Vysloukh, "Symmetry breaking and multipeaked solitons in inhomogeneous gain landscapes," *Phys. Rev. A*, 83, 041806, 2011.

50. D. A. Zezyulin, V. V. Konotop, and G. L. Alfimov, "Dissipative double-well potential: Nonlinear stationary and pulsating modes," *Phys. Rev. E*, 82, 056213, 2010.

51. C. H. Tsang, B. A. Malomed, and K. W. Chow, "Multistable dissipative structures pinned to dual hot spots," *Phys. Rev. E*, 84, 066609, 2011.

52. Y. V. Kartashov, V. V. Konotop, and V. V. Vysloukh, "Dissipative surface solitons in periodic structures," *EPL*, 91, 340003, 2010.

53. W. Zhu, Y. He, B. A. Malomed, and D. Mihalache, "Two-dimensional solitons and clusters in dissipative lattices," *J. Opt. Soc. Am. B*, 31, A1-A5, 2014.

54. D. A. Zezyulin, G. L. Alfimov, and V. V. Konotop, "Nonlinear modes in a complex parabolic potential," *Phys. Rev. A*, 81, 013606, 2010.

55. F. K. Abdullaev, V. V. Konotop, M. Salerno, and A. V. Yulin, "Dissipative periodic waves, solitons, and breathers of the nonlinear Schrödinger equation with complex potentials," *Phys. Rev. E*, 82, 056606, 2010.

56. V. Skarka, N. B. Aleksić, H. Leblond, B. A. Malomed, and D. Mihalache, "Varieties of stable vortical solitons in Ginzburg-Landau media with radially inhomogeneous losses," *Phys. Rev. Lett.*, 105, 213901, 2010.

57. V. E. Lobanov, Y. V. Kartashov, V. A. Vysloukh, and L. Torner, "Stable radially symmetric and azimuthally modulated vortex solitons supported by localized gain," *Opt. Lett.*, 36, 85, 2011.

58. O. V. Borovkova, V. E. Lobanov, Y. V. Kartashov, and L. Torner, "Rotating vortex solitons supported by localized gain," *Opt. Lett.*, 36, 1936, 2011.

59. O. V. Borovkova, Y. V. Kartashov, V. E. Lobanov, V. A. Vysloukh, and L. Torner, "Vortex twins and anti-twins supported by multiring gain landscapes," *Opt. Lett.*, 36, 3783, 2011.

60. C. Huang, F. Ye, B. A. Malomed, Y. V. Kartashov, and X. Chen, "Solitary vortices supported by localized parametric gain," *Opt. Lett.*, 38, 2177, 2013.

61. A. Ruschhaupt, F. Delgado, and J. G. Muga, "Physical realization of $\mathcal{PT}$-symmetric potential scattering in a planar slab waveguide," *J. Phys. A: Math. Gen.*, 38, L171–L175, 2005.

62. C. E. Rüter, K. G. Makris, R. El-Ganainy, D. N. Christodoulides, M. Segev, and D. Kip, "Observation of parity-time symmetry in optics," *Nat. Phys.*, 6, 192, 2010.

63. K. G. Makris, R. El-Ganainy, D. N. Christodoulides, and Z. H. Musslimani, "Beam dynamics in $\mathcal{PT}$- symmetric optical lattices," *Phys. Rev. Lett.*, 100, 103904, 2008.

64. S. Longhi, "Spectral singularities and Bragg scattering in complex crystals," *Phys. Rev. A*, 81, 022102, 2010.
65. Z. H. Musslimani, K. G. Makris, R. El-Ganainy, and D. N. Christodoulides, "Optical solitons in $\mathcal{PT}$-periodic potentials," *Phys. Rev. Lett.*, 100, 030402, 2008.
66. Z. Lin, H. Ramezani, T. Eichelkraut, T. Kottos, H. Cao, and D. N. Christodoulides, "Unidirectional invisibility induced by $\mathcal{PT}$-symmetric periodic structures," *Phys. Rev. Lett.*, 106, 213901, 2011.
67. X. Zhu, H. Wang, L.-X. Zheng, H. Li, and Y.-J. He, Gap solitons in parity-time complex periodic optical lattices with the real part of super lattices. *Opt. Lett.*, 36, 2680, 2011.
68. S. Nixon, L. Ge, and J. Yang, "Stability analysis for solitons in $\mathcal{PT}$-symmetric optical lattices," *Phys. Rev. A*, 85, 023822, 2012.
69. H. Cartarius, D. Haag, D. Dast, and G. Wunner, "Nonlinear Schrödinger equation for a $\mathcal{PT}$-symmetric delta-function double well," *J. Phys. A: Math. Theor.*, 45, 444008, 2012.
70. T. Mayteevarunyoo, B. A. Malomed, and A. Reoksabutr, "Solvable model for solitons pinned to a parity-time-symmetric dipole," *Phys. Rev. E*, 88, 022919, 2013.
71. B. A. Malomed, E. Ding, K. W. Chow, and S. K. Lai, "Pinned modes in lossy lattices with local gain and nonlinearity," *Phys. Rev. E*, 86, 036608, 2012.
72. N. K. Efremidis and D. N. Christodoulides, "Discrete Ginzburg-Landau solitons," *Phys. Rev. E*, 67, 026606, 2003.
73. K. Maruno, A. Ankiewicz, and N. Akhmediev, "Dissipative solitons of the discrete complex cubic-quintic Ginzburg-Landau equation," *Phys. Lett. A*, 347, 231.
74. N. K. Efremidis, D. N. Christodoulides, and K. Hizanidis, "Two-dimensional discrete Ginzburg-Landau solitons," *Phys. Rev. A*, 76, 043839, 2007.
75. N. I. Karachalios, H. E. Nistazakis, and A. N. Yannacopoulos, "Asymptotic behavior of solutions of complex discrete evolution equations: The discrete Ginzburg-Landau equation," *Discrete Contin. Dyn. Syst.*, 19, 711, 2007.
76. E. Kenig, B. A. Malomed, M. C. Cross, and R. Lifshitz, "Intrinsic localized modes in parametrically driven arrays of nonlinear resonators," *Phys. Rev. E*, 80, 046202, 2009.
77. C. Mejía-Cortés, J. M. Soto-Crespo, M. I. Molina, and R. A. Vicencio, "Dissipative vortex solitons in two-dimensional lattices," *Phys. Rev. A*, 82, 063818, 2010.
78. A. Christ, T. Zentgraf, J. Kuhl, S. G. Tikhodeev, N. A. Gippius, and H. Giessen, "Optical properties of planar metallic photonic crystal structures: Experiment and theory," *Phys. Rev. B*, 70, 125113, 2004.
79. Y. S. Bian, Z. Zheng, X. Zhao, Y. L. Su, L. Liu, J. S. Liu, J. S. Zhu, and T. Zhou, "Guiding of long-range hybrid plasmon polariton in a coupled nanowire array at deep-subwavelength scale," *IEEE Phot. Tech. Lett.*, 24, 1279, 2012.
80. B. A. Malomed, E. Ding, K. W. Chow and S. K. Lai, "Pinned modes in lossy lattices with local gain and nonlinearity," *Phys. Rev. E*, 86, 036608, 2012.
81. K. W. Chow, E. Ding, B. A. Malomed, and A. Y. S. Tang, "Symmetric and antisymmetric nonlinear modes supported by dual local gain in lossy lattices,"

    Eur. Phys. J. Special Topics, 223, 63, 2014.

82. E. Ding, A. Y. S. Tang, K. W. Chow, and B. A. Malomed, "Pinned modes in two-dimensional lossy lattices with local gain and nonlinearity," *Phil. Trans. Roy. Soc. A* (in press).

83. X. Zhang, J. Chai, J. Huang, Z. Chen, Y. Li, and B. A Malomed, "Discrete solitons and scattering of lattice waves in guiding arrays with a nonlinear $\mathcal{PT}$-symmetric defect," *Opt. Exp.* 22, 13927, 2014.

84. V. Skarka, N. B. Aleksić, H. Leblond, B. A. Malomed, and D. Mihalache, "Varieties of stable vortical solitons in Ginzburg-Landau media with radially inhomogeneous losses," *Phys. Rev. Lett.*, 105, 213901, 2010.

85. O. V. Borovkova, Y. V. Kartashov, V. A. Vysloukh, V. E. Lobanov, B. A. Malomed, and L. Torner, "Solitons supported by spatially inhomogeneous nonlinear losses," *Opt. Exp.*, 20, 2657, 2012.

86. G. Iooss and D. D. Joseph, *Elementary Stability and Bifurcation Theory*, Springer, 1980.

87. M. Vakhitov and A. Kolokolov, *Radiophys. Quantum. Electron*, 16, 783, 1973.

88. L. Berge, *Phys. Rep.*, 303, 259, 1998.

89. H. Sakaguchi and B. A. Malomed, "Solitons in combined linear and nonlinear lattice potentials," *Phys. Rev. A*, 81, 013624, 2010.

90. C. M. de Sterke and J. E. Sipe, "Gap solitons," *Progr. Opt.*, 33, 203, 1994.

91. D. N. Christodoulides and R. I. Joseph, "Slow Bragg solitons in nonlinear periodic structures," *Phys. Rev. Lett.*, 62, 1746, 1989.

92. A. B. Aceves and S. Wabnitz, "Self-induced transparency solitons in nonlinear refractive periodic media," *Phys. Lett. A*, 141, 37, 1989.

93. W. C. K. Mak, B. A. Malomed, and P. L. Chu, "Interaction of a soliton with a local defect in a fiber Bragg grating," *J.Opt. Soc. Am. B*, 20, 725, 2003.

94. I. V. Barashenkov, D. E. Pelinovsky, and E. V. Zemlyanaya, "Vibrations and oscillatory instabilities of gap solitons," *Phys. Rev. Lett.*, 80, 5117, 1998.

95. A. De Rossi, C. Conti, and S. Trillo, "Stability, multistability, and wobbling of optical gap solitons," *Phys. Rev. Lett.*, 81, 85, 1998.

96. F. Lederer, G. I. Stegeman, D. N. Christodoulides, G. Assanto, M. Segev, and Y. Silberberg, "Discrete solitons in optics," *Phys. Rep.*, 463, 1, 2008.

97. P. G. Kevrekidis, *The Discrete Nonlinear Schrödinger Equation: Mathematical Analysis, Numerical Computations, and Physical Perspectives*, Springer, 2009.

98. J. Yang, "Nonlinear Waves in Integrable and Nonintegrable Systems," SIAM/Philadelphia 2010.

99. N. Karjanto, W. Hanif, B. A. Malomed, and H. Susanto, "Interactions of bright and dark solitons with localized PT -symmetric potentials", Chaos 25, 023112, 2015

10 Exploring the frontiers of mode locking with fiber lasers

Philippe Grelu

10.1 INTRODUCTION

10.1.1 THE WONDER OF MODE LOCKING

The principle of the soliton has always been an important guideline for the development of short and ultrashort pulse sources [1]. However, the concept of passively mode-locked laser sources relies on the ability of the laser modes to self-organize in a highly competitive environment, excited and disturbed by noise, where gain and losses readily impact the development and the survival of any dynamical regime. For these reasons, it has become crystal clear that the dissipative soliton paradigm was the most adapted soliton concept to the case of mode locked lasers [2]. Contemplating the stable passive mode-locked dynamics is amazing: ultrashort pulses propagate through billions of kilometers, hour after hour, without, in general, any external feedback stabilization mechanism, while conserving their output features – pulse duration, peak power, spectral profile. This could probably never happen with conventional – conservative – solitons, as they generally require an ideal – lossless, uniform – nonlinear medium to exist.

The key feature in mode-locked lasers is the existence of stable focus dynamical attractors in specific domains of cavity parameters [2]. Such an attractor ensures that a balance can be reached between the various dissipative and dispersive physical effects that affect pulse propagation during every cavity roundtrip. Thus, in essence, mode-locked laser pulses are dissipative solitons. Attractors require dissipation to exist, and a stable focus attractor is a noise eater, which allows the attracting state to resist fluctuations coming in from the environment, to some extent. This is the origin of pulse stability over such long propagation distances. A stable focus attractor provides a fixed output temporal field profile. This is linked to the existence, at every location inside the laser cavity, of a fixed attracting temporal profile too, which spatially varies along the cavity according to the propagation dynamics through the various laser component parts.

Even though spontaneous emission noise and technical noise affect pulse features [3,4], these fluctuations are largely reduced owing to the attractor, so

235

that at every location, the average pulse profile will correspond to the given attracting profile. If the laser parameters slightly drift during the working time, due, for instance, to thermal effects, the attractor will move accordingly, so that the pulse regime remains, with a slightly altered field profile. This property is called robustness: there is always a certain extension in the domain of parameters where stable focus attractors can be found. Stability and robustness are the fundamental properties of mode-locked lasers that make them so popular for a wide range of applications. It is also known that these properties have limitations. When the cavity parameters are significantly shifted, they may leave the domain of stable focus attractors, so that, in general, mode locking stops, and the experimenter needs to readjust the cavity settings to retrieve the short pulse regime. Until recently, for the majority of mode-locked laser users, the operational regimes were split between "mode locked" and "not mode locked," the latter associated with a multimode quasi-continuous wave (cw) regime.

10.1.2   EXPLORING THE MODE-LOCKING FRONTIER

However, during the last 15 years or so, considerable interest has been driven by the investigation of the original dynamics lying in the territory between the "mode locked" and "not mode locked" classifications. Indeed, applying the tools of the nonlinear dynamics of dissipative systems, many types of bifurcations can be found by exploring the vicinity of stable stationary mode locking [5,6]. Nonstationary short pulse solutions can be found. These range from periodic pulsating dynamics to chaotic pulse dynamics. Periodic pulsations comprise short-period pulsations, where the periodicity of the dynamics corresponds to a small number of cavity roundtrips. For instance, this can result from a period-doubling or a period-tripling bifurcation [7]. Long-period pulsations are associated with Hopf-type bifurcations, leading to a limit cycle attractor [8]. In fact, various combinations of long and short pulsations can be found [6].

Since the dynamical system has an infinite number of degrees of freedom – the field values E(t) for all times t – the classification of bifurcations is most likely unlimited. Using universal models of nonlinear dissipative dynamics, such as the complex cubic-quintic Ginzburg–Landau equation, has helped tremendously in the exploration and the classification of original complex pulse dynamics [5]. Actually, there has been a mutual stimulation between theory and experiments. Sometimes, new experimental discoveries have challenged theoretical modeling and numerical simulations, such as with noise-like pulse laser emission [9,10] and soliton rain dynamics [11,12]. In other cases, theoretical predictions were confirmed by subsequent experiments, such as for stable soliton molecules [13–16] and dissipative rogue waves [17–19].

In fact, the majority of these investigations has involved fiber laser cavities. Compared to their bulk solid-state counterparts, such as titanium–sapphire lasers, fiber lasers involve a long propagation in the condensed fiber material, entailing a large amount of chromatic dispersion and nonlinearity per roundtrip. The distributed doped-fiber gain medium allows the integrated gain to be large, typically of the order of 10 dB – saturated gain – or larger. This procures a great flexibility in the design of a fiber laser cavity integrating multiple components and the efficient use of dissipation to tailor short pulse nonlinear dynamics. Thus, versatile fiber lasers have become the workbench for the investigation of complex nonlinear dynamics residing beyond the frontiers of conventional mode locking. At the same time, their key practical advantages in terms of power efficiency, compactness and beam delivery have opened the door, as efficient mode-locked laser sources, to major laboratory and industrial applications, from biomedical imaging and nonlinear spectroscopy to material processing [20–23].

Starting from stable single-pulse mode-locked dynamics, when the pumping power is increased, the pulse energy increases too. But above a certain pump threshold, an instability develops due to the excess of nonlinearity accumulated during pulse propagation. This is generally accompanied by an increase of the radiation of dispersive waves, and the pulse dynamics can also bifurcate. Besides pulsating dynamics, the formation of an additional pulse is often observed, and can be viewed as the best solution for the system to recover stable dynamics under intense pump power [24,25]. Keeping the pump power increasing, additional pulses are formed either one by one, or by multiple numbers. These pulses equally share the total cavity energy, since their profiles are identical, being shaped by a common dissipative soliton attractor. This way, multiple pulsing provides stepping stones of stability in the landscape of cavity parameters. However, multiple pulses will definitely interact, as they have virtually an infinite time to reveal the faintest possible interaction among them. Whereas long-ranged repulsive interaction procured by the fast depletion and slow recovery of the gain medium is known to generate a pattern of regularly spaced pulses all along the cavity, known as harmonic mode locking [26,27], there are other complex interaction processes leading to an amazing variety of multisoliton dynamics [2,28].

Likewise, single dissipative soliton dynamics, multisoliton dynamics in a fiber laser can lead to either stationary, pulsating or chaotic regimes. However, the amount of dynamical possibilities offered by combining multiple pulses and bifurcations appears unsurpassed [2]. Even the attempt to classify stationary multisoliton states is a daunting task, as pulses may combine in periodic or non-periodic bunches, with various possible separations and phase relationships, corresponding to the multiplicity of multisoliton attractors [29–31]. Stable soliton molecules and crystals are probably the most remarkable stationary multisoliton states [32–34]. They reflect the existence of coherent short range interactions between dissipative solitons, and lead in practice to the propagation of a tightly bound packet of pulses as a single entity around the laser cavity. These dynamical regimes, in particular self-phase-locked soliton molecules and crystals, can be viewed as generalized laser mode locking, since the spectral components have definite phase relationships.

10.1.3 PARTIALLY MODE-LOCKED REGIMES

As considered above, by shifting the cavity parameters, stationary single soliton or multisoliton mode locking may transform into complex nonstationary pulse dynamics. Soliton molecules may have pulsating states that are somehow comparable to vibrating molecules, hence strengthening the analogy between multisoliton complexes and the states of matter [35–37]. But chaotic bound states can be found for specific parameters [38]. Periodic pulsating dynamics can still be considered as generalized mode locking, since the longitudinal modes of an Nth-periodic pulsation are the modes of the extended cavity whose length is N times the original cavity length. But beyond that, complex nonstationary short-pulse dynamics can no longer be considered as genuine mode-locked states; instead, they can be viewed as weakly or partially mode-locked regimes.

During the past few years, there has been tremendous research activity devoted to the investigation of these weakly mode-locked regimes, which generate nonconventional short pulses of light ranging from partially coherent to incoherent. Understanding these complex laser dynamics is indeed a challenge, as they share the attributes of short or ultrashort pulse generation with conventional mode locking, whereas the effectiveness of laser modes tends to vanish when the pulses behave chaotically.

In the following sections, I shall illustrate the manifestation of partially mode-locked regimes found within fiber lasers when operated in the vicinity of conventional mode-locked pulse dynamics. The first peculiar regime, which has been named "soliton rain" dynamics, clearly illustrates the ability to gradually alter the coherence of multiple pulse laser dynamics [12]. Section 10.2 presents the experimental aspects and interpretations of soliton rain dynamics. The second type of regime consists of various chaotic multipulse packets, which can be dibbed noise-like pulses or chaotic pulse bunches according to the size of the temporal granularity involved [19,39]. Section 10.3 develops the subject of extended chaotic pulse bunches, whose temporal structure can be partially resolved in real time with fast acquisition electronics, and compact chaotic bunches, which have been dubbed "noise-like pulses." Their connection with the generation of optical rogue wave events is highlighted. Section 10.4 concludes by underlining the universal dynamical features that can be obtained using various laser cavity designs.

10.2 SOLITON RAIN DYNAMICS

10.2.1 INTRODUCTION

Despite the common understanding that mode locking appears as an abrupt transition between a noisy continuous wave regime and a clean short-pulse laser operation, recent investigations have highlighted, as discussed above, the existence of various partially mode-locked short-pulse dynamics. Among them, soliton rain dynamics is particularly striking, as it springs from the co-existence in the cavity of dissipative soliton pulses with a noisy cw background, both types of fields sharing the total cavity energy in comparable proportions. As a result, all these field components interact in a dramatic quasi-stationary fashion [12]. The coexistence of a strong cw background with soliton pulses seems to contradict the effectiveness of the saturable absorber, which should filter out the cw components. However, between mode locking and cw operation, there can be cavity settings resulting in a weak saturable absorption and a large amount of soliton radiation that feeds the cw. This cavity setting may not be easily achievable, as it relies on a precise shaping of the nonlinear transmission function associated with the saturable absorber. Fortunately, by using an effective saturable absorber effect based on nonlinear polarization evolution (NPE) [40, 41], we have direct intracavity access to several degrees of freedom readily affecting the nonlinear transmission function. This flexibility comes from the way nonlinear interferences can be shaped, which is not the case with most saturable absorbers relying on a specific saturable material, such as semiconductors, carbon nanotubes or graphene. For these reasons, NPE mode locking is very popular among investigators of mode-locked laser dynamics, and is most easily implemented in fiber lasers [42].

10.2.2 FIBER LASER SETUP

The experimental setup is sketched on Fig. 10.1. The erbium-doped fiber (EDF) laser, which emits at wavelengths around 1.5 μm, is an all-fiber ring cavity that series several optical fibers and fiber pigtailed integrated components. Unidirectional laser propagation is ensured by an optical isolator. A 2-m long EDF constitutes the gain medium, pumped by two counter-propagating laser diodes operating at 980 nm and coupling up to 1 W of total pump power. Pieces of standard fiber (SMF 28) and dispersion-compensating fiber (DCF) are used to control the amount of chromatic dispersion per cavity roundtrip. The presence of a polarizing beam splitter (PBS) along with polarization control allows for the NPE mode-locking mechanism to be implemented. Indeed, NPE takes place during pulse propagation in the fibers: this phenomenon results from the nonlinear interference of orthogonal polarizations during propagation in a Kerr medium. Transmission through the PBS transforms the polarization modulation into an amplitude modulation. The specific nonlinear transfer function is defined by the setting of the polarization controllers (PCs) that are located before and after the PBS. Since the nonlinear transfer function is based on the Kerr effect in silica, its response time is quasi-instantaneous. Here, each PC consists of three fiber loops whose plane can be manually oriented: each of these loops acts as an orientable waveplate with a fixed phase delay. So, numerous dynamical regimes become accessible besides standard mode locking by simply altering the orientation of the PCs, among which is soliton rain dynamics. One part of the laser output intensity is

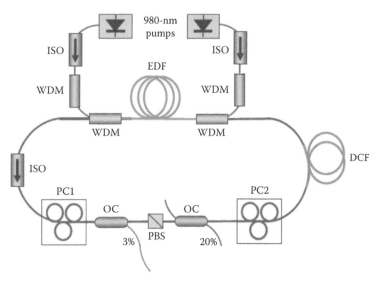

FIGURE 10.1: Sketch of the fiber ring laser experimental setup. EDF: erbium-doped fiber; WDM: wavelength-division multiplexer; ISO: polarization-insensitive optical isolator; OC: optical coupler; PC: fiber-optic polarization controller; DCF: (optional) dispersion-compensating fiber; PBS: polarizing beam splitter.

sent to a fast photodiode that is connected to a 2.5 GHz real-time oscilloscope, and another part is sent.

10.2.3   FIRST-ORDER MODE-LOCKING TRANSITION

Propagation in the anomalous dispersion regime combined with an intense pumping power – around 600 mW – ensures multiple-pulse mode locking operation with, typically, 10–100 interacting dissipative soliton pulses. However, the operational dynamical regime crucially depends on the setting of all the waveplates associated with the polarization controllers. We may start from a set of PC orientations where the laser is not mode locked. It is highly multimode, though, with an emission spectrum around 1 nm wide at −3 dB. The temporal trace corresponds to a noisy cw background – note that the observed noise level appears highly dependent on the electronic detection bandwidth, as a general rule. In the following, one single waveplate orientation is varied – here, the first paddle of PC1 – with its orientation initially at 74 degrees. The transition of regime from cw to mode locking is shown in Fig. 10.2, with the recording of optical spectra and temporal intensity oscilloscope traces for a succession of PC orientations. The cavity roundtrip time is here 66 ns. The change of the orientation of the waveplate (at 90 degrees) first produces

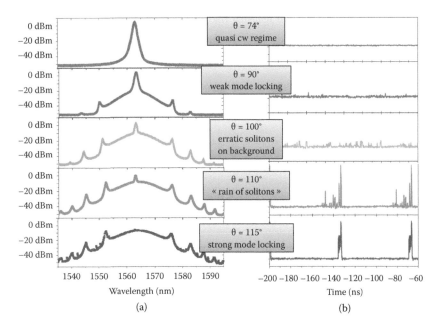

FIGURE 10.2: First-order mode-locking phase transition: variation of optical spectra (a) and temporal intensity (b) along with the orientation θ of the polarization controller (adapted from [12]).

the appearance of a weak but broad mode-locking spectrum, whereas the cw lines still dominate. The temporal trace displays higher noise, on top of which transient short pulses can be observed. Moving the waveplate further at 100 degrees, the mode-locking part of the spectrum has developed, but the cw lines are always there. Note the existence of additional quasi-cw sidebands accompanying the mode-locking spectrum: they result from the resonant radiation of dispersive waves by soliton pulses, while they experience a strong reshaping during their propagation along the various fiber cavity elements, a phenomenon well-known for mode-locked fiber lasers operated in the anomalous dispersion regime [43]. The oscilloscope trace shows many pulses moving erratically in the temporal domain. When the waveplate orientation reaches 110 degrees, there is an abrupt change in the collective temporal dynamics, as most of the pulses bunch together in a tight, sub-nanosecond packet. A noisy background level is still present on the oscilloscope trace, and, surprisingly, pulses pop out of this inhomogeneous background and drift toward the large pulse bunch. The optical spectrum (in log scale) indicates a relatively well balanced energy share between the cw component, and the soliton components. This is the particular "soliton rain" dynamics on which we concentrate in the following. Importantly, the fraction of the weakly coherent background could

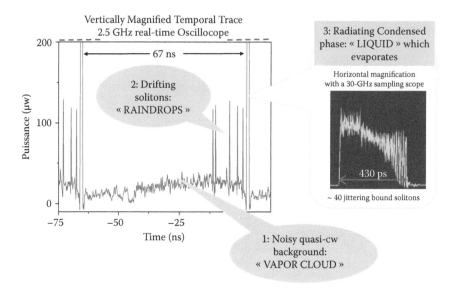

FIGURE 10.3: Analysis of the temporal trace of the optical intensity typical of soliton rain dynamics. Its major field ingredients are qualitatively highlighted, explaining the analogy with the cycle of water.

be gradually varied by tuning the cavity parameters, similar to the alteration of the proportions of mixed phases in the course of a first-order phase transition. Moving further the waveplate at 115 degrees, the popping out of solitons from the background ceases, as the background level is considerably reduced. A stable bunch comprising around 50 pulses remains. The optical spectrum is then a typical fiber laser soliton spectrum. Both spectral and temporal recordings are thus characteristic of a "strong" – i.e., coherent – mode-locking dynamics.

10.2.4   SOLITON RAIN DYNAMICS

Soliton rain is a complex self-organized dynamic, which takes place among large numbers of solitons and a substantial background. It involves interactions between three main field components, which are highlighted in Fig. 10.3.

One key component is a bunch of several tens of bound and jittering solitons. It is dubbed a "condensed soliton phase," since it appears analogous to a liquid thermodynamical phase [11,37]. This component appears as the main peak on the oscilloscope trace whenever the individual soliton constituents are not temporally resolved. Repeated at the cavity fundamental frequency, it provides the dynamical regime with a guise of mode locking. In practice, this peak serves as the triggering signal for the oscilloscope, so that other motions are analyzed in the moving frame of the condensed soliton phase.

The condensed soliton phase is accompanied by an inhomogeneous and noisy background spreading over the whole cavity. When fluctuations exceed a certain level, a new soliton is formed, such as a droplet would be formed from a vapor cloud. It then drifts to the condensed phase at a nearly constant relative velocity on the order of a few meters per second – negative relative velocity, which means it drifts from left to right on the oscilloscope – so that it can be followed in real time by the experimenter [11]. The condensed phase absorbs the incoming pulse, which does not traverse it as would be the case for elastic soliton collisions. This scenario repeats in a quasi-stationary fashion, with several drifting peaks at any given time.

In order for the condensed soliton phase not to grow indefinitely, radiation should be an important part of soliton rain dynamics. The arrival of new solitons causes, through collective rearrangements inside the condensed phase, the dissipation of a similar number of solitons. This dissipation in turn produces additional radiation and dispersive waves that contribute to the background, so that the cycle can be repeated. Consequently, like a liquid would evaporate, the condensed phase emits a large amount of radiation that moves to shorter times due to the conjunction of anomalous dispersion and spectral asymmetry of the radiated waves [12,28]. This radiation also superimposes with other pre-existing cw modes, altogether producing a noisy, inhomogeneous background with large fluctuations. As a major feature, there is a noise threshold above which fluctuations can be amplified to form the drifting solitons. This was tested by lowering the pumping power below the soliton rain threshold and triggering soliton rain dynamics by the injection of an external cw laser [11].

Since the discovery of soliton rain dynamics, very similar observations have been reported from diverse fiber laser configurations, showing that related dynamics belong to some universal class of complex short-pulse laser dynamics [44–46]. Modeling soliton rain dynamics represents a challenging task due to the extent of temporal scales and multiple interaction processes involved. It still remains to be undertaken. It is likely that a practical approach would tackle each process separately, at least in the first stage. Namely, it would address the formation of solitons from the noisy background, which is similarly a self-excitable process, the soliton drift under group velocity dispersion and among noise fluctuations, and, finally, the readjustment and radiation of a growing condensed soliton phase that becomes unstable.

10.3 CHAOTIC PULSE BUNCHES

10.3.1 EXTENDED CHAOTIC BUNCHES: DISSIPATIVE ROGUE WAVES

We have already seen that in the vicinity of soliton rain dynamics (see Fig. 10.2), we can find partially mode-locked states, which consist of fluctuating bunches of pulses that we dubbed the condensed – liquid – soliton phase traveling around the cavity. In this situation, the distribution of pulses inside the bunch fluctuates from one roundtrip to the next. We can anticipate that appropriate tuning of the cavity parameters would increase these fluctuations, to the extent that the traveling bunch of pulses becomes chaotic, whereas other parameter changes could lead to a freezing of these fluctuations, thus possibly forming a large soliton macromolecule [12].

In a chaotic bunch, the multiple constituent pulses are restless, so that they frequently collide inelastically. We can expect such collisions to be highly nonlinear. Nonlinearity and chaos in a high-dimensional system are conducive to the anomalously high occurrence of extreme-amplitude wave events, also popularized under the nicknames of "rogue waves" or "freak waves." These denominations have originated from oceanography, where the tales of mariners meeting giant waves have been reconsidered positively through recent surveys, systematic recordings and scientific modeling [47]. Waves whose amplitude is larger than twice the average of the one-third highest waves qualify as rogue waves, provided that they are formed and disappear unexpectedly, with the condition that their occurrence would be higher than the occurrence deduced from a classical statistical distribution such as a gaussian one. Tsunamis, for instance, do not meet all three criteria: since they can be traced over very long distances after being triggered, they are not considered rogue waves, properly speaking. Considering the human and financial costs at stake, the exploration of rogue wave dynamics has attracted considerable attention during the past decade. It has also been transferred into other fields of science, where rogue waves have been shown to be ubiquitous, and into optics in particular [48], where the fast dynamics have been shown to be particularly advantageous to collect significant statistics in a reduced amount of time.

Two independent theoretical works predicted the possibility of observing rogue waves among chaotic multiple pulse dynamics [17, 18]. These predictions were quickly followed by an experimental demonstration in 2012, using the experimental setup sketched in Fig. 10.1 [19]. Despite such success, it is essential to recall the inherent difficulties of the characterization of ultrafast non-stationary dynamics. Real time field intensity analysis is generally limited by the electronic detection bandwidth. This experiment, used a 20-GHz real time oscilloscope associated with a 40-GHz optoelectronic detector. This entails that structures internal to the bunch shorter than around 50 ps could not be resolved.

Figure 10.4 presents the experimental results of dissipative rogue wave observation [19]. Figure 10.4(a) and (b) displays simultaneous oscilloscope recordings of the output intensity. Using a low-bandwidth electronic detection of the output laser intensity in Fig. 10.4(a), the internal structure of the bunch of the pulse is not temporally resolved. The almost constant amplitude reflects the almost constant energy of the whole pulse bunch. This is an important indication showing that the rogue wave formation mechanism highlighted below does not originate from energy fluctuations, contrary to what was reported from a single chaotic pulse dynamic in a titanium–sapphire laser [49].

Using a 100-fold increase in the detection bandwidth in Fig. 10.4(b), we can see strong peak-intensity fluctuations. This corresponds to the fact that, as soon as the internal structure of the pulse bunch becomes partially resolved, the redistributions and collisions of pulses impact the recorded peak intensity amplitude [50]. A particular extreme amplitude event is shown on Fig. 10.4(c), which was shown to satisfy the rogue wave criteria [19]. Thus, the dynamics can be interpreted as the result from the nonlinear collisions taking place erratically inside a chaotic pulse bunch.

10.3.2   COMPACT CHAOTIC BUNCHES: NOISE-LIKE PULSES

In 1997, Horowitz et al. found an intriguing regime within a mode-locked fiber laser setup, which was dubbed "noise-like pulse" (NLP) emission [9]. NLP dynamics is characterized by the quasi-stationary circulation of an optical noise burst that is much shorter than the cavity roundtrip time. Even though the noise burst could not be analyzed and resolved in real time, its nature was extrapolated from the structure of the average optical autocorrelation trace, which featured a narrow (subpicosecond) coherence peak sitting on a large pedestal, the latter reflecting the temporal extension of the noise burst.

While NLP emission is not a coherent pulse regime, it generates a broad optical spectrum that can surpass the conventional gain medium bandwidth, with, for instance, spectra broader than 100 nm reported in erbium-doped fiber lasers [10, 51], This clearly involves the accumulation of a large non-linearity per cavity roundtrip, in a regime where a plain – coherent – pulse would be unstable. This statement is consistent with the fact that the burst of noise generally contains a large amount of energy – well above what would be allowed for a single stable pulse in a given cavity configuration. In addition, NLP regimes are found profusely in long (over 100 meters) fiber laser cavities, with the general trend that the longer the fiber laser, the longer the noise burst duration [53, 57].

While initially regarded as a dynamics curiosity subjected to a variety of interpretations, NLP emission has regained considerable interest over the past few years. NLP dynamics was first attributed to an instability provoked by the interplay between a large accumulated birefringence, normal dispersion, and the overdrive of nonlinear polarization evolution (NPE) mode locking [9]. However, NLP regimes have also been observed in short (on the order of 10 meters long) fiber laser cavities operated in the anomalous-path-averaged dispersion regime [39, 52]. The destabilizing role of the Raman effect on pulse propagation in long fiber cavities has been pointed out as promoting NLP dynamics [53–56]. But several numerical works have highlighted the existence of various routes leading to unstable pulse propagation and formation of NLPs, without involving the Raman effect or linear birefringence [17, 57].

10.3.3   SPECTRAL ROGUE WAVES

In order to be able to identify in each experimental situation the main destabilization mechanism, it is essential to acquire significantly more data than the information provided by the average autocorrelation and spectral measurements. One significant recent improvement concerns the use of the dispersive Fourier transform technique (DFT) associated with a fast real-time acquisition [58]. The DFT technique consists in sending the attenuated output pulses through a long dispersive line, in a linear propagation regime, so that the spectrum of each pulse is mapped into a temporal waveform owing to the large accumulated dispersion. Thus DFT can evidence shot-to-shot spectral fluctuations, as shown in Fig. 10.5, where the data were obtained with the fiber laser setup of Fig. 10.1 [39, 59]. The experiment produced NLP regimes in both anomalous and normal chromatic dispersions. The duration of the noise burst was from 20 to 40 ps, so that the internal temporal structure could not be resolved in real time due to the limitations of the electronic bandwidth.

However, the DFT technique reveals spectral fluctuations that are linked to the temporal chaotic dynamics. Indeed, Fig. 10.5 shows highly distorted single-shot optical spectra, which largely fluctuate roundtrip after roundtrip. The spectrum recorded by the OSA (optical spectrum analyzer) is smooth due to the averaging effect only. Note that such a major difference between single-shot and average spectra is also found in the case of supercontinuum generation. Taking the maximal spectral intensity for each single-shot spectrum, the long-tailed probability distribution function of Fig. 10.5 is obtained.

To see whether spectral wave events would satisfy the rogue wave criteria, the significant wave height (SWH) is calculated as the mean of the highest one-third population. Waves whose amplitude is higher than twice the SWH qualify as rogue wave events. We can see that a significant fraction of the long-tailed distribution satisfy the criteria: they were dubbed "spectral rogue waves." Thus the DFT technique has allowed us to connect NLP emission to rogue wave dynamics, or, more precisely, to show a significant overlap between the two domains of chaotic dynamics [39, 56].

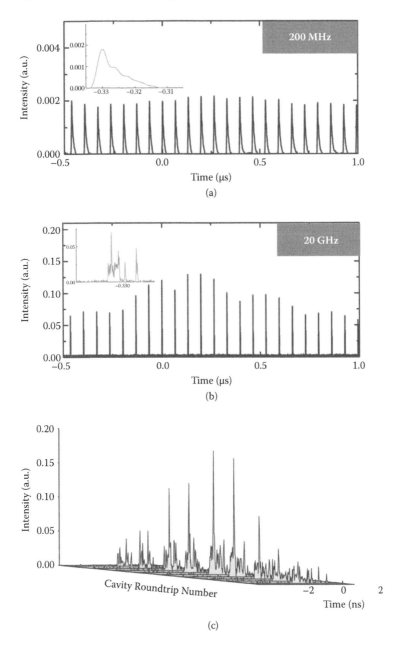

FIGURE 10.4: Temporal recordings of the output laser intensity, using (a) a low electronic bandwidth of 200 MHz, (b) a high bandwidth of 20 GHz. Using the 20-GHz bandwidth, an example of the stroboscopic recording of a rogue wave event is shown (c) (adapted from [19]).

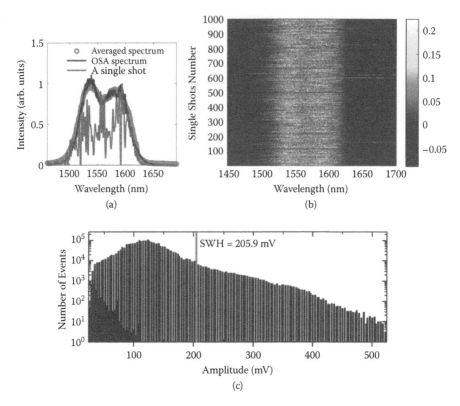

FIGURE 10.5: Observation of spectral rogue waves from noise-like pulse dynamics. (a) Example of a single-shot spectrum (dotted line) recorded through the use of the DFT measurement technique. The averaging of single shot spectra (circles) perfectly coincides with the spectrum recorded by the OSA (black line). In (b) consecutive single-shot spectra are displayed in a 2D contour plot, showing highlighting the level of fluctuations. (c) The L-shaped statistical distribution of spectral maxima is characteristic of rogue wave dynamics. It concerns events whose recorded amplitude exceeds twice the significant wave height (SWH) measure (Adapted from [39]).

10.4 CONCLUSION

The area of partially mode-locked laser dynamics is currently attracting most attention in fundamental research. The development of efficient, reliable, and—to some extent—pump-power scalable high-gain fiber lasers has allowed a quick generalization of experimental investigations, involving the propagation of multiple pulses and/or chaotic pulses in the cavity. The study of the interaction between solitons and a background field has been revived by live soliton rain dynamics, which also posed new modeling challenges yet to be solved, and provided vivid analogies between light dynamics and condensed matter physics. The keen interest in optical rogue waves that the scientific community has been manifesting since 2007 has also boosted research into chaotic pulse dynamics, entailing the discovery of dissipative optical rogue waves among chaotic multiple pulse dynamics, as well as a revisit of the "noise-like" pulse emission that, a decade ago, seemed like a curiosity. No doubt the area of partially mode-locked laser dynamics will keep growing at an accelerated pace, facilitated on one hand by technological improvements of characterization methods and devices, and on the other hand by the development of advanced statistical tools.

ACKNOWLEDGMENTS

This work was supported by the Agence Nationale de la Recherche, France (projects ANR-2010-BLANC-0417-01 and ANR-2012-BS04-0011) and by CE-FIPRA Indo-French agency through project 5104-2. I am grateful to my students and collaborators for their efforts and stimulating discussions. In particular, I wish to thank C. Lecaplain and S. Chouli for their decisive experimental contributions, N. Akhmediev and J.M. Soto-Crepos for their insightful modeling of complex dissipative soliton dynamics, F. Sanchez and his team for conducting stimulating parallel investigations, and K. Porsezian for his recent interest and interactions on dissipative soliton dynamics, along with his invitation to contribute to this chapter.

REFERENCES

1. L. Mollenauer and R. Stolen, "The soliton laser," *Opt. Lett.*, 9, 13, 1984.
2. Ph. Grelu, N. Akhmediev, "Dissipative solitons for mode locked lasers," *Nature Photonics*, 6, 84, 2012.
3. A. L. Schawlow and C. H. Townes, "Infrared and optical masers," *Phys. Rev.*, 112, 1940, 1958.
4. C. R. Menyuk, J. K. Wahlstrand, J. T. Willits, R. P. Smith, T. R. Schibli, and S. T. Cundiff, "Pulse dynamics in mode-locked lasers: relaxation oscillations and frequency pulling," *Opt. Express*, 15, 6677, 2007.
5. N. Akhmediev, J.M. Soto-Crespo, G. Town, "Pulsating solitons, chaotic solitons, period doubling, and pulse coexistence in mode-locked lasers: Complex Ginzburg-Landau equation approach," *Phys. Rev. E*, 63, 056602, 2001.

6. J. M. Soto-Crespo, M. Grapinet, Ph. Grelu, and N. Akhmediev, "Bifurcations and multiple-period soliton pulsations in a passively mode-locked fiber laser," *Phys. Rev. E*, 70, 066612, 2004.

7. D. Coté and H. M. van Driel, "Period doubling of a femtosecond Ti:sapphire laser by total mode locking," *Opt. Lett.*, 23, 715, 1998.

8. J. M. Soto-Crespo, N. Akhmediev, and A. Ankiewicz, "Pulsating, creeping, and erupting solitons in dissipative systems," *Phys. Rev. Lett.*, 85, 2937, 2000.

9. M. Horowitz, Y. Barad, and Y. Silberberg, "Noiselike pulses with a broadband spectrum generated from an erbium-doped fiber laser," *Opt. Lett.*, 22, 799, 1997.

10. J. Kang and R. Posey, "Demonstration of supercontinuum generation in a long-cavity fiber ring laser, *Opt. Lett.*, 23, 1375, 1998.

11. S. Chouli, Ph. Grelu, "Rains of solitons in a fiber laser," *Opt. Express*, 17, 11776, 2009.

12. S. Chouli and Ph. Grelu, "Soliton rains in a fiber laser: An experimental study," *Phys. Rev. A.*, 81, 063829, 2010.

13. B. Malomed, "Bound solitons in the nonlinear Schrödinger-Ginzburg-Landau equation," *Phys. Rev. A*, 44, 6954, 1991.

14. N. Akhmediev, A. Ankiewicz, J. M. Soto-Crespo, "Multisoliton solutions of the complex Ginzburg-Landau equation," *Phys. Rev. Lett.*, 79, 4047, 1997.

15. Ph. Grelu, F. Belhache, F. Gutty, and J. M. Soto-Crespo, "Phase-locked soliton pairs in a stretched-pulse fiber laser," *Opt. Lett.*, 27, 966, 2002.

16. D. Y. Tang, W. S. Man, H. Y. Tam, and P. D. Drummond, "Observation of bound states of solitons in a passively mode-locked fiber laser," *Phys. Rev. A*, 64, 033814, 2001.

17. J. M. Soto-Crespo, Ph. Grelu, and N. Akhmediev, "Dissipative rogue waves: Extreme pulses generated by passively mode-locked lasers," *Phys. Rev. E*, 84, 016604, 2011.

18. A. Zavyalov, O. Egorov, R. Iliew, and F. Lederer, "Rogue waves in mode-locked fiber lasers," *Phys. Rev. A*, 85, 013828, 2012.

19. C. Lecaplain, Ph. Grelu, J. M. Soto-Crespo, and N. Akhmediev, "Dissipative rogue waves generated by chaotic pulse bunching in a mode-locked laser," *Phys. Rev. Lett.*, 108, 233901, 2012.

20. D. J. Richardson, J. Nilsson, and W. A. Clarkson, "High power fiber lasers: Current status and future perspectives," (Invited) *J. Opt. Soc. Am. B*, 27, B63, 2010.

21. C. Xu and F. W. Wise "Recent advances in fiber lasers for nonlinear microscopy," *Nature Photonics*, 7, 875, 2013.

22. M. E. Fermann and I. Hartl, "Ultrafast fiber lasers," *Nature Photonics*, 7, 868, 2013.

23. J. M. Delavaux, Ph. Grelu, F. O. Ilday, and W. Pu (Eds.), Special Issue on Short-Pulse Fiber Lasers, *Opt. Fib. Technol.* (2014).

24. D. J. Richardson, R. I. Laming, D. N. Payne, V. J. Matsas, and M. W. Phillips, "Pulse repetition rates in passive, self-starting, femtosecond soliton fiber laser," *Electronics Letters*, 27, 1451, 1991.

25. F. Gutty, Ph. Grelu, G. Vienne, N. Huot, and G. Millot, "Stabilisation of modelocking in fiber ring laser through pulse bunching," *Electron. Lett.*, 37, 745, 2001.

26. J. N. Kutz, B. C. Collings, K. Bergman, and W. H. Knox, "Stabilized pulse spacing in soliton lasers due to gain depletion and recovery," *IEEE J. Quantum Electron.*, 34, 1749, 1998.

27. C. Lecaplain and Ph. Grelu, "Multi-gigahertz repetition-rate-selectable passive harmonic mode locking of a fiber laser," *Opt. Express*, 21, 10897, 2013.

28. F. Sanchez, Ph. Grelu, H. Leblond, A. Komarov, K. Komarov, M. Salhi, A. Niang, F. Amrani, C. Lecaplain, and S. Chouli, "Manipulating dissipative soliton ensembles in passively mode-locked fiber lasers," (Invited), *Optical Fiber Technology*, Vol. 20, 562–574 2014.

29. Ph. Grelu, F. Belhache, F. Gutty, and J. M. Soto-Crespo, "Relative phase locking of pulses in a passively mode-locked fiber laser," *J. Opt. Soc. Am. B*, 20, 863, 2003.

30. J. M. Soto-Crespo, N. Akhmediev, Ph. Grelu, and F. Belhache, "Quantized separations of phase-locked soliton pairs in fiber lasers," *Opt. Lett.*, 28, 1757, 2003.

31. N. Akhmediev, J. M. Soto-Crespo, M. Grapinet, and Ph. Grelu, "Dissipative soliton interactions inside a fiber laser cavity," (Invited) *Opt. Fib. Technol.*, 11, 209, 2005.

32. Ph. Grelu and J. M. Soto-Crespo, "Multisoliton states and pulse fragmentation in a passively mode-locked fiber laser," *Journal of Optics B: Quantum Semiclass. Opt.*, 6, Special issue on Optical Solitons, S271, 2004.

33. A. Haboucha, H. Leblond, M. Salhi, A. Komarov, and F. Sanchez, "Analysis of soliton pattern formation in passively mode-locked fiber laser," *Phys. Rev. A*, 78, 043806, 2008.

34. F. Amrani, M. Salhi Ph. Grelu, H. Leblond, F. Sanchez, "Universal soliton pattern formations in passively mode-locked fiber lasers," *Opt. Lett.*, 36, 1545, 2011.

35. M. Grapinet and Ph. Grelu, "Vibrating soliton pairs in a mode-locked laser cavity," *Opt. Lett.*, 31, 2115, 2006.

36. J. M. Soto-Crespo, Ph. Grelu, N. Akhmediev, and N. Devine "Soliton complexes in dissipative systems: Vibrating, shaking and mixed soliton pairs," *Phys. Rev. E*, 75, 016613, 2007.

37. F. Amrani, A. Haboucha, M. Salhi, H. Leblond, A. Komarov, and F. Sanchez, "Dissipative solitons compounds in a fiber laser: Analogy with the states of the matter," *Appl. Phys. B*, 99, 107, 2010.

38. D. Turaev, A. G. Vladimirov, and S. Zelik, "Chaotic bound state of localized structures in the complex Ginzburg-Landau equation," *Phys. Rev. E*, 75, 045601(R), 2007.

39. C. Lecaplain and Ph. Grelu, "Rogue waves among noise-like-pulse laser emission: An experimental investigation," *Phys. Rev. A*, 90, 013805, 2014.

40. P. D. Maker and R. W. Terhune, "Study of optical effects due to an induced polarization third order in the electric field strength," *Phys. Rev.*, 137, A801, 1965.

41. H. G. Winful, "Self-induced polarization changes in birefringent optical fibers," *Appl. Phys. Lett.*, 47, 213, 1985.

42. V. J. Matsas, T. P. Newson, D. J. Richardson, and D. N. Payne, "Self-starting passively mode-locked fiber ring soliton laser exploiting nonlinear polarization rotation," *Electron. Lett.*, 28, 1391, 1992.

43. J. P. Gordon, "Dispersive perturbations of solitons of the nonlinear Schrödinger equation," *J. Opt. Soc. Am. B*, 9, 91, 1992.

44. C. Bao, X. Xiao, and C. Yang, "Soliton rains in a normal dispersion fiber laser with dual-filter," *Opt. Lett.*, 38, 1875, 2013.

45. S. S. Huang, Y. G. Wang, P. G. Yan, G. L. Zhang, J. Q. Zhao, H. Q. Li, and R. Y. Lin, "Soliton rains in a graphene-oxide passively mode-locked ytterbium-doped fiber laser with all-normal dispersion," *Laser Phys. Lett.*, 11, 025102, 2014.

46. A. Niang, F. Amrani, M. Salhi, Ph. Grelu, and F. Sanchez, "Rains of solitons in a figure-of-eight passively mode-locked fiber laser," *Appl. Phys. B*, 2014, http://dx.doi.org/10.1007/s00340-014-5760-y.

47. N. Akhmediev and E. Pelinovsky (Eds.), "Rogue waves - towards a unifying concept," *Special Issue of Eur. Phys. J. Special Topics*, 185, 2010.

48. D. R. Solli, C. Ropers, P. Koonath, and B. Jalali, Optical rogue waves, *Nature*, 450, 1054, 2007.

49. M. Kovalsky, A. Hnilo, and J. Tredicce, Extreme events in the Ti: Sapphire laser, *Opt. Lett.*, 36, 4449, 2011.

50. C. Lecaplain, Ph. Grelu, J. M. Soto-Crespo, and N. Akhmediev, Dissipative rouge wave generation in multiple-pulsing mode-locked fiber laser, *J. Opt.*, 15, 064005, 2013.

51. L. M. Zhao, D. Y. Tang, T. H. Cheng, H. Y. Tam, and C. Lu, "120 nm bandwidth noise-like pulse generation in an erbium-doped fiber laser," *Optics Commun.*, 281, 157, 2008.

52. D. Y. Tang, L. M. Zhao, and B. Zhao, "Soliton collapse and bunched noised-like pulse generation in a passively mode-locked fiber ring laser," *Opt. Express*, 13, 2289, 2005.

53. A. Boucon, B. Barviau, J. Fatome, C. Finto, T. Sylvestre, M. Won Lee, Ph. Grelu, and G. Millot, "Noise-like pulses generated at high-harmonics in a partially-mode-locked km-long Raman fiber laser," *Appl. Phys. B*, 106, 283, 2012.

54. A. Bednyakova, S. Babin, D. Kharenko, E. Podivilov, M. Fedoruk, V. Kalashnikov, and A. Apolonski, "Evolution of dissipative solitons in a fiber laser oscillator in the presence of strong Raman scattering," *Opt. Express*, 21, 20556, 2013.

55. T. North, M. Rochette, "Raman-induced noiselike pulses in a highly nonlinear and dispersive all-fiber ring laser," *Opt. Lett.*, 38, 890, 2013.

56. A. Runge, C. Aguergaray, N. G. R. Broderick, M. Erkintalo, "Raman rogue waves in a partially mode-locked fiber laser", *Opt. Lett.*, 39, 319, 2014.

57. S. Kobtsev, S. Kukarin, S. Smirnov, S. Turitsyn, and A. Latkin, "Generation of double-scale femto/pico-second optical lumps in mode-locked fiber lasers," *Opt. Express*, 17, 20710, 2009.

58. K. Goda and B. Jalali, "Dispersive Fourier transformation for fast continuous single-shot measurements," *Nature Photonics*, 7, 102, 2013.

59. A. Runge, C. Aguergaray, N. G. R. Broderick, and M. Erkintalo, "Coherence and shot-to-shot spectral fluctuations in noise-like ultrafast fiber lasers," *Opt. Lett.*, 38, 4327, 2013.

11 Matter wave solitons and other localized excitations in Bose–Einstein condensates in atom optics

P. Muruganandam and *M. Lakshmanan*

11.1 INTRODUCTION

Solitons belong to a special class of localized solitary wave packets which maintain their shapes and amplitudes owing to self-stabilization against dispersion through a nonlinear interaction [1,2] which arise as the solutions of a variety of weakly nonlinear dispersive partial differential equations describing physical systems. Study of such localized structures has been a central theme for evolution dynamics of many physical systems. In this context, considerable interest has been shown in exploring localized excitations in the variable coefficient nonlinear Schrödinger (NLS) equation and its generalizations [3–5]. The motivation comes from the fact that the NLS equation and its variants appear in several branches/topics of physics, including nonlinear optics [6,7] and Bose–Einstein condensates (BECs) [8–10], etc. The NLS equation and its generalizations have also been widely used in the nonlinear optics literature to describe the evolution of coherent light in a nonlinear Kerr medium, envelope dynamics of quasi-monochromatic plane wave propagation in a weakly dispersive medium, high intensity pulse propagation in optical fibers and so on.

The experimental realization of BECs in dilute alkali-metal gases confined in magnetic traps has triggered an immense interest in studying the properties of ultra cold gases [11,12]. In particular, several attempts have been made by many experimenters worldwide to understand the properties of ultra cold matter and to exploit it for various purposes such as atom laser, atom interferometry, simulation of condensed-matter problems, quantum computing, information processing, and so on. An interesting dynamical feature is the formation of bright and dark matter wave solitons [13–26]. Matter-wave solitons in atom optics are expected to be useful for applications in atom lasers including atom interferometry, coherent atom transport, and so on [27]. The behavior of a BEC crucially depends on the sign of the atomic interactions:

253

dark (bright) solitons can be created in BECs with repulsive (attractive) interactions, resulting from the positive (negative) scattering length. The sign of the s-wave atomic scattering length changes by applying Feshbach resonance [28–30].

In this chapter, we briefly review critically the variety of matter wave solitons and their interactions in single- and multi-component Bose–Einstein condensates. In Section 11.2, we provide a brief overview of the mean Gross–Pitaevskii (GP) equation, derivation of the one-dimensional GP equation, and mapping of the 1D GP equation to the standard NLS equation. In Section 11.3 we discuss matter wave single and multi bright and dark solitons of Bose–Einstein condensates with varying trap potentials and atomic interactions. Then in section 11.4 we show various solitons solutions of multi-component BECs described by the two coupled GP equations. In this case, the coupled GP equation has been mapped onto a two coupled NLS equation (Manakov model). Then we illustrate the soliton interactions for appropriate choices of trap potential and intra- and inter-atomic interactions.

11.2   GROSS–PITAEVSKII EQUATION

The properties of a BEC at absolute zero temperature can be described by the mean-field GP equation, which is a generalized form of the ubiquitous NLS equation [1, 2, 31], for the macroscopic wave function of the condensate atoms.

We consider the GP equation describing the evolution of condensate wavefunction $\Psi(\mathbf{r}, t)$, of N bosons, each of mass m, at absolute zero temperature, of the form

$$i\frac{\partial \Psi(\mathbf{r}, t)}{\partial t} = \left[-\frac{1}{2}\nabla^2 + V(\mathbf{r}, t) + 4\pi a(t)N|\Psi(\mathbf{r}, t)|^2 \right] \Psi(\mathbf{r}, t), \qquad (11.1)$$

where $a(t)$ is the varying s-wave scattering length, which can be tuned or modulated via Feshbach resonance, and $a(t) < 0$ and $a(t) > 0$ for attractive and repulsive atomic interactions, respectively. In Equation (11.1) length is measured in units of characteristic harmonic oscillator length $l \equiv \sqrt{\hbar/m\omega}$, angular frequency of trap in units of ω, time t in units of ω^{-1}, and energy in units of $\hbar\omega$. The radial and axial trap frequencies are assumed as $\omega_\perp \equiv \omega$ and $\omega_z(t) \equiv \Omega(t)\omega$, respectively. Then the dimensionless trap potential is given by

$$V(\mathbf{r}, t) = \frac{1}{2}\left(x^2 + y^2\right) + \frac{1}{2}\Omega(t)z^2, \qquad (11.2)$$

and the normalization condition for the condensate wavefunction $\Psi(\mathbf{r}, t)$ will read as

$$\int_{-\infty}^{\infty} |\Psi|^2 d^3\mathbf{r} = 1. \qquad (11.3)$$

In the case of strong transverse confinement, that is, $\omega_z \ll \omega_\perp$, the GP equation (11.1) reduces to an effective one-dimensional (quasi-one-dimensional)

form. This can be achieved by assuming that the condensate remains confined to the ground state of the transverse trap and the wave function

$$\Psi(\mathbf{r}, t) = \frac{1}{\sqrt{2\pi}} \exp\left(-it - \frac{x^2 + y^2}{2}\right) \Phi(z, t).$$ (11.4)

An effective one-dimensional equation can be written by simply substituting
the above wave function (11.4) into Equation (11.1) as

$$i\Phi_t = -\frac{1}{2}\Phi_{zz} - \sigma R(t)|\Phi|^2\Phi + \frac{\Omega^2(t)}{2} z^2\Phi,$$ (11.5)

where $R(t) = 2a(t)$. The normalization condition can now be rewritten as

$$\int_{-\infty}^{\infty} |\Phi|^2 d^3\mathbf{r} = 2N.$$ (11.6)

The above equation (11.5) can be mapped onto the standard NLS equation
of the form

$$iU_T + U_{ZZ} + \sigma|U|^2 U = 0,$$ (11.7)

by applying the transformation,

$$\Phi = \Lambda U(Z, T),$$ (11.8)

where the new independent variables T and Z are chosen as

$$T = G(t), \quad Z = F(z, t),$$ (11.9)

and $\Lambda = \Lambda(z, t)$ along with the conditions that the functions Λ, F, G, Ω and
R should satisfy the following set of equations:

$$i\Lambda_t + \frac{1}{2}\Lambda_{zz} - \frac{\Omega(t)}{2} z^2\Lambda = 0,$$ (11.10a)

$$i\Lambda F_t + \frac{1}{2}(2\Lambda_z F_z + \Lambda F_{zz}) = 0,$$ (11.10b)

$$G_t = \frac{F_z^2}{2} = R|\Lambda|^2.$$ (11.10c)

The coefficient σ is included in the above equation (11.7) to account for the
nature of atomic interaction, that is, $\sigma < 1$ for attractive and $\sigma = -1$ for
repulsive interactions. The unknown functions Λ, F, G in the above equations
(11.10) can be solved by assuming the polar form

$$\Lambda = r(z, t) \exp[i\theta(z, t)],$$ (11.11)

One can immediately check from the relations (11.10c) that r is a function of t only, $r = r(t)$, since G and R are functions of t only. Then from

Equations (11.10) one can easily deduce the transformation function Λ given by (11.11) and the transformations $G(t)$ and $F(z,t)$ through the following relations:

$$r^2 = 2r_0^2 R, \tag{11.12a}$$

$$\theta = -\frac{1}{2}\frac{R_t}{R}z^2 + 2br_0^2 Rz - 2B^2 r_0^4 \int R^2 dt, \tag{11.12b}$$

$$F(z,t) = 2r_0 Rz - 4Br_0^3 \int R^2 dt \equiv Z, \tag{11.12c}$$

$$G(t) = 2r_0^2 \int R^2 dt \equiv T, \tag{11.12d}$$

where B, r_0 are arbitrary constants, and R and Ω^2 should satisfy the following Riccati-type equation [32–34]:

$$\frac{d}{dt}\left(\frac{\dot{R}}{R}\right) - \left(\frac{\dot{R}}{R}\right)^2 - \Omega^2(t) = 0, \tag{11.13}$$

where the overdot denotes differentiation with respect to t. One may also note that $R = R(t)$ represents the modulated scattering length and $\Omega(t) = \omega_z(t)/\omega_\perp$ represents the time-dependent trap frequency.

11.3   MATTER WAVE BRIGHT AND DARK SOLITONS

The NLS equation (11.7) has been known for quite some time for its novel soliton solutions. It admits Jacobian elliptic function solutions, bright and dark solitary wave solutions and trigonometric function solutions [1,35], depending on the sign of σ. For instance, soliton solutions of the standard NLSE (11.7) were given by Zakharov and Shabat by solving the Cauchy initial value problem through the inverse scattering transform method for $\sigma = +1$ (focusing or attractive case) [36] and for $\sigma = -1$ [37] (defocusing or repulsive case). The corresponding soliton solutions are called bright ($\sigma = +1$) and dark ($\sigma = -1$) solitons, respectively. In the following subsection, we will make use of the one-soliton solutions only and higher-order soliton solutions will be discussed later in Sec. 11.3.4. We obtain the one-soliton solutions of the effective one-dimensional GP equation (11.5) by making use of the one-soliton solution of the NLSE (11.7) through the relations (11.8) and (11.12).

11.3.1   ONE-SOLITON DYNAMICS

11.3.1(a) ATTRACTIVE INTERACTION ($\sigma = +1$)

The bright soliton solution of (11.7) for the case of $\sigma = +1$ takes the following form [37] (leaving aside the unimportant phase constants):

$$U^+(Z,T) = A\exp\left[i\left(\frac{C}{2}Z + \frac{2A^2 - C^2}{4}T\right)\right]\operatorname{sech}\left[\frac{A}{\sqrt{2}}(Z - CT)\right]. \tag{11.14}$$

Here the real parameters A and C correspond to the amplitude and velocity, respectively, of the bright soliton envelope and the plus sign in the superscript represents the $\sigma = +1$ case. Correspondingly, for the GP equation (11.5), using the relations (11.8) and (11.14), the bright soliton solution can be written as

$$\Phi^+(z,t) = \Lambda(z,t)\, U^+(Z,T). \tag{11.15}$$

Now substituting the forms of $\Lambda(z,t)$, Z and T from Equations (11.9), (11.11) and (11.12) in the above equation (11.15), we get a generalized expression for the bright soliton as

$$\Phi^+(z,t) = Ar_0\sqrt{2R(t)}\mathrm{sech}\left[\xi^+(z,t)\right]\exp\left[i\eta^+(z,t)\right]. \tag{11.16}$$

Here the amplitude of the generalized soliton (11.16) depends on the form of the atomic scattering length R, and the generalized wave variable ξ^+, which specifies the width and velocity of the bright soliton as a function of time and longitudinal distance, respectively, is given by

$$\xi^+(z,t) = \sqrt{2}Ar_0R\left[z - r_0(C + 2Br_0)\frac{1}{R}\int R^2 dt\right], \tag{11.17a}$$

so that the width is proportional to $1/\sqrt{R(t)}$ and the velocity is a function of time. Also, the generalized phase of the bright soliton is given by the expression

$$\eta^+(z,t) = r_0\left(C + 2Br_0\right)Rz - \frac{\dot{R}}{2R}z^2 - r_0^2\left[\frac{(C + 2Br_0)^2}{2} - A^2\right]\int R^2 dt. \tag{11.17b}$$

It may be noted that the amplitude of the bright soliton (11.16) can vary as a function of time depending upon the forms of the scattering length R.

Finally, the normalization condition (11.6) can be rewritten in terms of the soliton parameters as

$$\int_{-\infty}^{\infty} |\Phi^+(z,t)|^2 dz = 2\sqrt{2}Ar_0 = 2N, \tag{11.18}$$

which allows one to fix the constant r_0 in terms of the experimental BEC parameters as

$$r_0 = \frac{N}{\sqrt{2}A}. \tag{11.19}$$

11.3.1(b) REPULSIVE INTERACTION ($\sigma = -1$)

On the other hand, when $\sigma = -1$, one may write the dark one-soliton solution in the form [36]

$$U^-(Z,T) = \frac{1}{\sqrt{2}}\left[C - 2i\beta\tanh\beta(Z - CT)\right]\exp\left[-\frac{i}{2}(C^2 + 4\beta^2)T\right], \tag{11.20}$$

where the parameters β and C correspond to the depth and velocity of the dark soliton, respectively. Here the minus sign in the superscript means the $\sigma = -1$ case.

Using Equations (11.8) and (11.20), one can write the dark soliton solution of (11.5) as

$$\Phi^-(z,t) = \Lambda(z,t)U^-(Z,T). \tag{11.21}$$

Substituting $\Lambda(z,t)$, Z and T from Equations (11.9), (11.11) and (11.12) in the above equation (11.21), we get

$$\Phi^-(z,t) = r_0\sqrt{R}\left(C - 2i\beta\tanh\left[\xi^-(z,t)\right]\right)\exp\left[i\eta^-(z,t)\right]. \tag{11.22a}$$

Here the generalized wave variable ξ^- which specifies the width and velocity of the dark soliton as a function of time and longitudinal distance, respectively, is given by the expression

$$\xi^-(z,t) = 2\beta r_0 R\left[z - r_0(C + 2Br_0)\frac{1}{R}\int R^2 dt\right], \tag{11.22b}$$

and the generalized wave phase of the dark soliton is expressed as

$$\eta^-(z,t) = 2Br_0^2 Rz - \frac{\dot{R}}{2R}z^2 - r_0^2\left(C^2 + 2B^2r_0^2 + 4\beta^2\right)\int R^2 dt. \tag{11.22c}$$

Also, the depth of the generalized dark soliton is $2\beta r_0\sqrt{R}$, which depends on R. As in the case of the bright soliton solution, we can get stable dark soliton solutions by choosing the forms of R to balance each other so as to achieve a constant amplitude/depth. In this case also the width depends on R and can increase/decrease or remain constant with time. Again the normalization condition (11.6) of the dark soliton solution for the GP equation leads us to the condition

$$\int_{-\infty}^{\infty}\left[r_0^2 R(C^2 + 4\beta^2) - |\Phi^-(z,t)|^2\right]dz = 4\beta r_0 = 2N, \tag{11.23}$$

so that the parameter r_0 can be fixed in this case as

$$r_0 = \frac{N}{2\beta}. \tag{11.24}$$

We consider the trap frequencies to be of the form (i) $\Omega(t)^2 = \Omega_0^2$, (ii) $\Omega^2(t) = (\Omega_0^2/2)\left[1 - \tanh(\beta_0 t/2)\right]$ and (iii) $\Omega(t)^2 = 4\Omega_0^2[1 + 2\tan^2(2\Omega_0 t)]$, where Ω_0 is a constant. The effective scattering lengths for these three cases turn out to be (i) $R(t) = \text{sech}(\Omega_0 t + \delta)$, (ii) $R(t) = 1 + \tanh(\Omega_0 t/2)$ and (iii) $R(t) = 1 + \cos(2\Omega_0 t)$.

S. No.	Trap frequency (Ω^2)	Interaction strength (R)	Example
1.	Ω_0^2, constant	$\sec(\Omega_0 t + \delta)$, Bright soliton	Figure 11.1
2.	$-\Omega_0^2$, constant	$\mathrm{sech}(\Omega_0 t + \delta)$, Bright soliton	Figure 11.2
3.	$-\Omega_0^2$, constant	$\mathrm{sech}(\Omega_0 t + \delta)$, Soliton bound state	Figure 11.3
4.	$(\Omega_0^2/2)[\tanh$ $(\beta_0 t/2) - 1]$	$1 + \tanh(\Omega_0 t/2)$, Bright soliton	Figure 11.4

11.3.2 TIME-INDEPENDENT TRAP

Let us first focus our attention on the case of time-independent harmonic potential for which $\Omega^2(t)$ in (11.5) is a constant $(\Omega^2(t) = \pm\Omega_0^2)$. In this case, we shall identify the different soliton solutions for both confining $\Omega_0^2 > 0$, as well as expulsive, $\Omega_0^2 < 0$, potentials for various forms of scattering length $\tilde{R}(t)$. It is worth pointing out that bright matter wave solitons in an ultracold ^{7}Li have been observed in experiments, both the confining as well as expulsive potentials [20, 21]. Also the formation of dark solitons, their oscillations and interaction in single component BECs of ^{87}Rb atoms with confining harmonic potential have been experimentally demonstrated [38].

Substituting $\Omega(t)^2 = \Omega_0^2$ in the integrability condition (11.13), one may find that the time-dependent interaction term should be of the form $R(t) = \sec(\Omega_0 t + \delta)$, where δ is an integration constant. Figure 11.1 depicts the space-time plot of the condensate density $|\Psi^2|$ in a confining potential for a typical choice of parameters $A = 1$, $\Omega_0 = 0.1$, $\delta = 0$, $B = 0.1$, and $C = 0.2$. The bright matter wave soliton for $R = \sec(\Omega_0 t + \delta)$ shows periodic oscillation

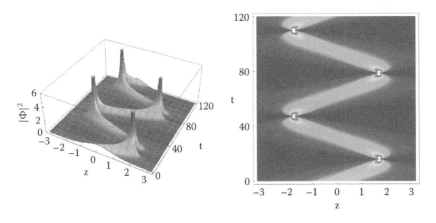

FIGURE 11.1: Bright soliton for the case of $\Omega^2(t) = \Omega_0^2$, and $R(t) = \sec(\Omega_0 t + \delta)$ and the corresponding contour plot. The parameters are chosen as $A = 1$, $\Omega_0 = 0.1$, $\delta = 0$, $B = 0.1$, and $C = 0.2$ in Equation (11.16).

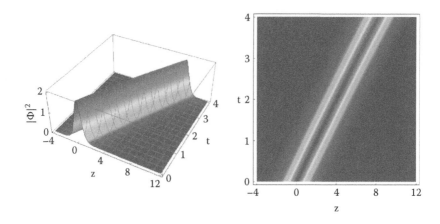

FIGURE 11.2: Bright soliton for the case of $\Omega^2(t) = -\Omega_0^2$, and $R(t) = \text{sech}(\Omega_0 t + \delta)$ and the corresponding contour plot. The parameters are chosen as $\Omega_0 = 0.1$, $\delta = 0$, $A = 1$, $B = 0.5$, and $C = 1$ in Equation (11.16).

in amplitude due to the presence of the periodic terms $\sec(\Omega_0 t + \delta)$. This particular soliton is used to provide a naive interpretation of wave collapse in quantum systems [39]. In this case, the dark soliton in the repulsive interaction case becomes singular at finite values of t and so is not physically interesting.

In the case of expulsive potential, $\Omega^2(t) = \Omega_0^2$, one may find a particular solution of physical interest, namely, $R(t) = \text{sech}(\Omega_0 t + \delta)$. The corresponding matter wave bright soliton profile is shown in Fig. 11.2 for $\Omega_0 = 0.1$, $\delta = 0$, $A = 1$, $B = 0.5$, and $C = 1$.

Another interesting structure that occurs when the propagation term is absent in the above soliton solution (that is, velocity is zero) for the parametric choice $B = 0$ and $C = 0$ is shown in Fig. 11.3. Similar plots for the case of dark solitons can also be deduced from (11.22a), which we do not present here. For details, see [33].

11.3.3   TIME-DEPENDENT TRAP

Next we consider the time-dependent trap frequency in the form $\Omega_0^2(t) = (\Omega_0^2/2)\,[1 - \tanh(\Omega_0 t/2)]$. For this choice, the relation (11.13) fixes the interatomic interaction term to be of the form $R(t) = 1 + \tanh(\beta_0 t/2)$ (see Fig. 11.4).

11.3.4   DARK AND BRIGHT MULTI-SOLITON DYNAMICS

In the previous sections, we analyzed the exact bright and dark one-soliton solutions of BECs for various forms of time-varying parameters. It is natural to extend the previous analysis in the case of multi-soliton solutions. For

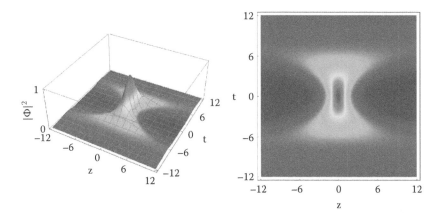

FIGURE 11.3: Bright soliton bound state for the case of $\Omega^2(t) = -\Omega_0^2$ and $R(t) = \mathrm{sech}(\Omega_0 t + \delta)$. The parameters are chosen as $\Omega_0 = 1/2$, $A = 1$, $B = 0$, and $C = 0$ in Equation (11.16).

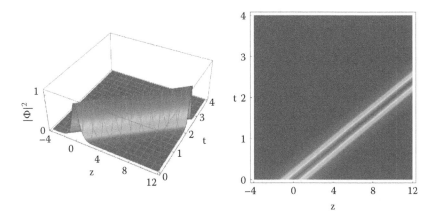

FIGURE 11.4: Bright soliton for the case of $\Omega^2(t) = (\Omega_0^2/2)\,[1 - \tanh\,(\Omega_0 t/2)]$, and $R(t) = 1 + \tanh\,(\beta_0 t/2)$. The parameters are chosen as $\Omega_0 = 0.01$, $r_0 = 1$, $A = 1$, $B = 2$, and $C = 1$ in Equation (11.16).

attractive type nonlinearity ($\sigma = 1$), the NLS equation admits N-multi bright soliton solutions.

The multi-soliton (N-soliton) solution of the NLS equation (11.7) the attractive type nonlinearity ($\sigma = 1$) can be written by defining the following $(1 \times N)$ row matrix C and $\mathbf{0}$, and $(N \times 1)$ column matrices ϕ, and the $(N \times N)$

identity matrix I:

$$C = -(\alpha_1, \alpha_2, \ldots, \alpha_N), \quad \mathbf{0} = (0, 0, \ldots, 0), \tag{11.25a}$$

$$\phi = \begin{pmatrix} \exp \eta_1 \\ \exp \eta_2 \\ \vdots \\ \exp \eta_N \end{pmatrix}, \quad I = \begin{pmatrix} 1 & 0 & \cdots & 0 \\ 0 & 1 & \cdots & 0 \\ \vdots & \vdots & \ddots & \vdots \\ 0 & 0 & \cdots & 1 \end{pmatrix}. \tag{11.25b}$$

Here α_j, $j = 1, 2, \ldots, N$, are arbitrary complex parameters and $\eta_j = \sqrt{2} k_j Z + 2i k_j^2 T$, $j = 1, 2, \ldots, N$, and k_j's are complex parameters. We write the multi-soliton solution of the NLS equation (11.7) as

$$U^+(Z, T) = g^+ / f^+, \tag{11.26}$$

where the plus sign in the superscript represents the $\sigma = +1$ case and

$$g^+ = \begin{vmatrix} A & I & \phi \\ -I & B^T & \mathbf{0}^T \\ \mathbf{0} & C & 0 \end{vmatrix}, \quad f = \begin{vmatrix} A & I \\ -I & B^T \end{vmatrix}. \tag{11.27a}$$

Here the matrices A and B are defined as

$$A_{ij} = \frac{\exp(\eta_i + \eta_j^*)}{k_i + k_j^*}, \quad B_{ij} = \frac{\alpha_j^\dagger \alpha_i}{k_j^* + k_i}, \quad i, j = 1, 2, \ldots, N. \tag{11.27b}$$

In Equation (11.27b), † represents the transpose conjugate. In particular, the one-soliton solution ($N = 1$ case) reads as

$$U^+(Z, T) = \begin{vmatrix} A_{11} & 1 & \exp \eta_1 \\ -1 & B_{11} & 0 \\ 0 & -\alpha_1 & 0 \end{vmatrix} \times \begin{vmatrix} A_{11} & 1 \\ -1 & B_{11} \end{vmatrix}^{-1} = \frac{\alpha_1 \exp \eta_1}{1 + \beta_0 \exp(\eta_1 + \eta_1^*)}, \tag{11.28}$$

where

$$\beta_0 = \frac{|\alpha_1|^2}{(k_1 + k_1^*)^2}, \quad \eta_1 = \sqrt{2} k_1 Z + 2i k_1^2 T. \tag{11.29}$$

The explicit form of the bright one-soliton solution can be written as

$$U^+(Z, T) = \frac{2\alpha_1 k_{1R}}{|\alpha_1|} \text{sech} \left[\sqrt{2} k_{1R} \left(Z - 2\sqrt{2} k_{1I} T \right) + \frac{\eta_0}{2} \right]$$

$$\times \exp \left[i \left(\sqrt{2} k_{1I} Z + 2(k_{1R}^2 - k_{1I}^2) T \right) \right], \tag{11.30}$$

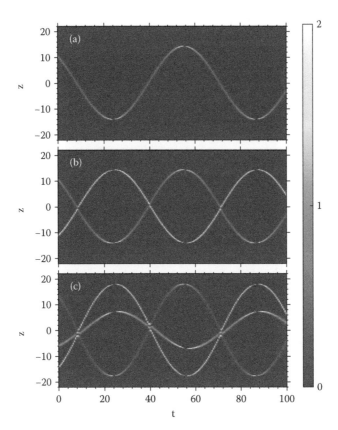

FIGURE 11.5: Density contour plot showing bright (a) one- (b) two- and (c) three-soliton solutions obtained from (11.26) for the time-independent confining harmonic potential $(\Omega^2(t) = \Omega_0^2)$ and $R(t) = \sec(\Omega_0 t + \delta)$ with $\Omega_0 = 0.1$, $\delta = -4.0$, $r_0 = 0.5$ and $b = 0$. The other parameters are (a) $k_1 = 1 + i$, $\alpha_1 = 1.0$, (b) $k_1 = 1 + i$, $k_2 = 2 - i$, $\alpha_1 = \alpha_2 = 1.0$, and (c) $k_1 = 0.75 + 1.25i$, $k_2 = 2 - 1.25i$, $k_3 = 1.25 - 0.52i$, $\alpha_1 = \alpha_2 = \alpha_3 = 1.0$.

where $\eta_0 = \log \beta_0$, k_{1R} and k_{1I} are real and imaginary parts of $k1$, η_{1R} and η_{1I} are real and imaginary parts of η_1. For $\eta_0 = 0$ (i.e., $\alpha_1 = 2k_{1R}$), the above equation (11.30) is the same as (11.16).

Fig. 11.5(a) shows the density contour plot of bright one soliton solution (11.30) for time independent confining harmonic trap.

For the $N = 2$ case, we can write the two-soliton solution from Equation

(11.26) as

$$U^+(Z,T) = \begin{vmatrix} A_{11} & A_{12} & 1 & 0 & \exp\eta_1 \\ A_{21} & A_{22} & 0 & 1 & \exp\eta_2 \\ -1 & 0 & B_{11} & B_{21} & 0 \\ 0 & -1 & B_{12} & B_{22} & 0 \\ 0 & 0 & -\alpha_1 & -\alpha_2 & 0 \end{vmatrix} \times \begin{vmatrix} A_{11} & A_{12} & 1 & 0 \\ A_{21} & A_{22} & 0 & 1 \\ -1 & 0 & B_{11} & B_{21} \\ 0 & -1 & B_{12} & B_{22} \end{vmatrix}^{-1},$$

(11.31)

with $j = 1, 2$. The explicit form of the two-soliton solution can be written as

$$U^+(Z,T) = \frac{1}{D} \sum_{l=1}^{2} [\alpha_l \exp\eta_l + \xi_l \exp(\eta_1 + \eta_2 + \eta_l^*)], \qquad (11.32)$$

where

$$D = 1 + \sum_{l,m=1}^{2} [\beta_{lm} \exp(\eta_l + \eta_m^*)] + \beta_3 \exp(\eta_1 + \eta_1^* + \eta_2 + \eta_2^*)$$

$$\xi_l = \frac{k_1 - k_2}{(k_1 + k_l^*)(k_2 + k_l^*)}(\alpha_1\beta_{2l} - \alpha_2\beta_{1l}),$$

$$\beta_3 = \frac{|k_1 - k_2|^2(\beta_{11}\beta_{22} - \beta_{12}\beta_{21})}{(k_2 + k_2^*)(k_2 + k_1^*)|k_1 + k_2^*|^2}, \quad \beta_{lm} = \frac{\alpha_l\alpha_m^*}{(k_l + k_m^*)^2}. \qquad (11.33)$$

Here

$$\eta_1 = \sqrt{2}k_1 Z + 2ik_1^2 T, \quad \eta_2 = \sqrt{2}k_2 Z + 2ik_2^2 T. \qquad (11.34)$$

Fig. 11.5(b) shows the density contour plot of bright two solitons (11.32) for time independent confining harmonic trap.

Similarly, the bright three- and N-soliton solutions of the NLS equation can be written explicitly.

Correspondingly for the GP equation (11.5), using the relation (11.8), the N-bright soliton solution can be written as

$$\Phi^+(z,t) = \Lambda(z,t)U^+(Z,T). \qquad (11.35)$$

Now substituting the forms of $\Lambda(z,t)$, Z and T from Equations (11.9), (11.11) and (11.12) in (11.35), one can get a generalized expression for the bright N-soliton as

$$\Phi^+(z,t) = r_0\sqrt{2R}\exp[i\theta(z,t)]U^+(Z,T). \qquad (11.36)$$

In Fig. 11.5(c) we show the density contour plot of bright three solitons for time independent confining harmonic trap.

11.3.5 *N*-DARK SOLITON SOLUTION

For repulsive type nonlinearity ($\sigma = -1$), the NLS equation admits the N-dark soliton solution. The general expression for the N-dark-dark soliton solution of the two-coupled NLS equation is obtained using Hirota's bilinear method [40]. Now we deduce the N-dark soliton solution of the scalar NLS equation from N-dark-dark soliton solutions of the NLS equation (Fig. 11.6).

The N-dark soliton solution of the NLS equation can be written as

$$U^-(Z,T) = g^-/f^-, \tag{11.37}$$

where the minus sign in the superscript represents the $\sigma = -1$ case and

$$g^- = \tau \exp(i\psi) \sum_{\alpha=0,1} \exp\left(\sum_{j=1}^N \alpha_j(\zeta_j + i\theta_j) + \sum_{i<j}^N a_{ij}\alpha_i\alpha_j\right), \tag{11.38}$$

$$f^- = \sum_{\alpha=0,1} \exp\left(\sum_{j=1}^N \alpha_j\zeta_j + \sum_{i<j}^N a_{ij}\alpha_i\alpha_j\right), \tag{11.39}$$

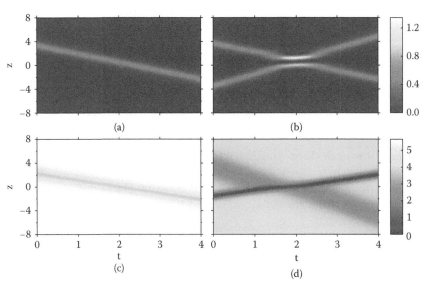

FIGURE 11.6: Density contour plots showing bright (a) one- (b) two-solitons, and dark (c) one- and (d) two-solitons obtained from (11.26) for the time-independent confining harmonic potential, $\Omega^2(t) = \Omega_0^2$, and $R(t) = \text{sech}(\Omega_0 t + \delta)$ with $\Omega_0 = 0.1$, $\delta = -0.2$, and $b = 0$. The other parameters are (a) $r_0 = 0.5$, $k_1 = 1 - i$, $\alpha_1 = 1.0$, (b) $r_0 = 0.5$, $k_1 = 1 + \sqrt{2}i$, $k_2 = 1 - i$, $\alpha_1 = \alpha_2 = 1.0$, (c) $r_0 = 1$, $a_1 = \sqrt{2}$, $\tau = \sqrt{2a_1}$, $l = 0.5$, and (d) $r_0 = 1$, $a_1 = 2$, $a_2 = \sqrt{2}$, $\tau = \sqrt{2}$, $l = 0.5$.

where

$$\psi = lT + (|\tau| - l^2)Z + \psi^{(0)}, \tag{11.40}$$

$$\zeta_j = -a_j T - \left(\sqrt{2|\tau|^2 - a_j^2} - 2l\right) a_j Z + \zeta_j^{(0)}, \tag{11.41}$$

in which a_j, l, $\psi^{(0)}$ and $\zeta_j^{(0)}$ are real constants and τ is a complex constant. Also,

$$\exp(i\theta_j) = \rho_j = \frac{-a_j + i\sqrt{2|\tau|^2 - a_j^2}}{a_j + i\sqrt{2|\tau|^2 - a_j^2}}, \tag{11.42}$$

and

$$\exp(a_{ij}) = \beta_{ij}^- = \frac{(a_i - a_j)^2 + \left(\sqrt{2|\tau|^2 - a_j^2} - \sqrt{2|\tau|^2 - a_j^2}\right)^2}{(a_i + a_j)^2 + \left(\sqrt{2|\tau|^2 - a_j^2} - \sqrt{2|\tau|^2 - a_j^2}\right)^2}. \tag{11.43}$$

For the $N = 1$ case, the dark one-soliton solution of the NLS equation can be written as

$$U^-(Z, T) = \tau \exp(i\psi)\frac{1 + \rho_1 \exp \zeta_1}{1 + \exp \zeta_1}. \tag{11.44}$$

For the $N = 2$ case, the dark two-soliton solution of the NLS equation can be deduced as

$$U^-(Z, T) = \tau \exp(i\psi)\frac{1 + \rho_1 \exp \zeta_1 + \rho_2 \exp \zeta_2 + \beta_{12}^- \rho_1 \rho_2 \exp(\zeta_1 + \zeta_2)}{1 + \exp \zeta_1 + \exp \zeta_1 + \beta_{12}^- \exp(\zeta_1 + \zeta_2)}. \tag{11.45}$$

Similarly, the dark three- and N-soliton solutions of the NLS equation can be written explicitly.

The N-dark soliton solution for the GP equation (11.5), using the relation (11.8) can be written as

$$\Phi^-(z, t) = \Lambda(z, t) U^-(Z, T). \tag{11.46}$$

Substituting the forms of $\Lambda(z, t)$, Z and T from Equations (11.9), (11.11) and (11.12) in (11.46), a generalized expression for the dark N-soliton of the GP equation (11.5) can be written as

$$\Phi^-(z, t) = r_0\sqrt{2R}\exp[i\theta(z, t)] U^-(Z, T). \tag{11.47}$$

11.4   MATTER WAVE SOLITONS IN MULTI-COMPONENT BECs

Multi-component BECs have been experimentally studied in mixtures of different hyperfine states of the same atomic species and mixtures of different atomic species. Experimental generation of two-component BECs of different hyperfine states of rubidium atoms in a magnetic trap [41] and of sodium atoms in an optical trap [42] stimulated theoretical studies devoted to the mean-field dynamics of multi-component condensates. In this connection, multi-component generalization of the soliton dynamics is very natural in the context of atomic BECs because of the several ways to create such systems. Matter wave solitons in multi-component BECs hold promise for a number of applications, including multi-channel signals and their switching, coherent storage and processing of optical fields.

In the case of BECs of cold atomic gases, the two vector components which evolve under the Gross–Pitaevskii equation are the macroscopic wave functions of Bose condensed atoms in two different internal states, which we shall denote as $|1\rangle$ and $|2\rangle$. In the presence of strong transverse trap potential, the properties of a one-dimensional two-component trapped BEC can be described by the mean-field GP equations for the wave functions Ψ_1 and Ψ_2 of the condensates $|1\rangle$ and $|2\rangle$:

$$i\frac{\partial \Psi_j}{\partial t} = -\frac{1}{2}\frac{\partial^2 \Psi_j}{\partial z^2} + \left[R\sum_{k=1}^{2} \sigma_{jk}|\Psi_k|^2 + V(z,t) \right]\Psi_j, \quad j = 1,2, \qquad (11.48)$$

where $V(z,t) = \Omega^2(t)z^2/2$ is the external time-varying trap potential, which is expulsive for $\Omega^2(t) > 0$ and confining for $\Omega^2(t) < 0$. σ_{jk}'s are the signs of the s-wave scattering lengths, which are negative for attractive and positive for repulsive interactions, and μ_j is the chemical potential of the j-th component.

Equation (11.48) can be mapped onto the two coupled nonlinear Schrödinger (Manakov) equations of the form

$$i\frac{\partial U_j}{\partial T} + \frac{\partial^2 U_j}{\partial X^2} + 2\left(\sum_{k=1}^{2} |U_k|^2 \right)U_j = 0, \quad j = 1,2, \qquad (11.49)$$

under the following transformation [33]:

$$\Psi_j(z,t) = \Lambda U_j(Z,T), \quad j = 1,2, \qquad (11.50)$$

where the new independent variables T and X are chosen as functions of the old independent variables as $T = G'(t)$ and $Z = F(z,t)$ subject to the conditions on the functions Λ, F, Ω and R as given earlier in Equations (11.10), and G' should satisfy

$$G'_t = \frac{F_x^2}{2} = \frac{R}{2}|\Lambda|^2. \qquad (11.51)$$

Following a similar procedure as in Sec. 11.2, one can see that the transformation functions Λ, F will have the same form as (11.11) and (11.12c). However, $G'(t)$ takes the form

$$G'(t) = 2r_0^2 \int R^2 dt \equiv T'. \qquad (11.52)$$

From the solutions of Equation (11.49), one can straightforwardly construct the one-, two-, three- and N-bright–bright soliton solutions for Equation (11.48), provided R and $\Omega(t)$ satisfy the Riccati-type equation (11.13).

11.4.1 DARK–BRIGHT SOLITONS

11.4.1(a) ONE-SOLITON DYNAMICS

The dark-bright components of the one-soliton solution of the defocusing Manakov system (11.49) can be given as

$$U_1(Z, T') = \left(\frac{\tau}{\rho^*}\right) \frac{(\rho^* - \rho)\chi \, \exp(\eta + \eta^*)}{1 + \chi \, \exp(\eta + \eta^*)} \exp\left(\mathrm{i}cZ - \mathrm{i}\left[c^2/2 + \tau^2\right] T'\right),$$

(11.53)

$$U_2(Z, T') = \frac{\exp \eta}{1 + \chi \, \exp(\eta + \eta^*)},$$

(11.54)

where

$$\eta = \kappa X + \mathrm{i}\left(\frac{\kappa^2}{2} - \tau^2\right) T', \quad \chi = \left[(\kappa + \kappa^*)^2 \left(\frac{|\tau|^2}{\rho\rho^*} - 1\right)\right]^{-1},$$

and $\kappa = a_1 + \mathrm{i}b_1$, $\rho = \kappa - \mathrm{i}c_1$. The parameters a_1 and b_1 correspond to the amplitude and velocity of the soliton envelope, respectively, and c_1 refers to the phase. The corresponding dark-bright one-soliton solution of the two coupled GP equation can be written as

$$\Psi_j(z, t) = \sqrt{2R(t)} \, U_j(Z, T') \exp(\mathrm{i}\left[\theta(z, t) + \mu t\right]), \quad j = 1, 2.$$

(11.55)

11.4.1(b) TWO-SOLITON DYNAMICS

The dark-bright two-soliton solution of the defocusing Manakov system (11.49) can be written as

$$U_1(Z, T') = \frac{\tau \exp(\mathrm{i}cZ - \mathrm{i}\left[c^2/2 + \tau^2\right] T')}{D}$$

$$\times \left(1 - \sum_{j,k=1}^{2} \frac{\rho_j \chi_{jk} \exp(\eta_j + \eta_k^*)}{\rho_k^*} + \frac{\rho_1 \rho_2 f \exp(\eta_1 + \eta_1^* + \eta_2 + \eta_2^*)}{\rho_1^* \rho_2^*}\right),$$

(11.56)

$$U_2(Z, T') = \frac{1}{D}\left(\sum_{j,k=1}^{2} \exp \eta_j - \sum_{j,k=1}^{2} \nu_{jk} \chi_{1j} \chi_{2j} \, \exp(\eta_j + \eta_j^* + \eta_k)\right),$$

(11.57)

where

$$D = 1 + \sum_{j,k=1}^{2} \chi_{jk} \exp(\eta_j + \eta_k^*) + f \, \exp(\eta_1 + \eta_1^* + \eta_2 + \eta_2^*),$$

with

$$\eta_j = \kappa_j X + i\left(\frac{1}{2}\kappa_j^2 - \tau^2\right)T', \quad \chi_{jk} = \left[(\kappa_j + \kappa_k^*)^2\left(\frac{|\tau|^2}{\rho_j\rho_k^*} - 1\right)\right]^{-1},$$

$$\nu_{jk} = (\kappa_j - \kappa_k)^2\left(\frac{|\tau|^2}{\rho_j\rho_k} + 1\right), \quad f = \chi_{11}\chi_{22}|\nu_{12}\chi_{12}|^2,$$

$\kappa_j = a_j + ib_j$, and $\rho_j = \kappa_j - ic$. The dark-bright two-soliton solution of the two coupled GP equation (11.48) can again be represented by Equations (11.55) with U_1 and U_2 of the forms (11.56) and (11.57). The procedure can be extended to N-soliton solutions also. However, we do not present their forms here. Depending on the form of the trap potential and interatomic interaction, a novel type of dark-bright matter wave solitons can be deduced using the above forms (11.55)–(11.57). In the following, we demonstrate them for two simple trap potentials. For the other choices, results can be similarly deduced; see, for example, [32]. In the present study we fixed the trap parameters as similar to those used in a recent experiment on dark-bright solitons in two component ^{87}Rb condensates [38].

The explicit form of the two-soliton solution can be written as

$$U_j(Z, T') = \frac{1}{D}\sum_{l=1}^{2}\left[\alpha_l^{(j)}\exp\eta_l + \xi_l^{(j)}\exp(\eta_1 + \eta_2 + \eta_l^*)\right], \qquad (11.58)$$

where

$$D = 1 + \sum_{l,m=1}^{2}\beta_{lm}\exp(\eta_l + \eta_m^*) + \beta_3\exp(\eta_1 + \eta_1^* + \eta_2 + \eta_2^*)$$

$$\xi_l^{(j)} = \frac{(k_1 - k_2)\left(\alpha_1^{(j)}\kappa_{2l} - \alpha_2^{(j)}\kappa_{1l}\right)}{(k_1 + k_l^*)(k_2 + k_l^*)}, \quad \beta_{lm} = \frac{\kappa_{lm}}{k_l + k_m^*},$$

$$\beta_3 = \frac{|k_1 - k_2|^2(\kappa_{11}\kappa_{22} - \kappa_{12}\kappa_{21})}{(k_2 + k_2^*)(k_2 + k_1^*)|k_1 + k_2^*|^2}, \quad \kappa_{lm} = \sum_{j=1}^{2}\frac{\alpha_l^{(j)}\alpha_m^{(j)*}}{k_l + k_m^*}. \qquad (11.59)$$

Similarly, the three-soliton and N-soliton solutions can be written down explicitly.

The N-soliton solution of the two-coupled GP equations can be written as

$$\Psi_j(z, t) = 2r_0\sqrt{R}\exp(i\theta)\,U_j(Z, T'), \qquad (11.60)$$

Figure 11.7 illustrates the one and two dark-bright solitons in the case of time-independent expulsive potential $\Omega^2(t) = -\Omega_0^2$ with corresponding $R(t) = \mathrm{sech}(\Omega_0 t + \delta)$.

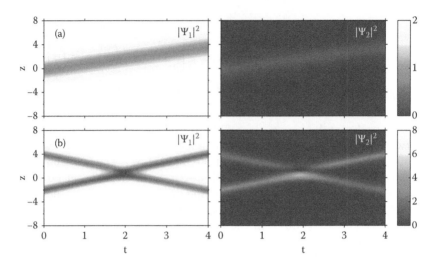

FIGURE 11.7: Density contour plots showing one and two dark-bright solitons obtained from (11.26) for the time-independent confining harmonic potential, $\Omega^2(t) = \Omega_0^2$, and $R(t) = \text{sech}\,(\Omega_0 t + \delta)$ with $\Omega_0 = 0.1$, $\delta = -0.2$, $r_0 = 1$ and $b = 0$. The other parameters are (a) and (b) $a_1 = 0.7$, $b_1 = 0.5$, $c_1 = c = 0$ and $\tau = 1$, (c) and (d) $a_1 = 1.4$, $a_2 = -1.4$, $b_1 = 0.7$, $b_2 = -0.7$, $c_1 = c_2 = c = 0$, and $\tau = 2$.

11.4.2 BRIGHT–BRIGHT SOLITONS

In the previous section we have analyzed the nature of interaction of the dark-bright solitons in two-component BECs with time-varying parameters. It is also of importance to bring out the exact bright–bright one-, two-, three- and N-soliton solutions in two-component BECs. In particular, we demonstrate the bright–bright shape changing/matter redistribution of two and three solitons under collision, while elastic collision occurs for a very special choice of parameters. We will also show the shape restoring property in the three-soliton interaction. This type of elastic and shape changing interactions of two and three solitons have been well studied in the context of optical computing, where the intensity of light pulses is transformable between two modes of Manakov type systems [43–46]. From the BEC point of view, the shape changing interaction can be interpreted as the transformation of the fraction of atoms between the components, which is the so called matter wave switch. Such matter wave switching phenomena can be used to manipulate matter wave devices such as switches, logic gates and atom chips. One of the long term perspectives of matter wave devices is their potential application to quantum information processing. In this section we show that such a matter wave switching phenomenon in two-component BECs is possible via shape changing soliton interactions by suitably tailoring the trap potential and atomic scattering length.

The 2CNLS (coupled nonlinear Schrödinger) equation (11.49) in contrast to the single component NLS system admits solutions which exhibit certain novel energy sharing (shape changing) collisions [43–46]. Recently, the general expression for the N-soliton solution of the Manakov system in the Gram determinant form has been given in [43] by using Hirota's bilinear method.

In order to write the multi-soliton (N-soliton) solution of the focussing Manakov system (11.49), we define the following ($1 \times N$) row matrix C_s, $s = 1, 2$, (2×1) column matrices ψ_j, and ($N \times 1$) column matrices ϕ, $j = 1, 2, \ldots, N$, and the ($N \times N$) identity matrix I:

$$C_s = - \left(\alpha_1^{(s)}, \alpha_2^{(s)}, \ldots, \alpha_N^{(s)} \right), \quad \mathbf{0} = (0, 0, \ldots, 0), \tag{11.61a}$$

$$\psi_j = \begin{pmatrix} \alpha_j^{(1)} \\ \alpha_j^{(2)} \end{pmatrix}, \phi = \begin{pmatrix} e^{\eta_1} \\ e^{\eta_2} \\ \vdots \\ e^{\eta_N} \end{pmatrix}, I = \begin{pmatrix} 1 & 0 & \cdots & 0 \\ 0 & 1 & \cdots & 0 \\ \vdots & \vdots & \ddots & \vdots \\ 0 & 0 & \cdots & 1 \end{pmatrix}. \tag{11.61b}$$

Here $\alpha_j^{(s)}$, $s = 1, 2$, $j = 1, 2, \ldots, N$, are arbitrary complex parameters and $\eta_i = k_i Z + i k_i^2 T'$, $i = 1, 2, \ldots, N$, are the wave variables and k_i's are complex parameters. We write the multi-soliton solution of the 2CNLS system [43] as

$$U_s(Z, T) = \frac{g^{(s)}}{f}, \quad s = 1, 2, \tag{11.62}$$

where

$$g^{(s)} = \begin{vmatrix} A & I & \phi \\ -I & B^T & \mathbf{0}^T \\ \mathbf{0} & C_s & 0 \end{vmatrix}, \quad f = \begin{vmatrix} A & I \\ -I & B^T \end{vmatrix}. \tag{11.63a}$$

Here the matrices A and B are defined as

$$A_{ij} = \frac{e^{\eta_i + \eta_j^*}}{k_i + k_j^*}, \quad B_{ij} = \frac{\psi_j^\dagger \psi_i}{k_j^* + k_i}, \quad i, j = 1, 2, \ldots, N. \tag{11.63b}$$

In Equation (11.63b), $\dagger$ represents the transpose conjugate. In particular, the one-soliton solution ($N = 1$ case) reads

$$U_j(Z, T) = \begin{vmatrix} A_{11} & 1 & e^{\eta_1} \\ -1 & B_{11} & 0 \\ 0 & -\alpha_1^{(j)} & 0 \end{vmatrix} \times \begin{vmatrix} A_{11} & 1 \\ -1 & B_{11} \end{vmatrix}^{-1} = \frac{\alpha_1^{(j)} e^{\eta_1}}{1 + \beta_0 e^{\eta_1 + \eta_1^*}}, \tag{11.64}$$

where $j = 1, 2$ and

$$\beta_0 = \frac{|\alpha_1^{(1)}|^2 + |\alpha_1^{(2)}|^2}{(k_1 + k_1^*)^2}. \tag{11.65}$$

Figure 11.8 depicts the bright–bright one-soliton for the two components $|\Psi_1|$ and $|\Psi_2|$ using equations (11.60) and (11.64) for the case with time independent confining harmonic potential, $\Omega^2(t) = \Omega_0^2$, and $R(t) = \sec(\Omega_0 t + \delta)$.

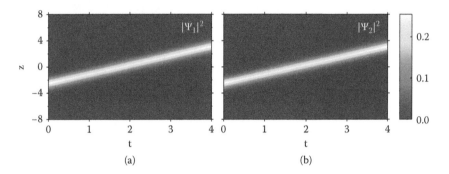

FIGURE 11.8: Density contour plots showing bright–bright one solitons obtained using equations (11.60) and (11.64) for the time independent confining harmonic potential $(\Omega^2(t) = \Omega_0^2)$ and $R(t) = \sec(\Omega_0 t + \delta)$ with $\Omega_0 = 0.1$, $\delta = -0.2$, $r_0 = 0.5$ and $b = 0$. The other parameters are $k_1 = 1 + i$, $\alpha_1^{(1)} = \alpha_1^{(2)} = 1.0$.

For the $N = 2$ case, we can write the two-soliton solution from Equation (11.62) as

$$U_j(Z,T) = \begin{vmatrix} A_{11} & A_{12} & 1 & 0 & e^{\eta_1} \\ A_{21} & A_{22} & 0 & 1 & e^{\eta_2} \\ -1 & 0 & B_{11} & B_{21} & 0 \\ 0 & -1 & B_{12} & B_{22} & 0 \\ 0 & 0 & -\alpha_1^{(j)} & -\alpha_2^{(j)} & 0 \end{vmatrix} \times \begin{vmatrix} A_{11} & A_{12} & 1 & 0 \\ A_{21} & A_{22} & 0 & 1 \\ -1 & 0 & B_{11} & B_{21} \\ 0 & -1 & B_{12} & B_{22} \end{vmatrix}^{-1},$$

$$\tag{11.66}$$

with $j = 1, 2$. The explicit form of the two-soliton solution can be written as

$$U_j(Z,T) = \frac{1}{D} \sum_{l=1}^{2} \left[\alpha_l^{(j)} e^{\eta_l} + \xi_l^{(j)} e^{\eta_1 + \eta_2 + \eta_l^*} \right], \tag{11.67}$$

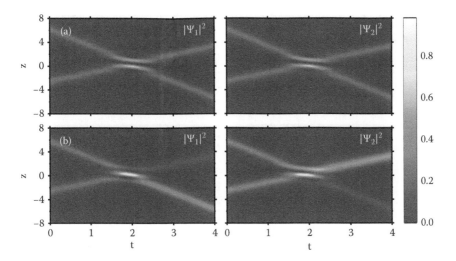

FIGURE 11.9: Density contour plots of bright–bright two-soliton solutions from Equations (11.60) and (11.67) showing (a) elastic collision and (b) inelastic collision for the time-independent confining harmonic potential, $\Omega^2(t) = \Omega_0^2$, and $R(t) = \sec(\Omega_0 t + \delta)$ with $\Omega_0 = 0.1$, $\delta = -0.2$, $r_0 = 0.5$, $b = 0$, $k_1 = 1 + i$, and $k_2 = 1 - 2i$. The other parameters are (a) $\alpha_1^{(1)} = \alpha_1^{(2)} = \alpha_2^{(1)} = \alpha_2^{(2)} = 1.0$, and (b) $\alpha_1^{(1)} = \alpha_2^{(1)} = \alpha_2^{(2)} = 1.0$, $\alpha_1^{(2)} = 1/3 + 2i$.

where

$$D = \sum_{l,m=1}^{2} \left[\beta_{lm} e^{\eta_l + \eta_m^*} \right] + \beta_3 e^{\eta_1 + \eta_1^* + \eta_2 + \eta_2^*},$$

$$\xi_l^{(j)} = \frac{k_1 - k_2}{(k_1 + k_l^*)(k_2 + k_l^*)} \left[\alpha_1^{(j)} \kappa_{2l} - \alpha_2^{(j)} \kappa_{1l} \right], \quad \kappa_{lm} = \sum_{j=1}^{2} \frac{\alpha_l^{(j)} \alpha_m^{(j)*}}{k_l + k_m^*}$$

$$\beta_3 = \frac{|k_1 - k_2|^2 (\kappa_{11}\kappa_{22} - \kappa_{12}\kappa_{21})}{(k_2 + k_2^*)(k_2 + k_1^*)|k_1 + k_2^*|^2}, \quad \beta_{lm} = \frac{\kappa_{lm}}{k_l + k_m^*} \tag{11.68}$$

Figure 11.9(a) shows the elastic interaction of the bright–bright two solitons for $k_1 = 1 + i$, $k_2 = 1 - 2i$, $\alpha_1^{(1)} = \alpha_1^{(2)} = \alpha_2^{(1)} = \alpha_2^{(2)} = 1.0$. Here the intensity of the two solitons in both the components is unchanged before and after interaction. Figure 11.9(b) illustrates the shape changing two-soliton interaction for $k_1 = 1 + i$, $k_2 = 1 - 2i$, $\alpha_1^{(1)} = \alpha_2^{(1)} = \alpha_2^{(2)} = 1.0$, and $\alpha_1^{(2)} = 1/3 + 2i$. Here the intensity of soliton 1 is enhanced while that of soliton 2 is suppressed after interaction in the component Ψ_1, whereas in the component Ψ_2 it is reversed.

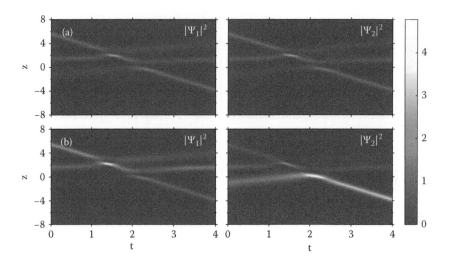

FIGURE 11.10: Density contour plots of bright–bright three-soliton solutions showing (a) elastic collision and (b) inelastic collision for the time-independent confining harmonic potential, $\Omega^2(t) = \Omega_0^2$, and $R(t) = \sec(\Omega_0 t + \delta)$ with $\Omega_0 = 0.1$, $\delta = -0.2$, $r_0 = 0.5$, $b = 0$, $k_1 = 1 + 0.5i$, $k_2 = 1.5$, and $k_3 = 2 - 1.5i$. The other parameters are (a) $\alpha_1^{(1)} = \alpha_2^{(1)} = \alpha_3^{(1)} = \alpha_1^{(2)} = \alpha_2^{(2)} = \alpha_3^{(2)} = 1.0$, and (b) $\alpha_1^{(1)} = 0.39 + 0.1i$, $\alpha_1^{(2)} = 0.3 + 0.1i$, $\alpha_2^{(1)} = 0.3 + 0.1i$, $\alpha_2^{(2)} = 1$, $\alpha_3^{(1)} = 1$, and $\alpha_3^{(2)} = 1$.

Similarly, the three-soliton solution can be written explicitly. The bright–bright N-soliton solution of the two-coupled GP equation (11.48) can be written as

$$\Psi_j(z,t) = 2r_0\sqrt{R}\exp(i\theta)\,U_j(Z, T'), \tag{11.69}$$

where Z and T' are given in Equations (11.12c) and (11.52). On examining Equation (11.69) for the choice of $\Omega^2(t) = \Omega_0^2$, and $R(t) = \sec(\Omega_0 t + \delta)$, one may find that elastic interaction occurs only for $\alpha_1^{(1)} = \alpha_1^{(2)}$, $\alpha_2^{(1)} = \alpha_2^{(2)}$ and $\alpha_3^{(1)} = \alpha_3^{(2)}$, and for all other choice of parameters, matter redistribution interaction occurs. Figure 11.10(a) shows the elastic interaction of the bright–bright three-soliton ($N = 3$ case) solution for $k_1 = 1 + 0.5i$, $k_2 = 1.5$, $k_3 = 2 - 1.5i$, and $\alpha_1^{(1)} = \alpha_2^{(1)} = \alpha_3^{(1)} = \alpha_1^{(2)} = \alpha_2^{(2)} = \alpha_3^{(2)} = 1.0$. Here the intensities of the three solitons in both the components are unchanged before and after interactions. Figure 11.10(b) shows the shape changing interactions of the bright–bright three-soliton solution for $\alpha_1^{(1)} = 0.39 + 0.1i$, $\alpha_1^{(2)} = 0.3 + 0.1i$, $\alpha_2^{(1)} = 0.3 + 0.1i$, $\alpha_2^{(2)} = 1$, $\alpha_3^{(1)} = 1$, and $\alpha_3^{(2)} = 1$. Here the intensity of soliton 1 is suppressed, while that of solitons 2 and 3 is enhanced after interaction in the component Ψ_1, whereas in the component Ψ_2 it is reversed. One can show similar one, two and three bright–bright solitons as in Figs. 11.8–11.10 for the case of expulsive trap potential $\Omega(t)^2 = -\Omega_0^2$ and with $R(t) = \text{sech}(\Omega_0 t + \delta)$.

Similar shape changing interactions of solitons during the interactions have been studied in the context of optical computing, where the intensities of light pulses are transformable between two modes [45, 46]. In the context of multicomponent BEC, the shape changing interactions may be interpreted as the transform of the fraction of atoms between the two components, which can be achieved experimentally by suitable tuning of atomic scattering length. These shape changing soliton interactions can be used in matter wave switching devices, logic gates and quantum information processing as in the case of optical computing.

11.5 SUMMARY

In summary, we have presented a brief overview of the exact bright and dark soliton solutions of single- and two-component Bose–Einstein condensates in an effective one-dimensional setting. In particular, by mapping the GP equation(s) onto the NLS equation(s) under suitable similarity transformation with appropriate conditions, we have deduced different kinds of bright, dark, dark-bright and bright–bright one-, two- and multi-soliton solutions and analyzed the interaction of two solitons for time-independent harmonic trapping potential. For other trapping potentials, interesting novel structures can also be deduced; see, for example, [32–34]. The present study provides an understanding of the possible mechanism for the localized excitations in Bose–Einstein condensates. The shape changing collisions of matter wave solitons can used for matter wave switching devices, logic gates and quantum information processing . These elastic and shape changing interactions can be realized in experiments by suitable control of time-dependent trap parameters and interatomic interaction.

ACKNOWLEDGMENTS

The work of PM forms a part of the Council of Scientific and Industrial Research Grant No. 03(1186)/10/EMR-II, and the Department of Science and Technology Grant No. INT/FRG/DAAD/P-220/2012, both Government of India funded research projects. The work of ML is supported by the Department of Science and Technology (DST), Government of India—DST-IRHPA project, and a DST Ramanna Fellowship. He is also supported by a DAE Raja Ramanna Fellowship.

REFERENCES

1. M. J. Ablowitz, P. A. Clarkson, *Nonlinear Evolution Equations and Inverse Scattering*, Cambridge University Press, 1991.

2. M. Lakshmanan, S. Rajasekar, *Nonlinear Dynamics: Integrability, Chaos and Patterns*, Springer–Verlag, 2003.

3. S. Ponomarenko, G. P. Agrawal, "Do solitonlike self-similar waves exist in nonlinear optical media?" *Physical Review Letters*, 97, 013901, 2006.

4. M. Centurion, M. A. Porter, P. G. Kevrekidis, D. Psaltis, "Nonlinearity management in optics: Experiment, theory, and simulation," *Physical Review Letters*, 97, 033903, 2006.

5. V. N. Serkin, A. Hasegawa, T. L. Belyaeva, "Nonautonomous solitons in external potentials," *Physical Review Letters*, 98, 074102, 2007.

6. A. Hasegawa, M. Matsumoto, *Optical Solitons in Fibers*, Springer, 2003.

7. D. R. Solli, C. Ropers, P. Koonath, B. Jalali, "Optical rogue waves," *Nature*, 450, 1054, 2003.

8. F. Dalfovo, S. Giorgini, L. P. Pitaevskii, S. Stringari, "Theory of Bose–Einstein condensation in trapped gases," *Review of Modern Physics*, 71, 463, 1999.

9. C. J. Pethick, H. Smith, *Bose–Einstein Condensation in Dilute Gases*," Cambridge University Press, 2002.

10. R. Carretero-Gonzalez, D. J. Frantzeskakis, P. G. Kevrekidis, "Nonlinear waves in Bose–Einstein condensates: Physical relevance and mathematical techniques," *Nonlinearity*, 21, R139-R202.

11. K. B. Davis, M. O. Mewes, M. R. Andrews, N. J. van Druten, D. S. Durfee, D. M. Kurn, W. Ketterle, "Bose–Einstein condensation in a gas of sodium atoms," *Physical Review Letters*, 75, 3969, 1995.

12. J. R. Ensher, D. S. Jin, M. R. Matthews, C. E. Wieman, E. A. Cornell, "Bose–Einstein condensation in a dilute gas: Measurement of energy and ground-state occupation," *Physical Review Letters*, 77, 4984, 1996.

13. P. A. Ruprecht, M. J. Holland, K. Burnett, M. Edwards, "Time-dependent solution of the nonlinear Schrödinger equation for Bose-condensed trapped neutral atoms," *Physical Review A*, 51, 4704, 1995.

14. V. M. Pérez-García, H. Michinel, H. Herrero, "Bose–Einstein solitons in highly asymmetric traps," *Physical Review A*, 57, 3837, 1998.

15. S. Burger, K. Bongs, S. Dettmer, W. Ertmer, K. Sengstock, A. Sanpera, G. V. Shlyapnikov, and M. Lewenstein, "Dark solitons in Bose-Einstein condensates," *Physical Review Letters*, 83, 5198 1999.

16. B. P. Anderson, P. C. Haljan, C. A. Regal, D. L. Feder, L. A. Collins, C. W. Clark, E. A. Cornell, "Watching dark solitons decay into vortex rings in a Bose–Einstein condensate," *Physical Review Letters*, 86, 2926, 2001.

17. J. Denschlag, J. E. Simsarian, D. L. Feder, C. W. Clark, L. A. Collins, J. Cubizolles, L. Deng, E. W. Hagley, K. Helmerson, W. P. Reinhardt, S. L. Rolston, B. I. Schneider, W. D. Phillips, "Generating solitons by phase engineering of a Bose–Einstein condensate," *Science*, 287, 97, 2000.

18. Th Busch, J. R. Anglin, "Motion of dark solitons in trapped Bose–Einstein condensates," *Physical Review Letters*, 84, 2298, 2002.

19. A. Trombettoni, A. Smerzi, "Discrete solitons and breathers with dilute Bose–Einstein condensates," *Physical Review Letters*, 86, 2353, 2000.

20. L. Khaykovich, F. Schreck, G. Ferrari, T. Bourdel, J. Cubizolles, L. D. Carr, Y. Castin, C. Salomon, "Formation of a matter-wave bright soliton," *Science*, 296, 1290, 2002.

21. K. E. Strecker, G. B. Partridge, A. G. Truscott, R. G. Hulet, "Formation and propagation of matter-wave soliton trains," *Nature*, 417, 150, 2002.

22. U. Al Khawaja, H. T. C. Stoof, R. G. Hulet, K. E. Strecker, G. B. Partridge, "Bright soliton trains of trapped Bose–Einstein condensates," *Physical Review Letters*, 89, 200404, 2002.

23. S. L. Cornish, S. T. Thompson, C. E. Wieman, "Formation of bright matter-wave solitons during the collapse of attractive Bose–Einstein condensates," *Physical Review Letters*, 96, 170401, 2006.

24. K. Bongs, S. Burger, D. Hellweg, M. Kottke, S. Dettmer, T. Rinkleff, L. Caccia-puoti, J. Arlt, K. Sengstock, W. Ertmer, "Spectroscopy of dark soliton states in Bose–Einstein condensates," *Journal of Optics B: Quantum and Semiclassical Optics*," 5, S124, 2003.

25. B. Eiermann, Th. Anker, M. Albiez, M. Taglieber, P. Treutlein, K. P. Marzlin, M. K. Oberthaler, "Bright Bose–Einstein gap solitons of atoms with repulsive interaction," *Physical Review Letters*, 92, 230401, 2004.

26. N. S. Ginsberg, J. Brand, L. V. Hau, "Observation of hybrid soliton vortex-ring structures in Bose–Einstein condensates," *Physical Review Letters*, 94, 040403, 2005.

27. P. Meystre, *Atom Optics*, Springer, 2001.

28. S. Inouye1, S. Andrews, M.R. Stenger, J. Miesner, H.-J. Stamper-Kurn, W. Ketterle. Observation of Feshbach resonances in a Bose–Einstein condensate, "*Nature*," 392, 151–154, 1998.

29. T. Köhler, K. Góral, P. S. Julienne, "Production of cold molecules via magnetically tunable Feshbach resonances," *Reviews of Modern Physics*, 78, 1311, 2006.

30. C. Chin, R. Grimm, P. Julienne, E. Tiesinga, "Feshbach resonances in ultracold gases," *Reviews of Modern Physics*, 82, 1225, 2010.

31. G. P. Agrawal, *Nonlinear Fiber Optics*, Academic Press Inc, 1995.

32. S. Rajendran, P. Muruganandam, M. Lakshmanan, "Interaction of dark-bright solitons in two-component Bose–Einstein condensates," *Journal of Physics B*, 42, 145307, 2009.

33. S. Rajendran, P. Muruganandam, M. Lakshmanan, "Bright and dark solitons in a quasi-1D Bose–Einstein condensates modelled by 1D Gross-Pitaevskii equation with time-dependent parameters," *Physica D*, 239, 366, 2010.

34. S. Rajendran, P. Muruganandam, M. Lakshmanan, "Matter wave switching in Bose–Einstein condensates via intensity redistribution soliton interactions," *Journal of Mathematical Physics B*, 52, 023515, 2011.

35. C. Sulem, P. L. Sulem, *The Nonlinear Schrödinger Equation*, Springer, 1999.

36. V. E. Zakharov, A. B. Shabat, "Exact theory of two-dimensional self-focusing and one-dimensional self-phase modulation of waves in nonlinear media," *Soviet Physics – JETP*, 34, 62, 1972.

37. V. E. Zakharov, A. B. Shabat, "Interaction between solitons in a stable medium," *Soviet Physics – JETP*, 37, 823, 1973.

38. C. Becker, S. Stellmer, P. Soltan-Panahi, S. Dorscher, M. Baumert, E. M. Richter, J. Kronjäger, K. Bongs, K. Sengstock,"Oscillations and interactions of

dark and dark-bright solitons in Bose–Einstein condensates," *Nature Physics*, 4, 496, 2008.

39. K. von Bloh, "A Wave Collapse in the Causal Interpretation of Quantum Theory," Wolfram Demonstrations Project. http://demonstrations.wolfram.com/AWaveCollapseInTheCausalInterpretationOfQuantumTheory/

40. R. Radhakrishnan, M. Lakshmanan, "Bright and dark soliton solutions to coupled nonlinear Schrödinger equations," *Journal of Physics A: Mathematical and General*, 28, 2683, 1995.

41. C. J. Myatt, E. A. Burt, R. W. Ghrist, E. A. Cornell, C. E. Wieman, "Production of two overlapping Bose–Einstein condensates by sympathetic cooling," *Physical Review Letters*, 78, 586, 1997.

42. D. M. Stamper-Kurn, M. R. Andrews, A. P. Chikkatur, S. Inouye, H. J. Miesner, J. Stenger, W. Ketterle, "Optical confinement of a Bose–Einstein condensate," *Physical Review Letters*, 80, 2027, 1998.

43. M. Vijayajayanthi, T. Kanna, M. Lakshmanan, "Multisoliton solutions and energy sharing collisions in coupled nonlinear Schrödinger equations with focusing, defocusing and mixed type nonlinearities," *European Physical Journal Special Topics*, 173, 57, 2009.

44. R. Radhakrishnan, M. Lakshmanan, J. Hietarinta, "Inelastic collision and switching of coupled bright solitons in optical fibers," *Physical Review E*, 56, 2213, 1997.

45. T. Kanna, M. Lakshmanan, "Exact soliton solutions, shape changing collisions, and partially coherent solitons in coupled nonlinear Schrödinger equations," "*Physical Review Letters*," 86, 5043, 2001.

46. T. Kanna, M. Lakshmanan, "Exact soliton solutions of coupled nonlinear Schrödinger equations: Shape-changing collisions, logic gates, and partially coherent solitons," *Physical Review E*, 67, 046617, 2003.

12 $\mathcal{PT}$-symmetric solitons

Chandroth P. Jisha and *Alessandro Alberucci*

12.1 INTRODUCTION

$\mathcal{PT}$- symmetric systems have become a topic which has generated considerable research interest. $\mathcal{PT}$-symmetric quantum mechanics originated after the work by Bender et al. [1,2], who conjectured that Hermiticity of the Hamiltonian, required in quantum mechanics for the existence of real eigenvalues, can be replaced by $\mathcal{PT}$ symmetry, i.e., mirror reflection (parity operator $\mathcal{P}$) and time inversion (time-inversion operator $\mathcal{T}$) even though the Hamiltonian is not Hermitian. Bender et. al. [1,3] made the connection between real eigenvalues of the non-Hermitian system and $\mathcal{PT}$ symmetry, thus paving the way for the development of $\mathcal{PT}$-symmetric quantum mechanics. Actions of the operators $\mathcal{P}$ and $\mathcal{T}$ correspond to the transformation $x \to -x$, $p \to -p$, $i \to i$ and $x \to x$, $p \to -p$, $i \to -i$, respectively: thus, the Bessis's Hamiltonian is neither symmetric with parity $\mathcal{P}$ nor time operation $\mathcal{T}$ applied alone, but it is symmetric with respect to the product operator $\mathcal{PT}$. However, the $\mathcal{PT}$-symmetry , corresponding to the commutation between the Hamiltonian and the $\mathcal{PT}$ operator, is neither a necessary nor a sufficient condition for a Hamiltonian to have real eigenvalues. If eigenfunctions of H are simultaneously eigenfunctions of the $\mathcal{PT}$ operator, symmetry is conserved; otherwise $\mathcal{PT}$ symmetry is spontaneously broken. A complex Hamiltonian of the form $H = \frac{p^2}{2m} + V(x)$ is $\mathcal{PT}$-symmetric if $V(x) = V^*(x)$, implying that the real part of $V(x)$ should be an even function and the imaginary part an odd function, respectively.

Fundamental ideas concerning $\mathcal{PT}$ systems have been successfully applied in optics. Indeed, optical systems have become an ideal workbench for the implementation of various quantum mechanical phenomena [4]: the paraxial Helmholtz equation can be considered formally equivalent to the Schrödinger equation and the condition for $\mathcal{PT}$ symmetry can be satisfied if the real part of the refractive index is an even function and the imaginary part of the refractive index is an odd function. The biggest advantage of considering optical systems is the ease in realizing complex Hamiltonians in the spatial [5] as well as in the temporal domain [6,7] and, consequently, the easy access to the experimental demonstration of theoretical findings.

In this chapter we will discuss nonlinear wave propagation in the presence of linear (i.e., independent from the power) potentials possessing $\mathcal{PT}$ symmetry; in particular we will consider the localized case, consisting of a bell-shaped trapping potential, and periodic potentials, infinitely extended across the transverse plane. After discussing the linear spectrum of the structures, we will discuss which types of solitons are supported by the system, comprising their stability [8]. Periodic systems possessing linear bandgaps are interesting as the bandgap can be tuned by nonlinearities and $\mathcal{PT}$-symmetric effects [9], allowing the observation of various phenomena such as $\mathcal{PT}$ phase transition, non-reciprocal behavior, double refraction [10,11], and also new kinds of gap solitons, i.e., self-trapped wavepackets existing inside the linear bandgap, both in continuous [12] and in discrete systems [13]. We will also show that $\mathcal{PT}$ systems predominantly exhibit oscillatory instability (OI), with the eigenvalues of the perturbation modes appearing in quartets, the latter being in the complex plane of the vertices of a square centered in the origin [14–17]. When the imaginary part of the potential is small enough, solitonic solutions can be quasi-stable, that is, the instability length is negligible if compared with the Rayleigh distance of the beam. We will show that the degree of instability is strictly related to the linear spectrum. In the case of periodic systems the effect of OI is the unidirectional transverse motion of the beam hopping from guide to guide (manifestation of PT breaking), with the field preferentially drifting toward the gain region. Conversely, for a localized potential, OI leads to exponentially growing modes which become highly localized into the gain region upon propagation.

12.2   RULING EQUATION

Propagation of a complex scalar field Ψ describing either the particle distribution for matter waves or the electromagnetic field for photons can be modeled using a modified nonlinear Schrödinger equation (NLSE) [18]. We assume that the material is linearly inhomogeneous, the inhomogeneity modeled by a potential $V(x)$ with a nonvanishing imaginary part [10]; furthermore, we consider a nonlocal nonlinear medium (nonlinear perturbation extending beyond the intensity profile), the nonlinear mechanism ascribable to some kind of diffusive process (heat conduction, diffusion, intermolecular interaction to cite only a few). Thus, in the mono-dimensional case the evolution of the field Ψ versus the evolution coordinate z (time for matter waves, propagation distance for light) will obey the following system of equations:

$$i\frac{\partial \Psi}{\partial z} = -\frac{1}{2}\frac{\partial^2 \Psi}{\partial x^2} + V(x)\Psi + V_{\mathrm{NL}}(|\Psi|^2)\Psi, \qquad (12.1)$$

$$V_{\mathrm{NL}} - \sigma\frac{\partial^2 V_{\mathrm{NL}}}{\partial x^2} = -n_2|\Psi|^2, \qquad (12.2)$$

where x is the transverse coordinate and V_{NL} is the nonlinear portion of the complete potential $V + V_{\mathrm{NL}}$. Material parameter n_2, equal to ± 1 in our dimensionless framework, determines the nature of the nonlinearity: focusing if $n_2 = 1$ and defocusing if $n_2 = -1$. Equation (12.1) is the well-known Gross–Pitaevskii equation (also called the nonlinear Schrödinger equation (NLSE) in optics), whereas Eq. (12.2) is a screened Poisson equation (or Yukawa equation). The parameter σ (defined as a positive quantity) determines the range of nonlocality: a large σ corresponds to a highly nonlocal response, whereas $\sigma = 0$ corresponds to a local Kerr response [19]. In particular, the width d of the Green function $G(x)$ of Eq. (12.2) is proportional to $\sqrt{\sigma}$, with $G(x) = -1/(2\sqrt{\sigma}) \exp(-|x|/\sqrt{\sigma})$. We stress that, in the local limit $\sigma = 0$, Eqs. (12.1) and (12.2) also describe temporal evolution of optical pulses in guided geometries if x is interpreted as a temporal coordinate [18]; moreover, the nonlocal model constituted by Eqs. (12.1) and (12.2) can be extended to account for noninstantaneous effects in the temporal domain by using a nonlinear response accounting for the principle of causality [20]. Recently, $\mathcal{PT}$ systems in the temporal domain have been created by means of fiber loops, harnessing various types of linear potentials [6,7].

Our aim is to search for propagation invariant solutions of Eqs. (12.1) and (12.2) in the form $\Psi(x,z) = \phi(x)\exp(-i\mu z)$, with $-\mu$ being the propagation constant for light waves or μ the chemical potential for matter waves. The substitution into (12.1) and (12.2) provides the nonlinear eigenvalue problem

$$\mu\phi = -\frac{1}{2}\phi_{xx} + [V_R(x) + iV_I(x)]\phi + V_{\mathrm{NL}}\phi, \qquad (12.3)$$

$$V_{\mathrm{NL}} - \sigma\frac{\partial^2 V_{\mathrm{NL}}}{\partial x^2} = -n_2|\phi|^2. \qquad (12.4)$$

12.3 $\mathcal{PT}$ LINEAR MODES

Let us first discuss solutions of Eqs. (12.3)–(12.4) in the linear case, i.e., when V_{NL} can be neglected. As mentioned in the introduction, a complex potential $V(x) = V_R(x) + iV_I(x)$ ($V_R(x)$ and $V_I(x)$ both real) will be $\mathcal{PT}$-symmetric if the potential satisfies the condition $V_R(x) = V_R(-x)$ and $V_I(x) = -V_I(-x)$. Both for localized and periodic potentials, there is a critical magnitude of V_I above which the $\mathcal{PT}$-symmetry is spontaneously broken, resulting in complex eigenvalues $\mu = \mu_R + i\mu_I$. A typical localized potential with a Gaussian real part, i.e., $V_R = -V_r \exp(-x^2)$ (so that $V_r > 0$ means a trapping potential) and an odd imaginary part $V_I = V_i x \exp(-x^2)$ is shown in Fig. 12.1 (a). The behavior of μ for the ground state (i.e., the state encompassing the lowest μ_R) is shown in Fig. 12.1(c) and (d). Eigenvalue μ is purely real for low V_i, becoming complex ($\mu_I \neq 0$) for large values of the imaginary potential: spontaneous symmetry breaking of $\mathcal{PT}$ symmetry takes place for V_i exceeding a given threshold value for V_i, the latter depending on $V_R(x)$ and on the shape of $V_I(x)$ [21].

The profile and bandgap spectrum for a periodic potential is shown in Fig. 12.1(b)–(e) with $V(x) = V_R(x) + iV_I(x) = V_r \sin^2(x) - iV_i \sin(2x)$, setting to π the period of the linear potential without any loss of generality. For

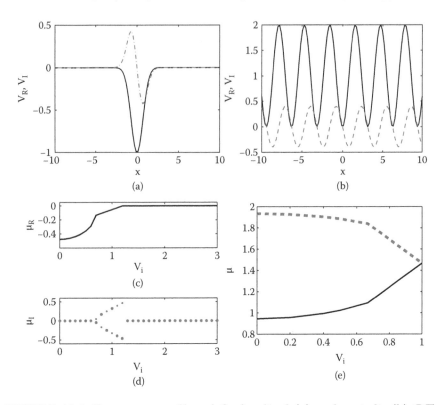

FIGURE 12.1: Transverse profiles of the localized (a) and periodic (b) $\mathcal{PT}$-symmetric potentials. The solid lines represent the real part and the dashed lines represent the imaginary part of the potential, respectively. The variation of the ground state eigenvalue with V_i for the localized potential with $V_r = 1$ is depicted, (c) showing the variation of μ_R and (d) showing the variation of μ_I, respectively. (e) Behavior of μ_R in correspondence to the lower (solid line) and upper (dashed line) bandgap edge versus V_i for $V_r = 2$; for high V_i the bandgap closes and the spectrum becomes complex above this point.

periodic potentials, the size of the bandgap decreases with increasing V_i and the eigenvalue spectrum remains real ($\mathcal{PT}$ symmetry is conserved) until $V_i = V_r/2 = V_c$, where the potential reduces to $V(x) = V_r[1 - \exp(2ix)]/2$: for $V_i = V_c$ the energy bandgaps disappear, resulting in a spectrum equivalent to that of a free particle, but shifted above by V_c, i.e., $\mu = K^2/2 + V_c$ [17,22,23]. In contrast, when $V_i > V_c$ (broken $\mathcal{PT}$ symmetry) the spectrum becomes complex; furthermore, there is no energy bandgap in the first Brillouin zone, as the first bandgap shifts to the second Brillouin zone. Such symmetry breaking in the spectrum is discussed in detail in [11,12,17,24].

12.3.1   PARTICLE CONSERVATION

The existence of a complex stationary solution can be explained by invoking the conservation of particle in the presence of a complex potential; from the complex Schrödinger equation it is possible to derive the particle conservation equation

$$\nabla \cdot j = -\frac{\partial \rho}{\partial z} + 2V_I(x)\rho, \tag{12.5}$$

where $\rho = |\psi|^2$ is the wave intensity and the particle flux j is given by $j = \frac{1}{2i}\left(\psi^* \frac{\partial \psi}{\partial x} - \psi \frac{\partial \psi^*}{\partial x}\right)$. Equation (12.5) states that the complex potential $V_I(x)$ physically acts as a source (when $V_I(x) > 0$) or as a sink (when $V_I(x) < 0$) for particles. Setting $\psi = \sqrt{\rho}e^{i\chi(x)}$, the flux reads $j = \rho\frac{\partial \chi}{\partial x}$. In the stationary case $\partial \rho/\partial z = 0$ Eq. (12.5) yields

$$\frac{\partial}{\partial x}\left(\rho\frac{\partial \chi}{\partial x}\right) = 2V_I(x)\rho. \tag{12.6}$$

When $V_I(x)$ is odd, according to Eq. (12.6), an even ρ corresponds to an odd χ, that is, an even real part and an odd imaginary part, respectively. Physically, particles are created within the gain regions and then they diffuse (transversely) toward the loss regions in order to keep the overall (i.e., integrated along x) particle number constant. It is noteworthy that in the case of single-hump potentials the flux j is an even function, that is, a unidirectional flow takes place [12]. Equation (12.6) also states that the larger V_I is the larger the anti-symmetric part is due to the increase in the transverse flux of particles necessary to compensate the inhomogeneous gain/loss. The flux j corresponds in optics to the transverse component of the Poynting vector; see [12].

Equation (12.6) explains qualitatively the dependence of the linear modes from the amplitude of the imaginary potential shown in Fig. 12.1. In fact, in the first approximation the flux j is proportional to the width of ρ, the latter determined by the real potential $V_R(x)$: when particle generation/destruction (proportional to V_i) is too large, flux is not strong enough to ensure the removal of the excess particle, leading to linear modes centered on the gain (lossy) region and undergoing amplification (attenuation). Such conclusions are supported by the behavior of the linear modes when the real potential $V_R(x)$ is kept fixed, whereas the shape of the imaginary part is varied: a lower spatial overlap between ρ and $V_I(x)$ leads to a breaking of $\mathcal{PT}$ symmetry occurring for larger amplitudes V_i. Figure 12.2 shows the ground state modes in the Gaussian case for V_i values below and above the $\mathcal{PT}$ breaking point: when the symmetry is not conserved, modes are centered in the gain (loss) region and are amplified (attenuated) in propagation, as shown in the bottom row [5].

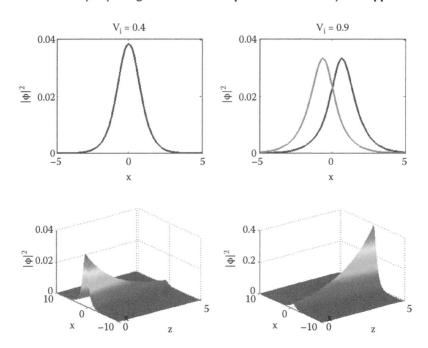

FIGURE 12.2: (Top row) Linear modes $|\phi|^2$ versus x for localized Gaussian potentials. Two representative examples below ($V_i = 0.4$) and above ($V_i = 0.9$) the $\mathcal{PT}$ breaking point are shown: when $\mathcal{PT}$ is broken linear modes are subject to amplification (decay) across z due to the complex nature of the associated eigenvalues (i.e., $\mu_I \neq 0$). (Bottom row) Evolution of the decaying mode localized in the lossy region and that of the amplifying mode localized in the gain region, respectively. In all the cases $V_r = 1$.

12.4   NONLINEAR MODES

We solved Eqs. (12.3) and (12.4) using standard relaxation techniques to solve both equations. Here, we used a pseudo-spectral method to calculate the differential matrix using Chebyshev polynomials to obtain the stationary profiles for both local and periodic potential. We will focus on the lowest order bright soliton (multi-hump solitons in Gaussian potential are investigated in [25]); moreover, in the periodic case we will concentrate on fundamental gap solitons (higher order solitons have been investigated in [23], whereas nonlinear delocalized Bloch waves are treated in [9]), that is, nonlinear localized states lying into the bandgaps of the linear spectrum. For periodic potential we will refer solely to the defocusing case, leaving for the reader the available literature for the focusing case. At variance with the standard case $V_i = 0$, all the solitons have a symmetric real part and an anti-symmetric imaginary part in order to fulfill the power conservation stated by Eq. (12.6).

12.4.1 GAUSSIAN POTENTIAL

Let us consider a localized linear potential in the form of a Gaussian for the real part; the shape of the odd imaginary part is taken to be equal to the first derivative of the real part. We will look for stationary solutions both in the focusing and defocusing regimes. Typical profiles of $|\phi|^2$ for solitons both in the focusing and defocusing regimes are depicted in Fig. 12.3; in the reported case the linear potential is exactly the same and its ground state conserves $\mathcal{PT}$ symmetry. The corresponding soliton phase, showing a gradual step profile, is reported as well: in accordance with Eq. (12.6), the larger V_i is the larger magnitude of the imaginary part of the solution, implying an increasing phase jump across the soliton.

For focusing nonlinearity solitons encompass a purely real eigenvalue ($\mu_I = 0$) and are found to exist for μ_R lower than the linear eigenvalue μ_{lin}: physi-

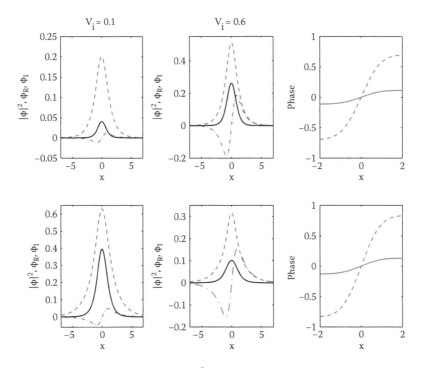

FIGURE 12.3: Stationary profiles $|\phi|^2$ (solid lines) together with the real part ϕ_R (dashed lines) and imaginary part ϕ_I (dashed-dotted lines) for (top row) local Kerr focusing nonlinearity for $\mu = -0.5$, and (bottom row) local Kerr defocusing nonlinearity for $\mu = -0.2$; linear potential is Gaussian with $V_r = 1$ and V_i equal to the marked value. The last column depicts the phase profile for the stationary solution with the solid line representing $V_i = 0.1$ and the dashed lines representing $V_i = 0.6$.

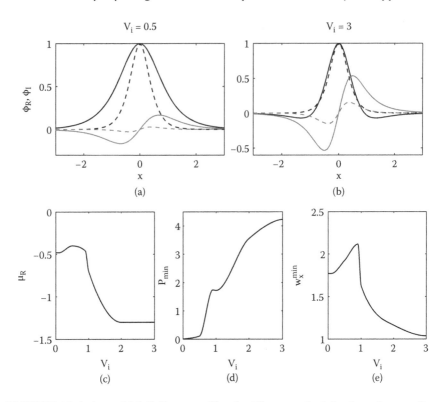

FIGURE 12.4: (a and b) Soliton profiles for V_i as marked for focusing nonlinearity in the presence of a Gaussian potential with $V_r = 1$. Symmetric solid curves represent the normalized real part of the profile for $\mu = -1.3$ (solid lines) and $\mu = -6$ (dashed lines). Anti-symmetric curves represent the imaginary part normalized to the peak of the real part: the higher the magnitude the higher the μ. The different panels show the features of the fundamental soliton for the largest allowed μ_R once the imaginary gain magnitude V_i is fixed: from left to right (c) variation of the real part of eigenvalue μ_R, (d) minimum power $P_{\min}$ and (e) width $w_x^{\min}$ of the soliton versus V_i are plotted.

cally, nonlinear effects increase wave confinement, thus decreasing the energy of the nonlinear eigenstate with respect to the linear value. For fixed μ and $V_R(x)$, soliton power P (defined as $\int |\phi|^2 dx$) increases with V_i in order to ensure a narrower soliton, in turn providing a larger transverse flux (first row in Fig. 12.3).

The most striking effect due to the nonlinearity is the *generation* of at least one nonlinear bound state featured by $\mu_I = 0$, no matter how large V_i is (Fig. 12.4(a) and (b)), including regions where linear ground states grow/decay exponentially: in other words, even if the linear spectrum is complex, nonlinearity causes a transition from broken $\mathcal{PT}$ to a $\mathcal{PT}$-symmetric state (for example, see Fig. 12.4(b): in the linear regime no bound states with $\mu_I = 0$

exist, see Fig. 12.1(c)) [9, 12]. Let us now discuss how solitonic profile ϕ depends on the nonlinear eigenvalue μ_R. For V_i below the exceptional point, lower μ_R corresponds to a more confined soliton, decreasing the overlap with $V_I(x)$: thus imaginary part ϕ_I is lower when the soliton is tightly trapped. Above $\mathcal{PT}$ transition behavior is similar, but now the soliton width at different μ_R is quite similar: nonlinearity shrinks the beam in order to decrease the transverse flux necessary to compensate particle generation/destruction [21].

Let us now turn our attention to the role played by the imaginary potential $V_I(x)$: when V_i is vanishing or small, solitons exist at very small power, with a profile almost matching the linear mode; initially the maximum μ_R increases with V_i, thus mimicking the behavior of the linear eigenvalue (see Fig. 12.4(c)). As V_i is further increased, the upper threshold for μ_R inverts its trend and starts to decrease monotonically with V_i owing to the effects of nonlinearity. Physically, the power has to be large enough to confine the field away from the peak of $V_I(x)$ in order to ensure that Eq. (12.6) is satisfied; in fact, for a given V_i the minimum power $P_{\min}$ necessary to create a soliton increases with V_i (Fig. 12.4(d)). This interpretation is confirmed by the width of the soliton corresponding to $P_{\min}$: it first slightly increases with V_i, then undergoes an abrupt fall when the decrease of the soliton eigenvalue is maximum (Fig. 12.4(e)).

The behavior with power is quite different for defocusing nonlinearity [26, 27]: first, from the second row in Fig. 12.3 it is clear that, as V_i is increased and μ_R kept fixed, soliton power diminishes owing to the repelling nature of the nonlinear potential $V_{\rm NL}$. For the same reason, solitary waves exist only for μ values larger than the corresponding linear value, as can be seen from Fig. 12.5. Moreover, solitons cease to exist near the $\mathcal{PT}$ breaking point due to self-repelling action, leading to a net decrease in the transverse flux: accordingly, second and third columns in Figure 12.5 show, respectively, the power P and width w_x versus V_i for solitons at the lower and upper boundaries of the existence interval in μ_R.

12.4.2   PERIODIC POTENTIAL

Bright solitons in periodic potentials show a vast phenomenology, much more heterogeneous than in homogeneous systems or in single-humped potentials. Solitons in lattice are based upon the inhibition of tunneling between adjacent traps, reached via the breaking of phase-matching by means of nonlinear phase modulation [28]. For this reason, bright solitons can exist both in the focusing and in the defocusing cases [29], with profiles usually extending over several waveguides. Types of solitons can be distinguished according to the position of their own eigenvalue: in particular, gap solitons lie inside the linear bandgaps, thus behaving as *defect* states generated by nonlinearity. Solitons in periodic media have interesting transverse profiles [30, 31] with solutions closer to the bandgap edges being essentially an envelope of Bloch waves: as the soliton eigenvalue gets deeper into the bandgap, solitons get more and more confined into a single waveguide, with decreasing tails.

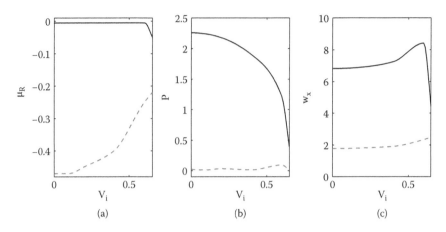

FIGURE 12.5: Features of the fundamental soliton vs V_i for a defocusing nonlinearity when $V_r = 1$. (a) Variation of real part of eigenvalue μ_R with dashed lines representing the lowest allowed μ_R and the solid lines representing the highest allowed μ_R. (b) Soliton power P and (c) width w_x corresponding to the two extremal cases plotted in panel (a). With increasing V_i the region of existence of the solitons gets narrower, with solitons ceasing to exist above the $\mathcal{PT}$-symmetry breaking.

We will deal with fundamental gap solitons in a defocusing nonlinear medium, considering the first bandgap only. For a focusing nonlinearity and a localized potential, nonlinearity is always able to create a gap soliton with $\mu_I = 0$. It is noteworthy that only on-site solitons exist [29,31] owing to the breaking of the left/right symmetry provoked by V_i [17]. Soliton profiles for $V_r = 2$ and $V_i = 0.5$ are depicted in Fig. 12.6. For the lowest μ, i.e., nonlinear states closer to the bandgap edge, solitons extend over several guides, resembling a modulated Bloch wave; furthermore, fields in adjacent guides are out of phase; thus a staggered soliton is formed [28]. As a soliton moves inside the bandgap (μ increases), excitation of adjacent guides drops off, in agreement with what happens in a real lattice [28,31]. As mentioned above, all the gap solitons show an imaginary component, with the largest gradient in the imaginary part ϕ_I occurring in correspondence to the peak of the real part ϕ_R so that the flux is large enough to allow particles to migrate from gain toward lossy regions (Eq. (12.6)). In Fig. 12.6 nonlinearity is varied from the local to the nonlocal case, moving from the top to the bottom row: it is evident that the soliton shape is scarcely affected by the degree of nonlocality. Instead, for a given μ solitons with high nonlocality have a larger power than for lower nonlocality owing to the decrease in nonlinearity associated with a larger σ.

Actually, in general, nonlocality affects the amplitude of solitonic tails as well. To understand this, we refer to the discrete soliton model [28]: a single-mode of each isolated guide is considered, taking the eigenmode ϕ of the

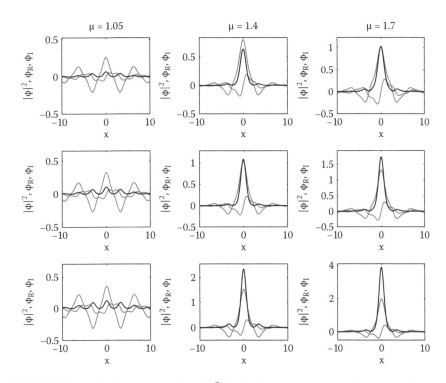

FIGURE 12.6: Stationary profiles $|\phi|^2$ (solid lines) together with the real part ϕ_R (dashed lines) and imaginary part ϕ_I (dashed-dotted lines) for $V_i = 0.5$ and for various μ values as marked in the figure: nonlocality is (top row) $\sigma = 0$, (middle row) $\sigma = 1$, and (bottom row) $\sigma = 10$ for $V_r = 2$. Notice that, when μ is closer to the lower bandgap edge, the solitonic profiles assume a Bloch-like trend.

complete periodic system as a superposition of these unperturbed eigenmodes. Each of these modes exchanges power with the modes corresponding to the adjacent guides (nearest neighbor approximation [28]) through its exponential tails, the amount of coupling depending on the nonlinearity, the latter affecting the phase matching condition by means of the Kerr effect. In the previous case ($V_r = 2$) single-modes of each isolated guide extend far beyond the corresponding trap; thus nonlocality does not affect the reciprocal coupling between guides. Conversely, when V_r is increased, the linear mode of the isolated guide gets narrower, and coupling between adjacent modes drops off. In this case, nonlocality is important because it permits the nonlinear perturbation to extend up to the adjacent guides.

Gap solitons for $V_R = 4$ are shown in Fig. 12.7: regardless of V_i, the larger the nonlocality is the lower the tails are, similar to the case $V_i = 0$ [31]. In fact,

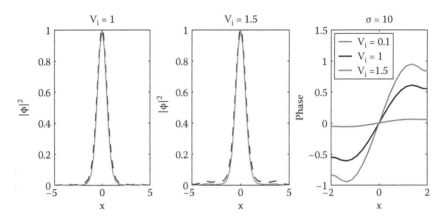

FIGURE 12.7: Stationary profiles $|\phi|^2$ for $V_r = 4$, $\sigma = 1$ (dashed lines) and $\sigma = 10$ (solid lines) for V_i as marked for $\mu = 3$ (closer to the upper bandgap edge). The phase of the solitons changing with V_i values is shown in the last column for $\sigma = 10$.

to exist, gap solitons require out-of-phase excitation of nearest-neighboring guides [29], with the nonlinear response providing the needed difference in μ (propagation constant in optics, energy for matter waves) between core and side channels. In the local Kerr case, a strong excitation is needed in the adjacent sites to reach the necessary nonlinear phase modulation, whereas in the nonlocal case the nonlinear perturbation induced by the mode in the central guide, spreading outwards owing to the finite size of the Green function $G(x)$, provides the required modulation. The phase of the solution due to the complex nature of the profiles depends on the magnitude of V_i and is also depicted in Fig. 12.7, showing an increase with V_i. Actually, such a simplified model is valid when the soliton eigenvalue is close to the upper edge of the bandgap: for solitons in the middle of the bandgap, behavior is opposite. Theoretically, this is due to the inadequacy of the coupled mode theory in this case, a Bloch wave approach being necessary for a correct interpretation of the physics.

Finally, let us discuss qualitatively what is going on in a focusing nonlinear material with a periodic potential written on it. Musslimani and collaborators [12] first showed that stable bright single-peak solitons exist in the semi-infinite bandgap, that is, for μ lower than that of the minimum of the linear spectrum (see also [23] for a deeper investigation). At the same time, they demonstrated that such solitons are stable only if the spectrum is completely real, undergoing exponential amplification otherwise; the role of nonlocality in focusing media has been discussed in [32]. Finally, ubiquitous instability of gap solitons for focusing nonlinearity in $(2+1)$D geometries has been numerically demonstrated in [33].

12.5 VARIATIONAL APPROACH FOR PERIODIC POTENTIAL AND DE-FOCUSING NONLINEARITY

Variational methods represent an important tool in nonlinear physics when a rough idea of the solution shape is a priori known, allowing us to address the qualitative (and sometimes quantitative) behavior of physically relevant quantities via analytical formulae. The semi-analytical approach used in variational methods has been successfully applied in solving nonlinear problems encompassing various kinds of nonlinearity and types of nonlinear waves, involving both spatial as well as temporal dynamics [34–40]. The method was first conceived by Whitham [41]: first a well suited trial solution has to be guessed, as similar as possible to the real wave field; then the Lagrangian density of the system has to be computed and minimized with respect to the free parameters. The variational method was applied to the NLSE first by Anderson [42], who used a Gaussian trial function and a Ritz optimization procedure to obtain approximate solutions for the evolution of pulse width, pulse amplitude and frequency chirp. The results obtained were in good agreement with exact solutions from inverse scattering and numerical simulations, at the same time ensuring deeper physical insight and lower resource usage. Since then the method has been used successively by various authors in several contexts [43,44]. The success of the method relies on the accurate guess of the trial solution. We adopt this method to predict semi-analytical solutions in a $\mathcal{PT}$- symmetric system [17].

The method proceeds by obtaining the Lagrangian of the investigated system, in our case Eqs. (12.1) and (12.2), for solutions of the form $\Psi(x, z) = \phi(x, z)\exp(-i\mu z)$; the Lagrangian density reads

$$
\begin{aligned}
L_C &= \frac{i}{2}(\phi_z^*\phi - \phi_z\phi^*) - \mu|\phi|^2 + \frac{1}{2}|\phi_x|^2 + V_R(x)|\phi|^2 \\
&+ n_2 V_{\mathrm{NL}}|\phi|^2 - \frac{\sigma}{2}\left(\frac{\partial V_{\mathrm{NL}}}{\partial x}\right)^2 - \frac{V_{\mathrm{NL}}^2}{2},
\end{aligned}
\tag{12.7}
$$

for the conservative part. For a purely local nonlinearity it takes the form

$$
L_C = \frac{i}{2}(\phi_z^*\phi - \phi_z\phi^*) - \mu|\phi|^2 + \frac{1}{2}|\phi_x|^2 + V_R(x)|\phi|^2 + n_2\frac{|\phi|^4}{2}.
$$

To illustrate the method, we proceed by considering the Lagrangian for the periodic system in a nonlocal medium. Application to the local case is straightforward. From the numerical solutions (see Figs. 12.6 and 12.7), we can infer that the width of the intensity profile does not vary appreciably for increasing μ, with the main portion of the wave packet being confined into a single channel with the tails extending into the neighboring channels decreasing for μ close to the upper bandgap edge. This allows us to choose an approximate trial function of the form

$$\phi = A \exp\left(-\frac{x^2}{w_b^2}\right) \exp\left[i\theta(z)f(x)\right], \qquad (12.8)$$

$$V_{\mathrm{NL}} = A_{\mathrm{NL}} \exp\left(-\frac{x^2}{w_{\mathrm{NL}}^2}\right), \qquad (12.9)$$

for the wave and nonlinear potential profile, with A corresponding to the field amplitude, θ the amplitude of the phase profile and $f(x)$ its spatial distribution along x; A_{NL} is the amplitude of the nonlinear potential and, finally, w_b and w_{NL} are the widths of the soliton and of the nonlinear perturbation V_{NL}, respectively. We assume $f(x)$ to be an odd function of x considering the fact that the real part of the solution is symmetric and that the imaginary part is anti-symmetric. In the limit of high nonlocality, the nonlinear potential well is parabolic-like in proximity to the soliton and proportional to the Green function of Eq. (12.2) away from it (i.e., $\exp(-|x|/d)$), but for simplicity we assume a Gaussian profile for the induced potential. The phase factor is accounted for in the θ term and, as shown below, can predict the observed behavior of the solutions to a reasonable extent.

The standard variational approach applied to conservative systems can be modified for systems with dissipation as [17, 37, 45]

$$\frac{d}{dz}\left(\frac{\partial\langle L_C\rangle}{\partial\beta_z}\right) - \frac{\partial\langle L_C\rangle}{\partial\beta} = 2\mathrm{Re}\int_{-\infty}^{\infty} Q\frac{\partial\phi^*}{\partial\beta}dx, \qquad (12.10)$$

where β stands for all the parameters which can be varied in our variational computation (A, A_{NL}, P, w_b, w_{NL}, θ) and collecting all the dissipative terms as $Q = iV_I(x)\phi$. Defining the reduced Lagrangian to be $\langle L_C\rangle = \int_{-\infty}^{\infty} L_C dx$, using Eqs. (12.8) and (12.9) for a defocusing nonlinearity we obtain

$$\begin{aligned}
\langle L_C\rangle &= \sqrt{\frac{2}{\pi}}\frac{P}{w_b}\theta_z\int_{-\infty}^{\infty} f(x)e^{-\frac{2x^2}{w_b^2}}dx - \mu P + \frac{P}{2w_b^2} \\
&\quad + \sqrt{\frac{2}{\pi}}\frac{P}{2w_b}\theta^2\int_{-\infty}^{\infty} f_x^2 e^{-\frac{2x^2}{w_b^2}}dx + \frac{V_r P(1-e^{-\frac{w_b^2}{2}})}{2} \\
&\quad + \frac{\sqrt{2}A_{\mathrm{NL}}Pw_{\mathrm{NL}}}{(2w_{\mathrm{NL}}^2 + w_b^2)^{1/2}} - \frac{\sqrt{\pi}\sigma A_{\mathrm{NL}}^2}{2\sqrt{2}w_{\mathrm{NL}}} - \frac{\sqrt{\pi}w_{\mathrm{NL}}A_{\mathrm{NL}}^2}{2\sqrt{2}}. \qquad (12.11)
\end{aligned}$$

Further simplifications to Eq. (12.11) can be made using the symmetries in the equation, for example, the first term on the right hand side vanishes owing to the anti-symmetry of $f(x)$. Using Eq. (12.10), variational equations are obtained for each variable parameter. The right hand side of Eq. (12.10) is nonzero only for $\beta = \theta$ because $f(x)$ is the only odd function; the substitution of Eq. (12.11) into (12.10) provides

$$\theta = -2V_i \frac{\text{Re}\left[\int_{-\infty}^{\infty} \sin(2x) f(x) e^{-\frac{2x^2}{w_b^2}} dx\right]}{\int_{-\infty}^{\infty} f_x^2 e^{-\frac{2x^2}{w_b^2}} dx}. \tag{12.12}$$

Equation (12.12) clearly shows the dependence between the phase associated with the solution (i.e., real and imaginary components in the stationary solution) and the imaginary part of the refractive index V_I: it vanishes when the potential is purely real, eventually providing a flat-phase soliton.

The effects of the complex potential are apparent (through the parameter θ) in the expression for w_b:

$$\frac{1}{w_b^4} = \frac{V_r}{2} e^{-\frac{w_b^2}{2}} + \sqrt{\frac{2}{4\pi w_b^2}} \theta^2 \frac{\partial}{\partial w_b} \left(\frac{1}{w_b} \int_{-\infty}^{\infty} f_x^2 e^{-\frac{2x^2}{w_b^2}} dx\right)$$

$$- \frac{2w_{\text{NL}}^3 P}{(2w_{\text{NL}}^2 + w_b^2)^2 (\sigma + w_{\text{NL}}^2)}, \tag{12.13}$$

and μ:

$$\mu = \frac{1}{2w_b^2} + \frac{V_r}{2}(1 - e^{-\frac{w_b^2}{2}}) + \sqrt{\frac{2}{4\pi w_b^2}} \theta^2 \int_{-\infty}^{\infty} f_x^2 e^{-\frac{2x^2}{w_b^2}} dx$$

$$+ \frac{2w_{\text{NL}}^3 P}{(2w_{\text{NL}}^2 + w_b^2)(\sigma + w_{\text{NL}}^2)}. \tag{12.14}$$

Similarly, the variation of Eq. (12.10) with respect to A_{NL} and w_{NL} yields

$$w_{\text{NL}}^2 = \frac{w_b^2 + 2\sigma + \sqrt{(w_b^2 + 2\sigma)^2 + 24\sigma w_b^2}}{4}. \tag{12.15}$$

Assuming a hyperbolic tangent profile for the transverse phase distribution, i.e., $f(x) = \tanh(x)$, using Eq. (12.13) we calculate the beam width versus μ for various σ: comparison with the numerical exact result is presented in Fig. 12.8(a). In both cases, soliton width increases with μ: in fact, larger μ correspond to larger soliton power (see Fig. 12.8(d)); thus the wave tends to self-repel owing to the defocusing character of nonlinearity. Furthermore, the variational approach is able to catch qualitatively the dependence on nonlocality σ: the soliton gets narrower as nonlocality increases. A qualitative agreement is achieved for the width of the nonlinear index well w_{NL} (Fig. 12.8(b)), larger discrepancies arising for low μ owing to the non-negligible tails of the soliton with the nearest-neighbor guides. The smooth step behavior of the solitonic phase profile is retrieved, together with its quasi-independence from nonlocality σ (Fig. 12.8(c)): in fact, the phase distribution depends mainly on $V_I(x)$ (see Fig. 12.7 and [17] for a direct comparison). Results for the propagation constant μ computed via Eq. (12.14) are reported in Fig. 12.8(d):

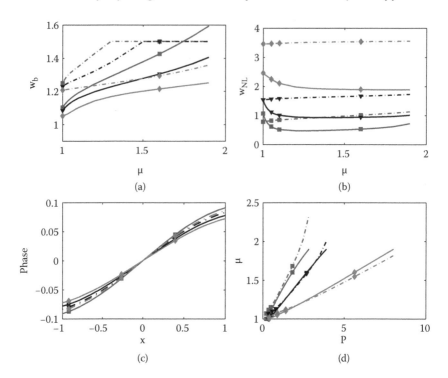

FIGURE 12.8: Graphs of (a) w_b versus μ, (b) w_{NL} versus μ, (c) phase profile and (d) μ versus P. In all panels $\sigma = 0$ ($\square$), $\sigma = 1$ (∇), $\sigma = 10$ ($\diamond$) and $V_i = 0.1$. Solid lines correspond to numerics and dotted lines are variational results.

agreement is very good, except for low σ and large μ. Summarizing, the variational method agrees with the numerical results fairly well, being able to describe the main dependence on physically relevant parameters such as non-locality and soliton power.

12.6   STABILITY ANALYSIS

In this section we address the stability of $\mathcal{PT}$- symmetric solitons by using a linear stability analysis (LSA) wherein small perturbations are added to the soliton solution and their effect analyzed. The stability of the solitons can be addressed by considering the effect of small perturbations in the form

$$\Psi(x,z) = [\phi(x) + p(x)e^{i\lambda_g z} + q(x)e^{-i\lambda_g^* z}]e^{-i\mu z}, \tag{12.16}$$
$$V_{NL}(x,z) = V_{NL}^{\mu}(x) + \Delta V_{NL}(x,z), \tag{12.17}$$

where V_{NL}^{μ} is the nonlinear potential computed via Eq. (12.4), corresponding to the soliton $\phi(x)$ once the propagation constant μ, the degree of nonlocality σ and the linear properties (i.e., V_r and V_i) of the structure are selected.

In the perturbative limit, i.e., neglecting nonlinear terms stemming from the added perturbation, we obtain from Eqs. (12.3) and (12.4) the following linear eigenvalue problem:

$$\lambda_g p = [\boldsymbol{L} - iV_i(x)]p - \phi \boldsymbol{D} \left(\phi^* p + \phi q^*\right), \tag{12.18}$$

$$\lambda_g q^* = [-\boldsymbol{L} - iV_i(x)]q^* + \phi^* \boldsymbol{D} \left(\phi q^* + \phi^* p\right), \tag{12.19}$$

where we defined the operator $\boldsymbol{L} = 0.5\partial_x^2 - V_{\mathrm{NL}}^\mu(x) - V_r(x) + \mu$. Moreover, we introduced $\boldsymbol{D} = (1 - \sigma\partial_x^2)^{-1}$, capable of computing the field intensity from the nonlinear potential V_{NL}; in other words, $\boldsymbol{D}$ is the convolution between the Green function of Eq. (12.2) and the intensity profile.

Solitons are stable if $\mathrm{Im}(\lambda_g) = 0$ holds for all the eigenvalues, i.e., if the system has only real eigenvalues. The existence of complex eigenvalues corresponds to an oscillatory instability : $\mathrm{Im}(\lambda_g) > 0$ (< 0) implies exponentially decaying (growing) modes together with intensity oscillations while the wave evolves along z. We solved the system of Eqs. (12.18-12.19) using pseudo-spectral techniques based on Chebyshev polynomials to compute both the diffraction operator and the operator $\boldsymbol{D}$. We chose a grid extending to 40 along x (much larger than the maximum degree of nonlocality used, i.e., $\sigma = 10$) to avoid artifacts. The grid consisted of 1001 points, the latter ensuring independence from the numerical resolution for both the eigenvalues and the eigenfunctions in the range of interest (see below). In general, eigenvalues associated with $\mathcal{PT}$- symmetric systems are complex irrespective of whether the system is composed of localized defects or periodic defects. Specifically, for $V_i \neq 0$ the eigenvalues appear in quartets, each eigenvalue being the vertex of a square centered in the origin: thus the system can be absolutely stable only if eigenvalues are purely real, the latter occurring only in the absence of an imaginary potential, i.e., $V_I(x) = 0$ everywhere. A nonvanishing real part of the complex eigenvalue corresponds to a different phase velocity for the perturbative mode with respect to the soliton, thus inducing periodic modulation across z due to the mutual interference: in a few words, the system is subject to oscillatory instability. The shape of the perturbation modes depends on the nature of the linear potential: hereafter we discuss the case of a Gaussian and of a periodic potential.

12.6.1 GAUSSIAN POTENTIAL

The stability spectrum for the Gaussian defect for both focusing and defocusing nonlinearity corresponding to the solutions depicted in Fig. 12.3 is shown in Fig. 12.9. As expected, for $V_i = 0.1$ solutions are quasi-stable with low magnitude for the eigenvalues (lower than 10^{-4}), that is, exponential amplification is negligible on several Rayleigh distances, thus playing no role in real experiments. Instability growth increases gradually until the $\mathcal{PT}$ transition for the linear spectrum is reached: in correspondence to that point, solitons become highly unstable. At $V_i = 0.6$, solitons are just below the $\mathcal{PT}$ breaking point and hence the corresponding stability spectrum is composed of a

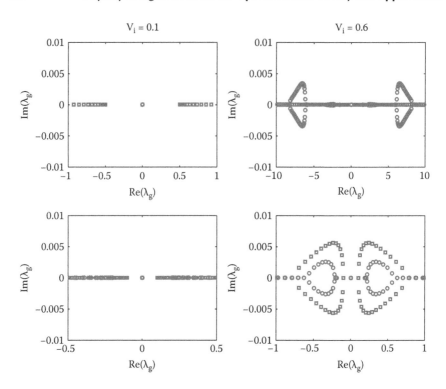

FIGURE 12.9: Eigenvalues λ_g for a local defect with (top row) focusing local Kerr nonlinearity for $\mu_R = -0.5$ ($\square$) and $\mu_R = -6$ ($\circ$) and with (bottom row) defocusing local Kerr nonlinearity for $\mu_R = -0.1$ ($\square$) and $\mu_R = -0.2$ ($\circ$) in the complex plane (the horizontal axis corresponds to the real part, the vertical axis to the imaginary one) for V_i as marked and $V_r = 1$.

large number of low frequency modes whose effect is manifested only at very large distances. In contrast, for V_i values above the $\mathcal{PT}$ breaking point a single dominant mode with an asymmetric transverse profile is observed having a dominant quartet of eigenvalues. Physically, the soliton wave is propagating inside a *sea* of linear highly unstable modes, comprising low-frequency modes (i.e., the bound states): every small perturbation couples with the linear modes, thus inducing the soliton breaking on small propagation distances. Finally, we note that, for a focusing nonlinearity, instability increases as $|\mu|$ increases.

For defocusing nonlinearity the behavior of perturbation eigenvalues λ_g with respect to V_i is quite similar to the focusing case (see second row in Fig. 12.9): the biggest difference is that now solitons are more stable for smaller $|\mu_R|$.

12.6.2 PERIODIC POTENTIAL

A typical stability eigenvalue spectrum for a periodic defect in a defocusing nonlinear medium is shown in Fig 12.10 for various degrees of nonlocality. First, stability undergoes small variations when σ takes nonvanishing values; conversely, it is about five times less when nonlinearity is local, i.e., $\sigma = 0$. Generally speaking, the solutions closer to the lower bandgap edge are more stable than the solutions closer to the upper bandgap edge. As with Gaussian potential, solitons are quasi-stable for V_i below the $\mathcal{PT}$ transitions, whereas they are subject to strong instability when V_i is increased. It can also be seen that for the same V_i/V_r, solutions of potentials with a lower magnitude of the real part of the potential V_r are more stable ($|\mathrm{Im}(\lambda_g)|$ on the order of 0.003) than higher V_r ($|\mathrm{Im}(\lambda_g)|$ on the order of 0.1) [17].

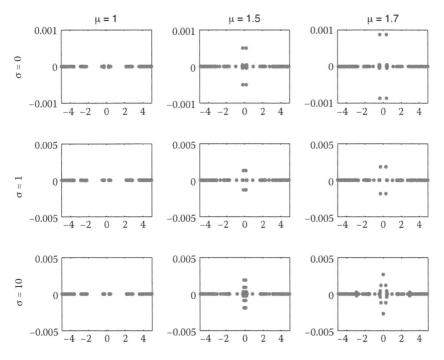

FIGURE 12.10: Eigenvalues λ_g for the periodic potential for various μ (i.e., soliton power) and nonlocality σ in the complex plane (the horizontal axis corresponds to the real part, the vertical axis to the imaginary one) for $V_i = 0.1$ and $V_r = 2$.

12.7   DYNAMICAL EVOLUTION OF THE SOLITON

We now proceed to study the stability properties of the soliton upon adding noise to the stationary solution. A standard beam propagation method using a Crank–Nicolson scheme for diffraction is employed and the evolution is studied by adding Gaussian random noise to the exact soliton profile.

12.7.1   GAUSSIAN POTENTIAL

For localized linear potentials, oscillatory instability (OI) manifests as an exponential growth of the beam together with high localization of the beam in the gain region: the effect is more relevant as V_i is increased, reaching appreciable magnitude with respect to the Rayleigh distance in proximity to the $\mathcal{PT}$ breaking point, in agreement with LSA discussed above. Figure 12.11

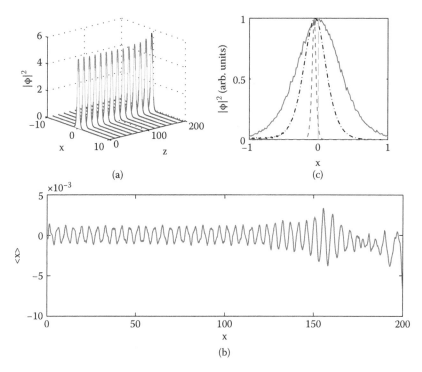

FIGURE 12.11: (a) Dynamic evolution of a soliton beam with 1% noise added to the amplitude and (b) trajectory of the evolution for $V_r = 1$ and $V_i = 0.5$ for a local Gaussian defect. (c) Transverse profiles of the solution for $V_i = 1$ and $V_r = 1$ at $z = 0$ (solid line), $z = 100$ (dotted line) and $z = 110$ (dashed line) showing narrowing and peak oscillation upon propagation. The profiles are normalized so that beam maximum in each section is one. In all cases $\mu = 3.4$.

illustrates this effect for a Gaussian potential and focusing nonlinearity: the soliton shows appreciable exponential growth on propagation [Fig. 12.11(a)] when $V_i = 0.5$, i.e., near the $\mathcal{PT}$ breaking point. Corresponding evolution along z of the *center of mass* of the field $\langle x \rangle = \int |\psi|^2 x dx / \int |\psi|^2 dx$ is depicted in Fig. 12.11(b). OI induces oscillatory motion in the soliton path due to the interference between the dominant perturbation modes and the soliton itself, with a period determined by the difference between the real part of the corresponding eigenvalue λ_g and soliton eigenvalue μ_R. The transverse profiles of the beam for various z values are shown in Fig. 12.11(c) for $V_i = 1$, which is well beyond the $\mathcal{PT}$ breaking point and enlightening the final step of the instability insurgence, consisting in the localization of the beam in the gain region. It is to be noted that narrow width corresponds to higher power: in the figure profiles are normalized with respect to power to improve visibility.

12.7.2   PERIODIC POTENTIAL

LSA predicts quasi-stable solutions for $V_i = 0.1$ and $V_r = 2$ (corresponds to $V_i/V_r = 0.05$) for μ_R inside the bandgap: distances greater than $z = 1000$ are necessary to observe exponential growth in the local case (see Fig. 12.10); conversely, solitons can be stable for distances over $z = 500$ for nonlocality σ larger than 1. To address the influence of instability on soliton propagation, the typical diffraction length of these solitons is $z = 20$ in the linear regime: thus soliton stability is ensured on length at least ten times larger than the typical diffraction length, such a result being confirmed by numerical simulations.

In Fig. 12.12 we compare the evolution profile for two different values of V_r, but fixing the effective magnitude of the imaginary part to be $V_i/V_r = 0.05$. We remind the readers that $\mathcal{PT}$ breaking occurs at $V_i = V_r/2$. As expected from the predictions of LSA, solutions near the lower bandgap edge are more stable than solutions near the upper bandgap edge; furthermore, solutions for $V_r = 2$ remain stable over larger propagation distances than for $V_r = 4$.

Numerical simulations also show that instability is enhanced when V_i becomes larger: in full analogy with the Gaussian potential, gap solitons undergo strong instability when the system in the linear regime is beyond the $\mathcal{PT}$ transition [17]. Different from the Gaussian case, now instability produces a transverse motion of a light beam through the lattice, with the transverse momentum of the wave pointing toward the gain region (i.e., toward negative x in our case), i.e., the gain is breaking the left/right symmetry. Such behavior is confirmed by the transverse profile of the dominant instability modes, peaked on the nearest neighbor guide with respect to the unperturbed gap soliton.

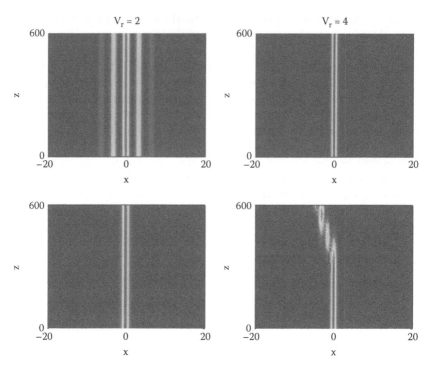

FIGURE 12.12: Dynamic evolution of a soliton beam with 1% noise added to the amplitude for solitons near the lower and upper band edge for two different V_r but keeping the ratio $V_i/V_r = 0.05$ and $\sigma = 10$. Left: Evolution for $V_i = 0.1$ and $V_r = 2$ with $\mu = 1$ (top) and $\mu = 1.9$ (bottom) showing stable propagation. Right: Evolution for $V_i = 0.2$ and $V_r = 4$ with $\mu = 1.9$ (top) and $\mu = 3$ (bottom) showing unstable propagation for large μ. In both cases predictions from LSA agree very well with the dynamics.

12.8   CONCLUSION

To conclude, we discussed the existence and stability of bright solitons in the presence of $\mathcal{PT}$-symmetric linear potentials. We dealt with two fundamental types of linear potentials: a localized single-hump potential with a trapping character and a periodic lattice. We showed how their properties can be explained considering the particle conservation on the transverse plane: stationary solutions can exist when the flux is large enough to enable migration of particles from the gain toward the lossy regions. For both cases, solitons are always unstable: however, self-trapped waves are quasi-stable for low values of V_i, meaning that the effects of instability are apparent for very large propagation distances, making the soliton effectively stable for all practical purposes. Conversely, when V_i is such that even in the linear regime $\mathcal{PT}$ is broken, solitons are subject to strong instability. The onset of instability assumes different forms according to the type of linear potential: single-hump potential leads to field localization in the gain region, eventually leading to exponential growth; in periodic potentials gap solitons undergo unilateral transverse motion toward the gain region, jumping continuously from guide to guide. Finally, we illustrated a variational method adapted for $\mathcal{PT}$-symmetric potentials, leading to a correct qualitative description of the physics.

In optics, $\mathcal{PT}$-symmetric systems pave the way for the design of a novel kind of laser, where the transverse light confinement is ensured by means of a linear waveguide and inhomogeneous gain/loss can be achieved by space-dependent charge injection, easily realizable using current technology, for example, in AlGaAs systems: we expect that the Kerr effect plays a fundamental role in these set-ups owing to the high intensities associated with cavities, permitting new functionalities as the switching between different emission states as the gain of the system is varied. Finally, the beam drift in a waveguide array can find application in the design of novel integrated optical circuits, showing both linear and nonlinear tunability.

ACKNOWLEDGMENT

JCP gratefully acknowledges POPH-QREN and ISE Fundancia a Technologia (FCT, Portugal), SFRH/BPD/77524/2011 for financial support.

REFERENCES

1. C. M. Bender and S. Boettcher, "Real spectra in non-Hermitian Hamiltonians having $\mathcal{PT}$ symmetry," *Phys. Rev. Lett.*, 80, 5243, 1998.
2. C. M. Bender, D. C. Brody and H. F. Jones, "Complex extension of quantum mechanics," *Phys. Rev. Lett.*, 89, 270401, 2002.
3. C. M. Bender, "Introduction to $\mathcal{PT}$-symmetric quantum theory," *Contemp. Phys*, 46, 277, 2005.
4. S. Longhi, "Quantum-optical analogies using photonic structures," *Laser Phot. Rev.*, 3, 243, 2009.
5. C. E. Rüter, K. G. Makris, R. El-Ganainy, D. N. Christodoulides, M. Segev and D. Kip, "Observation of parity-time symmetry in optics," *Nat. Phys.*, 6, 192, 2010.
6. A. Regensburger, C. Bersch, M. A. Miri, G. Onishchukov, D. N. Christodoulides and U. Peschel, "Parity-time synthetic photonic lattices," *Nature*, 488, 167, 2012.
7. A. Regensburger, M. A. Miri, C. Bersch, J. Näger, G. Onishchukov, D. N. Christodoulides and U. Peschel, "Observation of defect states in $\mathcal{PT}$-symmetric optical lattices," *Phys. Rev. Lett.*, 110, 223902, 2013.
8. F. K. Abdullaev, Y. V. Kartashov, V. V. Konotop and D. A. Zezyulin, "Solitons in $\mathcal{PT}$-symmetric nonlinear lattices," *Phys. Rev. A*, 83, 041805, 2011.
9. Y. Lumer, Y. Plotnik, M. C. Rechtsman and M. Segev, "Nonlinearly induced $\mathcal{PT}$ transition in photonic systems," *Phys. Rev. Lett.*, 111, 263901, 2013.

10. K. G. Makris, R. El-Ganainy, D. N. Christodoulides and Z. H. Musslimani, "Beam dynamics in PT symmetric optical lattices," *Phys. Rev. Lett.*, 100, 103904, 2008.

11. K. G. Makris, R. El-Ganainy, D. N. Christodoulides and Z. H. Musslimani, "$\mathcal{PT}$-symmetric optical lattices," *Phys. Rev. A*, 81, 063807, 2010.

12. Z. H. Musslimani, K. G. Makris, R. El-Ganainy and D. N. Christodoulides, "Optical solitons in $\mathcal{PT}$ periodic potentials," *Phys. Rev. Lett.*, 100, 030402, 2008.

13. S. V. Dmitriev, A. A. Sukhorukov and Yu. S. Kivshar, "Binary parity-time-symmetric nonlinear lattices with balanced gain and loss," *Opt. Lett.*, 35, 2976, 2010.

14. I. V. Barashenkov, M. M. Bogdan and V. I. Korobov, "Stability diagram of the phaselocked solitons in the parametrically driven, damped nonlinear Schrödinger equation," *Europhysics Letters*, 15, 113, 1991.

15. A. De Rossi, C. Conti and S. Trillo, "Stability, multistability, and wobbling of optical gap solitons," *Phys. Rev. Lett.*, 81, 85, 1998.

16. I. V. Barashenkov, D. E. Pelinovsky and E. V. Zemlyanaya, "Vibrations and oscillatory instabilities of gap solitons," *Phys. Rev. Lett.*, 80, 5117, 1998.

17. C. P. Jisha, A. Alberucci, Valeriy A. Brazhnyi and G. Assanto, "Nonlocal gap solitons in $\mathcal{PT}$-symmetric periodic potentials with defocusing nonlinearity," *Phys. Rev. A*, 89, 013812, 2014.

18. Y. S. Kivshar, G. P. Agrawal, *Optical Solitons*, Academic Press, 2003.

19. A. Alberucci and G. Assanto, "Propagation of optical spatial solitons in finite size media: Interplay between nonlocality and boundary conditions," *J. Opt. Soc. Am. B*, 24, 2314, 2007.

20. C. Conti, M. A. Schmidt, Philip St. J. Russell and F. Biancalana, "Highly noninstantaneous solitons in liquid-core photonic crystal fibers," *Phys. Rev. Lett.*, 105, 263902, 2010.

21. C. P. Jisha, L. Devassy, A. Alberucci and V. C. Kuriakose, "Influence of the imaginary component of the photonic potential on the properties of solitons in symmetric systems," *Phys. Rev.*, 2014.

22. K. G. Makris, R. El-Ganainy, D. N. Christodoulides and Z. H. Musslimani, "$\mathcal{PT}$-symmetric periodic optical potentials," *Int. J. Theor. Phys.*, 50, 1019, 2011.

23. S. Nixon, Lijuan Ge and J. Yang, "Stability analysis for solitons in $\mathcal{PT}$-symmetric optical lattices," *Phys. Rev. A*, 85, 023822, 2012.

24. B. Midya, B. Roy and R. Roychoudhury, "A note on the PT invariant periodic potential," *Phys. Lett. A*, 374, 2605, 2010.

25. S. Hu, X. Ma, D. Lu, Z. Yang, Y. Zheng and W. Hu, "Solitons supported by complex $\mathcal{PT}$-symmetric gaussian potentials," *Phys. Rev. A*, 84, 043818, 2011.

26. Z. Shi, X. Jiang, X. Zhu, H. Li, "Bright spatial solitons in defocusing Kerr media with $\mathcal{PT}$-symmetric potentials," *Phys. Rev. A*, 84, 053855, 2011.

27. Z. Shi, H. Li, X. Zhu and X. Jiang, "Nonlocal bright spatial solitons in defocusing Kerr media supported by pt symmetric potentials," *Eur. Phys. Lett.*, 98, 64006, 2012.

28. F. Lederer, G. I. Stegeman, D. N. Christodoulides, G. Assanto, M. Segev and Y. Silberberg, "Discrete solitons in optics," *Phys. Rev.*, 463, 1, 2008.

29. Y. S. Kivshar, "Self-localization in arrays of defocusing waveguides," *Opt. Lett.*, 18, 1147, 1993.

30. D. Mandelik, R. Morandotti, J. S. Aitchison and Y. Silberberg, "Gap solitons in waveguide arrays," *Phys. Rev. Lett.*, 92, 093904, 2004.

31. Y. Y. Lin, C. P. Jisha, Ching-Jen Jeng, Ray-Kuang Lee and B. A. Malomed, "Gap solitons in optical lattices embedded into nonlocal media," *Phys. Rev. A*, 81, 063803, 2010.

32. H. Li, X. Jiang, X. Zhu and Z. Shi, "Nonlocal gap solitons in parity-time symmetric optical lattices," arXiv:1109.4987 (2011).

33. J. Zeng and Y. Lan, "Two-dimensional solitons in PT linear lattice potentials," *Phys. Rev. E*, 85, 047601, 2012.

34. Y. S. Kivshar, "Switching dynamics of solitons in fiber directional couplers," *Opt. Lett.*, 18, 7, 1993.

35. B. A. Malomed, "Variational methods in nonlinear fiber optics and related fields," *Progress in Optics*, 43, 71, 2002.

36. D. P. Caetano, S. B. Cavalcanti, J. M. Hickmann, A. M. Kamchatnov, R. A. Kraenkel and E. A. Makarova, "Soliton propagation in a medium with Kerr nonlinearity and resonant impurities: A variational approach," *Phys. Rev. E*, 67, 046615, 2003.

37. A. Ankiewicz, N. Akhmediev and N. Devine, "Dissipative solitons with a Lagrangian approach," *Opt. Fiber Techn.*, 13, 91, 2007.

38. C. P. Jisha, V. C. Kuriakose and K. Porsezian, "Modulational instability and beam propagation in photorefractive polymer," *J. Opt. Soc. Am. B*, 25, 674, 2008.

39. A. Alberucci, G. Assanto, D. Buccoliero, A. S. Desyatnikov, T. R. Marchant and N. F. Smyth, "Modulation analysis of boundary-induced motion of optical solitary waves in a nematic liquid crystal," *Phys. Rev. A*, 79, 043816, 2009.

40. C. P. Jisha, V. C. Kuriakose and K. Porsezian, "Variational method in soliton theory," *Eur. Phys. J. Spec. Top*, 173, 341, 2009.

41. G. B. Whitham, *Linear and Nonlinear Waves*, John Wiley and Sons, 1974.

42. D. Anderson, "Variational approach to nonlinear pulse propagation in optical fibers," *Phys. Rev. A*, 27, 3135, 1983.

43. Z. Rapti, P. G. Kevrekidis, A Smerzi and A. R. Bishop, "Variational approach to the modulational instability," *Phys. Rev. E*, 69, 017601, 2004.

44. C. P. Jisha, V. C. Kuriakose and K. Porsezian, "Variational approach to spatial optical solitons in bulk cubic-quintic media stabilized by self-induced multiphoton ionization," *Phys. Rev. E*, 71, 056615, 2005.

45. S. Chvez Cerda, S. B. Cavalcanti and J. M. Hickmann, "A variational approach of nonlinear dissipative pulse propagation," *Eur. Phys. J. Spec. Top*, 1, 313, 1998.

13 Suspended core photonic crystal fibers and generation of dual radiation

Samudra Roy, Debashri Ghosh, and *Shyamal K. Bhadra*

13.1 INTRODUCTION

In the middle of the 1800s, experiments demonstrated that light can travel through a curved stream of water or a curved glass rod by total internal reflection (Daniel Colladon in 1841, Jacques Babinet in 1842, John Tyndall in 1854). However, the first all-glass optical fiber which had a higher index core surrounded by a lower index cladding was realized almost one hundred years later. Kao was the first who identified the critical specifications of the fibers that are needed for low loss long range communication [1–3]. The advent of new manufacturing processes led to the achievement of the theoretical minimum of loss value (0.2 dB/m at 1.5 μm) by the end of 1980 [4]. The loss reduction in fibers enhanced the scope for the study of different nonlinear processes which required a longer propagation path length at the available power levels in the early 1970s. During the 1980s most of the experiments related to nonlinear phenomena in optical fibers were performed at Bell Laboratories. Stimulated Raman scattering (SRS) and stimulated Brillouin scattering (SBS) were studied first [5,6] followed by the optical Kerr effect [7], parametric four-wave mixing (FWM) [8] and self-phase modulation (SPM) [9]. The basics of nonlinear effects in optical fiber are described in Chapter 1. In a classic paper, Zakharov and Shabat first demonstrated the mathematical formalism based on the inverse scattering theory to solve the nonlinear Schrödinger equation (NLSE) [10]. In 1973, the theoretical prediction of optical solitons as the interplay of fiber dispersion and fiber nonlinearity was made by Hasegawa and Tappert [11] and soliton propagation was demonstrated experimentally seven years later in a single-mode optical fiber by Mollenauer [12]. The advent of mature techniques to tailor structures on the micro and nano scales in the late 1980s provided the opportunity to investigate the interaction of light with structured matter. Within this framework, the photonic crystal emerged and became an extensively studied scientific area in 1987 when Eli Yablonovitch and S. John proposed the idea of "photonic bandgap structures" [13–15]. 2D and 3D photonic crystal structures are artificially generated in periodic dielectric structures in which light behaves the same way as electron waves in nat-

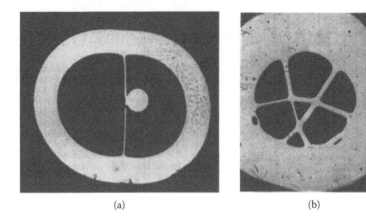

(a)   (b)

FIGURE 13.1: Cross-section of the (a) single material multimode fiber and
(b) three-ring multi-core single material fiber, after [16].

ural crystals. Under proper periodic arrangements of the dielectric medium,
the propagation of light can be totally inhibited at certain frequency bands,
regardless of propagation direction and polarization. This frequency band is
generally called the photonic bandgap (PBG).

Single material optical fibers with air holes running along their axes were
first fabricated by Kaiser et al. in 1974 at Bell Labs [16] and then in 1982 by
Okoshi et al. [17]. The single material fibers consisted of small diameter rods
supported on thin plates in the center of large diameter protective tubes. The
formation of three-ring multi-core single material fiber was also attempted
during this work by Kaiser et al., as shown in Figure 13.1. However, those
structures were quite premature and after almost 20 years, with available
technology, the so- called photonic crystal fiber (PCF) was first demonstrated
by Knight et al. in 1996 [18]. Typically, the PCF can be subdivided into two
basic types — a hollow core and a solid core PCF [19–21]. The guiding mechanisms of these two types of PCFs are also different. In a hollow core PCF
the light is confined in the hollow core region. When the high index cladding
which is made of periodic dielectric layers (essentially form Bragg planes) are
in resonance with the core mode, light is relayed outward in the transverse
direction by the high index layers. On the other hand, when the cladding
layers are not in resonance (in antiresonance) with the core mode, light is
rejected and forced to be confined in the low index hollow core. The solid core
PCF is realized by considering a central defect in the form of a solid silica
rod in the periodic air hole lattice (Figure 13.2). The effective refractive index
of cladding formed by the microstructured air holes is essentially less than
that of solid silica in the core, and light is confined in the core by the conventional mechanism of total internal reflection (TIR). Since the meticulous
arrangement of air holes in the cladding is not essential for solid core PCFs

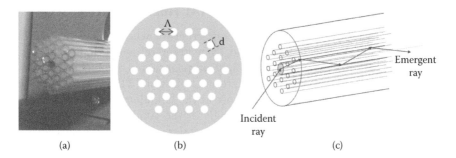

FIGURE 13.2: (a) Macroscopic structure of the core and the periodic cladding in terms of capillary arrangement. (b) Schematic structure of a hexagonal lattice. (c) Schematic picture of the modified TIR process where solid arrows represent the light rays trapped inside the solid fiber core.

and the guiding mechanism is also not based on typical photonic band structure, it is logical to term the solid core PCFs microstructured optical fibers (MOFs).

Cladding air-hole diameter and distance between two consecutive air holes, namely pitch, are the two important structural parameters for MOFs. MOFs offer unprecedented dispersion tailoring and high nonlinearity through modification of the design parameters. Due to the reduced effective core area and large refractive index contrast between core and cladding, light is confined very tightly inside the core of the MOFs, resulting in high nonlinearity. Light with a narrow spectral width can undergo extreme spectral broadening during propagation inside such nonlinear MOFs. This spectral broadening results in supercontinuum generation (SCG), which was observed in bulk glass in 1970 [22]. SCG is a complex phenomenon which involves different higher order nonlinear processes like SRS, FWM, SPM, etc. [23, 24]. Dispersion also plays a dominant role in controlling such wide band spectra [25]. SRS and SPM play a major role in generating low frequency components, whereas phase matched resonant radiation is the key phenomenon behind the generation of high frequencies. The details of SCG in optical fiber have been described by different authors in a book edited by Dudley and Taylor [26]. It is evident from previous studies [27] that the dispersion and nonlinearity of MOF are the key factors in tailoring supercontinuum output. In the present chapter we will focus our discussion on describing a special kind of MOF whose core is suspended through the specially designed air-hole structure adjacent to the core. This special geometric structure of the air holes reduces the effective core area and hence increases nonlinearity. The dispersion properties of such MOFs also critically depend on the geometrical design parameter called the suspension factor (SF).

13.2   SOLID CORE PHOTONIC CRYSTAL FIBER: A BRIEF OUTLINE

Optical waveguides constitute a general structure where typically a higher refractive index region is surrounded by a lower refractive index dielectric material. This arrangement ensures light propagation in the higher refractive index part of the waveguide by total internal reflection (TIR). The light propagation occurs when the angle of incidence at the boundary of the two dielectric materials is higher than the critical angle of the total internal reflection. If the higher index material has a refractive index of n_1 and the surrounding low index one n_2, then the critical angle of incidence for TIR can be given by $\theta_c = \sin^{-1}(n_1/n_2)$. Solid core PCFs are formed by an array of microstructured hollow channels running along the longitudinal axis of the fiber with a defect in the core, as shown in Fig. 13.1(a). The hollow channels are commonly arranged such that they form a lattice providing a six-fold symmetry. The light is guided in a silica core formed by one missing strand at the center of the photonic crystal. The guidance mechanism in this type of structure can be understood by utilizing a simple step-index fiber configuration, where light is confined in a high index core surrounded by a low index cladding by means of modified TIR. The fiber geometry is defined by the cladding hole diameter d and the hole-to-hole distance, Λ (pitch), as illustrated in Fig. 13.1(b). d and Λ strongly affect the guidance properties of the MOF, including the number of modes and the spectral dependence of the group velocity. The refractive index of the cladding is taken to be the effective index of the fourth-order dispersion (FOD), which is defined as the mode with the largest value of propagation constant (β) which would propagate in an infinite cladding structure without any defect. The cladding refractive index in MOFs is hence not a constant quantity but is strongly dependent on the wavelength. For these reasons (absence of a finite boundary at the core-cladding interface and strong dependence of the cladding refractive index on wavelength), the guidance mechanism in MOFs is termed "modified" TIR.

Effective area is another important parameter for an MOF, especially for nonlinear applications. According to the definition, the effective area can be described as [23]

$$A_{eff} = \frac{\left[\int_{-\infty}^{\infty}\int_{-\infty}^{\infty}|E(x,y)|^2 dxdy\right]^2}{\int_{-\infty}^{\infty}\int_{-\infty}^{\infty}|E(x,y)|^4 dxdy} \tag{13.1}$$

where $E(x,y)$ is the transverse electric field distribution. The mode area can be tailored to a large extent by varying air-hole diameter and pitch. It can be increased by reducing the hole diameter which is required for high power transmission where nonlinear effects are unwanted. However, one of the major implications of the MOF is to enhance the nonlinear effect during pulse propagation. The nonlinear coefficient (γ) of the MOF can simply be written as [23]

$$\gamma = \frac{2\pi n_2}{\lambda A_{eff}} \qquad (13.2)$$

where n_2 is the nonlinear index coefficient of the material used in the MOF and related to third order susceptibility. Typically for silica, $n_2 = 2.7 \times 10^{20}$ m^2/W at the wavelength of 1.06 μ m. The nonlinear coefficient is inversely related to A_{eff}, which means with proper geometric arrangement one can enhance the nonlinearity. The highest nonlinearity available in conventional step-index fiber is $\gamma = 20$ W^{-1}Km^{-1} at 1550 nm [23]. By comparison, a solid core PCF with a core diameter of 1 μm has a nonlinearity of $\gamma = 550$ W^{-1}Km^{-1} at 1550 nm [28]. Reducing the core diameter and increasing the air-filling fraction is the standard procedure to fabricate a highly nonlinear PCF.

13.3   GROUP VELOCITY DISPERSION

Frequency dependent group velocity leads to temporal pulse broadening and is generally termed group velocity dispersion (GVD). In single-mode optical fibers dispersion occurs due to two processes. One is the intrinsic material or chromatic dispersion, D_m, of the glass which originates from the variation of refractive index with wavelength, and the other is due to waveguide dispersion, D_w, which depends on the waveguide geometry. The effective refractive index of a confined mode in a fiber varies as a function of frequency, w. The individual speeds experienced by different frequencies are different and can be written as $c/n(w)$ where c is the velocity of light and $n(w)$ is the refractive index. During the propagation of an optical pulse envelope, different frequency components travel at different speeds. The pulse envelope will spread out as it propagates, since the slower frequencies lag behind the faster ones. This process is known as chromatic dispersion. To capture the effect of pulse reshaping due to dispersion, the propagation constant, β, is often expressed as a Taylor series expansion about the center frequency, w_0:

$$\beta(w) \approx \beta_0 + \beta_1(w - w_0) + \frac{\beta_2}{2!}(w - w_0)^2 + \frac{\beta_3}{3!}(w - w_0)^3 ... \qquad (13.3)$$

where, β_m is written for $d^m \beta/dw^m$ evaluated at the center frequency, w_0. β_0 is the propagation constant at the center frequency and β_1 is the group delay, defined as the time it takes a pulse to travel a given distance, or, equivalently, the inverse of the group velocity (v_g). The expression is given as

$$\beta_1 = (v_g)^{-1} = \frac{n_g}{c} = \frac{1}{c}\left(n + w\frac{dn}{dw}\right) \qquad (13.4)$$

Here n_g is called the group index. Higher order terms denote GVD and account for reshaping of the pulse. Depending on the sign of the GVD,

Kerr-based phase modulation can be counterbalanced (in the case of anomalous/negative dispersion) or reinforced (in the case of normal/positive dispersion). The expression for the second order dispersion term β_2 is

$$\beta_2 = \frac{1}{c}\left(2\frac{dn}{d\omega} + \omega\frac{d^2n}{d\omega^2}\right) = -\frac{\lambda^2}{2\pi c}D \qquad (13.5)$$

Here c is the speed of light in a vacuum, λ is the operating wavelength and D is the dispersion parameter commonly used because it has practical units of ps/km/nm which can be directly related to the amount of pulse spreading (in ps) per propagation length (in km) per bandwidth (in nm). On the other hand, β_2 has units of ps^2/km. Both β_2 and D are used when referring to GVD. The high refractive index difference between silica and air as well as the flexibility of changing the air hole sizes and patterns, which makes the cladding index strongly wavelength dependent, offer a variety of unusual dispersion behaviors in MOFs. By properly changing the geometric characteristics of the air holes in the MOF cross-section, the waveguide contribution to the dispersion can be significantly altered, thus obtaining unusual positions of the zero dispersion wavelength (ZDW) which can be tailored over a very wide range, or particular values of dispersion slope which can be engineered to be ultra-flattened. For example, the ZDW can be shifted to the visible by reducing the core size and increasing the air-filling fraction. On the other hand, very flat dispersion curves can be obtained in certain wavelength ranges in MOFs with small air holes, i.e., with a low air-filling fraction. MOFs with a high air-filling fraction can also be designed to compensate the anomalous dispersion ($\beta_2 < 0$) of single-mode fibers. The dispersion parameter (D) of MOFs is computed using the real part of n_{eff} as

$$D = -\frac{\lambda}{c}\frac{d^2[Re(n_{eff})]}{d\lambda^2} \qquad (13.6)$$

The total dispersion is approximated as the sum of the material dispersion, D_m, and the waveguide dispersion, D_w, as $D(\lambda) = D_m(\lambda) + D_w(\lambda)$. The chromatic dispersion and consequently $D_m(\lambda)$ become inherent properties of the system for air-silica MOFs. The effective index of the guided mode, n_{eff}, is calculated taking into account the material dispersion of the structure through the Sellmeier equation, and the total dispersion at a particular wavelength is obtained by adding material and waveguide contributions, as shown in Fig. 13.3(a).

The amazing flexibility of tailoring the dispersion leads to MOFs with two ZDWs (sometimes even more!) which open up new and interesting features of pulse propagation such as dual radiations. The dispersion of MOFs can be tailored over a large extent simply by changing the MOF design parameters. In Fig. 13.3(b) we show how the dispersion profile changes significantly with changing geometric parameters of an MOF. For a pulse propagating in a fiber, we can define a characteristic length over which the effects of dispersion

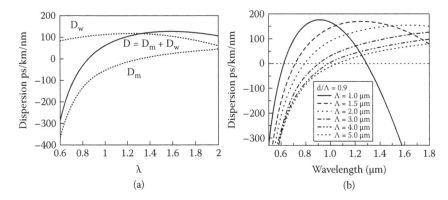

FIGURE 13.3: (a) The individual contribution of the material (D_m) and waveguide (D_w) dispersion for an MOF with an air-filling fraction of 80 is shown in the figure. The solid line indicates the overall dispersion. Wavelengths are in microns. (b) Dispersion curves with a constant air-filling fraction (d/Λ) but varying pitch (Λ).

become important. This length is known as the dispersion length, L_D, and is given by $L_D = t_0^2/|\beta_2|$, where t_0 is the input pulse width. It is straightforward to note that dispersive effects become more important for shorter pulses. The physical explanation for this is that a shorter transform limited pulse will have a broader bandwidth than a pulse of longer temporal duration, and will therefore experience a greater degree of spreading.

13.4   FABRICATION OF SUSPENDED CORE PCFs

Nonlinear MOFs are fabricated by the stack and draw method, which is very versatile and cost effective and allows easy control over core shape and size as well as the index profile of the cladding region. A schematic of the fiber drawing tower is shown in Fig. 13.4. The preform was formed by stacking high purity silica capillaries of suitable diameter in a hexagonal array of rings around a central silica rod. The individual capillary tubes in the preform stack are completely sealed at one end to pressurize all the capillaries uniformly and prevent their collapse. Furthermore, a precise evacuation system is necessary in order to obtain perfect microstructured cladding. This was done by attaching a special evacuation cap to the top end of the stack. The evacuation cap is machined to accommodate the stack geometry at one end and to fit the output tubing that evacuates the air from the preform at the other end. The preform formed by the stacked capillaries and rod are first drawn to intermediate canes (2–3 mm diameter). Drawn cane should be absolutely free from any interstitial holes. Once the cane is drawn, it is inserted into a thick silica tube and the end outside the furnace is again fused or pres-

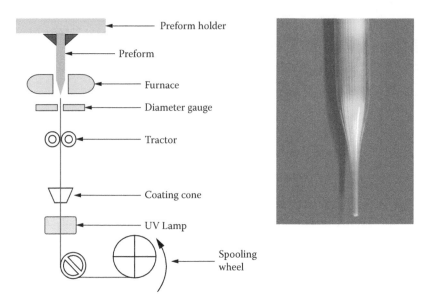

FIGURE 13.4: Schematic diagram of the fiber drawing tower and a preform end is shown after drawing cane in a hexagonal configuration. The cane was drawn at Fiber Optics and Photonics Division, CSIR-CGCRI, Kolkata.

surized for controlling the air-filling fraction. Then it is finally drawn to fibers of desired dimensions. This step allows a very high increase of the air-filling fraction as well as diminished core size, which is a prerequisite for enhancing the nonlinearity of the MOF. During drawing, it was observed that the core suspension can be varied according to the requirement by suitably adjusting the furnace temperature and the fiber drawing speed under specific conditions of differential pressure inside the preform structure. High speed drawing ensured higher SF. In order to minimize complexities during the MOF drawing process, the pitch (Λ) had comparatively higher values (between 3 and 7 μm) and d/Λ values 0.9. A d/Λ value up to 0.97 could be achieved with the help of the customized pressurization and evacuation system attached to the conventional fiber drawing tower with certain special arrangements. The drawing parameters are optimized in order to obtain the desired core suspension, and the optimized parameters are given in Table 13.1. A cross-section of the core and microstructured region of one such fabricated MOF is shown in Fig. 13.5(a), while Fig. 13.5(b) schematically reveals the difference between an ideally circular structure and the fabricated structure with a suspended core. The egg shape of the air holes in the first ring can be clearly seen in the fabricated structure which effectively reduces the core area and enhances nonlinearity.

TABLE 13.1

Optimized values of the drawing parameters for cane and MOF drawing

Cane drawing		Fiber drawing	
Drawing parameters	Optimized values	Drawing parameters	Optimized values
Furnace temperature	1950–1975°C	Furnace temperature	2000–2020°C
Feed rate	10–12 mm/min	Feed rate	6–8 mm/min
Draw speed	1.5–1.6 m/min	Draw speed	25–35 m/min
Evacuation pressure	300–600 mm Hg		

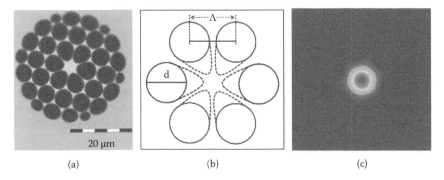

(a)   (b)   (c)

FIGURE 13.5: (a) Cross-section showing the microstructured region of the MOF fabricated at Fiber Optics and Photonics Division, CSIR-CGCRI, Kolkata. (b) Schematic diagram showing the ideal (circular) and fabricated (egg-shaped) structures. (c) Field distribution of the fundamental mode for the fabricated fiber at 1 μm wavelength; the modal profile is simulated using commercial COMSOL software.

13.5   CHARACTERISTICS OF SUSPENDED CORE PCF

In this section the different properties of suspended core PCFs are studied. We consider $d/\Lambda \geq 0.8$ in order to have high nonlinearity. A commercial full-vector finite element method (FEM) based mode solver, COMSOL Multiphysics, is used to theoretically study modal propagation in suspended core PCFs and investigate dispersion properties as well as the nature of the variation of different parameters such as effective mode area (A_{eff}) and nonlinear parameter (γ) as a function of suspension factor (SF). In order to reduce the effective

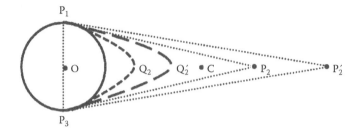

FIGURE 13.6: Schematic of elongated air holes formed by Bézier curves.

core area, each air hole in the first ring, instead of being perfectly circular, has been suitably elongated and tapered toward the center of the core to give it an egg-shaped form. The elongation and tapering is controlled by means of a parameter called suspension factor (SF) introduced by Mohsin et al. [29]. Each air hole in the first ring is formed by a combination of a circle and a second degree Bézier curve [29, 30], which is mathematically defined by Eq. (13.7) where P_1, P_2, P_3 are the control points and t is the varying parameter.

$$B(t) = (1-t)^2 P_1 + 2(1-t)t P_2 + t^2 P_3 \quad t \in [0,1] \qquad (13.7)$$

The formation of such egg-shaped air holes is shown in Fig. 13.6, where O is the center of the circular air hole, C is the center of the entire MOF geometry and P_1, P_2, P_3 are the control points of the Bézier curve, as mentioned in Eq. (13.7). P_2' denotes a different position of P_2.

The suspension factor (SF) of the core is defined as OP_2/OC [29]. Different suspended conditions for the core can be obtained by manipulating the control point P_2 of the Bézier curve. For example, the curve $P_1 Q_2 P_3$ denotes a particular suspended condition with control point P_2, while the curve $P_1 Q_2' P_3$ denotes a different suspended condition with control point P_2'. The SF of the core cannot have any arbitrary values; its value is delimited by a lower and upper bound determined by the structural integrity. A perfectly circular air hole corresponds to the minimum value of SF, while the maximum value is limited by the width of silica between two adjacent holes. The SF can be increased only up to the point up at which the individuality of the holes is preserved and they do not collapse or merge with each other. Using the commercial mode solver, the minimum and maximum values of SF (SF$_{min}$ and SF$_{max}$, respectively) for $0.75 \leq d/\Lambda \leq 0.95$ are individually determined and are listed in Table 13.2. The maximum value of SF is further plotted as a function of d/Λ in Fig. 13.7, and it is observed that they are linearly related to each other.

Hence, from Table 13.2 and Fig. 13.7, the mathematical expressions for SF_{min} and SF_{max} in terms of d/Λ are determined and given by Eq. (13.8)

TABLE 13.2

Minimum and maximum allowable values of SF for various d/Λ

d/Λ	SF_{min}	SF_{max}
0.75	0.75	1.75
0.80	0.80	1.70
0.85	0.85	1.65
0.90	0.90	1.60
0.95	0.95	1.55

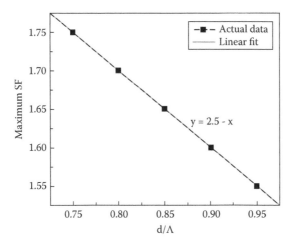

FIGURE 13.7: Maximum value of the suspension factor plotted as a function of d/Λ.

and Eq. (13.9), respectively.

$$SF_{min} = \frac{d}{\Lambda} \tag{13.8}$$

$$SF_{max} = 2.5 - \frac{d}{\Lambda} \tag{13.9}$$

Once the allowable values of SF are determined, the variations of the MOF parameters are studied with respect to the permissible range of SF. The two key parameters of MOF that affect SCG are the nonlinear parameter (γ) which was defined in Eq. (13.2) and dispersion [23, 31].

Variation of A_{eff} and γ is studied with respect to SF for different MOF designs at a wavelength of 1.0 μm, as pump sources close to this wavelength are easily available and frequently used for SCG experiments. The d/Λ value

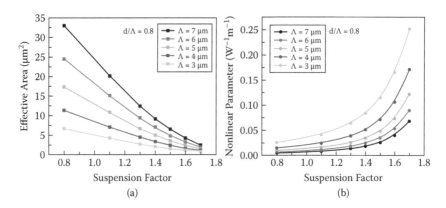

FIGURE 13.8: Effective area (a) and nonlinearity (b) as functions of SF for $\Lambda = 3 - 7\mu$m for $d/\Lambda = 0.8$.

is fixed at 0.8 while the value of Λ is chosen on the basis of practical feasibility and ranges from 3 μm to 7 μm. It should be noted that A_{eff} was calculated using the standard expression given in Eq. (13.1) where $E(x, y)$ is the fundamental mode field. The variation of effective area (A_{eff}) and nonlinear parameter (γ) with SF is shown in Figs. 13.8(a) and (b), respectively, for $d/\Lambda = 0.8$. As expected, A_{eff} decreases with the increase in SF, thereby enhancing the nonlinearity. The nature of the graphs are similar for the other higher d/Λ values and hence are not shown here. It is important to note here that the value of SFmax increases for lower d/Λ according to Eq. (13.9) (also seen in Table 13.2) thereby reducing the effective core area. Hence, by properly tuning the SF, one can obtain greater nonlinearity for low d/Λ than that possible for higher d/Λ.

Dispersion is another important property that affects SCG in MOFs. The shape of the dispersion curve determines which frequencies are phase-matched and can therefore undergo coherent nonlinear processes; it also determines which frequencies are matched in group velocity. As mentioned, one of the biggest advantages of MOFs is the freedom of tailoring dispersion properties over a very wide wavelength range. Flexibility in tuning the ZDW in MOFs allows different lasers to be used for SCG experiments. Following Eq. (13.6), dispersion D is computed from the real part of the fundamental mode index (n_{eff}) [23]. The values of n_{eff} at different wavelengths are obtained from COMSOL Multiphysics, taking Sellmeier's equation into consideration. In addition to the geometric parameters like Λ and d/Λ, it is found that SF can also be a guiding parameter for tailoring the dispersion properties of MOFs. Hence, depending on the availability of laser sources for SCG, the zero dispersion wavelength (Λ_{ZDW}) for a particular MOF design can be tuned easily by merely varying the SF. The dispersion curves for the range of allowable SF for $\Lambda = 3$ μm and $d/\Lambda = 0.8$ are shown in Fig. 13.9.

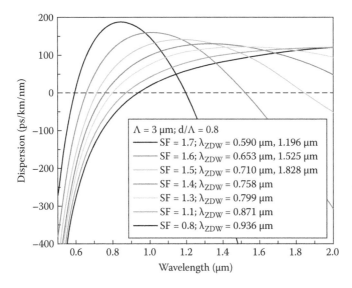

FIGURE 13.9: Dispersion curves for various SF values for $\Lambda = 3$ μm and $d/\Lambda = 0.8$. It should be noted that the two zero dispersion points can be achieved by modifying the SF.

It is seen that the ZDW shifts toward shorter wavelengths with an increase in SF. This trend of the dispersion curves complies with the established idea that ZDW shifts toward shorter wavelengths with a decrease in core size, which in this case has been obtained by increasing the SF. It is interesting to note that for higher values of SF, the dispersion curves bend sharply to give two ZDWs while for MOF structures with perfectly circular air holes only a single ZDW is obtained. It is also worth noting that the second ZDW moves toward shorter wavelengths at a faster rate than the first ZDW with the increase in SF. The curves shown in Fig. 13.9 clearly bring out the advantage of the SF in offering a wider range of tunability of dispersion.

13.6 DUAL-RESONANT RADIATION

In this section we try to show that the suspension factor critically affects the dispersion profile of MOFs. Under reasonable tolerance one can have two zero dispersion points and tailor the dispersion profile by pushing the first zero dispersion point in the lower wavelength regime. Dispersive wave (DW) generation [32, 33] is a phase matching (PM) phenomenon and largely depends on the dispersion profile. In the schematic in Fig. 13.10, we show how single and dual phase matched radiations are generated for different dispersion profiles. The established theory [25] already predicts that the higher order dispersion (HOD) plays a crucial role in generating DWs. Generally a DW is not phase matched with a fundamental soliton because the soliton wave number lies in a

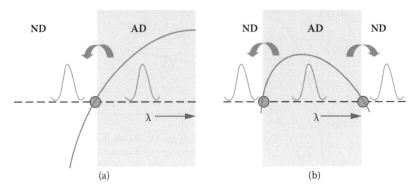

FIGURE 13.10: (a) Schematic diagram of a single radiation. The solid line and the dot represent the dispersion profile (as a function of wavelength) and zero dispersion points, respectively. The pulse is launched at the anomalous dispersion (AD) domain, and one can have radiation beyond the zero dispersion point at the lower wavelength side. (b) Schematic diagram of dual radiation. Here two normal dispersion (ND) regions are separated by an intermediate AD regime. If the pulse is launched at the AD domain, then the radiation will fall in ND regimes of both sides.

range forbidden for a linear DW. The presence of HOD terms, however, leads to a phase matched situation in which energy is transferred from the soliton to a DW at specific frequencies. In a dimensionless notation, the frequencies of DWs can be calculated by using a relatively simple PM condition, which arises from the equality of the soliton and radiation propagation constant [25],

$$\sum_{n=2}^{\infty} \delta_m \Omega^m = \frac{1}{2}(2N-1)^2, \qquad (13.10)$$

where $\delta_m = \beta_m/m!|\beta_2|t_0^{m-2}$, $\Omega = (\omega_d - \omega_s)t_0$, ω_d, ω_s are carrier frequencies associated with the DW and the soliton, respectively. The right hand side of the equation corresponds to the nonlinear phase shift. Here the mth order dispersion coefficient is denoted by β_m. N and t_0 correspond to the soliton order and input pulse width. Equation (13.10) is basically a polynomial of the detuned frequency Ω whose solutions represent radiation frequencies. Normally, the dispersion profile contains only one zero dispersion point and one can expect a single radiation, as depicted schematically in Fig. 13.10(a). Here the major contribution of dispersion comes through third order dispersion. In the two zero dispersion case, however, the major contribution will come from the fourth order dispersion term, and one can have two real solutions representing dual radiation.

Considering only the fourth order dispersion term (and neglecting third order dispersion), one can expand the polynomial as given in Eq. (13.10). The solution of the polynomial gives us the frequency of the radiation (Ω_{res}), which is

$$\Omega_{res} = \pm \left[\frac{-\delta_2 + \sqrt{\delta^2 + 4\delta_4\epsilon}}{2\delta_4} \right]^{1/2}. \tag{13.11}$$

Here ϵ corresponds to the nonlinear phase shift effect. The solution suggests that one can expect two symmetric radiations across the operating frequency Ω_0. Note that, with negligible nonlinear effect ($epsilon \approx 0$) one shouldn not have any radiation (mathematically $\Omega_{res} \approx 0$) if the pump pulse is launched at the normal dispersion regime, i.e., $\delta_2 > 0$. However, as shown in Fig. 10(b), dual radiation is still there if the pulse is launched at the anomalus dispersion regime ($\delta_2 < 0$), with the radiation frequency $\Omega_{res} = \pm\sqrt{\delta_2/\delta_4}$.

We try to give a complete picture of the evolution of dual radiation in Fig. 13.11, where we solve the generalized nonlinear Schrödinger equation numerically. In the simulation we include the well-known Raman and shock

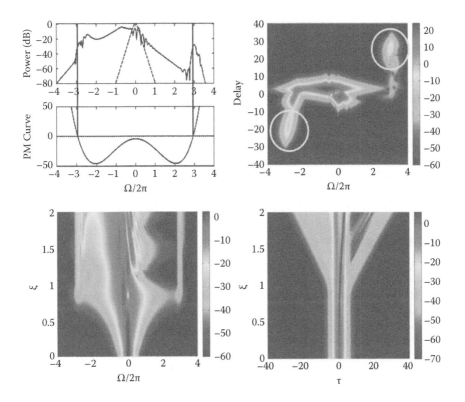

FIGURE 13.11: Complete picture of dual radiation. In the upper panel we show the phase matching curve based on Eq. 13.10 which efficiently predicts the radiation frequencies. The XFROG is also shown where the two radiations are evident and marked by white circles. In the lower panel we show the frequency and time evolution over distance for the two ZDW condition. The dual radiation in terms of dispersive wave is evident in the simulated figures.

effects and find that the dual radiations are assymetric in nature. This assymetry is generated because of the inclusion of the Raman and shock terms.

CONCLUSION

A practical design of highly nonlinear photonic crystal fiber is still an important issue for generating stable high power supercontinuum light. In this context perhaps suspended core photonic crystal fiber would be an ideal solution. The suspension factor can be tailored suitably depending upon the need and many such PCFs (MOFs) can be fabricated with two zero dispersion points in the perceptable wavelength regime for generating new radiation. The dispersion properties of such fibers with different suspension factors are discussed while extending the effect of higher order disperion terms on group velocity dispersion. The experimental process of drawing suspended core MOF is described and the optimized parameters are tabulated. Generation of dual radiation is one of the interesting consequences of MOFs with two zero dispersion wavelengths that is studied theoretically in this chapter. Further experimental work is necessary in order to get perfect highly nonlinear MOFs for stable supercontinuum output power over a wide range of wavelenths extending into deep blue region. The effect of higher order disperion in supercontinuum generation is interesting to study for nonlinear MOFs having two zero dispersion wavelengths.

ACKNOWLEDGMENT

All the staff members of the Fiber Optics and Photonics Division, CSIR-CGCRI, have given full support while carrying out this work and we are thankful to our Director, CSIR-CGCRI, for encouragement and support.

REFERENCES

1. A. C. S. Van Heel, "A new method of transporting optical images without aberrations," *Nature*, 173, 39, 1954.
2. H. H. Hopkins, N. S. Kapany, "A flexible fibrescope, using static scanning," *Opt. Lett.*, 15, 1443, 1990.
3. K. C. Kao, G. A. Hockham, "Dielectric-fiber surface waveguides for optical frequencies," *IEEE Proc.*, 113, 1151, 1966.
4. T. Miya, Y. Terunuma, T. Hosaka, T. Miyashita, "Ultimate low-loss single-mode fiber at 1.55 Îijm," *Electron. Lett.*, 15, 106, 1979.
5. R. H. Stolen, E. P. Ippen, A. R. Tynes, "Raman oscillation in glass optical waveguide," *Appl. Phys. Lett.*, 20, 62, 1972.
6. E. P. Ippen, R. H. Stolen, "Stimulated Brillouin scattering in optical fibers," *Appl. Phys. Lett.*, 21, 539, 1972.
7. R. H. Stolen, A. Askin, "Optical Kerr effect in glass waveguide," *Appl. Phys. Lett.* 22, 294, 1973.

8. R. H. Stolen, J. E. Bjorkholm, A. Ashkin, "Phase-matched three-wave mixing in silica fiber optical waveguides," *Appl. Phys. Lett.*, 24, 308, 1974.

9. R. H. Stolen, C. Lin, "Self-phase-modulation in silica optical fibers," *Phys. Rev. A*, 17, 1448, 1978.

10. V.E. Zakharov, A.B. Shabat, "Exact theory of wave two-dimensional self-focusing and one-dimensional self-modulation in nonlinear mediums," *Sov. Phys. JETP* 34, 62, 1971.

11. A. Hasegawa, F. D. Tappert, "Transmission of stationary nonlinear optical physics in dispersive dielectric fibers I: Anomalous dispersion," *Appl. Phys. Lett.*, 23, 142, 1973.

12. L. F. Mollenauer, R. H. Stolen, J. P. Gordon, "Experimental observation of picosecond pulse narrowing and solitons in optical fibers," *Phys. Rev. Lett.*, 45, 1095, 1980.

13. E. Yablonovitch, "Inhibited spontaneous emission in solid-state physics and electronics," *Phys. Rev. Lett.*, 58, 2059, 1987.

14. S. John, "Strong localization of photons in certain disordered dielectric superlattices," *Phys. Rev. Lett.* 58, 2486, 1987.

15. E. Yablonovitch," Photonic band-gap structures," *J. Opt. Soc. Am. B*, 10, 283, 1993.

16. P. Kaiser and H. W. Astle, "Low-loss single-material fibers made from pure fused silica," *The Bell Systems Technical Journal*, 53, 1021, 1974.

17. T. Okoshi, K. Oyamada, M. Nishimura, H. Yokata, "Side-tunnel-fiber: An approach to polarization maintaining optical waveguide scheme," *Electron Lett.*, 18, 824, 1982.

18. J. C. Knight, T. A. Birks, P. St. Russel, D. M. Atkin, "All-silica single-mode optical fiber with photonic crystal cladding," *Optics Letters.*, 21, 1547, 1996.

19. P. S. J. Russell,"Photonic crystal fiber," *Science*, 229, 358, 2003.

20. P. S. J. Russell, "Photonic crystal fiber," *J. Lightwave Technology*, 24, 4729, 2006.

21. F. Benabid, "Hollow-core photonic band gap fiber: New light guidance for new science and technology," *Phil. Trans. R. Soc. A.*, 364, 3439, 2006.

22. R. R. Alfano, S. L. Shapiro, "Observation of self-phase modulation and small-scale filaments in crystals and glasses," *Phys. Rev. Lett.*, 24, 592, 1970.

23. G. P. Agrawal, *Nonlinear Fiber Optics*, Academic Press, 5th ed., 2012.

24. J. M. Dudley, G. Genty, S. Coen, "Supercontinuum generation in photonic crystal fiber," *Rev. Mod. Phys.*, 78, 1135, 2006.

25. S. Roy, S. K. Bhadra, G. P. Agrawal, "Effect of higher-order dispersion on resonant dispersive waves emitted by solitons," *Optics Letters*, 34, 2072, 2009.

26. J. M. Dudley, J. R. Taylor, *Supercontinuum Generation in Optical Fiber*, Cambridge University Press, 2010.

27. S. K. Bhadra, A. Ghatak, *Guided Wave Optics and Photonic Devices*, Taylor and Francis Group, CRC Press, 2013.

28. P. Petropoulos, H. Ebendorff-Heidepriem, V. Finazzi, R. Moore, K. Frampton, D. J. Richardson, and M. Monro," Highly nonlinear anomalously dispersive lead silicate glass holey fibers," *Optics Express* 11, 3568, 2003.

29. K. M. Mohsin, M. S. Alam, D. M. N. Hasan, M. N. Hossain, "Dispersion and nonlinear properties of a chalcogenide As_2Se_3 suspended core fiber," *Appl. Opt.*, 50, E102, 2011.

30. G. Farin, *Curves and Surfaces for Computer-Aided Geometric Design*, 4th ed., Elsevier, 1997.

31. F. Zolla, G. Renversez, A. Nicolet, B. Kuhlmey, S. Guenneau, D. Felbacq, *Foundations of Photonic Crystal Fibers*, Imperial College Press, 2005.

32. S. Roy, S. K. Bhadra, G. P. Agrawal, "Dispersive wave generation in supercontinuum process inside nonlinear microstructured fiber," *Current Science*, 100, 312, 2011.

33. S. Roy, S. K. Bhadra, G. P. Agrawal, "Dispersive waves emitted by solitons perturbed by third-order dispersion inside optical fibers." *Phy. Rev. A*, 79, 023824, 2009.

14 Parabolic similaritons in optical fibers

Finot Christophe and *Boscolo Sonia*

14.1 INTRODUCTION

Parabolic pulses in optical fibers have several remarkable properties which have stimulated an increasing number of applications in nonlinear optics. Here, we review the physics underlying the generation of such pulses as well as the results obtained in a wide range of experimental configurations. For more than three decades, optical fibers have been recognized as a versatile testbed for the investigation of a large variety of nonlinear concepts. The unique dispersive and nonlinear properties of optical fibers lead to various scenarios of the evolution of short pulses propagating in the fiber which result in particular changes of the pulse temporal shape, spectrum and phase profile. A fascinating example is the formation of optical solitons [1] in the anomalous dispersion regime of a fiber as a result of cooperation between linear group-velocity dispersion (GVD) and nonlinear self-phase modulation (SPM). Recent analogies between optical supercontinuum generation and the occurrence of oceanic rogue waves [2] have led to the first experimental demonstration of a peculiar hydrodynamic solution analytically predicted decades ago: the Peregrine soliton [3], later followed by the experimental observation of its counterpart in deep water hydrodynamics [4].

The fundamental interest in optical fiber systems is not limited to pulse propagation in the anomalous GVD region. Indeed, recent developments in nonlinear optics have brought to the fore of intensive research an interesting class of pulses with a parabolic intensity profile and a linear instantaneous frequency shift or chirp [5, 6]. Parabolic pulses propagate in optical fibers with normal GVD in a self-similar manner, holding certain relations (scaling) between pulse power, duration and chirp parameter, and can tolerate strong nonlinearity without distortion or wave breaking [7]. These solutions, which have been dubbed similaritons, were demonstrated theoretically and experimentally in fiber amplifiers in 2000 [8]. Similaritons in fiber amplifiers are, along with solitons in passive fibers, the most well-known classes of nonlinear attractors for pulse propagation in optical fiber [8–11], so they take on major fundamental importance. The unique properties of parabolic similaritons have stimulated numerous applications ranging from high-power ultrashort pulse generation to optical nonlinear processing of telecommunication signals.

In this chapter, we review several features of parabolic similariton formation as well as recent results obtained in a wide range of experimental configurations. The chapter is organized as follows: Section 14.2 introduces the key model governing optical pulse propagation in a nonlinear optical fiber and its quasi-classical solutions in the normal regime of dispersion, and discusses the nonlinear reshaping of a pulse into a parabolic intensity profile. Section 14.3 outlines the key properties of parabolic similaritons. Section 14.4 is devoted to the experimental techniques that have been proposed to generate a parabolic intensity profile in both amplifying fibers and passive devices. Section 14.5 highlights some novel applications that have taken advantage of the intrinsic properties of parabolic pulses.

14.2   SHORT-PULSE DYNAMICS IN NORMALLY DISPERSIVE FIBERS

14.2.1   MODEL AND SITUATION UNDER INVESTIGATION

In a single-mode fiber amplifier, gain, dispersion and nonlinearity interact to strongly modify the properties of short optical pulses. The resultant longitudinal evolution of the complex electric field envelope $\psi(z,t)$ can be quantitatively modeled in terms of the nonlinear Schrödinger equation (NLSE) with a linear gain term:

$$i\frac{\partial \psi}{\partial z} - \frac{1}{2}\beta_2 \frac{\partial^2 \psi}{\partial t^2} + \gamma |\psi|^2 \psi = i\frac{g(z)}{2}\psi \qquad (14.1)$$

where z is the propagation distance, t is the reduced time, β_2 and γ are the respective GVD and Kerr nonlinearity coefficients of the fiber, and $g(z)$ is the gain profile along the fiber.

This equation neglects higher-order linear effects such as third-order dispersion (TOD) as well as higher-order nonlinear effects such as self-steepening or intra-pulse Raman scattering. Moreover, neglecting higher-order gain effects such as gain bandwidth and saturation is well suited to describing experiments that use high-gain, broadband rare-earth fiber amplifiers [8]. The impact of higher-order effects on pulse evolution will be discussed in Section 14.3.2 in the context of self-similar propagation.

Depending on the dispersion regime of the fiber, different pulse dynamics can be observed. In the anomalous dispersion regime ($\beta_2 < 0$), ultrashort pulses experience soliton-like compression [1] before splitting as a result of perturbed higher-order soliton dynamics [12]. On the other hand, high-power pulses that propagate in the normal GVD ($\beta_2 > 0$) region are susceptible to strong distortions and breakup owing to optical wave breaking [13–16], which is a consequence of excessive nonlinear phase accumulation combined with dispersion.

For the purpose of illustration, in the numerical simulations of Eq. (14.1) described in this chapter we consider the propagation of an initial transform-limited Gaussian pulse with a root-mean-square (rms) temporal duration of

1 ps in a highly nonlinear fiber (HNLF) with the following parameters at telecommunication wavelengths: $\beta_2 = 1\,\mathrm{ps}^2\,\mathrm{km}^{-1}$, $\gamma = 10\,(\mathrm{W\,km})^{-1}$. Taking advantage of the well-known scaling laws of the NLSE [16,17], the insights into the pulse dynamics gained from these simulations can be easily extended to a wide range of fiber and pulse parameters. To quantify the pulse evolution, here we use the excess kurtosis parameter and the parameter of misfit M_p between the pulse intensity profile $|\psi|^2$ and a parabolic fit of the same energy [18]:

$$M_p^2 \;=\; \int \Big[|\psi(t)|^2 \;-\; I_p(t) \Big]^2 dt \;/\; \int |\psi(t)|^4 \, dt, \qquad (14.2)$$

with $I_p(t)$ corresponding to a compactly supported pulse with a parabolic intensity profile:

$$I_p(t) \;=\; P_p \left(1 - \frac{t^2}{T_p^2} \right) \quad \text{if} \quad |t| < T_p, \quad I_p = 0 \quad \text{otherwise} \qquad (14.3)$$

where P_p is the peak power of the pulse and T_p the zero-crossing points defining effective pulse width. In the following sections, we will use the rms pulse duration, which for a parabolic pulse is given by $T_{\mathrm{rms}} = 2\,T_p/\sqrt{5}$.

14.2.2 PARABOLIC WAVEFORM AS A TRANSIENT STAGE OF EVOLUTION IN A PASSIVE FIBER

In this section, we discuss the main features of the pulse evolution in an HNLF in the absence of gain and in the nonlinearity-dominant regime of propagation, that is, when the length scale over which nonlinear effects become important for pulse evolution, $L_{\mathrm{NL}} = 1/(\gamma P_0)$, is much shorter than the dispersion length $L_{\mathrm{D}} = T_0^2/|\beta_2|$ (T_0 and P_0 are some characteristic temporal value and the peak power of the input pulse). Figure 14.1 shows the temporal and spectral evolution of the initial Gaussian pulse with $P_0 = 20\,\mathrm{W}$. The initial pulse form undergoes significant changes during the first stage of propagation, as it is highlighted by the rapid variation of the excess kurtosis (panel (c1)). We also observe an initial drop of the misfit parameter M_p (panel (c2)), indicating that the Gaussian pulse tends to reshape into an inverted parabola [15,17].

This is confirmed by the resultant pulse profile after the initial stage (after 20 m propagation distance here; panel (b2)). The combination of SPM and normal GVD makes the pulse develop a frequency chirp (equaling the first derivative of the phase with respect to time) with a linear variation over the pulse center (panel (b1)). However, the nonmonotonic nature of the chirp causes the red- (blue-) shifted light near the leading (trailing) edge to travel faster (slower) than the unshifted pulse parts, i.e., the pulse center and low-intensity wings. As a result, the parabolic shape formed is not maintained with increasing propagation distance, and the pulse changes shape toward an almost trapezoidal pulse form with a linear frequency chirp variation over most of the pulse (panels (b1) and (c1)). Ultimately, when the shifted light overruns the pulse tails, the wave breaks; oscillations appear in the wings of

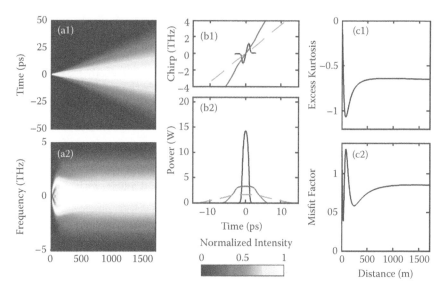

FIGURE 14.1: Evolution of an initially unchirped Gaussian pulse with $P_0 = 20\,\text{W}$ propagating in a passive normally dispersive fiber, as obtained by numerical integration of the NLSE. (a) Longitudinal evolution of the temporal (subplot 1) and spectral (subplot 2) intensity profiles (the maximum of the profile is normalized to one at each propagation step). (b) Temporal intensity (subplot 1) and chirp (subplot 2) profiles at different propagation distances in the fiber: $20\,\text{m}$ (solid black), $250\,\text{m}$ (solid grey) and $500\,\text{m}$ (dashed grey). (c) Longitudinal evolution of the excess kurtosis (subplot 1) and the misfit parameter M_p (subplot 2).

the pulse because of interference and, concomitantly, side lobes appear in the pulse spectrum [14–16].

Soon after the distance at which wave breaking occurs, the spectral expansion of the pulse starts to saturate [16]. With the accumulation of a parabolic spectral phase induced by dispersion, the temporal and spectral contents of the pulse become increasingly close to each other and the temporal intensity profile does not evolve anymore, as it is highlighted by the constant kurtosis or misfit factor in panels (c). This long-term far-field evolution corresponds to the formation of nonlinear structures of a spectronic nature in the passive system [19–21].

As highlighted by Fig. 14.1, the parabolic waveform that is formed after the first propagation stage represents **a transient state** of the nonlinear dynamic evolution of the pulse in the passive fiber medium with a life distance that depends on the initial conditions [22]. Note that a change in the initial conditions can lead to different transient states, such as that featuring a triangular-profiled pulse [22–24]. While no analytic exact formulae exist to describe these reshaping processes, some insights can be obtained from approximate analyses based on the method of moments [25, 26].

The situation is different with a high-intensity initial pulse with a parabolic intensity profile, where such a pulse can be used to create strongly frequency-swept pulses without the degrading effect of wave breaking. Indeed, in 1993 Anderson et al. showed that a sufficient condition to avoid wave breaking is that a pulse acquires a monotonic frequency chirp as it propagates, and that wave-breaking-free solutions of the NLSE (with only nonlinearity and GVD) exist when the GVD is normal [7]. These solutions of the NLSE in the quasi-classical or Wentzel–Kramers–Brillouin (WKB) limit (i.e., the limit of high amplitude or small dispersion such that $\beta_2|(|\psi|)_{tt}|/(2\gamma|\psi^3|) \ll 1$) exist when the GVD is normal [7] and take the form of linearly chirped downward-opened parabolae:

$$\psi_p(t) = \sqrt{I_p(t)} \ \exp\left(-i\left(C_p\frac{t^2}{2} + \phi\right)\right). \qquad (14.4)$$

with C_p is the linear chirp coefficient and ϕ a phase offset. While all the parameters of this solution evolve longitudinally, such a pulse remarkably maintains its shape, and is always a scaled version of itself; i.e., it evolves self-similarly. Note that it is yet possible to generate these parabolic similaritons from conventional pulses in a passive manner via nonlinear propagation in a two-segment normally dispersive fiber system where the nonlinear and dispersive characteristics of the second fiber are specially adjusted relative to the first fiber [17, 22]. An experimental demonstration will be presented in Section 14.4.1.

14.2.3 PARABOLIC WAVEFORM AS AN ASYMPTOTIC ATTRACTING STATE OF PULSE EVOLUTION IN A FIBER AMPLIFIER

Now we consider the impact of distributed gain on pulse evolution. Similar plots to those of Fig. 14.1 are presented in Fig. 14.2 except that the Gaussian pulse with $P_0 = 4\,\text{W}$ propagates in an active fiber with constant gain $g = 20\,\text{dB/m}$. Both the pulse duration and bandwidth now increase monotonically in the gain fiber. In contrast to the passive propagation described in the previous section, a parabolic intensity profile is formed in the amplifier after some initial transition stage and is maintained with further propagation, as indicated by the longitudinal decrease of the misfit parameter down to very low values (panel (c2)) and highlighted by the pulse shapes recorded at different propagation distances (panel (b2)). Consequently, in the context of a fiber amplifier, the parabolic profile is not a transient stage at all [8], as also confirmed in Fig. 14.2(b2), where the pulse shape is clearly parabolic for different propagation distances.

More details of the temporal and spectral shapes are given in Fig. 14.3. The output pulse from the amplifier has a highly parabolic form with compact support and a highly linear chirp over its entire duration in both the

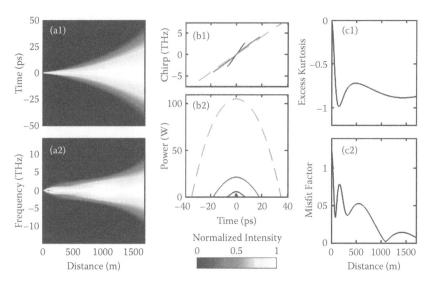

FIGURE 14.2: The same figure as Fig. 14.1 except that the Gaussian pulse propagates in a fiber with linear gain. The temporal intensity and chirp profiles in panel b are recorded at the propagation distances: 500 m (solid black), 1000 m (solid grey) and 1500 m (dashed grey).

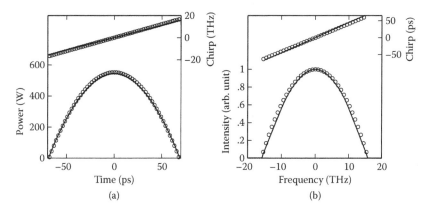

FIGURE 14.3: (a) Temporal and (b) spectral intensity and chirp profiles of the pulse at the end of a 2-km long normally dispersive gain fiber from NLSE numerical simulation (solid) compared to the theoretical predictions in the asymptotic analysis (circles).

temporal and spectral domains, and is in excellent agreement with the theoretical predictions obtained from the asymptotic similarity analysis of Eq. (14.1) [8, 9, 27]. In the normal GVD region, the dynamic evolution of any pulse can be accurately described by Eq. (14.4) in the limit $z \to \infty$, where the dynamic equations in the case of constant gain for the pulse peak amplitude P_{ps}, characteristic width T_{ps}, and chirp parameter C_{ps} have the explicit solutions (here, U_{in} is the initial pulse energy)

$$\sqrt{P_{ps}}(z) = \frac{1}{2} \left(\frac{g\,U_{\text{in}}}{\sqrt{\gamma\beta_2/2}} \right)^{\frac{1}{3}} \exp\left(\frac{g}{3}z\right), \qquad (14.5)$$

$$T_{ps}(z) = 3 \left(\frac{(\gamma\beta_2/2)\,U_{\text{in}}}{g^2} \right)^{\frac{1}{3}} \exp\left(\frac{g}{3}z\right), \qquad (14.6)$$

$$C_{ps} = \frac{g}{3\,\beta_2}. \qquad (14.7)$$

For a sufficiently high value of the chirp coefficient, the pulse waveform is imaged in the frequency domain, with the spectral width ν_{ps} given by

$$\nu_{ps}(z) = C_{ps}\,T_{ps}(z), \qquad (14.8)$$

and the spectral chirp coefficient $1/C_{ps}$. This can be explained using the stationary phase method, i.e., the cancellation of oscillating contributions with rapidly varying phase in the Fourier transform integral.

14.3 PROPERTIES OF SELF-SIMILAR PULSES AND EXTENSION TO OTHER CONFIGURATIONS

14.3.1 PARABOLIC SIMILARITON PROPERTIES

Pulses with a parabolic intensity profile and a linear frequency chirp variation have been the first class of optical pulses to be clearly identified as experiencing self-similar dynamics in a nonlinear optical system. This has led to the term "similariton" [5] that is now widely used to indicate such pulses. However, other pulse waveforms can also maintain their typical shape unchanged during propagation in nonlinear media with longitudinally tailored linear and nonlinear parameters, including chirped hyperbolic secant and hyperbolic tangent profiled pulses [28–33]. For the remainder of the chapter, we will limit our discussion to parabolic pulse profiles.

An intense optical pulse with an initial parabolic profile keeps its shape and acquires a linear chirp upon propagation in a passive normally dispersive fiber, while its amplitude decays monotonically and its width consequently increases monotonically [7]. On the contrary, as we outlined in the previous section, an arbitrary pulse inserted into a gain fiber is nonlinearly attracted to an asymptotically evolving parabolic pulse [8] subject to exponential growth of the amplitude and width at the same rate (Eq. (14.5) and (14.6) as well as Fig. 14.4(a)). Note that the self-similar parabolic solution is found for any gain profile, and the functional form of $g(z)$ determines only the self-similar scaling of the propagating pulse in the amplifier [34, 35].

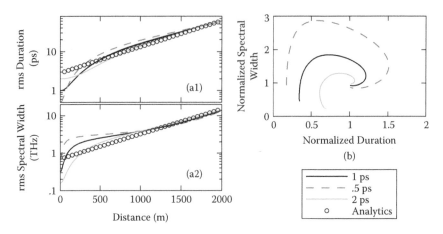

FIGURE 14.4: Pulse evolution toward the asymptotic attracting solution in a fiber amplifier as obtained by numerical integration of the NLSE. (a) Evolution of the rms temporal (subplot 1) and spectral (subplot 2) widths for initial Gaussian pulses with the same energy and different durations: 0.5 ps (dashed grey), 1 ps (solid black), and 2 ps (solid grey). The results are plotted on a logarithmic scale and compared to the theoretical predictions (circles). (b) Phase diagrams representing the combined evolution of the rms temporal and spectral widths of the pulses. The parameters are normalized to the asymptotic values predicted by Eq. (14.6) and (14.8).

The output pulse parameters are entirely determined by the energy of the initial pulse and the parameters of the fiber. In other words, the asymptotic state is a global attractor to the system for arbitrary initial conditions [8, 9, 11]. This attracting nature of amplifier similaritons is illustrated by the phase diagram of Fig. 14.4(b), and has also been experimentally confirmed in long distributed Raman amplifiers where initial pulses with significantly different temporal widths but with the same energy converge after a sufficient propagation length toward the theoretically predicted asymptotic state [10].

The challenge is for the pulse to reach this solution in a fiber length compatible with efficient system design. The problem of choosing the optimum initial conditions in order to shorten the propagation length required to enter into the asymptotic parabolic regime has been the subject of several studies [16, 18, 27, 36–38] which have highlighted the interactions between dispersion and nonlinearity in the initial stages of propagation. Note that some slowly decaying wings may exist in the spectrum of the amplifier similariton, and are residues of the pre-asymptotic evolution [9, 11, 39].

We have mentioned in the previous section that similaritons are intrinsically resistant to the deleterious effects of optical wave breaking [7]. The normal GVD of the fiber effectively linearizes the accumulated phase of the pulse during nonlinear propagation, allowing for the spectral bandwidth to increase without destabilizing the pulse. This self-similar evolution can stabilize high-energy pulses in fiber lasers, as we will describe in Section 14.3.3.2. It is worth mentioning that the strictly linear chirp coefficent developed during self-similar amplification (Eq. (14.7) and Fig. 14.3) does not depend on either the nonlinear coefficient or the input pulse parameters. Regarding the spectral chirp, a practical consequence of its high linearity is that it can be fully canceled by various means that we will detail in Section 14.5.1 and thus lead to high-quality, temporally compressed pulses with high energies.

14.3.2 IMPACT OF HIGHER-ORDER EFFECTS

The description of the self-similar pulse shaping regime given so far is based on the assumption that pulse evolution in a fiber can be described by Eq. (14.1). In fact, higher-order dispersion gain and nonlinear effects may become important for pulse propagation and ultimately limit the life distance of similariton solutions.

Among these effects, the Raman response of silica has been shown to have noticeable impact on parabolic pulse dynamics [40, 41]. Further, self-similar evolution is disrupted if the pulse encounters a limitation to its bandwidth. This must be reconciled with the tendency of a pulse to fill in available gain bandwidth [42–45].

Because the spectral bandwidth of the pulse may increase substantially in self-similar amplification, the inherent TOD in the fiber can become critically important for similariton propagation and ultimately lead to an intra pulse shock within the similariton. Further, in passive dispersion-decreasing fibers (DDFs; see Section 14.3.3.2) the GVD decreases asymptotically to zero, making the relative contribution of higher-order dispersion significant. The effect of TOD on self-similar evolution and parabolic pulses has been studied theoretically in several works [46–48]. In recently developed fibers with flattened dispersion characteristics around the pump wavelength the effect of TOD is significantly reduced below that of fourth-order dispersion (FOD). It has been theoretically demonstrated that the progressive nonlinear reshaping of a parabolic pulse driven by FOD in a passive normally dispersive fiber can result in a temporal triangular intensity profile, as well as a broad, compact, flat spectral profile at a particular point in the evolution [49]. The effects of fiber birefringence on parabolic pulses in fiber amplifiers have also been investigated theoretically [50, 51].

Theoretical and experimental studies of the dynamics of a pair of optical similaritons with the same or different central wavelengths [52] have revealed that similaritons are robust against collisions [53], while their interaction leads to the generation of high-repetition rate dark soliton trains [54]. The propagation of dark structures on a compact parabolic background in nonlinear waveguides has also been the subject of recent theoretical studies [55–58].

14.3.3   EXTENSION TO OTHER CONFIGURATIONS

Parabolic self-similar pulse dynamics can occur in other optical systems than normally dispersive fiber amplifiers or passive fibers with a fixed dispersion.

14.3.3.1   In dispersion-tailored fibers

In addition to fiber amplifiers, similariton formation with nonlinear attraction can be achieved in passive fibers provided a suitable longitudinal variation of the dispersion is introduced [59–61]. This approach is based on the observation that a longitudinal decrease of the normal dispersion is formally equivalent to linear gain. Continuously tapered DDFs with a dispersion profile exponentially decreasing down to zero can indeed mimic the effect of an active gain.

In the context of applications where high signal power is not necessarily required and where the most valuable features are the specific parabolic shape and phase of the pulse such as, e.g., optical telecommunications, passive approaches to the generation of parabolic pulses are of strong interest to overcome some of the drawbacks of an amplifying process, including the need for adequate rare earth dopants, availability and cost of pump lasers, and bandwidth-limited gain. However, TOD and optical attenuation have been found to severely affect similariton propagation performance [62–64]: as we mentioned before, the decrease of the ratio between the second and third dispersion coefficients favors the emergence of optical shock-type instabilities [62, 64]. Hybrid configurations using tapering down to nonzero (normal) dispersion values with possibly additional Raman pumping can overcome those limitations [64, 65].

14.3.3.2   In fiber-based cavities

The possibility of exploiting self-similar propagation in the cavity of a fiber laser was demonstrated numerically as early as 2001 [52]. Further theoretical and experimental work has led to the development of a new class of high-energy ultrashort pulse fiber lasers [66–70]. Fiber lasers that use self-similar pulse shaping in the normal dispersion regime have been demonstrated to achieve pulse energy and peak power performances much higher than those of prior approaches and that can directly compete with solid-state systems [71, 72] but with the major practical advantages of fiber.

The approach taken by Ilday et al. [66] to stabilize parabolic similaritons in an oscillator was to design a laser that supports self-similar evolution in a passive fiber segment which is subsequently amplified in the presence of minimal dispersion and nonlinearity. In such a laser, a short segment of gain fiber decouples gain filtering from the nonlinear evolution in the long segment of passive fiber, and the chirp accumulated in the passive fiber is compensated by linear, anomalously dispersive delay. More recently, the formation of a self-consistent solution in a laser cavity that supports self-similar evolution

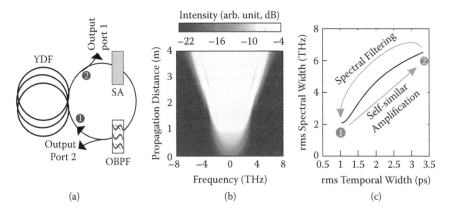

FIGURE 14.5: (a) Schematic of a laser relying on nonlinear self-similar attraction in a fiber amplifier YDF, ytterbium-doped fiber; SA, saturable absorber; OBPF, optical bandpass filter. (b) Evolution of the spectrum over one rountrip in the fiber cavity. (c) Phase diagram representing the combined evolution of the rms temporal and spectral widths. More details on the configuration in study can be found in [75].

in the gain fiber segment of the cavity has been demonstrated [67–70]. Unlike the passive similariton regime, the amplifier similariton laser relies on a local nonlinear attraction to stabilize the pulse in the cavity, while spectral filtering plays a critical supporting role. Oktem et al. built a laser where the parabolic pulse, following spectral filtering, gradually evolves into a soliton in an anomalous-dispersion segment, which allows for a short transform-limited pulse to return to the input of the gain fiber [67]. In the all-normal-dispersion laser experimentally demonstrated by Renninger et al. and theoretically investigated by Bale et al., (Figure 14.5) strong spectral filtering compensates for both the broad pulse duration and bandwidth after the gain segment and, hence, facilitates the creation of a self-consistent cavity [68, 69, 73]. The inclusion of a passive linear segment in the laser cavity has enabled further enrichment of the spectral content of the pulse [74]. Aguergaray et al. built a Raman oscillator with kilometers of gain fiber, which provides enough propagation length for the asymptotic solution to be achieved [70].

14.3.3.3   In nonlinear waveguides

The concept of a parabolic similariton has spread to other nonconventional nonlinear waveguides where engineering the waveguide dimensions offers accurate control of the dispersive properties, so that gain-free operating conditions similar to those in Section 14.3.3.1 become possible. Theoretical works have therefore demonstrated the nonlinear reshaping of a pulse in a tapered silicon fiber or nanowire [76, 77] and have discussed the limitations that may be associated with this new material.

Another variety of parabolic self-similar structure has been discussed in gradient index waveguides with this time the demonstration of spatial parabolic similaritons having a parabolic transverse beam profile evolving self-similarly upon propagation [78], with potential applications to the field of plasmonics [79]. Propagation of 2D spatial beams has also been investigated [80, 81].

14.4 EXPERIMENTAL GENERATION OF PARABOLIC PULSE SHAPE

14.4.1 IN PASSIVE SEGMENTS

Parabolic intensity profiles generated as a transient stage of passive propagation (see Section 14.2.2) were first reported in [82], where an initial sinusoidal beating having a convenient peak power evolves in a well-chosen length of normally dispersive fiber into a parabolic pulse train. In other experimental configurations, well-separated picosecond input pulses are reshaped into a parabola and a second stage involving an HNLF enables the stabilization of the profile [17]. This all-fiber architecture is particularly well suited for high-repetition rate low-duty cycle sources [83]. An illustration of this approach is provided in Fig. 14.6: a 40-GHz Gaussian pulse train is first reshaped into a train of a highly-chirped parabola that then evolves self-similarly in an HNLF. A standard single-mode fiber then achieves an efficient compression of the pulses [84] (see Section 14.5.1).

Regarding the generation of similaritons in a passive dispersion tailored fiber (see Section 14.5.1), experimental demonstrations have also been achieved at telecommunication wavelengths [65, 85] relying on specifically drawn fibers. A potential drawback of the method based on DDFs is that such dedicated tapered fibers are not to date commercially available. Fortunately, the use of comb-like architectures made of several segments with fixed dispersion has been found to efficiently mimic a continuous dispersion decrease [86, 87].

14.4.2 IN FIBER AMPLIFIERS

The first clear experimental demonstration of the generation of a parabolic similaration and the validation of the theoretical predictions of Eq. (14.4) have been carried out in the framework of a few meters long ytterbium-doped fiber amplifier [8]. A precise intensity and phase measurement based on the frequency resolved optical gating definitively confirmed the parabolic intensity profile of the emerging picosecond pulses, the rapid decrease of the wings as well as the high linearity of the chirp. Since then, several other experiments have benefited from the parabolic features in ytterbium-doped fibers with increasing output power levels and various initial femtosecond seeds from passively mode-locked fiber lasers with repetition rates of a few tens of MHz [8] to vertical external-cavity surface-emitting lasers (VECSELs) with GHz repetition rates [88]. Since the pulse reshaping process is not restricted to a specific

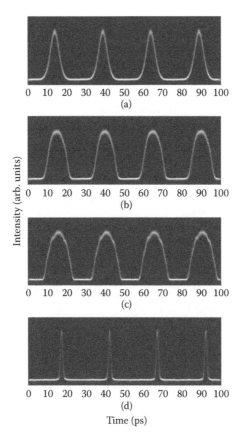

FIGURE 14.6: Passive generation, stabilization and compression of parabolic profiles at telecommunication wavelengths. Temporal intensity profiles are experimentally recorded using an optical sampling oscilloscope. (a) Train of incoming Gaussian pulses at a repetition rate of 40 GHz. (b) Pulses after propagation in a passive dispersion-shifted fiber with normal dispersion. (c) Pulses after propagation in an HNLF. (d) Pulses after cancellation of their linear chirp in a standard fiber with anomalous dispersion. More details on the experiment can be found in [83, 84].

rare-earth dopant, generation in erbium-doped fiber amplifiers at telecommunication wavelengths has also been successfully tested [39, 89–91]. Indeed, it should be noted that as early as 1996, experiments in erbium-doped fiber done by Tamura and Nakazawa provided indirect evidence of the chirp linearity of pulses amplified in a normally dispersive fiber [89]. At this time, however, the self-similar nature of the parabolic pulses and their underlying fundamental properties were not clearly pointed out.

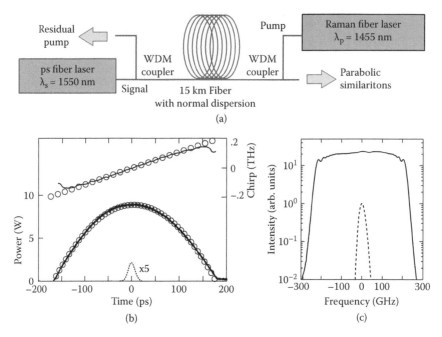

FIGURE 14.7: Generation of parabolic similaritons through Raman amplifica-tion at telecommunication wavelengths. (a) Experimental setup. (b) The tem-poral intensity profiles of the initial (dotted) and output (solid) pulses WDM, wavelength division multiplexing recorded in real time by a high-speed pho-todetector and oscilloscope are compared with a parabolic fit (circles). Chirp of the output pulse is compared with a linear fit (circles). (c) Spectrum of the initial (dashed) and output (solid) pulses. More details on the experiments can be found in [93].

Other gain mechanisms are also fully suitable for asymptotic parabolic generation, and configurations based on the use of the Raman amplification effect have been successfully investigated [38, 92, 93]. As an illustration, Fig. 14.7 presents the setup and results of an experiment using a Raman amplifier operating at telecommunication wavelengths. In this all-fiber setup using only standard devices, initial 14-ps pulses are adiabatically amplified in a 15-km normally dispersive fiber. The emerging 200-ps pulses clearly exhibit the ex-pected parabolic intensity profile combined with a highly linear chirp. The output optical spectrum also highlights the clear reshaping of the pulse, with a low ripple in the central part and a rapid decrease of the wings.

This then confirms that parabolic pulses can be efficiently generated over a wide range of initial pulse durations and fiber configurations. However, and to the best of our knowledge, no experimental demonstration of parabolic similaritons has been reported in parametric amplifiers up to now.

14.4.3 THROUGH LINEAR PULSE SHAPING

All the previously described generation methods are based on the intrinsic nonlinear reshaping of an optical pulse propagating in an active or passive normally dispersive fiber. It is of course also possible to benefit from the growing field of linear pulse shaping. While the resulting parabolic pulses do not present all features described in Section 14.3, using linear approaches removes the need for a high initial peak power, which is especially beneficial in the context of telecommunication applications described in Section 14.5.3, where the required average power could be prohibitive. We note, however, that whereas the nonlinear passive reshaping does not induce losses and generate a spectrum much broader than the incoming pulse spectrum, most of the linear techniques rely on the dissipative shaping of an ultrashort incoming pulse in order to fit the parabolic target.

Several approaches have been reported to efficiently synthetize optical parabola. The first set of approaches relies on the use of a superstructured fiber Bragg grating to tailor the initial spectrum of picosecond pulses [94]. Contrary to nonlinear methods, it is here possible to record chirp-free parabolic intensity profiles. A second flexible approach takes advantage of the line by line shaping in an arrayed waveguide grating fibre bragg grating (FBG) sensors [95,96] to produce periodic bright or dark parabolic pulse trains. With the progress of liquid crystal on silicon (LCOS), technology reconfigurable pulse shapers have been successfully studied at telecommunication wavelengths [97]. Other shaping schemes include the use of long period fiber grating filters [99] or of acousto-optic devices [100]. The wavelength-to-time mapping has also been explored to dynamically generate parabolic profiles [101].

14.5 APPLICATIONS OF PARABOLIC PULSES

Parabolic pulses are not only of theoretical interest but they have very wide practical potential. We provide here an overview of the numerous applications both in ultrashort high-power pulse generation and in optical pulse processing that have been stimulated by their remarkable features.

14.5.1 HIGH-POWER PULSE AMPLIFICATION AND ULTRASHORT PULSE GENERATION

The inital interest in self-similar dynamics was initially strongly driven by the field of ultrashort high-power pulse generation [8]. Indeed, the direct nonlinear amplification of pulses avoids the necessity of a prechirping stage as required in the linear chirped pulse amplification (CPA) approach [102] where the ultimate goal is to avoid any nonlinearity by stretching the initial pulse and therefore reducing the initial peak power. The approach based on parabolic amplification is intrinsically different, as it takes advantage of the nonlinear optical spectrum expansion of the pulse. Indeed, a strong frequency chirp

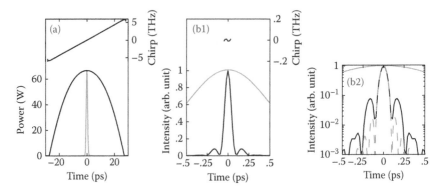

FIGURE 14.8: (a) Temporal intensity and chirp profiles of a parabolic pulse obtained after propagation in 1200 m of distributed fiber amplifier (same conditions as in Section 14.2.3; seed pulse is plotted using gray line). (b) Temporal intensity and chirp profiles after cancelation of the linear spectral chirp of the pulse. Panel (b2) provides a logarithmic plot of the compressed intensity profile. The results of numerical simulations (black solid lines) are compared with a first-order Bessel function fit (gray dashed line) and with the similariton at the output of the amplifier (gray solid line).

develops during amplification and new spectral components are generated. Thanks to the high level of linearity of this chirp, it is possible to efficiently compensate for the positive chirp slope by using a pair of diffraction gratings. Contrary to the CPA method and as illustrated by Fig. 14.8, the temporal duration of the compressed pulses is much shorter than the initial seed pulse and the resulting pulses exhibit a high-quality temporal profile. Indeed, the theoretical pulse shape obtained after compression of a perfectly parabolic pulse is a first-order Bessel function of the first kind that exhibits noticeably low sidelobes [18].

Therefore, several experimental works have reported high performance results [39,103–105], with temporal widths below 100 fs and peak powers exceeding 4 MW [44] or pulse energy at the microjoule level [104]. Improvements in the technique have led to the use of cubicon pulses where the effect of residual TOD can be overcome by an initial pulse shaping [106,107]. Other approaches have included an initial stage of parabolic preshaping before SPM-driven high-power amplification [108,109] or CPA device [110,111].

Parabolic amplification can also sustain GHz repetition rates [88,91] and can be integrated in an all-fiber setup with the judicious use of hollow-core photonic bandgap fibers employed as compression stages [39]. Additional nonlinear higher-order soliton compression has enabled the generation of pulses with temporal durations as short as 20 fs (i.e., four optical cycles at 1550 nm) [112]. If the primary goal is not to reach high peak powers, a length of anomalously dispersive fiber can be used [83,84,94,99].

14.5.2   HIGHLY COHERENT CONTINUUMS FOR OPTICAL COMMUNICA-TIONS

Highly coherent pulse generation has also taken advantage of the parabolic pulses' ability to resist the effects of optical wave breaking. Given the spectral properties of highly chirped parabolic pulses (see Fig. 14.3(b)), a continuum with low spectral ripple and high-energy spectral density in the central part can indeed be achieved [91, 94, 97]. The temporal coherence of a continuum obtained in the normal dispersion regime has been found superior to the case of a continuum generated in the anomalous regime in the presence of solitonic effects or modulational instability [12, 113]. This remarkable high flatness and stability can be of great interest for applications such as pulse shaping [114] or multiwavelength picosecond sources. Examples of low-noise multichannel sources running at 10 GHz and covering the whole C-band of optical telecommunications have therefore been demonstrated based on km-long erbium-doped fiber amplifiers [91] or on passive propagation in a highly nonlinear fiber (HNLF) of a parabolic pulse synthetized thanks to FBG-or LCOS-based technologies [94, 97] (Figure 14.9). Note that self-similar lasers having enhanced conherence properties have also stimulated interest in the generation of frequency combs [115].

14.5.3   ULTRAFAST ALL-OPTICAL SIGNAL PROCESSING

In the high-bit-rate return-to-zero telecommunication context, using parabolic pulses has also enabled noticeable improvements of existing optical pulse processing techniques based on the quasi-instantaneous Kerr response of HNLF. Various applications targeting the optical regeneration of an optical on-off-keyed signal impaired by amplitude or timing jitter have been proposed.

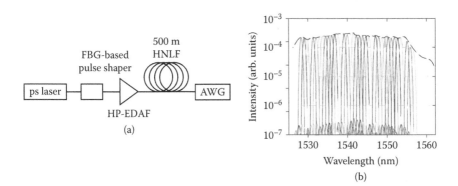

FIGURE 14.9: (a) Experimental setup of the generation of a flat continuum based on the spectral expansion of a parabolic pulse obtained through linear reshaping of a ps pulse train in an FBG. (b) Resulting continuum and channels obtained after slicing of this continuum. More details can be found in [94].

In an extension of the technique initially proposed by P.V. Mamyshev [116], distributed Raman amplification in a normally dispersive fiber followed by offset spectral filtering has been shown to improve the ability of a 2R regenerator to eliminate spurious noise pulses as well as to simultaneously reduce any fluctuations in the signal 1 bit level [117, 118].

Given their typical parabolic intensity profiles, the cross-phase modulation (XPM) induced by a parabolic pulse onto another pulse is also strictly parabolic. This feature is of great interest for several optical high-speed applications where electro-optic phase modulators are not available or are unable to provide a sufficient phase shift. Therefore, taking advantage of the perfectly linear chirp that can be generated through XPM induced by a parabolic pulse, it is possible to increase the level of tolerable temporal jitter that an optical retiming system can accommodate [119]. More recently, parabolic phase profiles induced by XPM have also been shown to be beneficial for the restoration of the intensity profile of pulses degraded by linear dispersive effects. Indeed, the time-domain optical Fourier transformation approach is based on imparting a highly linear chirp to the distorted data bits [120], and the use of parabolic pulses represents an efficient solution to generate the required chirp for picosecond pulse signals [121, 122]. We also anticipate that in the near future other applications, such as those based on time-lens effects, generation of Nyquist pulses [123], spectral self-imaging [124] or ultrashort pulse characterization [125, 126] will benefit from parabolic features.

14.5.4   SPECTRAL COMPRESSION

As a final application of parabolic pulses, we would like to mention their use in the context of spectral compression. If in textbooks SPM is generally associated with spectral broadening, it is, however, not always the case with chirped input pulses. Indeed, for a pulse having a chirp typical of the one brought by anomalous dispersion, the SPM-induced nonlinear phase shift results in a frequency downshift of the leading edge and an upshift of the trailing edge of the pulse, i.e., a positive chirp. Consequently, for a negatively chirped input pulse the linear and nonlinear phases cancel each other, leading to a redistribution of long and short wavelengths toward the center wavelength and consequently to spectral compression [127, 128]. This efficient process, which does not induce energy loss, has been reported for various parameters [129, 130]. However, for the usual intensity profiles, the resulting pulses are not compressed to the Fourier transform limit because the SPM-induced chirp does not exactly cancel in the input chirp for all times. Some residual side-lobes are therefore present in the wings of the compressed spectrum, as illustrated in Fig. 14.10.

An elegant solution to overcome this limitation is to take advantage of the highly linear chirp induced by the SPM of a pulse having a temporal parabolic intensity profile. Experimental demonstrations at telecommunication wavelengths using an HNLF and a LCOS-based shaping [98] or at the

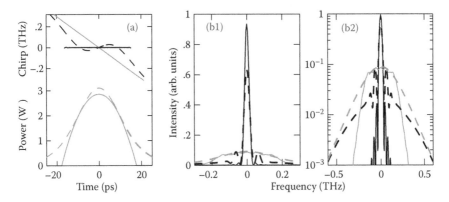

FIGURE 14.10: Nonlinear spectral compression of an hyperbolic secant (dashed) and a parabolic (solid) pulse. (a) Temporal chirp and intensity profiles. (b) Spectral intensity profiles plotted on linear (subplot 1) and logarithmic (subplot 2) scales. The pulses obtained after passive propagation in a 500-m fiber with parameters described in Section 14.2.2 are compared with the linearly chirped input pulses.

wavelengths around 1 μm using a photonic crystal fiber and an acousto-optics shaper [100] have confirmed all the benefits of preshaped pulses for achieving high-quality compressed spectra with very low substructures and enhanced brightness. Similar experiments were used to highlight the existence of a $\pi/2$ spectral phase shift at the point of spectral focusing [131], therefore constituting a spectral analog in nonlinear optics of the well-known Gouy phase shift existing in the spatial linear focusing of Gaussian beams.

14.6 CONCLUSION

Parabolic self-similar pulses represent a new class of ultrashort optical pulses that have opened new avenues of fundamental and applied research in nonlinear science. The success of self-similarity analysis in nonlinear fiber optics is motivating more general studies into the dynamics of guided wave propagation as well as related self-similar evolution in physical systems such as Bose–Einstein condensates. In parallel, a wide range of active and passive approaches have become available to generate parabolic pulses, which have enabled new techniques or improvements in existing techniques in diverse application fields such as ultrashort high-power pulse generation, highly coherent continuum sources, and optical signal processing for high-bit-rate telecommunications, among others.

ACKNOWLEDGMENTS

We would like to acknowledge the significant scientific contributions of K. Hammani, J. Fatome, B. Kibler, I. El Mansouri, S. Pitois, G. Millot (University of Burgundy), S. Wabnitz (University of Brescia), P. Dupriez, F. Parmigiani, L. Provost, P. Petropoulos, D.J. Richardson (Optoelectronics Research Centre), S. Turitsyn and B.G. Bale (Aston University), W. Lu (Zhejiang Forestry University), E.R. Andresen, H. Rigneault (Fresnel Institute) and J.M. Dudley (University of Franche-ComtÃ¯).

REFERENCES

1. L. F. Mollenauer, R.H. Stolen, J.P. Gordon, "Experimental observation of picosecond pulse narrowing and solitons in optical fibers," *Phys. Rev. Lett*, 45, 1095, 1980.

2. N. Akhmediev, E. Pelinovsky, "Discussion & database: Rogue waves – Towards a unifying concept?," *Eur. Phys. J. Spec. Top*, 185, 1, 2010.

3. B. Kibler, J. Fatome, C. Finot, "The Peregrine soliton in nonlinear fiber optics, *Nat. Phys.*, 6, 790, 2010.

4. A. Chabchoub, N.P. Hoffmann, N. Akhmediev, "Rogue wave observation in a water wave tank," *Phys. Rev. Lett.*, 106, 204502, 2011.

5. J.M. Dudley, C. Finot, G. Millot, "Self-similarity in ultrafast nonlinear optics. *Nat. Phys.*," 3, 597, 2007.

6. C. Finot, J.M. Dudley, B. Kibler, "Optical parabolic pulse generation and applications," *IEEE J. Quantum Electron.*, 45, 1482, 2009.

7. D. Anderson, M. Desaix, M. Karlson, "Wave-breaking-free pulses in nonlinear optical fibers," *J. Opt. Soc. Am. B*, 10, 1185, 1993.

8. M.E. Fermann, V.I. Kruglov, B.C. Thomsen, "Self-similar propagation and amplification of parabolic pulses in optical fibers," *Phys. Rev. Lett.*, 84, 6010, 2000.

9. S. Boscolo, S.K. Turitsyn, V.Y. Novokshenov, "Self-similar parabolic optical solitary waves," *Theor. Math. Phys.*, 133, 1647, 2002.

10. C. Finot, G. Millot, J.M. Dudley, "Asymptotic characteristics of parabolic similariton pulses in optical fiber amplifiers," *Opt. Lett.*, 29, 2533, 2004.

11. V.I. Kruglov, J.D. Harvey, "Asymptotically exact parabolic solutions of the generalized nonlinear Schrödinger equation with varying parameters," *J. Opt. Soc. Am. B*, 23, 2541, 2006.

12. J.M. Dudley, G. Genty, S. Coen, "Supercontinuum generation in photonic crystal fiber," *Rev. Modern Physics*, 78, 1135, 2006.

13. H. Nakatsuka, D. Grischkowsky, A.C. Balant, "Nonlinear picosecond-pulse propagation through optical fibers with positive group velocity dispersion," *Phys. Rev. Lett.*, 47, 910, 1981.

14. W.J. Tomlinson, R.H. Stolen, A.M. Johnson, "Optical wave-breaking of pulses in nonlinear optical fibers," *Opt. Lett.*, 10, 457, 1985.

15. D. Anderson, M. Desaix, M. Lisak, "Wave-breaking in nonlinear optical fibers," *J. Opt. Soc. Am. B*, 9, 1358, 1992.

16. C. Finot, B. Kibler, L. Provost, "Beneficial impact of wave-breaking on coherent continuum formation in normally dispersive nonlinear fibers" *J. Opt. Soc. Am. B*, 25, 1938, 2008.

17. C. Finot, L. Provost, P. Petropoulos, "Parabolic pulse generation through passive nonlinear pulse reshaping in a normally dispersive two segment fiber device," *Opt. Express*, 15, 852, 2007.

18. C. Finot, F. Parmigiani, P. Petropoulos, "Parabolic pulse evolution in normally dispersive fiber amplifiers preceding the similariton formation regime," *Opt. Express*, 14, 3161, 2006.

19. A. Zeytunyan, G. Yesayan, L. Mouradian, "Nonlinear-dispersive similariton of passive fiber," *J. Europ. Opt. Soc. Rap. Public*, 4, 09009, 2009.

20. S.O. Iakushev, O.V. Shulika, I.A. Sukhoivanov, "Passive nonlinear reshaping towards parabolic pulses in the steady-state regime in optical fibers," *Opt. Commun.*, 285, 4493, 2012.

21. I.A. Sukhoivanov, S.O. Iakushev, O.V. Shulika, "Femtosecond parabolic pulse shaping in normally dispersive optical fibers," *Opt. Express*, 21, 17769, 2013.

22. S. Boscolo, A.I. Latkin, S.K. Turitsyn, "Passive nonlinear pulse shaping in normally dispersive fiber systems" *IEEE J. Quantum Electron.*, 44, 1196, 2008.

23. H. Wang, A.I. Latkin, S. Boscolo, "Generation of triangular shaped optical pulses in normally dispersive fiber," *J. Opt.*, 12, 035205, 2010.

24. N. Verscheure, C. Finot, "Pulse doubling and wavelength conversion through triangular nonlinear pulse reshaping," *Electron. Lett.*, 47, 1194, 2011.

25. B. Burgoyne, N. Godbout, S. Lacroix, "Nonlinear pulse propagation in optical fibers using second order moments," *Opt. Express.*, 15, 10075, 2007.

26. C.K. Rosenberg, D. Anderson, M. Desaix, "Evolution of optical pulses towards wave breaking in highly nonlinear fibers," *Opt. Commun.*, 273, 272, 2007.

27. V.I. Kruglov, A.C. Peacock, J.D. Harvey, "Self-similar propagation of parabolic pulses in normal-dispersion fiber amplifiers," *J. Opt. Soc. Amer. B*, 19, 461, 2002.

28. S.A. Ponomarenko, G.P. Agrawal, "Interactions of chirped and chirp-free similaritons in optical fiber amplifiers," *Opt. Express*, 15, 2963, 2007.

29. V.I. Kruglov, D. Mechin, J.D. Harvey, "Self-similar solutions of the generalized Schrödinger equation with distributed coefficients," *Opt. Express*, 12, 6198, 2004.

30. D. Mechin, S.H. Im, V.I. Kruglov, "Experimental demonstration of similariton pulse compression in a comblike dispersion-decreasing fiber amplifier," *Opt. Lett.*, 31, 2106, 2006.

31. S. Chen, L. Yi, G. Dong-Sheng, "Self-similar evolutions of parabolic, Hermite-Gaussian, and hybrid optical pulses: Universality and diversity," *Phys. Rev. E*, 72, 016622, 2005.

32. S.A. Ponomarenko, G.P. Agrawal, "Do solitonlike self-similar waves exist in nonlinear optical media?" *Phys. Rev. Lett.*, 97, 013901, 2006.

33. L. Wu, J.F. Zhang, L. Li, "Similaritons in nonlinear optical systems," *Opt. Express*, 16, 6352, 2008.

34. V.I. Kruglov, A.C. Peacock, J.M. Dudley, "Self-similar propagation of high-power parabolic pulses in optical fiber amplifiers," *Opt. Lett.*, 25, 1753, 2000.

35. C. Finot, "Influence of the pumping configuration on the generation of optical similaritons in optical fibers," *Opt. Commun.*, 249, 553, 2005.

36. S. Wabnitz, "Analytical dynamics of parabolic pulses in nonlinear optical fiber amplifiers," *IEEE Photon. Technol. Lett.*, 19, 507, 2007.

37. Y. Ozeki, Y. Takushima, K. Taira, "Clean similariton generation from an initial pulse optimized by the backward propagation method." In: Conference on Lasers and Electro-Optics (CLEO US), 2004. OSA Trends in Optics and Photonics Series, CTuBB51113-1114.

38. K. Hammani, S. Boscolo, C. Finot, "Pulse transition to similaritons in normally dispersive fiber amplifiers," *J. Opt.* 15, 025202, 2013.

39. C. Billet, J.M. Dudley, N. Joly, "Intermediate asymptotic evolution and photonic bandgap fiber compression of optical similaritons around 1550 nm," *Opt. Express*, 13, 3236, 2005.

40. G. Chang, A. Galvanauskas, H.G. Winful, "Dependence of parabolic pulse amplification on stimulated Raman scattering and gain bandwith," *Opt. Lett.*, 29, 2647, 2004.

41. D.B.S. Soh, J. Nilsson, A.B. Grudinin, "Efficient femtosecond pulse generation using parabolic amplifier combined with a pulse compressor. I. Stimulated Raman-scattering effects," *J. Opt. Soc. Amer. B*, 23, 1, 2006.

42. A.C. Peacock, R.J. Kruhlak, J.D. Harvey, "Solitary pulse propagation in high gain optical fiber amplifiers with normal group velocity dispersion," *Opt. Commun.*, 206, 171, 2002.

43. D.B.S. Soh, J. Nilsson, A.B.Grudinin, "Efficient femtosecond pulse generation using a parabolic amplifier combined with a pulse compressor. II. Finite gain-bandwidth effect," *J. Opt. Soc. Am. B*, 23, 10, 2006.

44. D.N. Papadopoulos, Y. Zaouter, M. Hanna, "Generation of 63 fs 4.1 MW peak power pulses from a parabolic fiber amplifier operated beyond the gain bandwith limit," *Opt. Lett.*, 32, 2520, 2007.

45. J.F. Zhang, L. Wu, L. Li, "Self-similar parabolic pulses in optical fiber amplifiers with gain dispersion and gain saturation," *Phys. Rev. A*, 72, 055801, 2008.

46. B.G. Bale, S. Boscolo, "Impact of third order dispersion on the evolution of parabolic optical pulses," *J. Opt. A*, 12, 015202, 2010.

47. V.I. Kruglov, C. Aguergaray, J.D. Harvey, "Quasi-parabolic pulse propagation and breakup in fiber amplifiers with third-order dispersion," *Opt. Lett.*, 35, 3084, 2010.

48. V.I. Kruglov, C. Aguergaray, J.D. Harvey, "Propagation and breakup of pulses in fiber amplifiers and dispersion-decreasing fibers with third-order dispersion," *Phys. Rev. A*, 84, 023823, 2011.

49. B.G. Bale, S. Boscolo, K. Hammani, "Effects of fourth-order fiber dispersion on ultrashort parabolic optical pulses in the normal dispersion regime," *J. Opt. Soc. Am. B*, 28, 2059, 2011.

50. G. Chang, H.G. Winful, A. Galvanauskas, "Incoherent self-similarities of the coupled amplified nonlinear Schrödinger equations," *Phys. Rev. E*, 73, 016616, 2006.

51. V.I. Kruglov, D. Mechin, J.D. Harvey, "Parabolic and quasiparabolic two-component coupled propagating regimes in optical amplifiers," *Phys. Rev. A*, 77, 033846, 2008.

52. A.C. Peacock, V.I. Kruglov, B.C. Thomsen, "Generation and interaction of parabolic pulses in high gain fiber amplifiers and oscillators." In: AMERICA,

O. S. O., ed. Optical Fiber Communication Conference (OFC), March 17–22 2001, Anheim, California. Optical Society of America, 1–3.

53. C. Finot, G. Millot, "Collisions of optical similaritons," *Opt. Express*, 13, 7653, 2006.

54. C. Finot, G. Millot, "Interactions of optical similaritons," *Opt. Express*, 13, 5825, 2005.

55. L. Wu, J.F. Zhang, L. Li, "Similaritons interaction in nonlinear graded-index waveguide amplifiers," *Phys. Rev. A*, 78, 053807, 2008.

56. L. Wu, J.F. Zhang, C. Finot, "Propagation of dark similariton on the compact parabolic background in dispersion-managed optical fibers," *Opt. Express*, 17, 8278, 2009.

57. L. Li, X. Zhao, Z. Xu, "Dark solitons on an intense parabolic background in nonlinear waveguides," *Phys. Rev. A*, 78, 063833, 2008.

58. X. Zhao, L. Li, Z. Xu, "Dark-soliton stripes on a paraboloidal background in a bulk nonlinear medium," *Phys. Rev. A*, 79, 043827, 2009.

59. T. Hirooka, M. Nakazawa, "Parabolic pulse generation by use of a dispersion-decreasing fiber with normal group-velocity dispersion," *Opt. Lett.*, 29, 498, 2004.

60. B. Nagaraju, R.K. Varshney, G.P. Agrawal, "Parabolic pulse generation in a dispersion-decreasing solid-core photonic bandgap Bragg fiber," *Opt. Commun.*, 283, 2525, 2010.

61. N. Vukovic, N.G.R. Broderick, F. Poletti, "Parabolic pulse generation using tapered microstructured optical fibers," *Advances in Nonlinear Optics*, 2008, 1, 2008.

62. A. Latkin, S.K. Turitsyn, A. Sysoliatin, "On the theory of parabolic pulse generation in tapered fiber," *Opt. Lett.*, 32, 331, 2007.

63. C. Zhang, G. Zhao, A. Luo, "Third-order dispersion role on parabolic pulse propagation in dispersion-decreasing fiber with normal group-velocity dispersion," *Appl. Phys. B*, 94, 227, 2008.

64. S. Wabnitz, C. Finot, "Theory of parabolic pulse propagation in nonlinear dispersion decreasing optical fiber amplifiers," *J. Opt. Soc. Am. B*, 25, 614, 2008.

65. C. Finot, B. Barviau, G. Millot, "Parabolic pulse generation with active or passive dispersion decreasing optical fibers," *Opt. Express*, 15, 15824, 2007.

66. F. Ilday, J.R. Buckley, W.G. Clark, "Self-similar evolution of parabolic pulses in a laser," *Phys. Rev. Lett.*, 92, 213902, 2004.

67. B. Oktem, C. Ulgudur, F. Ilday, "Soliton-similariton fiber laser," *Nat. Photonics*, 4, 307, 2010.

68. W.H. Renninger, A. Chong, F.W. Wise, "Self-similar pulse evolution in an all-normal-dispersion laser," *Phys. Rev. A*, 82,021805(R), 2010.

69. B.G. Bale, S. Wabnitz, "Strong spectral filtering for a mode-locked similariton fiber laser," *Opt. Lett.*, 35, 2466, 2010.

70. C. Aguergaray, D. Mechin, V.I. Kruglov, "Experimental realization of a mode-locked parabolic Raman fiber oscillator," *Opt. Express*, 18, 8680, 2010.

71. F. W. Wise, A. Chong, W.H. Renninger, "High-energy femtosecond fiber lasers based on pulse propagation at normal dispersion," *Laser & Photonics Rev.*, 2, 58, 2008.

72. W. Renninger, A. Chong, F.W. Wise, "Pulse shaping and evolution in normal-dispersion mode-locked fiber lasers," *IEEE J. Sel. Top. Quantum Electron.*, 18, 389, 2012.

73. W.H. Renninger, A. Chong, F.W. Wise, "Amplifier similaritons in a dispersion-mapped fiber laser," *Opt. Express*, 19, 22496, 2011.

74. A. Chong, H. Liu, B. Nie, "Pulse generation without gain-bandwidth limitation in a laser with self-similar evolution," *Opt. Express*, 20, 14213, 2012.

75. S. Boscolo, S.K. Turitsyn, C. Finot, "Amplifier similariton fiber laser with nonlinear spectral compression," *Opt. Lett.*, 37, 4531, 2012.

76. A.C. Peacock, N. Healy, "Parabolic pulse generation in tapered silicon fibers," *Opt. Lett.*, 35, 1780, 2010.

77. S. Lavdas, J.B. Driscoll, H. Jiang, "Generation of parabolic similaritons in tapered silicon photonic wires: Comparison of pulse dynamics at telecom and mid-infrared wavelengths," *Opt. Lett.*, 38, 3953, 2013.

78. L. Wu, L. Li, J.F. Zhang, "Controllable generation and propagation of asymptotic parabolic optical waves in graded-index waveguide amplifiers," *Phys. Rev. A*, 78, 013838, 2008.

79. A.R. Davoyan, S.K. Turitsyn, Y.S. Kivshar, "Self-similar parabolic plasmonic beams," *Opt. Lett.*, 38, 428, 2008.

80. G. Chang, H.G. Winful, A. Galvanauskas, "Self-similar parabolic beam generation and propagation," *Phys. Rev. E*, 72, 016609, 2005.

81. S. Chen, J.M. Dudley, "Spatiotemporal nonlinear optical self-similarity in three dimensions," *Phys. Rev. Lett.*, 102, 233903, 2009.

82. S. Pitois, C. Finot, J. Fatome, "Generation of 20-Ghz picosecond pulse trains in the normal and anomalous dispersion regimes of optical fibers," *Opt. Commun.*, 260, 301, 2006.

83. C. Finot, J. Fatome, S. Pitois, "All-fibered high-quality low duty-cycle 20-GHz and 40-GHz picosecond pulse sources," *IEEE Photon. Technol. Lett.*, 19, 1711, 2007.

84. I. El Mansouri, J. Fatome, C. Finot, "All-fibered high-quality stable 20- and 40-GHz picosecond pulse generators for 160-Gb/s OTDM applications," *IEEE Photon. Technol. Lett.*, 23, 1487, 2011.

85. A. Plocky, A.A. Sysoliatin, A.I. Latkin, "Experiments on the generation of parabolic pulses in waveguides with length-varying normal chromatic dispersion," *JETP Letters*, 85, 319, 2007.

86. B. Kibler, C. Billet, P.A. Lacourt, "Parabolic pulse generation in comb-like profiled dispersion decreasing fiber," *Electron. Lett.*, 42, 965, 2006.

87. D. Ghosh, M. Basu, S. Sarkar, "Generation of self-similar parabolic pulses by designing normal dispersion decreasing fiber amplifier as well as its staircase substitutes," *Journal of Lightwave Technology*, 27, 3880, 2009.

88. P. Dupriez, C. Finot, A. Malinowski, "High-power, high repetition rate picosecond and femtosecond sources based on Yb-doped fiber amplification of VECSELS," *Opt. Express*, 14, 9611, 2006.

89. K. Tamura, M. Nakazawa, "Pulse compression by nonlinear pulse evolution with reduced optical wave breaking in erbium-doped fiber amplifiers," *Opt. Lett.*, 21, 68, 1996.

90. J.W. Nicholson, A. Yablon, P.S. Westbrook, "High power, single-mode, all-fiber source of femtosecond pulses at 1550 nm and its use in supercontinuum generation," *Opt. Express*, 12, 3025, 2004.

91. Y. Ozeki, Y. Takushima, K. Aiso, "High repetition-rate similariton generation in normal dispersion erbium-doped fiber amplifiers and its application to multi-wavelength light sources," *IEICE Trans. Electron.*, 88, 904, 2005.

92. C. Finot, G. Millot, C. Billet, "Experimental generation of parabolic pulses via Raman amplification in optical fiber," *Opt. Express* 11, 1547, 2003.

93. K. Hammani, C. Finot, S. Pitois, "Real time measurement of long parabolic optical similaritons," *Electron. Lett.*, 44, 1239, 2008.

94. F. Parmigiani, C. Finot, K. Mukasa, "Ultra-flat SPM-broadened spectra in a highly nonlinear fiber using parabolic pulses formed in a fiber Bragg grating," *Opt. Express*, 14, 7617, 2006.

95. T. Hirooka, M. Nakazawa, K. Okamoto, "Bright and dark 40 GHz parabolic pulse generation using a picosecond optical pulse train and an arrayed waveguide grating," *Opt. Lett.*, 33, 1102, 2008.

96. K. Kashiwagi, H. Ishizu, T. Kurokawa, "Fiber transmission characteristics of parabolic pulses generated by optical pulse synthesizer," *Japanese Journal of Applied Physics*, 50, 2501, 2011.

97. A.M. Clarke, D.G. Williams, M.A.F. Roelens, "Reconfigurable optical pulse generator employing a Fourier-domain programmable optical processor," *J. Lightw. Technol.*, 28, 97, 2010.

98. J. Fatome, B. Kibler, E.R. Andresen, "All-fiber spectral compression of picosecond pulses at telecommunication wavelength enhanced by amplitude shaping," *Appl. Opt.*, 51, 4547, 2012.

99. D. Krcmarik, R. Slavik, Y. Park, "Nonlinear pulse compression of picosecond parabolic-like pulses synthesized with a long period fiber grating filter," *Opt. Express*, 17, 7074, 2009.

100. E.R. Andresen, J.M. Dudley, C. Finot, "Transform-limited spectral compression by self-phase modulation of amplitude shaped pulses with negative chirp," *Opt. Lett.*, 36, 707, 2011.

101. D. Nguyen, M.U. Piracha, D. Mandridis, "Dynamic parabolic pulse generation using temporal shaping of wavelength to time mapped pulses," *Optics Express*, 19, 12305, 2011.

102. D. Strickland, G. Mourou, "Compression of amplified chirped optical pulses," *Opt. Commun.*, 56, 219, 1985.

103. J.P. Limpert, T. Schreiber, T. Clausnitzer, "High-power femtosecond Yb-doped fiber amplifier," *Opt. Express*, 10, 628, 2002.

104. T. Schreiber, C.K. Nielsen, B. Ortac, "Microjoule-level all-polarization-maintaining femtosecond fiber source," *Opt. Lett.*, 31, 574, 2006.

105. A. Malinowski, A. Piper, J.H.V. Price, "Ultrashort-pulse Yb3+ fiber-based laser and amplifier system producing 25 W average power," *Opt. Lett.*, 29, 2073, 2004.

106. L. Shah, Z. Liu, I. Hartl, "High energy femtosecond Yb cubicon fiber amplifier," *Opt. Express*, 13, 4717, 2005.

107. S. Zhou, L. Kuznetsova, A. Chong, "Compensation of nonlinear phase shifts with third-order dispersion in short-pulse fiber amplifiers," *Opt. Express*, 13, 4869, 2005.

108. S. Pierrot, F. Salin, "Amplification and compression of temporally shaped picosecond pulses in Yb-doped rod-type fibers," *Opt. Express*, 21, 20484, 2013.

109. S. Wang, B. Liu, C. Gu, "Self-similar evolution in a short fiber amplifier through nonlinear pulse preshaping," *Opt. Lett.*, 38, 296, 2013.

110. D.N. Schimpf, J. Limpert, A. Tĩijnnermann, "Controlling the influence of SPM in fiber-based chirped-pulse amplification systems by using an actively shaped parabolic spectrum," *Opt. Express* 15, 16945, 2007.

111. D. Nguyen, M.U. Piracha, P.J. Delfyett, "Transform-limited pulses for chirped-pulse amplification systems utilizing an active feedback pulse shaping technique enabling five time increase in peak power," *Opt. Lett.*, 37, 4913, 2012.

112. B. Kibler, C. Billet, P.A. Lacourt, "All-fiber source of 20-fs pulses at 1550 nm using two-stage linear-nonlinear compression of parabolic similaritons," *IEEE Photon. Technol. Lett.*, 18, 1831, 2006.

113. K. Hammani, B. Kibler, J. Fatome, "Nonlinear spectral shaping and optical rogue events in fiber-based systems," *Opt. Fiber. Technol.*, 18, 248, 2012.

114. C. Finot, G. Millot, "Synthesis of optical pulses by use of similaritons," *Opt. Express*, 12, 5104, 2004.

115. T.R. Schibli, I. Hartl, D.C. Yost, "Optical frequency comb with submillihertz linewidth and more than 10W average power," *Nat. Photonics*, 2, 355, 2008.

116. P.V. Mamyshev, "All-optical data regeneration based on self-phase modulation effect." In: European Conference on Optical Communication, ECOC'98, Institute of Electrical and Electronics Engineering, Madrid, Spain, 475, 1998.

117. C. Finot, S. Pitois, G. Millot, "Regenerative 40-Gb/s wavelength converter based on similariton generation," *Opt. Lett.*, 30, 1776, 2005.

118. C. Finot, J. Fatome, S. Pitois, "Active Mamyshev regenerator," *Optical Review*, 18, 257, 2011.

119. F. Parmigiani, P. Petropoulos, M. Ibsen, "Pulse retiming based on XPM using parabolic pulses formed in a fiber Bragg grating," *IEEE Photon. Technol. Lett.*, 18, 829, 2006.

120. T. Hirooka, M. Nakazawa, "Optical adaptative equalization of high-speed signals using time-domain optical Fourier transformation," *J. Lightw. Technol.*, 24, 2530, 2006.

121. T. Hirooka, M. Nakazawa, "All-optical 40-GHz time-domain fourier transformation using XPM with a dark parabolic pulse," *IEEE Photon. Technol. Lett.*, 20, 1869, 2008.

122. T. T. Ng, F. Parmigiani, M. Ibsen, "Compensation of linear distortions by using XPM with parabolic pulses as a time lens," *IEEE Photon. Technol. Lett.*, 20, 1097, 2008.

123. A. Vedadi, M.A. Shoaie, C.S. Brs, "Near-Nyquist optical pulse generation with fiber optical parametric amplification," *Opt. Express*, 20, 558, 2012.

124. R. Maram, J. Azaa,"Spectral self-imaging of time-periodic coherent frequency combs by parabolic cross-phase modulation," *Opt. Express*, 21, 28824, 2013.

125. T. Mansuryan, A. Zeytunyan, M. Kalashyan, "Parabolic temporal lensing and spectrotemporal imaging: a femtosecond optical oscilloscope," *J. Opt. Soc. Am. B*, 25, A101, 2008.

126. A. Zeytunyan, A. Muradyan, G. Yesayan, "Generation of broadband similaritons for complete characterization of femtosecond pulses," *Opt. Commun.*, 284, 3742, 2011.

127. M. Oberthaler, R.A. Hãũpfel, "Spectral narrowing of ultrashort laser pulses by self-phase modulation in optical fibers," *Appl. Phys. Lett.*, 63, 1017, 1993.

128. A.V. Zohrabian, L.K. Mouradian, "Compression of the spectrum of picosecond ultrashort pulses," *Quantum Electronics*, 25, 1076, 1995.

129. E.R. Andresen, J. Thogersen, S.R. Keiding,"Spectral compression of femtosecond pulses in photonic crystal fibers," *Opt. Lett.*, 30, 2025, 2005.

130. J.P. Limpert, N. Deguil-Robin, I. Manek-Hÿnninger, "High-power picosecond fiber amplifer based on nonlinear spectral compression," *Opt. Lett.*, 30, 714, 2005.

131. E.R. Andresen, C. Finot, D. Oron, "Spectral analog of the Gouy phase shift," *Phys. Rev. Lett.*, 110, 143902, 2013.

15 Brillouin scattering: From characterization to novel applications

Victor Lambin Iezzi, *Sébastien Loranger*, and *Raman Kashyap*

15.1 INTRODUCTION

The early 20th century was one of the richest eras of scientific discoveries as was the mid-19th century for industrial development with the invention of the steam engine. To name some of the most important discoveries, one can think of quantum mechanics, relativity, but also diverse scattering mechanisms of light from matter, such as Rayleigh, Raman and Brillouin scattering. Today, these mechanisms are used extensively in many areas, as a spectroscopic tool, a sensing tool, in laser processes, and more. This chapter is dedicated mostly to recent advances in Brillouin scattering, especially as a means to an end, i.e., as a tool to enable new devices, but also as a characterization technique to probe acousto-optic properties of new materials.

The Brillouin scattering mechanism was first demonstrated theoretically by Leon Brillouin in 1922, when he showed that energy (inelastic scattering) from an incident photon could be lost through a coupling to an acoustic phonon in a material, leading to a lower energy photon emission called a Stokes wave [1]. This three wave interaction is now called spontaneous Brillouin scattering. Although it was first studied in gases [2] and liquids [3] as well as in some bulk glasses [4], it took almost half a century to see major progress in this field, with the invention of the laser in 1960 by Theodore Maiman [5] and the development of low loss optical fiber first proposed by Kao and Hockham in 1966 [6], which led to the experimental discoveries of the stimulated process of Brillouin scattering (SBS). This novel medium of transmission, the optical fiber, opened up a whole new era not only for research in communication systems, but also of nonlinear effects such as SBS that could be easily generated in this long low loss medium.

SBS attracted interest in the scientific community rapidly due to its low threshold in long lengths of optical fiber and its counterpropagating nature which severely limited optical transmission systems, especially with narrow bandwidth sources. Over the years, many methods have been developed to suppress SBS efficiently, some of which will be discussed in Sec. 15.2. This section also presents the basic concepts of SBS, such as frequency shift, Brillouin gain spectrum, threshold, temperature and strain dependence, etc. Section 15.3 covers the use of SBS as a lasing tool, from multi Stokes comb generation to narrow linewidth lasers, distributed feedback lasers (DFB) or pulsed laser sources for telecommunications applications. Finally, with improvements in research and technology, developments in the area of SBS, not only as a way of mitigation, have become increasingly important, typically in two fields of study outlined in Sec. 15.4: temperature and strain sensing using SBS, and the basic concepts behind this effect as well as progress in distributed sensing with SBS are presented.

15.2   BASIC CONCEPTS

15.2.1   GENERALITIES

SBS is one phenomenon in inelastic scattering in the broad spectrum of energy transfer, just as Raman scattering is. On the other hand, Rayleigh scattering is an elastic mechanism which reemits scattered light at the same frequency as the incident wave. Raman and Brillouin scattering can either give energy to an incident optical photon (anti-Stokes photons are emitted) or take energy from the photon, leading to an optical loss mechanism (generation of phonons) via a Stokes process. The Stokes emission, governed by Maxwell–Boltzmann statistics, occurs more often than the anti-Stokes emission since excited states are less populated than the fundamental ground state in thermodynamic equilibrium. The difference between Raman and Brillouin scattering comes from the type of coupling. Raman will couple light with an optical phonon (excitation of vibrational modes of non-neutral electronic particles excited by the incident electric field) which is of high energy (~ 10 THz) while Brillouin couples an incident photon with an acoustic phonon (general pressure vibration of a material due to local atom instabilities with one another) which is of lower energy (~ 10 GHz). These concepts are illustrated in Fig. 15.1.

The scattering process does not imply an electronic transition, which is why it is represented by the dashed line in Fig. 15.1 and called a pseudo-electronic state. It is an instantaneous process that does not let the molecule (atom) store energy during scattering and reemission of energy by a radiative process.

15.2.2   SPONTANEOUS BRILLOUIN SCATTERING

Using Maxwell's equations, it is possible to describe the way an electromagnetic wave propagates in an homogeneous medium as

$$\nabla^2 \mathbf{E} - \frac{n^2}{c^2} \frac{\partial^2 \mathbf{E}}{\partial t^2} = 0, \tag{15.1}$$

where $\mathbf{E}$ is the electric field of the incident wave, c the speed of light, $n = \sqrt{\frac{\epsilon_r}{\epsilon_0}} = \sqrt{1 + \chi}$ is the refractive index of the medium and ϵ_r is the relative

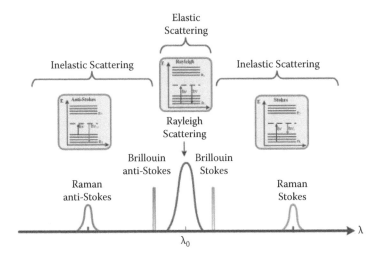

FIGURE 15.1: Schematic representation of the principal scattering mechanisms.

dielectric constant. The same propagation occurs in a homogeneous medium as in the air except that the speed of the wave becomes c/n. However, in an inhomogeneous medium, the density variation has to be taken into account, which adds a perturbation term, as shown in Eq. (15.2):

$$\nabla^2 \mathbf{E} - \frac{n^2}{c^2} \frac{\partial^2 \mathbf{E}}{\partial t^2} = \mu_0 \frac{\partial^2 \mathbf{P}}{\partial t^2}, \tag{15.2}$$

where μ_0 is the vacuum magnetic permittivity and $\mathbf{P}$ the polarization field showing the material response with an incident electromagnetic wave. $\mathbf{P}$, in a linear medium, can be expressed as an induced field as

$$\mathbf{P} = \epsilon_0 \chi \mathbf{E} = \mathbf{P}_0, \tag{15.3}$$

where χ is the medium dielectric susceptibility tensor and ϵ_0 the vacuum permittivity. To observe scattering, the medium must contain inhomogeneities. In this context, polarization will be modified according to temporal and spatial fluctuations of the dielectric susceptibility $\Delta\chi$ as presented below:

$$\mathbf{P}' = \epsilon_0 \chi \mathbf{E} + \epsilon_0 \Delta\chi \mathbf{E} = \mathbf{P}_0 + \mathbf{P}^D, \tag{15.4}$$

which enables us to rewrite the wave equation as ($\mathbf{P}^D = \mathbf{P}' - \mathbf{P}_0$):

$$\nabla^2 \mathbf{E} - \frac{n^2}{c^2} \frac{\partial^2 \mathbf{E}}{\partial t^2} = \mu_0 \frac{\partial^2 \mathbf{P}^D}{\partial t^2}, \tag{15.5}$$

in which the right hand term leads to scattering mechanisms since it corresponds to temporal and spatial variations of the medium's dielectric properties. In spontaneous Brillouin scattering, if we consider a compressible medium such as an optical fiber of volume V, which undergoes a change in volume, ΔV, depending on the pressure variation Δp, we can represent this as

$$\Delta p = -\kappa \frac{\Delta V}{V}, \tag{15.6}$$

where κ is the volume compressibility modulus of the material. The pressure variation will perturb the entire volume V according to classical mechanics, which is represented by the acoustic wave equation, similar to the wave equation, and described as follow:

$$\nabla^2 p - \frac{\rho}{\kappa} \frac{\partial^2 p}{\partial t^2} = 0, \tag{15.7}$$

where ρ is the medium density. It is then possible to find the velocity, V_A, of the acoustic wave in the medium as

$$V_A = \sqrt{\frac{\kappa}{\rho}}, \tag{15.8}$$

where $V_a \approx 5000$ m/s in silica (for instance, optical fiber) compared to the speed of sound in the air of ≈ 350 m/s since the density of silica is $\rho \approx 2210$ kg/m^3 and its compressibility modulus, $\kappa \approx 73.7$ GPa, is higher than in air [7]. It is possible to correlate this pressure wave to a similar wave of density variation $\Delta\rho$ which is dependent on forces $\mathbf{F}$ applied on a finite element of volume ΔV given by [8]

$$V_a^2 \nabla^2 \Delta\rho - \frac{\partial^2 \Delta\rho}{\partial t^2} - 2\Gamma_B \frac{\partial \Delta\rho}{\partial t} = \nabla \mathbf{F}, \tag{15.9}$$

where Γ_B is the damping coefficient characteristic of any resonant phenomenon. In an amorphous medium, the wave of pressure will generate a variation in density such as $\Delta p = \kappa \Delta\rho/\rho$, leading to a modification of the dielectric susceptibility $\Delta\chi$:

$$\Delta\chi = \frac{\gamma_e}{\epsilon_0} \frac{\Delta\rho}{\rho}, \tag{15.10}$$

where γ_e is the electrostrictive constant of the medium. Thus the variation of the dielectric susceptibility refers to the moving of the acoustic wave in the medium which modulates periodically the refractive index linked to $\Delta\chi$ just as if it were a Bragg grating moving at a certain speed V_A. Thereby, a diffraction of the optical wave (ω_p, k_p) by an acoustic wave Ω_B, k_a will generate another optical wave Ω_s, k_s when the phase matching conditions are respected, i.e., when there is conservation of momentum and energy as presented below:

$$\Omega_B = \omega_p - \omega_s; \mathbf{k}_a = \mathbf{k}_p - \mathbf{k}_s. \tag{15.11}$$

Using $|k_p| \approx |k_s|$ since these two waves are nearly identical (very close in frequency), it is possible to associate the pump wave vector k_p to the vector of the scattered Stokes wave k_s by the Bragg relationship

$$\Omega_B = V_a|\mathbf{k}_a| = 2V_a|\mathbf{k}_p|sin(\theta/2), \qquad (15.12)$$

where θ is the angle between the scattered wave k_s and the pump wave k_p. Since in an optical fiber the only two relevant directions are forward ($\theta = 0$) or backward ($\theta = \pi$), Brillouin scattering can only occur in the backward direction, with ($\theta = \pi$), forcing the acoustic wave vector (k_a) to be in the propagation axis of the fiber; thus the Brillouin frequency shift, corresponding to the coupling of the light with the acoustic phonon, can be written as

$$v_B = \frac{\Omega_B}{2\pi} = \frac{2n_{eff}V_a}{\lambda_p}, \qquad (15.13)$$

where v_B is the Brillouin frequency shift, n_{eff} is the effective refractive index of the mode in the optical fiber, V_A the velocity of the acoustic phonon and λ_p the pump wavelength. One of the ways of changing the frequency shift is by modifying the fiber's refractive index profile. In fact, changing the radius of the core or the cladding or the dopant type and concentration will modify the intrinsic Brillouin frequency shift, as shown in Fig. 15.2, where the SBS shift is compared for four different fibers.

It is also possible to determine the spectral broadening Δv_B of Brillouin scattering, which is linked to the phonon lifetime, dependent on the damping coefficient $T_B = \Gamma_B^{-1}$ and therefore

$$\Delta v_B = \frac{\Gamma_B}{2\pi} = \frac{1}{2\pi T_B}. \qquad (15.14)$$

Typically, SBS has a Brillouin gain spectrum bandwidth of around tens of MHz compared to stimulated Raman scattering (SRS) involving optical phonons, on the order of tens of THz. The difference is due to the damping effect of an acoustic wave related to the acoustic phonon lifetime, which is longer than that of an optical phonon. SBS can now be discussed in greater detail.

15.2.3   STIMULATED BRILLOUIN SCATTERING

Electrostriction is the tendency of a medium to compress under the temporally varying intensity of an electric field. When two optical waves at different frequencies beat together, they create an acoustic wave through electrostriction, as seen in Fig. 15.3.

First, a small fraction of the incident light is scattered by acoustic phonons with the right energy at the Brillouin frequency: this is the spontaneous scattering process. The scattered wave v_s (Stokes wave) interferes with the pump wave v_p since it is frequency shifted, creating a beat frequency at v_B to produce

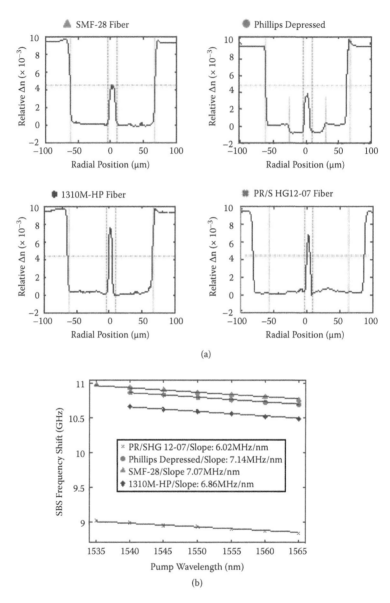

FIGURE 15.2: (a) Fiber profiles of four different fibers measured by the refracted near field technique. The dashed lines serve as a visual indication of
the differences in core radii, dopant concentrations and cladding radii compared to the reference, SMF-28 fiber. (b) SBS frequency shift as a function of
wavelength for these fibers, as described by Eq. (15.13) (from [9]).

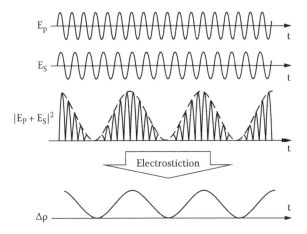

FIGURE 15.3: Creation of a pressure wave by electrostriction induced by the beating of two slightly frequency shifted optical waves (from [8]).

an acoustic wave of frequency v_B by electrostriction. This is the stimulated process. The generated acoustic wave amplifies the number of phonons at this energy, which in turn increases the intensity of the diffracted pump light to the Stokes wave, experiencing an optical gain by the stimulated process.

15.2.4 BRILLOUIN GAIN

After tedious algebraic manipulations, it is possible to arrive at the differential coupled mode equations of SBS [8] describing the interaction of the intensity of the pump wave with the Stokes wave:

$$
\begin{aligned}
\frac{\partial I_p}{\partial z} &= -\eta_p \, g_B(\nu) I_p I_S - \alpha I_p, \\
\frac{\partial I_S}{\partial z} &= -\eta_p \, g_B(\nu) I_p I_S + \alpha I_S,
\end{aligned}
\tag{15.15}
$$

where η_p is the polarization overlap coefficient, I_p is the intensity of the pump field, I_s is the intensity of the Stokes field, α is the loss in the propagating medium and $g_B(v)$ is the Brillouin gain spectrum (BGS) and has, in most cases, a Lorentzian shape, as described by Eq. (15.16) It can be seen from the previous equation that the Stokes wave experiences a gain while the pump is depleted due to energy transfer through electrostriction:

$$
g_B(\nu) = g_B \frac{(\Delta\nu_B/2)^2}{\Delta\nu^2 + (\Delta\nu_B/2)^2},
\tag{15.16}
$$

in which Eq. (15.14) is representing the full width at half maximum (FWHM) Δv_B of the BGS given by the previous equation where g_B is described as

$$g_B = \frac{4\pi\gamma_e^2}{n_{eff}\lambda_p^2\rho_0 c V_a \Delta v_B} = \frac{2\pi n_{eff}^7 P_{12}^2}{\lambda_p^2 c\rho_0 V_a \Delta v_B}, \qquad (15.17)$$

where γ_e is the electrostrictive constant, ρ_0 is the mean density of the medium, c is the light velocity and P_{12} is the longitudinal elasto-optic Pockels coefficient. The next equation gives the Brillouin gain coefficient G_A:

$$G_A = \exp\left(g_B P_0 L_{eff}/A_{eff} - \alpha L\right), \qquad (15.18)$$

where P_0 represents the initial pump power, L_{eff} the effective length , A_{eff} the effective area and αL the fiber loss coefficient, which will be discussed further in the next sub-section.

The most commonly known technique to measure the BGS is the so-called pump-probe technique first introduced by Niklès et al. in 1997 [10] in which they split the laser source into two arms, one as a pump and the other as a probe, via an optical coupler. This technique enables the user to avoid electrical locking of a seed laser to the pump laser, which is often difficult to do since it generally requires a high frequency loop to control the frequency drift around the center wavelength. This method is presented in Fig. 15.4.

Typically observed Lorentzian shaped BGS as described by Eq. (15.16) and caused by the phonon characteristic relaxation decay time of $\exp\left(-\Gamma_B t\right)$ is seen. For a standard telecomm SMF-28 fiber the BGS linewidth is 10–20 MHz [12]. Figure 15.5 shows five different BGS at various frequency shifts determined by the properties of the optical fiber used for the measurements:

Depending on the material properties, (type of dopant, concentration of dopant, profile of refractive index of core, etc.), fibers experience different frequency shifts, as seen in Fig. 15.5. However, the strongest influence on the BGS is the type of material used for the fiber. In fact, in Fig. 15.5(a), the ZBLAN fluoride fiber is known for its low phonon energy [14], which leads to a broadened Brillouin linewidth and smaller frequency shift. In a typical telecommunications fiber, it is possible to find two or more BGS peaks which can be caused by an inhomogeneous distribution of GeO_2 within the core [15] or by multiple supported acoustic modes in the fiber which will generate their own BGS since the modes have different velocities V_A [12,16,17]. One way to control the Brillouin frequency shift is to change the dopant concentration within the core of the fiber, which leads to a negative variation with concentration, as shown by Niklès et al. [10], and presented in Fig. 15.6.

Niklès et al. showed in 1997 that the Brillouin frequency shift varied as 94 MHz/wt% of the core dopant concentration. They also showed that there is a relationship of 1.4 MHz/wt% of the core dopant concentration and the linewidth of the BGS, which indicates an increase in the damping effect of the acoustic wave. To perform the experiments, they used a 1.32-μm wavelength

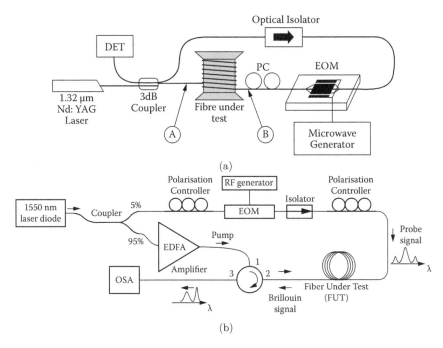

(a)

(b)

FIGURE 15.4: Experimental setup of the pump-probe technique for Brillouin gain spectrum measurement using an electro-optic modulator (EOM) to generate the probe wave at 1.32 μ m with the Nd:YAG laser to a detector (DET)(From [10]). (b) Modern experimental setup for measurement of gain at a telecommunication wavelength of 1.55 μ m in which a higher gain can be achieved by injecting a probe of lower intensity, amplified by a high power pumped Brillouin signal. (From [11]). EDFA : Erbium Doped Fiber Amplifier. PC : polarization controller. OSA : Optical Spectrum Analyzer." RF: Radio Frequency. YAG:Yttrium Aluminium Garnet.

source, which confirmed and specified the previously reported results from [15, 18].

The very first measurement of the Brillouin gain was made as early as 1950 in a bulk material [19]. In 1979, Heiman et al. [20] showed for the first time that Δv_B was dependent on the Brillouin frequency shift and varies as v_b^2, which leads to a dependence on the pump wavelength of λ_p^{-2} for Δv_B compared to the dependence of λ_p^{-1} for v_B, as seen in Eq. (15.13). However, the Brillouin gain peak value remains almost entirely independent of wavelength since the dependence of g_B in Eq. (15.17) cancels the narrowing of the Brillouin gain profile with increasing pump wavelength. Typical silica fiber parameters inserted in Eq. (15.17) lead to a gain of $3 - 5 \times 10^{-11}$ mW^{-1} [7], which is nearly three orders of magnitude larger than standard SRS gain. Fur-

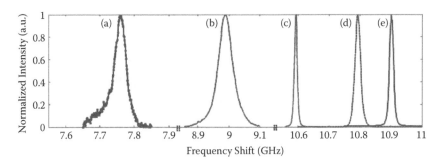

FIGURE 15.5: Brillouin gain spectra of five different fibers experiencing different frequency shifts at 1.55 μm. (a) Fluoride ZBLAN fiber, (b) PR/SHG 12-07 fiber (large cladding fiber), (c) 1310MH-HP fiber (small core fiber), (d) depressed cladding fiber, (e) SMF-28e fiber (from [9, 13]).

thermore, chalcogenide fibers have seen increasing interest lately since their Brillouin gain can reach 6.1×10^9 m/W [21, 22].

The main characteristic of BGS is its very narrow linewidth, which can act as a narrow band amplifier useful for telecom channel amplification in comparison to an EDFA or a Raman amplifier, which have a THz amplification bandwidth.

One last aspect regarding the Brillouin gain spectrum is its dependence on pump power. In fact, even though Eq. (15.16) refers to a Lorentzian spectral shape, Villafranca et al. in 2005 presented in their work that under different pump regimes, the BGS could have the shape of a narrower Gaussian form [23]. In fact, depending on the pump power, the spectrum will have a Lorentzian shape in the low gain regime and a Gaussian shape in the high gain limit, as can be seen in Fig. 15.7, where in (a), experimental results of such an effect are presented from [23], in (b) correlated results in ZBLAN fibers are observed and in (c), a hybrid case is analyzed, again from [23].

15.2.5   POWER THRESHOLD

The power threshold is a good indicator of how easily a nonlinear process in a medium can be observed. To measure the power threshold, one must know the effective mode area in the medium. The effective area indicates how the energy is spread spatially across a certain medium in which a mode is guided. Typically, for a step-index single-mode fiber, such as SMF-28 fiber from Corning, the beam shape is close to a Gaussian form and the effective mode area can be approximated as:

$$A_{eff} = \pi w_0^2. \tag{15.19}$$

However, if the fiber index profile is more complex, such as with a gradient index profile, it is necessary to perform an overlap integral to arrive at the

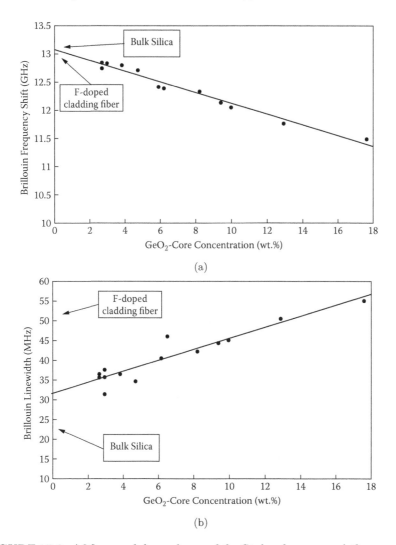

FIGURE 15.6: a) Measured dependence of the Stokes frequency shift v_B on the GeO_2–core concentration at 1.32μm. b) Measured dependence of the Brillouin linewidth Δv_B on the GeO_2–core concentration (From .[10])

proper effective area value. This effective area is directly linked to the power threshold, presented in Eq. (15.20) [7]:

$$P_{th} \approx \frac{21 A_{eff}}{g_B L_{eff}} \left[1 + \frac{\Delta \nu_L}{\Delta \nu_B} \right] \approx \frac{21 A_{eff}}{g_B L_{eff}}, \qquad (15.20)$$

where the number 21 is a numerical approximation reported for the first time by Smith in 1972 [24], who, at the time, accounted for a loss of 20 dB/km

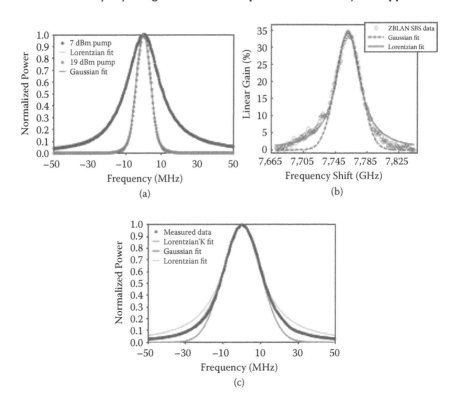

FIGURE 15.7: (a) SBS gain spectrum for a low pump (curve) and a high pump (curve) leading to, respectively, a Lorentzian and Gaussian shape [23]. (b) Same observation as (a) and (c) but for a ZBLAN fiber of 10.4 m instead of a standard silica fiber. The asymmetry comes from the fact that the frequency shift is closer to the pump wavelength than in a standard silica fiber [13]. (c) Intermediate gain region, in which the gain is a hybrid between Lorentzian and Gaussian with a fitting factor of k = 1.6 [23].

in an optical fiber in calculating this coefficient. Presently, this coefficient is closer to $17 - 19$, instead of 21. L_{eff} is the effective length and is given by Eq. (15.21). The term in brackets is a factor that shows the dependence on the linewidth of the pump and the Stokes wave (BGS) over the SBS power threshold. When the pump is very narrow, the term in brackets tends to unity, which means that it is easier to generate SBS when the pump linewidth is narrower than the BGS. If the pump bandwidth is larger than the BGS, as could be the case for a multimode laser or a single-mode pump laser whose phase varies quickly on a time scale shorter than the phonon lifetime P_B, the SBS threshold is increased [25–27]. The definition of the power threshold in the literature is not formally defined. However, all definitions have the same conceptual form, which corresponds to the moment when the Stokes wave acquires a certain fraction of energy from the pump. The reason why this fraction is not well defined is because of the fact that, close to this threshold, the growth of the Stokes wave is exponentially dependent on the pump power and thus the fraction required for the threshold is not a limiting factor in the calculation. The SBS power threshold is therefore dependent mostly on fiber parameters such as the effective length, defined as

$$L_{eff} = \frac{1 - e^{-\alpha L}}{\alpha}. \qquad (15.21)$$

The effective length is used is because it includes the loss term, really simplifying the coupled mode equations and all the calculations arising from them. For example, instead of measuring the variation of intensity of the waves in the fiber relative to the loss of the fiber on a certain length L, with the introduction of L_{eff}, it is possible to keep the intensity constant over its entire length (shorter than L), which eases calculations tremendously. The threshold can be easily approximated from Eq. (15.20), in which, if standard conditions are used, such as a fiber length of 10 km with $L_{eff} \approx 1/\alpha$ since $\alpha L \gg 1$, $A_{eff} = 50\mu m^2$, $\alpha = 0.2$ dB/km and $g_B = 5 \times (10^{-11})$ m/W, of $P_{th} \approx 4$ mW, a consequence of the long low loss fiber length. Figure 15.8a (a) presents the evolution of the transmitted saturated signal versus the reflected growth of the Brillouin–Stokes waves. In Fig. 15.8b (b), Stokes wave growth is presented as a function of the length of fiber.

15.2.6   BRILLOUIN STRAIN AND TEMPERATURE DEPENDENCE

Brillouin scattering is also determined by external parameters. Thus, when a mechanical strain is applied to an optical fiber, an increase in the frequency shift is observed, mainly due to the increase in the phonon velocity through the variation in the modulus of elasticity and Poisson coefficient [8]. This dependence on strain was first proposed by Tateda et al. in 1989 [28]. The same correlation can be made with temperature; the frequency shift is caused by the variation in the density and the increase in the modulus of elasticity proposed the same year by Culverhouse et al. [29]. Many groups have shown similar results over the years which all corroborate the observations, which Niklès et al. [10] investigated more thoroughly and concluded that it was of an order of magnitude around ~ 595 MHz/ %ϵ and 1.36 MHz/°C in a standard silica fiber, as shown in Fig. 15.9. These results are for silica based fibers, but they can vary significantly depending on the type of fiber used, shown as early as 1990 by Kurashima et al. [30].

The measurements presented in Fig. 15.9(a)–15.9(d) were made at a wavelength of 1.32 μm and show the dependence of SBS over a strain of $0 < \epsilon < 1\%$ and a temperature variation $-30°$ C $< T < 100°$ C. Niklès et al. [10] also showed a nonlinear decrease in the Brillouin linewidth with temperature which converges at a higher temperature to a constant value for all fiber types, as

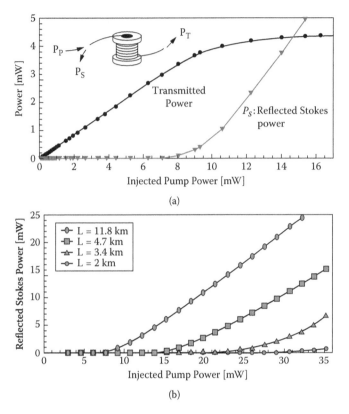

FIGURE 15.8: (a) Reflected Stokes power P_S as a function of the input power P_P and of the saturation of the transmitted power P_T. (b) Effect of the length over the power threshold, e.g., growth of the Stokes wave (from [8]).

shown in Fig. 15.9(d). As mentioned above, the maximum gain follows an increase with temperature proportional to Δv^{-1} while the linewidth diminishes accordingly, which means that the product $g_B \times \Delta v$ remains invariant to temperature and thus it demonstrates the independence on temperature of the electrostrictive constant responsible for SBS. In other words, the increase in the gain with temperature is only due to its spectral narrowing and thus lower phonon absorption at higher temperatures. As for the strain dependence on the Brillouin linewidth, after running experiments, Niklès et al. concluded that it has no effect at all, which could help decorrelate temperature from strain effect for point sensing devices using SBS, which will be further explained in Sec. 15.4.

Le Floch and Cambon [31] also analyzed the behavior of the Brillouin linewidth over this temperature range. Le Floch and Cambon found that there was a maxima in the linewidth caused by a maximum absorption at $110°K$,

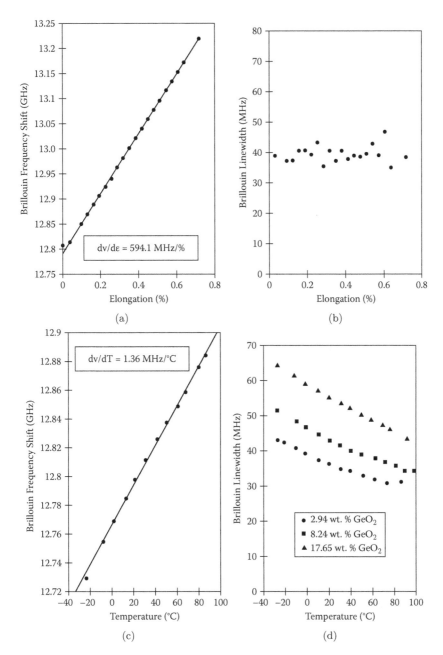

FIGURE 15.9: (a) SBS frequency shift variation induced by strain applied to an optical fiber. (b) Brillouin linewidth as a function of elongation. (c) SBS frequency shift caused by temperature variation. (d) Brillouin linewidth as a function of temperature for three different GeO_2 core dopant concentrations (from [10]).

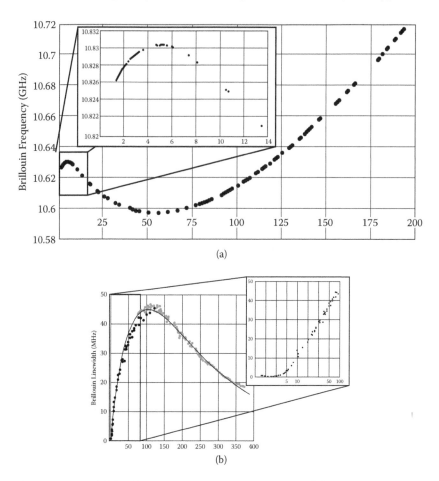

FIGURE 15.10: (a) Brillouin frequency shift variation. (b) Brillouin linewidth with temperature in a range from 1.4K to 370K (from [31]).

shown in Figure 15.10(b), as was previously proposed by Niklès et al. but so far not demonstrated. The change of color in Fig. 15.10 (a) and 15.10 (b) (from black to gray) represents a different approach used to perform temperature measurements. They explained this behavior by considering three states, as demonstrated by Rau et al. for oxide glasses [33]. At very low temperatures, the Jäckle perturbation theory [34] predominates; this explains that only a tunneling effect occurs, so that the phonon energy is very weak and thus the linewidth reaches a plateau of about 1 MHz for temperatures below 5K. However, at increased temperatures, the model is no longer valid due to the strong relaxation (more thermal agitation occurs); instead incoherent tunneling can explain this in which a competition between a two-level system governing the

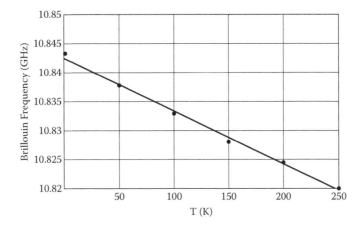

FIGURE 15.11: Measurement of the Brillouin frequency as a function of pressure (from [31]).

relaxation dynamics adequately is allowed and in which only one defect state can be excited at any time. This model fits the data up to 20K where at still higher temperatures, the behavior is governed by an Arrhenius-like relaxation function.

Finally, Le Floch and Cambon concluded their analysis by showing the influence of hydrostatic pressure on the SBS frequency shift — a very weak dependence of −91 kHz/bar for a range of 1 to 250 bars of pressure. They did their measurement in a pressure vessel in which the water pressure was slowly increased via a piston. The results obtained are presented in Fig. 15.11.

This ambient pressure had no effect on the Brillouin linewidth. However, what is of interest in this discovery is the use for oceanographic optical fiber bathythermographs that must be able to perform simultaneous measurement of temperature and depth. Since the linewidth is not dependant on pressure, it helps discriminate one factor from another.

15.2.7 BRILLOUIN MITIGATION

With the development of low loss optical fiber in the early 1970s and the increasing length of telecommunication fibers, researchers rapidly discovered that SBS was a real threat to the global network and much emphasis was placed on mitigating this effect. This type of research was really the first in-depth study of SBS since it was such an easy phenomenon to observe as the power threshold is so low, especially in the long fiber lengths being used. Thus this dramatically limited the transmitted input power [24], which saturates, reflecting the energy when the SBS threshold is reached.

15.2.8   PUMP LASER MODULATION

The first technique to mitigate the effects of SBS consisted of modulating the phase of the electric field of the incident wave in the optical fiber [35]. This technique is based on the beat frequency phenomenon which occurs when two electrical fields oscillate at discrete frequencies, but close to one another. The objective is to broaden the pump bandwidth by either using two narrowband single-mode lasers with a very small frequency difference or by allowing two longitudinal modes to oscillate within the gain bandwidth of the pump laser. In this way, the pump bandwidth Δv_L becomes broader, which increases the Brillouin power threshold, as seen in Eq. (15.20). In such a scenario, the term Δv_L is determined by the spacing between the two laser sources (respectively, the longitudinal modes). Thus a reduction in the Brillouin gain is observed since the phase variation of the beating optical waves occurs faster than the dephasing of the spontaneous acoustic wave, which limits the ability of this wave to generate SBS through electrostriction. Nowadays, many encoding techniques, such as amplitude-shift keying, frequency-shift keying or phase-shift keying by default use this pump modulation scheme and significantly limit the generation of SBS.

15.2.9   SEGMENTED FIBERS

Since Brillouin gain is dependent on the length of the propagation medium, one can use this fact to advantage. A multi-segmented configuration, using different types of fiber, can lead to an increase in the Brillouin threshold [36–38], thus increasing the fiber length in a fiber network, since each fiber experiences a different frequency shift, as shown in Fig. 15.12.

Thus, if BGS before and after some location L_1 along the whole length L, $(L_1 + L_2)$ do not overlap, the Stokes power $P_s(L_1)$ generated in segment $[L_1, L_2]$ is not further amplified, but attenuated in the segment $[0, L_1]$, while the Stokes power n section $[0, L_1]$ grows. The same behavior has been observed in nonuniform fibers [40], which leads to equivalent results, increasing the threshold power significantly.

15.2.10   MODULATION VIA TEMPERATURE OR STRAIN

By heating (or straining) sections of the fiber, the SBS frequency shift can be changed locally, which has the same effect as having segmented fiber sections. This principle is shown in Fig. 15.13. Research groups around the world have shown a Brillouin threshold increase with this technique, e.g., an increase of 8 dB by using a staircase ramp strain distribution in 580 m of dispersion shifted fiber [41].

Other works reported that changing the core radius along the fiber length increased the power threshold by 3.5 dB [42], while changing the dopant of the core of the fiber has shown a nonlinear increase in the power threshold by up to 10 dB [43, 44].

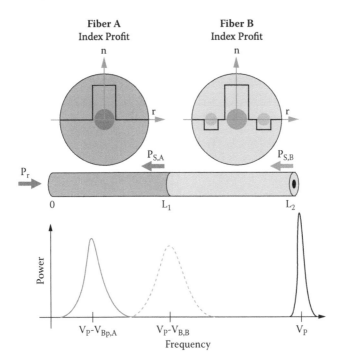

FIGURE 15.12: Schematic of the Brillouin frequency shift variation induced by the segmented portions of different optical fibers to increase the power threshold of SBS in a communication link. Adapted from [39]. The pump laser is at the for right, the fiber A Stokes wave is at the for left the fiber B Stokes wave is in the middle. Both Stokes waves are seeded from noise since their gain bandwidths do not overlap.

15.2.11   POLARIZATION AND SELF-INDUCED EFFECTS

When speaking of Brillouin scattering, it is important to mention the influence of polarization. Indeed, for SBS amplification to occur, the pump and Stokes wave must be co-polarized. If an initial Stokes signal to be amplified is orthogonal to a pump, there can be no mode coupling and therefore no amplification. This is why, in many systems, a polarization maintaining fiber is used to ensure the correct co-polarization between the pump and Stokes waves. However, SBS generated from noise occur with the correct polarization, but at the end of the fiber where SBS is seeded. The main issue of polarization arises when using fiber with residual fluctuating birefringence, which is the case for any non-polarization-maintaining fiber. In such a case, the SBS threshold is increased because of random changes in polarization, therefore momentarily losing gain at some points. It can be shown that this increase in threshold is 50% for a long length of random birefringence fluctu-

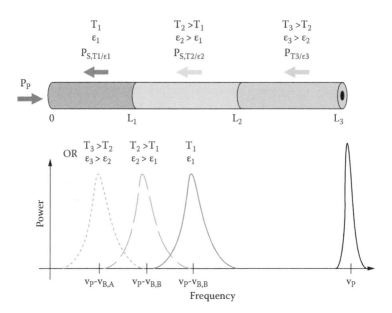

FIGURE 15.13: Schematic of the SBS frequency shift variation with temperature (or strain) producing the equivalent effect of having a segmented fiber link. The pump laser is presented in, while Stokes waves influenced by temperature are in (T1), (T2>T1) and (T3>T2). Adapted from [39].

ation [7]. This said, steady control of polarization is not necessary to generate SBS, but can increase the effect by 50% and is required when studying the stability of relaxation dynamics, where changes in polarization can affect the stability of the measurement. Many schemes presented in this chapter show the use of polarization maintaining (PM) fiber and polarization controllers, while some others do not require it.

SBS also has the interesting property of slightly changing the refractive index from Kramers–Kronig relations due to the presence of the gain curve. This refractive index change was only recently measured by Zhou et al. [45], but its effect has been observed indirectly by a resulting phenomenon: SBS induced slow light, observed for the first time by Okawachi et al. [46] and Song et al. [47, 48]. Indeed, from this refractive index change, an arriving signal within an SBS gain bandwidth (so counter-propagating a pump beam and with the correct frequency shift for resonance) will experience a delay in the ns range. Since then, many groups have improved the delay time of SBS slow light, which could be used for an optical buffer in all-optical communication systems.

As a final general remark on SBS, this nonlinear effect is not limited to single-mode fibers. Indeed, it was actually first observed in solid-state and gas materials [2–4], although its applications in such materials is so far quite limited, but recent progress has demonstrated it. However, another structure which might prove interesting in the future is its observation in multimode fibers. Indeed, there is increasing interest in using multimode fiber for telecommunications to encode information in various modes. It was recently shown that SBS can be generated in multimode fibers [11], and that the Brillouin frequency shift changes with the modes. Although this study did not show distinct and stable mode gain curves, it does bring the possibility of using SBS as a means to separate modal information if the effective indices of the modes are sufficiently separated, thus separating the gain curves.

15.3   BRILLOUIN FIBER LASER

Since SBS has such a low threshold, it has led over the years to a tremendous number of applications other than in the characterization of new materials. One such application is in the development of new lasers via SBS, which is the theme of this section. Due to the gain provided by SBS in optical fibers, it can be used as a laser when placed inside a cavity. The first SBS laser was reported in 1976 [49–51] and today it is more than ever a hot topic in the scientific community. The technique to fabricate such fiber lasers is either by using a ring cavity or a Fabry–Perot cavity, each with its own advantages.

15.3.1   CONTINUOUS WAVE SBS LASERS

With the present low loss fibers, it can be shown that the threshold in a ring cavity configuration can be written as

$$R_m \exp(g_B P_{th} L_{eff}/A_{eff} - \alpha L) = 1, \qquad (15.22)$$

where L is the ring cavity length, R_m is the fraction of Stokes power re-injected after each round trip and P_{th} is the threshold value of the pump power and in which the boundary condition $P_S(L) = R_m P_S(0)$ is taken into account. Using a ring cavity reduces the threshold factor of 21 in Eq. (15.20) to typically something between 0.1 and 1, depending on the value of R_m and the fiber loss, which is negligible for most of these types of lasers, since the fiber length is relatively short (¡ 100 m). With the improvement in fiber loss over the years with purer materials developed by the chemical vapor deposition process, the threshold required for such lasers was reduced to half a miliwatt in 1982 [52] compared to what was first reported (100 mW) in 1976 [49], a 200× improvement since the loss per round trip was reduced from 70% to only 3.5%.

The main difference between a Fabry–Perot (FP) cavity and the first proposed ring cavity setup comes from the fact that in a Fabry–Perot cavity, the pump and Stokes field propagate simultaneously in the forward and backward directions compared to the ring cavity, in which the pump and the Stokes

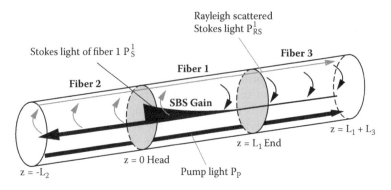

FIGURE 15.14: Cascaded fiber composed of three types of fibers with different Brillouin frequency shifts.

wave can be discriminated by their direction of propagation. The longitudinal modes allowed in both configurations will also be slightly different, as presented in Eq. (15.23)

$$\text{Ring Cavity: } \Delta\nu_L \quad = \quad c/n_{eff}L, \tag{15.23}$$

$$\text{FP Cavity: } \Delta\nu_L \quad = \quad c/2n_{eff}L, \tag{15.24}$$

where ν_L is the free spectral range (spacing between longitudinal modes), c is the speed of light, n_{eff} the effective refractive index of the transverse mode and L the length of the cavity. The factor of 2 comes from the fact that in a FP cavity, the length is considered for a full roundtrip (back and forth). To configure a stable CW laser with single longitudinal mode oscillation, $\Delta\nu_L$ must be such that $\Delta\nu_L > \Delta\nu_B$ where $\Delta\nu_B$ is the Brillouin gain bandwidth (typically 20 MHz). In contrast, if many longitudinal modes are present within the Brillouin bandwidth (($\Delta\nu_L \ll \Delta\nu_B$), i.e., L is long), it will typically require an active intracavity stabilisation to operate continuously, since the output can become periodic or even chaotic under certain conditions, as first reported in 1981 in [53]. For more information on the instability issues, the reader is directed to [53]–[56] or Section 9.5.2 in [7]. An interesting application of CW Brillouin lasers is in sensitive laser gyroscopes [54,57,58] which detect the frequency shift variation caused by a rotation ring cavity.

Recently, Pang et al. [59] demonstrated a novel technique to ensure a single frequency narrow linewidth laser in three cascaded fibers experiencing different Brillouin gains (as shown in Fig. 15.14). They showed that when the Brillouin gain of the middle fiber exceeds the effective loss of the Brillouin Stokes light, lasing with a narrow bandwidth 3 kHz is observed, as shown in Fig. 15.15. Rayleigh scattering, enhanced by the presence of the other two sections of fiber, acts as incoherent mirrors and is the contributing factor to this narrow bandwidth [60].

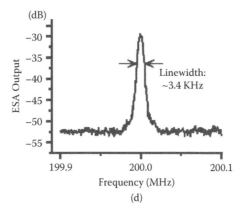

FIGURE 15.15: Measured power spectrum of Brillouin lasing by using self-heterodyne (from [60]). ESA: electrical spectrum analyzer

15.3.2 DFB BRILLOUIN LASER

In early 2012, the very first Brillouin distributed feedback fiber laser (DFB) was demonstrated by Abedin et al. [61], who used a 12.4 cm long DFB laser cavity made with a fiber Bragg grating incorporating a ϕ phase shift offset in the grating center, allowing different mirror reflectivities from either end. Such a DFB cavity has only one possible longitudinal mode, since the mirrors are relatively close to one another. The interest in a Brillouin DFB laser, compared to a standard single longitudinal mode Brillouin fiber laser, comes from its capability to easily operate at a single frequency without the need to match a mode within the BGS. This is attractive for optical microwave signal processing [62, 63] and amplitude/phase noise suppression of optical radiation [64, 65]. Since the Q-factor of DFB cavities can be extremely high, i.e., the number of effective roundtrips a Stokes wave has to accomplish to escape the cavity, such a DFB cavity is equivalent to a long length of fiber, and therefore has a low SBS threshold.

Typically, to fabricate a single frequency Brillouin laser, different approaches have been proposed, all with the common feature of using an external feedback system [52, 64, 65]. Abedin et al. [61] proposed, however, using the stimulated Brillouin gain inside the fiber Bragg grating (FBG). They obtained a DFB Brillouin laser in a 12.4 cm grating with a coupling coefficient of $\kappa \sim 90$ m^{-1} with a pump power as low as 30 mW. A conversion efficiency from the pump to the Stokes wave of 27% and an extinction ratio with the higher SBS orders of more than 20 dB were also observed. By controlling the offset of the ÏÅ-phase shift they could control the output direction (backward or forward) of the Stokes wave, which exhibited a very high signal to noise ratio of >50 dB. The experimental setup, the DFB grating and the laser output spectrum are represented in Fig. 15.16.

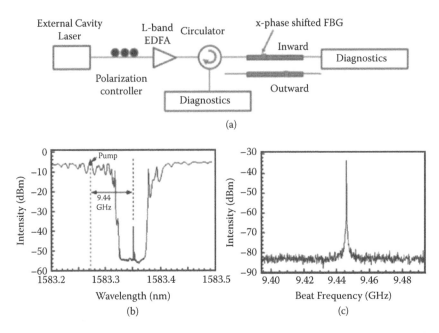

FIGURE 15.16: (a) Experimental setup of the first DFB Brillouin laser, (b) fiber Bragg grating spectrum and (c) output laser linewidth of the Stokes wave obtained (from [61]).

In 2013, Winful et al. [66] improved the analysis made on the first DFB Brillouin laser by Abedin et al. [61] by rigorously simulating a dynamic model of the phenomenon and by confirming these analytical results with experiments. They also provided a route toward a sub-mW threshold and a cm-long DFB Brillouin laser in chalcogenide fiber. This novel field of research in the Brillouin domain is promising on many levels, including the recent research on nanophotonic structures that leads to forward SBS [67, 68], which could help generate an integrated narrow linewidth DFB Brillouin on a silicon chip compatible with complementary metal-oxide-semiconductor (CMOS) signal processing technologies [66].

15.3.3   MULTI-STOKES ORDER COMB

Since the Fabry–Perot and ring configurations both lower the Brillouin threshold, it becomes easy to observe higher order Stokes waves generated through a cascaded SBS process in which each successive Stokes component pumps the next order Stokes wave after reaching the SBS threshold. As for the anti-Stokes components, these are assisted through a four wave mixing (FWM) process involving a pump and Stokes co-propagating waves.

The interest in developing very narrow spaced frequency combs via cascaded Brillouin–Stokes waves comes not only from the telecommunications industry but also from the sensing industry. Both need such a tool, one for multiplexing/demultiplexing transmitted communication channels and the other for optical characterization and measurement. In these applications, many parameters have to be taken into account such as the equalization of the peak intensities, tunability of the comb wavelengths, the equalization in the spacing of the frequencies of the comb as well as the number of peaks being generated.

Since SBS is a counter-propagating phenomenon, in a ring cavity configuration even and odd order Stokes waves propagate in different directions, while the pump follows the odd Stokes waves. The even or odd Stokes waves alone lead to a comb of equally spaced frequencies separated by $2\nu_B$. On the other hand, a Fabry–Perot configuration enables both even and odd Stokes waves to propagate in both directions, leading to a spacing in frequencies by the natural Brillouin frequency shift of ν_B.

One of the first demonstrations of using a Sagnac loop interferometer (SLI) in a ring cavity configuration was reported in 2004 in which the authors used the SLI in an end mirror configuration to allow each Stokes wave to propagate in the same direction, leading to a frequency spacing of ν_B. A tunable filter in the cavity also allowed the generation of the comb over a large bandwidth with the use of an EDFA [69]. The authors demonstrated 12 wavelength channels tuned over 14.5 nm, while in 2005 a tuning range of over 60 nm was demonstrated by the same group by the use of a combination of a dual fiber loop with optical circulators as mirrors to form a Fabry–Perot-like cavity with an 8.8 km SMF-28 fiber as the Brillouin gain medium. The schematic design is shown in Fig. 15.17. Song et al. in 2005 improved the design of a ring cavity configuration using an SLI (16 cm of PM fiber to create interference) as a mirror and showed a 120 line spectra all spaced by 11 GHz in a 5 km long SMF-28 fiber [70], shown in Fig. 15.18. Since the setup used self-generation of SBS by the amplified spontaneous emission (ASE) in the intra-cavity EDFA, the emission wavelengths are dependent on the intra-cavity modes linked to the maximum gain of the amplifier, and thus the emission wavelengths are not controlled and cannot be tuned properly.

Following the work done inter alia by Song et al. using a Sagnac loop reflector, in 2006 Fok and Shu [72] proposed a way to use an SLI also as a mirror, but to allow the tuning of the spacing between Stokes lines for comb generation, which obviated the use of a Brillouin pump (BP) to select the emission wavelengths. Thus, with a very similar cavity to that of Song et al., the BP was replaced by a polarization controller within the Sagnac loop, as shown schematically in Fig. 15.19.

Due to constructive/destructive interference, the Sagnac loop dictates the emission wavelength as well as the spacing between reflected modes generating a periodic output with frequency spacing in the comb as follows:

$$\Delta\lambda = \lambda^2/\Delta nL. \tag{15.25}$$

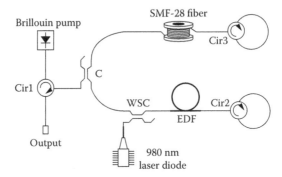

FIGURE 15.17: Experimental setup of a Brillouin fiber laser with two fiber loops acting as mirrors in a Fabry–Perot cavity (from [71]). WSC: wavelength selective coupler. EDF: erbium doped fiber.

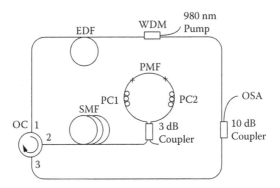

FIGURE 15.18: Self-seeded Brillouin fiber laser using ASE to start the cascaded SBS process and using an SLI to double the gain seen by the multiple Stokes waves (from [70]). WDM: wavelength division multiplexer. OSA: optical spectrum analyzer. PC: polarization controller.

This system uses the loop mirror filter's (LMF)'s wavelength dependent interference pattern, caused by the birefringence, to reflect only some equally spaced wavelength, where this frequency spacing is controlled by the birefringence in the loop. Although the SBS gain frequency shift is determined by the SMF characteristic, which is typically fixed at 0.088 nm, a multi-wavelength source with a mode spacing of $0.080nm$ was obtained with this configuration with the help of the LMF.

The authors extended the proof of concept as shown in Fig.15.19(c), in which they demonstrated a spacing of $0.16, 0.6, 0.8, 1.00$ and 1.20 nm using different lengths of polarization maintain fiber (PMF) of $50.0, 13.4, 10.0, 8.0$ and 6.7 m (from top to bottom Fig.15.19(c)), changing the interference pat-

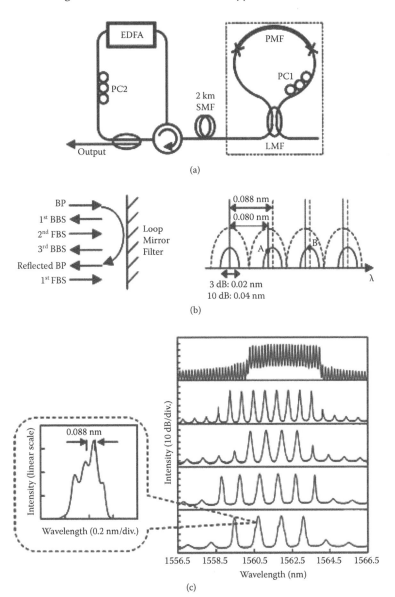

FIGURE 15.19: (a) Experimental setup of a Brillouin ring cavity for tuneably spaced multi-Stokes frequencies. LMF: birefringent loop mirror filter using 100 m of polarization maintaining fiber (PMF). (b) On the left illustration of the growth of the SBS components, while on the right, generation of multi-wavelength components with a 0.08 nm LMF spacing. Points A and B show two frequencies tuned by the LMF. (c) Optical spectra obtained with different frequency spaced combs of the LMF as a function of the EDFA output power (power decreasing from top to bottom). Adapted from [72].

terns by changing the length of the PMF. They also claimed that it is possible to achieve variable patterns with other techniques as well [73–75].

In 2008, Nasir et al. [76] demonstrated the utility of incorporating a bandpass filter in the ring cavity to enable an easier and more reproducibly tunable SBS frequency comb over the entire EDFA bandwidth. They used a bandpass filter to limit the gain region so the intra-cavity modes of the EDFA cannot win over SBS Stokes peaks through mode competition, since the gain is locked to the desired bandwidth in the EDFA gain spectrum. Thus they were able to demonstrate a frequency comb of 14 Stokes peaks equally spaced by 10.5 GHz over 32 nm in the C-band, and also equalized the power parameters from one spectrum to another.

Other configurations have also been developed over the years, some to generate even from odd Stokes waves. This is the case of Hambali et al. in 2011 [77], in which a combined SBS comb was generated with Raman gain in the C-band enabling a 22 GHz spacing with 16 even Stokes waves at the output. Shargh et al. [78] proposed another model using no physical mirrors to generate a very long comb of over 460 Stokes lines by combining Brillouin and Rayleigh scattering as virtual mirrors at the end of the cavity, as shown in Fig. 15.20. However, the use of virtual mirrors was not a new concept, as many before them had already demonstrated this principle [79–81].

In 2011, Pant et al. [82] showed for the first time, the possibility generating a Brillouin frequency comb in a few centimeters of a chalcogenide waveguide. Instead of using the typical few kilometers of fiber to generate the comb, they realized a 3 peak comb in 4 cm of a chalcogenide ridge waveguide, as shown in Fig. 15.21. The nonlinear coefficient of chalcogenide is over 2 orders of magnitude higher than silica. This demonstration opens up a whole new domain of research and is promising for compact and low cost telecommunication devices in the near future.

Finally, more recently, in 2013, Alimi et al. [83] showed a frequency comb of 150 Stokes lines in the C-band . They combined the FWM process with cascaded Brillouin–Stokes generation in a highly nonlinear fiber, just as Stepanov and Cowle [84] had shown in the past, generating anti-Stokes lines. They also demonstrated the tuneability of their device over 40 nm with a spacing of 0.078 nm. However, the amplitude of the channels was highly nonuniform.

In the light of these different results regarding multi-channel lasers, one can note that it is rather complicated to address all the desired conditions, especially for telecommunication applications, e.g., keeping the amplitude constant for each Stokes line, equal spacing with a high signal to noise ratio as well as tunable bandwidth over a wide span. However, great progress has been made and several commercial applications have resulted in multiple areas.

15.3.4 MODE-LOCKED BRILLOUIN LASER

Mode-locked lasers have been investigated extensively over several decades. The first report of a mode-locked laser in the literature was a ruby laser

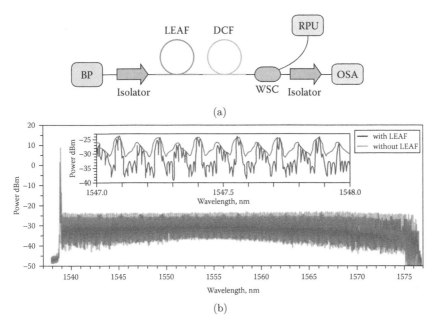

FIGURE 15.20: (a) Experimental setup for multiwavelength comb laser generation in a linear Brillouin–Raman fiber laser using large core fiber and dispersion compensating fiber as gain in the media. (b) Optimized flat amplitude spectrum of a linear Brillouin-Raman laser with 460 Brillouin–Stokes lines is shown with and without LEAF fiber. Employing LEAF fiber in this multiwavelength laser enhances the OSNR (optical signal-to-noise ratio), owing to its large core size, which suppresses the turbulent wave effect due to the interaction with self-lasing cavity modes along the optical fiber (from [78]).

made by Gürs and Müller [85, 86], as well as by Statz and Tang in an He-Ne laser [87]. However, the mode-locking process in a laser was first clearly identified by DiDomenico [88], Hargrove et al. [89] and Yariv [90]. The mode-locking principle allows multiple longitudinal modes to oscillate within a cavity together by forcing them to be in phase, thus creating a constructive interference for a short period of time, which enables a train of pulses at the output from a semi-transparent mirror. Mode-locking of a laser can be achieved in two ways: either passively or actively. The active technique typically uses an acousto-optic or electro-optic modulator inside a resonant cavity, which is driven by an RF (radio frequency) electrical signal to amplitude modulate the optical waves at the roundtrip frequency of the cavity. As for passive mode-locking, the use of a saturable absorber or Kerr lens [91–93] in the cavity is necessary. The saturable absorber enables the modes that are in phase to grow in the cavity, reaching a certain threshold at which they form a short pulse of high intensity light which saturates the absorption of a dye. For more information

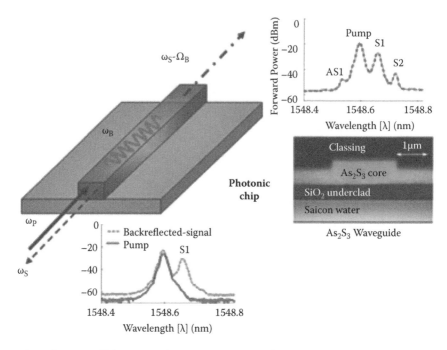

FIGURE 15.21: Schematic showing the principle of the on-chip cascaded SBS process where the backward and forward scattered spectrum (see inset) shows the generation of first and second Stokes shift due to the cascaded SBS process (from [82]).

on the complete theory of mode-lock lasers, the reader is directed to books published on the subject [94].

Mode-locking in the context of SBS, which is of interest in this chapter, was first reported by Hill et al. in 1976 [50] as a potential for generating pulses with 14 cascaded Stokes lines in a fiber Fabry–Perot resonator using a xenon laser pump operating at 535.5 nm. The active mode locking of the modes within the Brillouin gain bandwidth was first used in 1978 [95]. The authors used an acousto-optic modulator placed in the cavity leading to a train of pulses of 8 ns at a repetition rate 8 MHz limited by the length of the cavity.

In 1985, Bar-Joseph et al. showed that an energy transfer over time existed between the pump and the Stokes wave, leading to steady oscillations in time due to the relaxation characteristics of an SBS process [96]. They demonstrated oscillations depending on the length of their linear cavity of L = 50 m at 514.5 nm. An acousto-optic modulator was used to provide feedback and therefore lock the oscillating modes together, enabling a steady temporal behavior.

Another scheme was proposed in 1991 by Picholle et al. [97] as a way to generate quasi-solitonic pulses in a ring cavity using a cw argon pump laser to obtain pulses of 20 ns every 414 ns by periodically blocking the pump at a precise time using an acousto-optic modulator (AOM). The gating time of the AOM was set to be longer than the spontaneous acoustic damping time of the acoustic phonon present in the 83 m of the polarization maintaining fiber (PMF) used in the ring cavity so the Stokes wave could build up in energy and stabilize itself. Above a certain pump power threshold, the energy was high enough to create a Brillouin mirror, leading to a stabilized pulse train of 20 ns every 414 ns (15.22 (b)) while if the power was lower than the threshold, an unstable pulse regime occurred. Schematic representations of the experimental setup as well as experimental results of the stable regime are presented in Fig. 15.22.

Mode-locked SBS lasers were able to generate only nanosecond pulses since the gain bandwidth of the Brillouin line is 20 MHz; however, in 1993, Mirtchev and Minkovski [98] achieved a 200 ps pulse duration with an Nd:YAG laser of 300 ns pulse duration in a fibered ring cavity system. The way the system worked was quite simple; the length of the resonant cavity, e.g., the roundtrip time of a pulse, matched the pulse repetition rate of the pump laser, leading to a cumulative growth of the acoustic wave pumped by multiple pump pulses. Individually, these pulses are so short that they are unable to excite the acoustic wave significantly, but combined together within a time frame, they prevent the acoustic wave from decaying and thus build up in amplitude. The net result is a Stokes pulse generated through transient SBS with the passage of each pump pulse narrowing the output pulse compared to the pump.

In 2005, Botineau et al. [99] demonstrated that even though SBS could be used to passively lock spontaneous modes as Picholle et al. had shown, they demonstrated an active way to lock the modes together using a phase modulator directly inside the cavity, thereby decreasing the pulse width further. They explained that in a passive scheme, only the longitudinal modes close to the maximum of the gain bandwidth were allowed, limiting the pulse width as dictated by the Fourier transform. Thus, by externally modulating the phase modulator at a given frequency determined by the free spectral range of the cavity (which was composed of a PMF of 250 m with a pump of $1.319\ \mu m$), as seen in Eq. (15.16), they obtained pulses four times shorter than by locking it passively, since they were able to lock all the modes present within the gain bandwidth.

Botineau et al. also showed in another paper [100] that a simulation model based only on the intensity equations was not sufficient to explain the soliton-like self-pulsing phenomena of SBS due to the incoherence of the model. However, by combining the amplitude and the phase in a coherent interaction model and taking into account the damping effect of the acoustic wave, the simulations were in good agreement with the experimental observations. A PMF is typically used in such a soliton SBS laser since polarization dynamics in a Brillouin fiber ring laser may lead to unstable behavior, as was reported

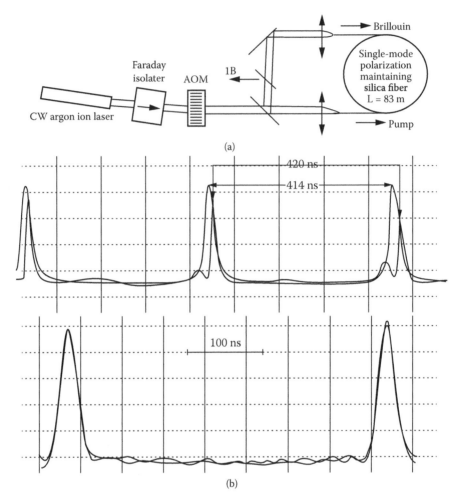

FIGURE 15.22: (a) Experimental setup for obtaining soliton-like pulses via a mode-locked SBS laser. (b) Results obtained in the stable regime in which very stable pulses are seen when the pump power is above the threshold (from [97]).

by Randoux and Zemmouri [101], and therefore the stability issues are more easily compensated. However, to confirm agreement of experimental results with simulations, Botineau et al. tried experiments with and without a PMF fiber. They determined the whole soliton localization domain as a function of the pump intensity Ip and the feedback R which corresponds to the intensity

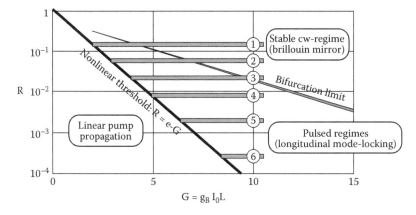

FIGURE 15.23: Bifurcation diagram of the Brillouin fiber ring laser with the experimentally explored areas (tinted areas) (from [100]).

reinjected into the cavity as

(1) Pulsed Brillouin soliton regime, for $R_{thres} < R < R_{pul}$,

(2) Oscillatory regime, for $R_{pul} < R < R_{crit}$, and

(3) Stationary Brillouin mirror regime, for $R > R_{crit}$. (15.26)

These conditions can also be represented schematically, as shown in Fig. 15.23. To observe the self-structuring of soliton pulses via a Brillouin–Stokes wave, the feedback must be below a critical feedback R_{crit} [102, 103] when the length of the cavity is large enough to admit many longitudinal modes under the Brillouin gain bandwidth $\left[L \gg (n\Delta\nu_B/c)^{-1}\right]$. Thus, for a given gain $G = g_B I_p L_{eff}$, the three conditions above can be obtained in which R_{thres} is determined by the relation $R = e^{-G}$, which is linear on the logarithmic scale in Fig. 15.23. This corresponds to the demarcation between the Stokes losses and the gain, with the pump propagating along the fiber in the linear regime. Above the bifurcation line, the laser works in a stationary regime, on a single longitudinal mode or as the so-called Brillouin mirror, in which the pump power is transferred to the Stokes wave. Finally, between these two regimes, the laser experiences pulsing on the order of 100 ns, which is the result of the competition between nonlinear interactions and acoustic damping [102].

Depending on different parameters, such as the gain, these pulses have a repetition rate close to the FSR of the cavity, but exhibit unstable pulse behavior that can be compensated for by the use of an acousto-optic modulator placed within the cavity to actively lock the modes together.

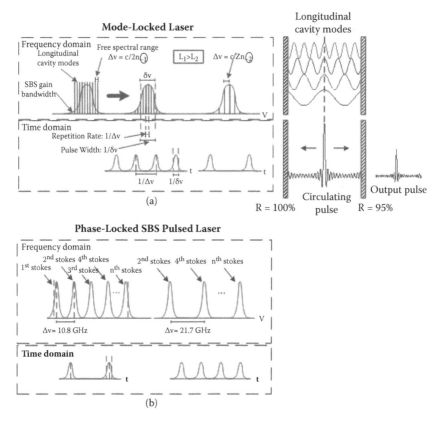

FIGURE 15.24: (a) Demonstration of the duality between frequency and time domain of a mode-lock laser. (b) The same duality demonstration, but for a self-phase-locked SBS laser.

15.3.5 SELF-PHASE-LOCKED BRILLOUIN LASER

Recently, in 2012 for the first time, Loranger et al. [104] demonstrated a novel technique using the locking in phase of a multi-Stokes Brillouin comb to generate pulses on the order of picoseconds at a tens of GHz repetition rate. Strictly speaking, since it is not a mode-locked laser, the best way to describe it is by presenting a comparison with a Brillouin mode-locked laser and the phase-locked SBS laser, as shown in Fig. 15.24.

The gain bandwidth in Fig. 15.24(a) is the Stokes linewidth of typically 10–20 MHz [12] in a standard optical fiber in which the modes within the FSR are equally spaced and spread within it and are dependent on the length L of the cavity. However, in Fig. 15.24(b) there are multiple Stokes waves all cascaded from the previous one and thus all are locked in phase to each other, since SBS is a parametric nonlinear effect, even though the phase drifts linearly between them. This leads to pulses in the time domain. By Fourier transform, the pulse duration is dependent on the total bandwidth of all Stokes waves, typically spaced apart by 10 GHz in the frequency domain, while the repetition rate is dictated by the spacing between each Stokes wave ($\sim$10 GHz). In comparison with the Brillouin mode-locked laser, which has a pulse duration inversely proportional to the Stokes linewidth, emitting pulses on the order of μs-ns, the phase-locked stimulated Brillouin scattering laser can tune its pulse width by simply modifying the number of Stokes waves being generated, leading to pulses on the order of a few ps. Since the pulses are not generated by locked modes, but by in-phase Stokes waves, it makes the mode-locking Brillouin lasers quite distinct from traditional mode locking.

For example, generating 9 Stokes orders leads to a span of 100 GHz in the frequency domain which corresponds to a 10 ps pulse duration at a repetition rate of 10 GHz. The temporal coherence of this laser will be given by the bandwidth of each Stokes wave in the Brillouin comb and leads to a coherence length of 10 m for a Stokes bandwidth of 10 MHz. Figure 15.25 presents experimental results of the phase-locked Brillouin laser in two configurations, one taking all Stokes orders together in the same direction by reflecting them with a Sagnac loop interferometer (SLI) mirror or a gold-tipped fiber (left-hand side), giving rise to a frequency spacing of 10.85 GHz. In the second one, only the odd or even Stokes waves (right-hand side) are selected, thus giving rise to a repetition frequency of 21.7 GHz. Results in the frequency and time domains are also shown in Fig. 15.25.

The main difference between the two configurations shown in Fig. 15.25 is the repetition rate while all the other parameters remain unchanged. Dianov et al. [105] suggested that pulses could not be generated using a Brillouin multi-Stokes comb fiber laser since the locking conditions, i.e., equidistant spectral components in a frequency comb, are not respected because of the Stokes wave dependence on the wavelength rather than frequency. However, Loranger et al. [104] argued that this frequency mismatch was less than 0.025% (2.8 MHz over 10.85 GHz in an SMF-28 fiber) which for less than 50 Stokes waves does not significantly perturb the locking of the laser.

Another interesting result of this novel laser also from the same paper [104] is its wide band tunability, as the output depends only on the seed wavelength. Tunability over the C-band in the frequency domain and pulse width in the time domain were also demonstrated [9]. Tuning the emission wavelength of the seed laser and a slight change in the grating filter center frequency to select the desired operating region is all that is required to tune the laser. This is much simpler to do than most commercially available ps laser sources. The spectra in frequency and time over the C-band region are shown in Fig. 15.26.

The authors also demonstrated that changing the pump power in the cavity controls the number of Stokes peaks and consequently the pulse width. They showed tunability from 15.4 ps to 3.65 ps by generating 5 to 28 Stokes waves shown in Fig. 15.27. Since the bandwidth increases in finite increments

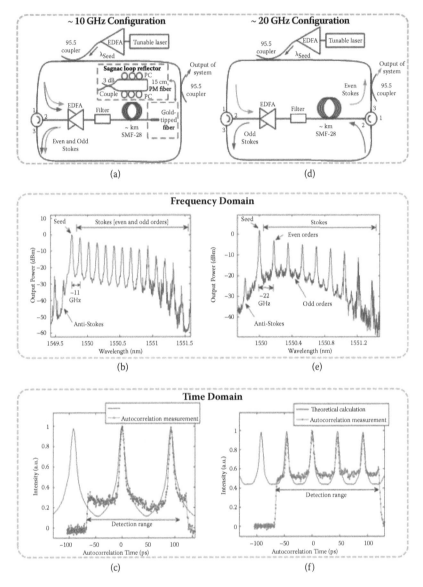

FIGURE 15.25: (a) Experimental setup for self-phase-locked SBS pulse generation for a unique cavity system using all Stokes waves together (left-hand side) or the even/odd Stokes waves separately (right-hand side). (b) Typical experimental results in the frequency domain of such a laser for both configurations and (c) the time domain results from an autocorrelation trace with the theoretical fast Fourier transform analysis for both configurations, using a 10 km SMF-28 fiber bundle as the gain medium. Adapted from [104].

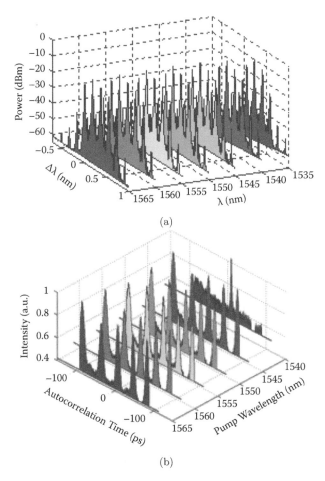

(a)

(b)

FIGURE 15.26: (a) Demonstration of the wavelength tunability of the self-phase-locked Brillouin laser. Trace obtained with an OSA for 1535 nm $< \lambda <$ 1565 nm operation. (b) Time measurement traces observed at an autocorrelator for wavelengths 1540 nm $< \lambda <$ 1565 nm (from [9]).

(multiple of Stokes waves generated), a criterion of -20 dB was fixed to count the number of peaks within the bandwidth.

Finally, a method was also proposed to modify the repetition rate of the laser, i.e., by changing the type of the in-cavity fiber and thus the Brillouin frequency shift. This technique allows the tuning of the repetition rate of the pulsed laser. So far, the main limitation of this laser is that the pulses are not fully depleted and a high CW or slow pulsating background is still present.

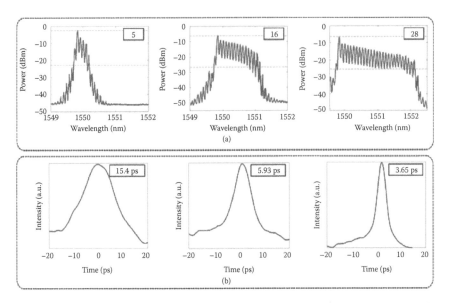

FIGURE 15.27: Demonstration of the effect of the spectral width on the pulse duration for the unique cavity configuration in which all the Stokes waves travel in the same direction. (a) From left to right, increasing number of Brillouin peaks generated and (b) decreasing pulse duration as a result of the number of Stokes waves generated.

After the demonstration by Loranger et al. [104] of the pulse emission of an SBS frequency comb by phase locking, Büttner et al. [55] proposed a theoretical model incorporating FWM to explain the temporal dynamics of such an SBS comb. To confirm the theoretical model, Büttner et al. created a comb in a short piece of chalcogenide fiber (higher nonlinear coefficient) in a Fabry–Perot configuration. They were able to acquire a very high extinction ratio for the emitted pulses by seeding the proper initial phase by using a quasi-CW seed laser. They obtained similar results as Loranger et al., presented in Fig. 15.28, but with a better control of the laser dynamics.

15.4 BRILLOUIN SCATTERING FOR SENSORS

In this section, we will look at the most developed application of BS in sensing. Such an application was introduced by Culverhouse et al. [29] in 1989. They showed for the first time that BS was temperature dependent. In the same year, Tateda et al. [28] showed the equivalent strain dependence of BS and proposed a point-sensing method in a fiber, which led to distributed sensing technologies. These developments therefore opened the field of all-fiber temperature and strain sensing, in which any type of fiber could be used as a sensor. The effect of temperature on BS is not trivial. Contrary to Raman

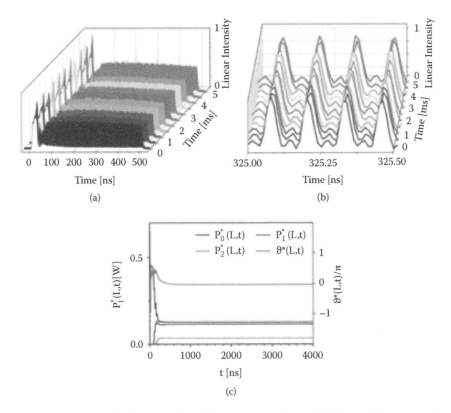

FIGURE 15.28: (a) Temporal stability over 500 ns, (b) inlet of the time domain showing the high extinction ratio of the pulses of 40 ps at an 8 GHz
repetition rate and (c) constant relative phase relationship (top line) of the
laser showing a high degree of stability over time (adapted from [55]).

scattering, where the temperature changes the number of phonons and therefore the ratio of anti-Stokes vs Stokes intensity, the Stokes intensity is not
used for temperature sensing in BS. It is rather the frequency shift, defined
as Eq. (15.13), that changes slightly with temperature and strain. This frequency shift depends on the phonon velocity, not on the number of phonons.
Therefore, in this case the change is rather induced by the material's acoustic
properties, such as density, hence the conjugated dependence on strain which
also affects the material density in the same way. The temperature and strain
dependence can be modeled by the following empirical formula:

$$\nu_B(T, \epsilon) = \frac{2n_{eff}V_A}{\lambda} + C_T(T - T_0) + C_\epsilon(\epsilon - \epsilon_0), \qquad (15.27)$$

where T_0 and the ϵ_0 are temperature and strain reference, respectively, for
the given n_{eff} and V_A, and C_T and C_ϵ are temperature and strain variation

coefficients measured experimentally. Typically, the dependence on temperature is in the range of $C_T \approx 1$ MHz/$^\circ C$ for silica fiber while the dependence of strain is on the order of $C_\epsilon \approx 594$ MHz/% ϵ. Considering that the SBS gain bandwidth is on the order of 10–20 MHz in optical fiber, this gives a rough resolution of 10-20° C. Fortunately, there are methods that allow this resolution to be improved by reducing the gain bandwidth. It is important here to note the "slight" variation in frequency shift with temperature and strain. Indeed, the temperature/strain variation is on the order of MHz, while the total Stokes shift is on the order of 10 GHz. Therefore, the temperature/strain variation cannot be detected by conventional optical spectroscopic techniques and requires interference and beating methods with a probe to measure the variation, similar to the techniques used to measure the BS gain bandwidth, as shown previously.

Considering that BS requires at least a few tens of meters in silica fiber to be detectable and its basic resolution is quite poor, it may be difficult to see the usefulness of this type of sensor, especially for temperature sensing where dozens of other technologies, miniaturizable and more precise, exist. However, the advantage of BS sensing does not lie in size or precision, but in another type of sensing where very few technologies have proven themselves: distributed sensing. Perfect distributed sensing requires a long single sensor which can measure a variable, such as temperature or strain, at any point within the sensor with a specified spatial resolution. Since SBS is an effect that occurs in a fiber, it can be easily implemented as a distributed sensor. The main challenge is to find a way to locally measure (within meters or tens of meters) the temperature and strain in a long section of fiber (several kilometers). Fortunately, this challenge was addressed the same year as the discovery of BS sensing by Tateda et al. [28]. Since then, it has been widely developed and is grouped into two methods: Brillouin optical time domain reflectometry (BOTDR) and Brillouin optical time domain analysis (BOTDA), which will be discussed in the next sections. There are also other variations of these techniques which are in the scope of this article.

15.4.1 BOTDR

The simplest method to implement Brillouin scattering in distributed sensing is to use a pulsed pump to generate a counter-propagating pulsed Stokes wave and to measure its time-of-flight hence the term time domain reflectometry [106]. By knowing the arrival time of the "reflected" Stokes wave, one can determine the position where that Stokes wave is generated with the pulse width determining the spatial resolution, as shown in Fig. 15.28(a). As mentioned before, temperature and strain will affect the Brillouin frequency shift; therefore, a local change in temperature or strain will locally shift the Brillouin gain spectrum by a few MHz, as shown in Fig. 15.28(b). In order to determine the local temperature, the reflected Stokes signal must be analyzed to measure this change in the spectrum. Different methods can be used, but

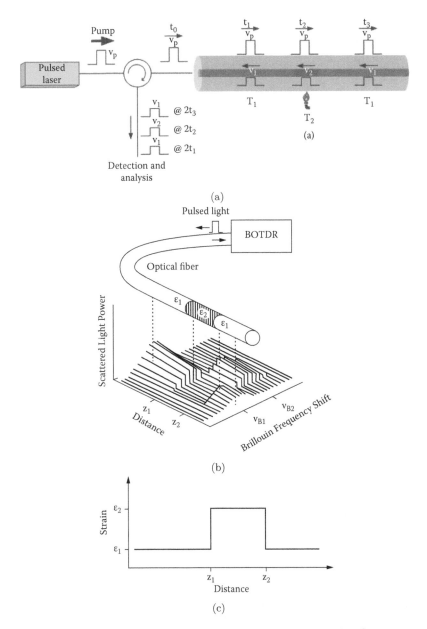

(a)

(b)

(c)

FIGURE 15.29: Examples of a BOTDR system [107].

since the frequency change in temperature is very small, interferometric methods are generally used, such as beating the signal with a reference. Examples of a proposed system are shown in Figs. 15.29 and 15.30.

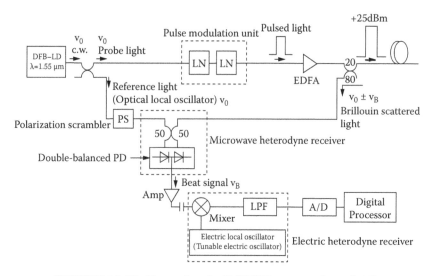

FIGURE 15.30: Example of a BOTDR system from [108].

Although simple in concept, the BOTDR signal is not simple to analyze, since two temporal characteristics of the signal must be determined simultaneously: the time of arrival (which gives the position) and the frequency shift (which gives the temperature/strain), generally obtained by measuring the beat frequency with a reference source. These complications are mostly in the electrical scheme, but they tend to limit the performance of BOTDR.

Despite these complications, BOTDR has the main advantage of a single physical system to send, receive and analyze optical signals only from one end of the fiber, thus making it a single ended technique. This implies that it requires only one fiber along the testing length and it is insensitive to accidental fiber rupture. Indeed, if the fiber is ruptured, the portion before this rupture will still function, therefore not only allowing the device to remain functional, but also allowing the operator to locate the rupture point.

15.4.2   BOTDA

The other technique for Brillouin distributed sensing is slightly more complex. In this method, we no longer send only one signal and retrieve the "reflected" Stokes wave, but send signals from both ends of the fiber to improve the possibilities for analysis, which is why "time domain reflectometry" is replaced by time domain analysis," a more general term. In a BOTDA scheme, the pump and a probe signal are sent from opposite ends of the test fiber, much as the pump-probe technique (Fig. 15.4) to determine the gain spectrum. To discriminate the signal spatially, either the pump or the probe can be pulsed

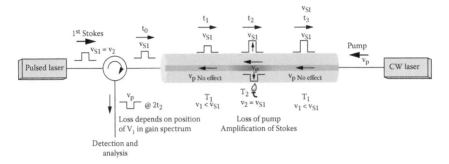

FIGURE 15.31: Schematic concept of the BOTDA distributed sensor.

(in most schemes, the probe is pulsed, while the pump is maintained in CW operation). The temperature is determined by the interaction between the pump and probe, and the position is determined by the time in which this interaction occurs, as shown in Fig. 15.30. Since we now have an extra degree of freedom with the control of the probe frequency shift from the pump, this gives more control of the performance of the distributed sensor compared to BOTDR.

For a higher temperature range, the probe frequency shift is scanned, while at each scan point several pulses are read to determine which parts of the fiber are at the corresponding temperature. The parts that are close to this corresponding temperature (within the gain spectrum) will allow interaction between pump and probe, while interactions are reduced in other regions. For higher temperature sensitivity, the intensity of this interaction can be measured, thus allowing the determination of the detuning of the probe from the peak of the gain spectrum.

When there is interaction between the pump and the probe, the probe (Stokes wave) is amplified, while the pump is depleted. To measure this interaction one can either measure the Stokes gain [28, 109–112] or the pump loss [113], which is the case shown in Fig. 15.31. The closer the probe is to the maximum of the BS gain spectrum at the corresponding temperature, the stronger is the interaction, therefore leading to a higher pump loss and higher Stokes gain. When the temperature is too far off (tens of degrees), the probe is completely out of the BS gain spectrum and there is no more interaction, thus the need to change the probe frequency to maintain the sensing capability.

Two examples are shown in Fig. 15.32. In the first (top of the figure), both the pump and probe are generated from the same laser (to avoid frequency jittering). The pump is pulsed, while the probe is modulated at a frequency corresponding to the BS frequency shift. In this case, the probe gain is measured. In the second case, pump and probes are generated separately, but frequency jitter is measured by beating both, which becomes the current fre-

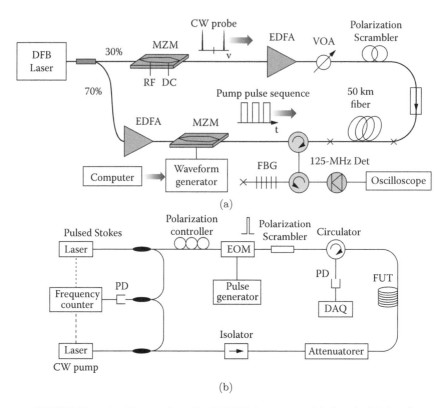

FIGURE 15.32: Examples of a BOTDA system. (a) [114], (b) [115].

quency shift reference. The probe is pulsed and it is the pump loss that is measured.

Because of the higher flexibility in the measurement, BOTDA offers a higher signal to noise ratio, higher resolution and longer fiber testing lengths than BOTDR. However, the physical measurement system needs to be connected to both ends of the test fiber. This requires twice the length of fiber for a testing length and implies that the system will suffer from total failure if the fiber is ruptured, with no way of knowing where that rupture occurred unless another technique was used simultaneously with it.

15.4.3 ADVANCES IN DISTRIBUTED SENSING

The first Brillouin distributed measurement of strain was performed in a 1.3 km long underwater cable [110], while the first distributed temperature measurement was observed in a Fabry–Perot system [29]. Later on, BOTDA was demonstrated over a distance of 1.2 km in a single-mode fiber with a resolution of 100 m and 3°C sensitivity [111]. In only 3 years, performance improved the reach to 22 km in distance with a resolution of 5 m and 1°C sensitivity [112]. Later, to reach longer distances, the probe gain measurement technique was put aside for a pump-loss measurement to be able to inject more power while avoiding gain saturation. Such a change allowed a record length of 50 km [113]. For BOTDR, the largest observed distance is 57 km, with a spatial resolution of 20 m and 3°C in temperature [116]. Recently, by combining BOTDR and Raman distributed sensing, a record length of 150 km was achieved with a resolution of 50 m and 5.2°C sensitivity [117]. By applying this same principle to BOTDA, a distance of 120 km was shown for a resolution of less than 2 m [118, 119].

Another variant of BOTDA was used by combining different fibers with different gain spectra one after the other. This gives a much better signal to noise ratio, but requires more post-analysis time to recover the measurement from the signal. With three segments of 25 km each of different fibers, a resolution of 1 m and 1° C/20 $\mu\epsilon$ sensitivity was achieved in 75 km of fiber [120].

The most recent upgrade uses two (or more) pulses with different pulse widths. The pulse width determines spatial resolution: the smaller the pulse, the smaller the resolution. However, a small pulse implies a small interaction length, therefore a weaker signal, which limits the total length of fiber used for distributed sensing. However, to overcome this limitation, Li et al. [115] proposed comparing the signal from two long pulses, but with different widths. With the correct analysis, the resolution is then given by the difference in pulse widths, which can be small (a few ns), while the interaction signal itself remains strong from the long pulses (50–100 ns), giving an exceptionally high signal to noise ratio for a very high resolution. With such a scheme, a resolution of 2 cm was demonstrated over a distance of 2 km. Other schemes have been proposed to pre-excite the acoustic wave with a weak CW probe-pump configuration to enable a higher signal during pulse probing, also giving rise to cm resolutions [121, 122].

15.4.4 STRAIN SENSING VS. TEMPERATURE SENSING

A big challenge in Brillouin sensing is the ability to separate strain measurements from effects of temperature. It is unfortunately impossible to separate the two with a single sensor. If one wants to measure temperature, the structure under test must not undergo strain variation, and vice versa. Sometimes, knowing both can be useful, since they can both indicate failure of a structure under test. If it is necessary to separate strain and temperature, two sensors are required to separate the variables:

$$S_1 = C_{T1}T + C_{\epsilon 1}\epsilon, \tag{15.28}$$

$$S_2 = aC_{T1}T + bC_{\epsilon 1}\epsilon. \tag{15.29}$$

Then the condition to extract the temperature and strain information is given by

$$a \neq b. \tag{15.30}$$

Two Brillouin fibers with different materials can be used in parallel. In such a case, the thermal expansion coefficient and the elastic coefficient must change from one material to the other. Another way to separate both variables is to combine two modalities, such as Raman scattering distributed sensing and Brillouin scattering distributed sensing. This will be successful so long as the dependence on temperature and strain is different from one modality to the other.

15.4.5 IMPROVING SENSITIVITY

Much effort has been placed in improving spatial resolution. However, due to physical restrictions, not much has been done to improve temperature and strain sensitivity. The only aspect that can physically be changed in terms of sensitivity is the fiber material, in which the phonon lifetime may change, therefore affecting the width of the gain spectrum. However, changing the material does not affect the sensitivity by more than a small factor and may require expensive speciality fibers, which are not suitable for long-distance distributed sensing. The other way to improve sensitivity is to improve the equipment and techniques to get a better signal to noise ratio, but that only allows one to approach the physical sensitivity limit.

A new idea was proposed by the authors of this article, Lambin Iezzi et al. [123] in 2014 to radically improve this physical limit by using higher order Stokes waves, i.e., with SBS. Indeed, distributed Brillouin sensing is normally operated below the SBS threshold to avoid gain saturation, and only the first Stokes wave is generated. However, it was recently shown that a higher order Stokes wave has its temperature/strain sensitivity of the frequency shift multiplied by the order of the Stokes wave in a cascaded frequency comb. The experimental setup is shown Fig. 15.33(a), in which a generator proposed by Loranger et al. [104] is built into a cross heterodyne detection scheme to beat together each pair of Stokes waves from two different generators. The authors showed the potential of this technique in improving temperature sensitivity, obtaining a 6× sensitivity increase with the sixth Stokes order, as shown in Fig. 15.33(b) and 15.33(c) in which the beat spectrum can be interpreted following the relationship:

$$\nu_{Beat} = \nu_{ref}(T_0, \epsilon_0) - \nu_{test}(T, \epsilon) = \sum_{n} \{C_{2n,T}\Delta T + C_{2n,\epsilon}\Delta\epsilon\}, \tag{15.31}$$

where $C_{2n,T}$ and $C_{2n,\epsilon}$ are the respective coefficients of temperature and strain dependence related to the nth Stokes order in which the second Stokes wave of the reference generator will beat with the second order Stokes wave of the test generator, and so on, being displayed at the output by a low frequency electrical spectrum analyzer.

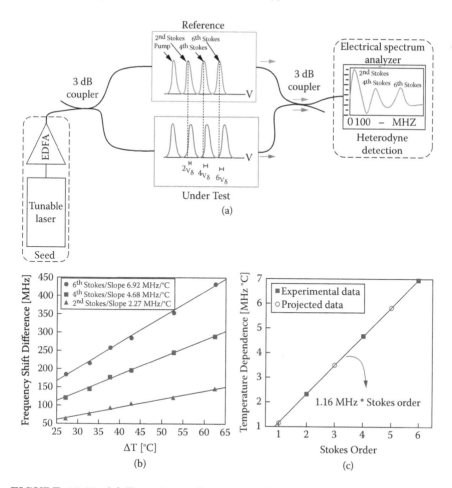

FIGURE 15.33: (a) Experimental setup to detect the increase in temperature sensitivity of a higher order stimulated Brillouin based sensor by cross-heterodyne detection, (b) demonstration of the increase in sensitivity with temperature with a higher order stimulated Stokes wave and (c) correlation of results obtained in (b) which correspond to a linear improvement in sensitivity with the Stokes order generated.

However, not demonstrated was this relationship of increased sensitivity with strain or in a distributed sensing system, which is the focus of continuing research [123].

15.4.6 OTHER TECHNIQUES AND LIMITATIONS

To better understand distributed BS sensing, it is important to analyze the competing optical technologies. Before fully distributed sensing was developed, quasi-distributed sensing was demonstrated with fiber Bragg gratings (FBG), which, placed at regular intervals, can measure temperature or strain and where this interval defines the spatial resolution. Although this scheme can give very good resolution (gratings of mm in length) and extreme sensitivity, it is rather expensive to make a high resolution FBG quasi-distributed sensor over multi-kilometer lengths, because thousands of FBG have to be fabricated. Therefore, this method is rather limited by cost.

Perfect distributed sensors have also been demonstrated with Rayleigh and Raman scattering in optical fibers. The Rayleigh method uses random defects in a fabricated optical fiber, which brings scattered reflected light back to the detector. The time-domain method yields a low resolution and sensing is done by the temperature dependence of Rayleigh loss, which depends on the material. A technique utilizing polarization can also be used to improve resolution and length, but it then also becomes dependent on strain. The more recent Fourier-domain technique yields a much better resolution, capable of actually seeing the Rayleigh defects, but requires high frequency and high cost acquisition equipment. As with an FBG, when the temperature or strain changes, the spacing between defects also changes, enabling the observer to correlate a local temperature or strain change in the observed Rayleigh scattering pattern. The systems require high spatial resolution (mm range) to be able to see these defects, which greatly limits the total length which can be measured (a few tens to a few hundreds of meters). The actual spatial resolution of sensing is longer since it must correlate as many defects as possible to get an acceptably accurate temperature measurement.

The Raman scattering distributed sensing method uses the high sensitivity of anti-Stokes intensity with temperature, relative to the Stokes shift. Therefore, by comparing anti-Stokes and Stokes intensity along the fiber using the time-domain reflectometry method, a temperature measurement can be made. However, as it is well known that an anti-Stokes wave has a very small intensity, a very intense pulse is required to generate an acceptable signal. This low signal therefore limits the length of the sensor and requires more sensitive detection and higher energy pump pulses, thus higher costs.

The characteristics of each technology have been summarized in Tables 15.1–15.5. As can be seen, the main advantage of Brillouin sensing is in the resolution/distance ratio. Indeed, Brillouin distributed sensing is the only optical technology which can reach hundreds of km in length while maintaining a resolution of meters (or less) as well as a low cost. However, it is not the most appropriate method for short lengths, where other lower cost (Rayleigh) or higher sensitivity (FBG, Raman) technologies can easily outpace a Brillouin sensor.

This said, sensing using Brillouin has had growing interest for its advantages for long-distance distributed sensing. Indeed, Brillouin is a rather simple technique to implement considering what is gained in precision and distance, although to push the limits of distributed sensing, complex schemes

TABLE 15.1

Method vs Variables. (after Bao Sensors 2012, 12, 8601–8639; doi:10.3390/s120708601)

Method	Variables
FBG Quasi-Distributed	Temperature and strain
Rayleigh Time Domain	Temperature and/or strain
Rayleigh Fourier Domain	Temperature and strain
Raman Time Domain	Temperature
Brillouin Time Domain	Temperature and strain

TABLE 15.2

Method vs Sensing Resolution (after Bao Sensors 2012, 12, 8601–8639; doi:10.3390/s120708601)

Method	Sensing Resolution
FBG Quasi-Distributed	Small (10–100 mK)
Rayleigh Time Domain	Large (2–5°C)
Rayleigh Fourier Domain	Large (3–5°C)
Raman Time Domain	Large (2–5°C)
Brillouin Time Domain	Medium ($<$ 1 to 2°C)

TABLE 15.3

Method vs Spatial Resolution (after Bao Sensors 2012, 12, 8601–8639; doi:10.3390/s120708601)

Method	Spatial Resolution
FBG Quasi-Distributed	Very Small (mm)
Rayleigh Time Domain	Large (10 m)
Rayleigh Fourier Domain	Small (cm)
Raman Time Domain	Medium-Large (1–10 m)
Brillouin Time Domain	Small-Large (cm-m)

TABLE 15.4

Method vs Distance (after Bao Sensors 2012, 12, 8601–8639; doi:10.3390/s120708601)

Method	Distance
FBG Quasi-Distributed	Limited by high cost
Rayleigh Time Domain	Medium-Short (0.1–1 km)
Rayleigh Fourier Domain	Short (100 m)
Raman Time Domain	Medium (10 km)
Brillouin Time Domain	Medium-Long (10–100 km)

TABLE 15.5

Method vs Complexity/Cost (after Bao Sensors 2012, 12, 8601–8639; doi:10.3390/s120708601)

Method	Complexity/Cost
FBG Quasi-Distributed	Low to High (Signal Analysis)
Rayleigh Time Domain	Simple
Rayleigh Fourier Domain	High (High Frequency)
Raman Time Domain	High (Low Frequency)
Brillouin Time Domain	Average (Signal Analysis)

of BOTDA have been demonstrated. Their sensitivity to temperature and to strain and the possibility of applying them with the most simple fiber makes Brillouin sensing adaptable to many industrial applications.

15.5   CONCLUSION

This chapter was dedicated to the analysis of Brillouin scattering in optical fiber as a characterization technique for new materials, but mostly as an all-fiber laser and a sensing tool for various purposes and applications. Some basics of SBS were described and applications schemes were discussed.

General remarks were introduced to understand the physics behind this third order nonlinear scattering mechanism in Sec. 15.2, while more specific results on Brillouin lasers were seen in Sec. 15.3. More specifically, CW operation via SBS as a narrow linewidth laser source, even DFB Brillouin lasers, were shown with an even narrower linewidth, to multi-wavelength emissions for telecommunication applications (WDM schemes) without forgetting the pulsed regime described in this section either by mode-locking or by the self-phase locking technique. The latter method is a new way to generate high repetition rate ps pulses with high tunability. Finally, Sec. 15.4. showed the evolution over the last two decades in the sensing area using Brillouin scattering. Two techniques were introduced, Brillouin optical time domain reflectometry and Brillouin optical time domain analysis, both of which allow sensing of a variety of temperatures and/or strains with high resolution over multiple kilometers of fibers, but with some difference in their conceptual schemes. These techniques were compared with other existing technologies, such as FBG quasi-distributed sensing and Raman and Rayleigh scattering sensing, showing that Brillouin scattering is a very useful tool for sensing since it has no true weaknesses compared to other techniques.

REFERENCES

1. L. Brillouin, "Diffusion de la lumière et des rayons X par un corps transparent homogène. Influence de l'agitation thermique," *Ann. Phys.(Paris)*, 17, 88, 1922.
2. E. Hagenlocker and W. Rado, "Stimulated Brillouin and Raman scattering in gases," *Applied Physics Letters*, 7, 236, 1965.
3. R. D. Mountain, "Thermal relaxation and Brillouin scattering in liquids," *J. Res. Natl. Bur. Stand. A*, 70, 207, 1966.
4. J. Budin, A. Donzel, J. Ernest, and J. Raffy, "Stimulated Brillouin scattering in glasses," *Electronics Letters*, 3, 31, 1967.
5. T. H. Maiman, "Stimulated optical radiation in ruby," 1960.
6. K. Kao and G. A. Hockham, "Dielectric-fiber surface waveguides for optical frequencies," in *Proceedings of the Institution of Electrical Engineers*, 1151, 1966.
7. G. P. Agrawal, *Nonlinear Fiber Optics*, Springer, 2000.
8. A. Küng, L'émission laser par diffusion Brillouin stimulée dans les fibers optiques," 1997.
9. V. L. Iezzi, S. Loranger, and R. Kashyap, "Wavelength tunable GHz repetition rate picosecond pulse generator using an SBS frequency comb," in SPIE OPTO, 86231S, 2013.
10. M. Niklès, L. Thevenaz, and P. A. Robert, "Brillouin gain spectrum characterization in single-mode optical fibers," *Journal of Lightwave Technology*, 15, 1842, 1997.
11. V. Lambin Iezzi, S. Loranger, A. Harhira, R. Kashyap, M. Saad, A. Gomes et al., "Stimulated Brillouin scattering in multi-mode fiber for sensing applications," in *Fiber and Optical Passive Components (WFOPC)*, 2011 7th Workshop, 2011.
12. A. Yeniay, J.-M. Delavaux, and J. Toulouse, "Spontaneous and stimulated Brillouin scattering gain spectra in optical fibers," *Journal of Lightwave Technology*, 20, 1425, 2002.
13. V. Lambin-Iezzi, S. Loranger, M. Saad, and R. Kashyap, "Stimulated Brillouin scattering in SM ZBLAN fiber," *Journal of Non-Crystalline Solids*, 359, 65, 2013.
14. C. E. Mungan, M. I. Buchwald, and G. L. Mills, "All-solid-state optical coolers: History, status, and potential," Defense Technical Information Center Document, 2007.
15. R. W. Tkach, A. R. Chraplyvy, and R. Derosier, "Spontaneous Brillouin scattering for single-mode optical-fiber characterisation," *Electronics Letters*, 22, 1011, 1986.

16. Y. Koyamada, S. Sato, S. Nakamura, H. Sotobayashi, and W. Chujo, "Simulating and designing Brillouin gain spectrum in single-mode fibers," *Journal of Lightwave Technology*, 22, 631, 2004.

17. J. H. Lee, T. Tanemura, K. Kikuchi, T. Nagashima, T. Hasegawa, S. Ohara et al., "Experimental comparison of a Kerr nonlinearity figure of merit including the stimulated Brillouin scattering threshold for state-of-the-art nonlinear optical fibers," *Optics Letters*, 30, 1698, 2005.

18. N. Shibata, K. Okamoto, and Y. Azuma, "Longitudinal acoustic modes and Brillouin-gain spectra for GeO_2-doped-core single-mode fibers," *JOSA B*, 6, 1167, 1989.

19. A. Melloni, M. Frasca, A. Garavaglia, A. Tonini, and M. Martinelli, "Direct measurement of electrostriction in optical fibers," *Optics Letters*, 23, 691, 1998.

20. D. Heiman, D. Hamilton, and R. Hellwarth, "Brillouin scattering measurements on optical glasses," *Physical Review B*, 19, 6583, 1979.

21. K. S. Abedin, "Observation of strong stimulated Brillouin scattering in single-mode As_2Se_3 chalcogenide fiber," *Optics Express*, 13, 10266, 2005.

22. K. Y. Song, K. S. Abedin, K. Hotate, M. González Herráez, and L. Thávenaz, "Highly efficient Brillouin slow and fast light using As_2Se_3 chalcogenide fiber," *Optics Express*, 14, 5860, 2006.

23. A. Villafranca, J. Lázaro, and I. Garcés, "Stimulated Brillouin scattering gain profile characterization by interaction between two narrow-linewidth optical sources," *Optics Express*, 13, 7336, 2005.

24. R. G. Smith, "Optical power handling capacity of low loss optical fibers as determined by stimulated Raman and Brillouin scattering," *Applied Optics*, 11, 2489, 1972.

25. G. C. Valley, "A review of stimulated Brillouin scattering excited with a broadband pump laser," *IEEE Journal of Quantum Electronics*, 22, 704, 1986.

26. P. Narum, M. D. Skeldon, and R. W. Boyd, "Effect of laser mode structure on stimulated Brillouin scattering," *IEEE Journal of Quantum Electronics*, 22, 2161, 1986.

27. E. Lichtman and A. Friesem, "Stimulated Brillouin scattering excited by a multimode laser in single-mode optical fibers," *Optics communications*, 64, 544, 1987.

28. M. Tateda, T. Horiguchi, T. Kurashima, and K. Ishihara, "First measurement of strain distribution along field-installed optical fibers using Brillouin spectroscopy,"*Journal of Lightwave Technology*, 8, 1269, 1990.

29. D. Culverhouse, F. Farahi, C. Pannell, and D. Jackson, "Potential of stimulated Brillouin scattering as sensing mechanism for distributed temperature sensors," *Electronics Letters*, 25, 913, 1989.

30. T. Kurashima, T. Horiguchi, and M. Tateda, "Thermal effects of Brillouin gain spectra in single-mode fibers," *IEEE Photonics Technology Letters*, 2, 718, 1990.

31. S. Le Floch and P. Cambon, "Study of Brillouin gain spectrum in standard single-mode optical fiber at low temperatures (1.4-370 K) and high hydrostatic pressures (1-250 bars)," *Optics Communications*, 219, 395, 2003.

32. R. Vacher, E. Courtens, and M. Foret, "Anharmonic versus relaxational sound damping in glasses. II. Vitreous silica," *Physical Review B*, 72, 214205, 2005.

33. S. Rau, C. Enss, S. Hunklinger, P. Neu, and A. Würger, "Acoustic properties of oxide glasses at low temperatures," *Physical Review B*, 52, 7179, 1995.

34. J. Jäckle, "On the ultrasonic attenuation in glasses at low temperatures," *Zeitschrift fÅijr Physik*, 257, 212, 1972.

35. D. Cotter, "Suppression of stimulated Brillouin scattering during transmission of high-power narrowband laser light in monomode fiber," *Electronics Letters*, 18, 638, 1982.

36. X. Mao, R. Tkach, A. Chraplyvy, R. Jopson, and R. Derosier, "Stimulated Brillouin threshold dependence on fiber type and uniformity," *IEEE Photonics Technology Letters*, 4, 66, 1992.

37. C. De Oliveira, C.-K. Jen, A. Z. Shang, and C. Saravanos, "Stimulated Brillouin scattering in cascaded fibers of different Brillouin frequency shifts," in *Fibers*, 92, 1993.

38. K. Shiraki, M. Ohashi, and M. Tateda, "SBS threshold of a fiber with a Brillouin frequency shift distribution,"*Journal of Lightwave Technology*, 14, 50, 1996.

39. A. Kobyakov, M. Sauer, and D. Chowdhury, "Stimulated Brillouin scattering in optical fibers," *Advances in Optics and Photonics*, 2, 1, 2010.

40. A. Kobyakov, M. Sauer, and J. E. Hurley, "SBS threshold of segmented fibers," in *Optical Fiber Communication Conference*, 2005, p. OME5.

41. J. Boggio, J. Marconi, and H. Fragnito, "Experimental and numerical investigation of the SBS-threshold increase in an optical fiber by applying strain distributions," *Journal of Lightwave Technology*, 23, 3808, 2005.

42. K. Shiraki, M. Ohashi, and M. Tateda, "Suppression of stimulated Brillouin scattering in a fiber by changing the core radius," *Electronics Letters*, 31, 668, 1995.

43. M. Ohashi and M. Tateda, "Design of strain-free-fiber with nonuniform dopant concentration for stimulated Brillouin scattering suppression," *Journal of Lightwave Technology*, 11, 1941, 1993.

44. K. Shiraki, M. Ohashi, and M. Tateda, "Performance of strain-free stimulated Brillouin scattering suppression fiber," *Journal of Lightwave Technology*, 14, 549, 1996.

45. D.-P. Zhou, L. Chen, and X. Bao, "Stimulated Brillouin scattering induced refractive index changes measurement in an optical fiber," in *OFS2012 22nd International Conference on Optical Fiber Sensor*, 2012, pp. 8421CC-8421CC-4.

46. Y. Okawachi, M. S. Bigelow, J. E. Sharping, Z. Zhu, A. Schweinsberg, D. J. Gauthier et al., "Tunable all-optical delays via Brillouin slow light in an optical fiber," *Physical Review Letters*, 94, 153902, 2005.

47. K. Y. Song, M. G. Herráez, and L. Thévenaz, "Observation of pulse delaying and advancement in optical fibers using stimulated Brillouin scattering," *Optics Express*, 13, 82, 2005.

48. M. Gonzalez-Herraez, K.-Y. Song, and L. Thévenaz, "Optically controlled slow and fast light in optical fibers using stimulated Brillouin scattering," *Applied Physics Letters*, 87, 081113, 2005.

49. K. Hill, B. Kawasaki, and D. Johnson, "CW Brillouin laser," *Applied Physics Letters*, 28, 608, 1976.

50. K. Hill, D. Johnson, and B. Kawasaki, "cw generation of multiple Stokes and anti-Stokes Brillouin-shifted frequencies," *Applied Physics Letters*, 29, 185, 1976.

51. K. Hill, Y. Fujii, D. C. Johnson, and B. Kawasaki, "Photosensitivity in optical fiber waveguides: Application to reflection filter fabrication," *Applied Physics Letters*, 32, 647, 1978.

52. L. Stokes, M. Chodorow, and H. Shaw, "All-fiber stimulated Brillouin ring laser with submilliwatt pump threshold," *Optics Letters*, 7, 509, 1982.

53. D. R. Ponikvar and S. Ezekiel, "Stabilized single-frequency stimulated Brillouin fiber ring laser," *Optics Letters*, 6, 398, 1981.

54. S. Smith, F. Zarinetchi, and S. Ezekiel, "Narrow-linewidth stimulated Brillouin fiber laser and applications," *Optics Letters*, 16, 393, 1991.

55. T. F. Büttner, I. V. Kabakova, D. D. Hudson, R. Pant, C. G. Poulton, A. C. Judge et al., "Phase-locking and pulse generation in multi-frequency Brillouin oscillator via four wave mixing," *Scientific Reports*, 4, 2014.

56. V. Lecoeuche, S. Randoux, B. Segard, and J. Zemmouri, "Dynamics of stimulated Brillouin scattering with feedback," *Quantum and Semiclassical Optics: Journal of the European Optical Society Part B*, 8, 1109, 1996.

57. S. Huang, L. Thévenaz, K. Toyama, B. Kim, and H. Shaw, "Optical Kerr-effect in fiber-optic Brillouin ring laser gyroscopes," *IEEE Photonics Technology Letters*, 5, 365, 1993.

58. P.-A. Nicati, K. Toyama, S. Huang, and H. J. Shaw, "Temperature effects in a Brillouin fiber ring laser," *Optics Letters*, 18, 2123, 1993.

59. M. Pang, S. Xie, X. Bao, D.-P. Zhou, Y. Lu, and L. Chen, "Rayleigh scattering-assisted narrow linewidth Brillouin lasing in cascaded fiber," *Optics Letters*, 37, 3129, 2012.

60. H. Cao, "Review on latest developments in random lasers with coherent feedback," *Journal of Physics A: Mathematical and General*, 38, 10497, 2005.

61. K. S. Abedin, P. S. Westbrook, J. W. Nicholson, J. Porque, T. Kremp, and X. Liu, "Single-frequency Brillouin distributed feedback fiber laser," *Optics Letters*, 37, 605, 2012.

62. A. Loayssa, D. Benito, and M. J. Garde, "Optical carrier Brillouin processing of microwave photonic signals," *Optics Letters*, 25, 1234, 2000.

63. S. Norcia, S. Tonda-Goldstein, D. Dolfi, J.-P. Huignard, and R. Frey, "Efficient single-mode Brillouin fiber laser for low-noise optical carrier reduction of microwave signals," *Optics Letters*, 28, 1888, 2003.

64. J. Geng, S. Staines, and S. Jiang, "Dual-frequency Brillouin fiber laser for optical generation of tunable low-noise radio frequency/microwave frequency," *Optics Letters*, 33, 16, 2008.

65. J. Geng, S. Staines, Z. Wang, J. Zong, M. Blake, and S. Jiang, "Actively stabilized Brillouin fiber laser with high output power and low noise," in *Optical Fiber Communication Conference*, 2006, p. OThC4.

66. H. G. Winful, I. V. Kabakova, and B. J. Eggleton, "Model for distributed feedback Brillouin lasers," *Optics Express*, 21, 16191, 2013.

67. H. Shin, W. Qiu, R. Jarecki, J. A. Cox, R. H. Olsson III, A. Starbuck et al., "Tailorable stimulated Brillouin scattering in nanoscale silicon waveguides," *Nature communications*, 4, 2013.

68. P. T. Rakich, C. Reinke, R. Camacho, P. Davids, and Z. Wang, "Giant enhancement of stimulated Brillouin scattering in the subwavelength limit," *Physical Review X*, 2, 011008, 2012.

69. Y. Song, L. Zhan, S. Hu, Q. Ye, and Y. Xia, "Tunable multiwavelength Brillouin-erbium fiber laser with a polarization-maintaining fiber Sagnac loop filter,"*IEEE Photonics Technology Letters*, 16, 2015, 2004.

70. Y. Song, L. Zhan, J. Ji, Y. Su, Q. Ye, and Y. Xia, "Self-seeded multiwavelength Brillouin-erbium fiber laser," *Optics Letters*, 30, 486, 2005.

71. M. Al-Mansoori, M. K. Abd-Rahman, F. Mahamd Adikan, and M. Mahdi, "Widely tunable linear cavity multiwavelength Brillouin-erbium fiber lasers,"*Optics Express*, 13, 3471, 2005.

72. M. P. Fok and C. Shu, "Spacing-adjustable multi-wavelength source from a stimulated Brillouin scattering assisted erbium-doped fiber laser," *Optics Express*, 14, 2618, 2006.

73. R. M. Sova, C.-S. Kim, and J. U. Kang, "Tunable all-fiber birefringence comb filters," in *Optical Fiber Communication Conference and Exhibit*, 2002, pp. 698–699.

74. M. Fok, K. Lee, and C. Shu, "Waveband-switchable SOA ring laser constructed with a phase modulator loop mirror filter," *IEEE Photonics Technology Letters*, 17, 1393, 2005.

75. K. Lee, M. Fok, S. Wan, and C. Shu, "Optically controlled Sagnac loop comb filter," *Optics Express*, 12, 6335, 2004.

76. M. M. Nasir, Z. Yusoff, M. Al-Mansoori, H. A. Rashid, and P. Choudhury, "Broadly tunable multi-wavelength Brillouin-erbium fiber laser in a Fabry-Perot cavity," *Laser Physics Letters*, 5, 812, 2008.

77. N. A. Hambali, M. Al-Mansoori, M. Ajiya, A. Bakar, S. Hitam, and M. Mahdi, "Multi-wavelength Brillouin-Raman ring-cavity fiber laser with 22-GHz spacing," *Laser Physics*, 21, 1656, 2011.

78. R. Sonee Shargh, M. Al-Mansoori, S. Anas, R. Sahbudin, and M. Mahdi, "OSNR enhancement utilizing large effective area fiber in a multiwavelength BrillouinâĂŘRaman fiber laser," *Laser Physics Letters*, 8, 139, 2011.

79. A. Zamzuri, M. Mahdi, A. Ahmad, M. Md Ali, and M. Al-Mansoori, "Flat amplitude multiwavelength Brillouin-Raman comb fiber laser in Rayleigh-scattering-enhanced linear cavity," *Optics Express*, 15, 3000, 2007.

80. K.-D. Park, B. Min, P. Kim, N. Park, J.-H. Lee, and J.-S. Chang, "Dynamics of cascaded Brillouin Rayleigh scattering in a distributed fiber Raman amplifier," *Optics Letters*, 27, 155, 2002.

81. A. Zamzuri, M. Md Ali, A. Ahmad, R. Mohamad, and M. Mahdi, "Brillouin-Raman comb fiber laser with cooperative Rayleigh scattering in a linear cavity," *Optics Letters*, 31, 918, 2006.

82. R. Pant, E. Li, D.-Y. Choi, C. Poulton, S. Madden, B. Luther-Davies et al., "Cavity enhanced stimulated Brillouin scattering in an optical chip for multi-order Stokes generation," *Optics Letters*, 36, 3687, 2011.

83. A. Al-Alimi, M. Yaacob, A. Abas, M. Mahdi, M. Mokhtar, and M. Al-Mansoori, "150-Channel four wave mixing based multiwavelength Brillouin-erbium doped fiber laser," *IEEE Photonics Journal*,5, 1501010, 2013.

84. D. Y. Stepanov and G. J. Cowle, "Properties of Brillouin/erbium fiber lasers," *IEEE Journal of Selected Topics in Quantum Electronics*, 3, 1049, 1997.

85. K. Gürs and R. Müller, "Breitband-modulation durch steuerung der emission eines optischen masers (Auskoppelmodulation)," *Physics Letters*, 5, 179, 1963.

86. K. Gürs and R. Müller, "Beats and modulation in optical ruby lasers," *Quantum Electronics*, 1964.

87. H. Statz and C. Tang, "Phase locking of modes in lasers," *Journal of Applied Physics*, 36, 3923, 1965.

88. M. DiDomenico Jr., "Small-signal analysis of internal (coupling-type) modulation of lasers," *Journal of Applied Physics*, 35, 2870, 1964.

89. L. Hargrove, R. Fork, and M. Pollack, "Locking of He-Ne laser modes induced by synchronous intracavity modulation," *Applied Physics Letters*, 5, 4, 1964.

90. A. Yariv, "Internal modulation in multimode laser oscillators," *Journal of Applied Physics*, 36, 388, 1965.

91. U. Morgner, F. Kärtner, S.-H. Cho, Y. Chen, H. A. Haus, J. G. Fujimoto et al., "Sub-two-cycle pulses from a Kerr-lens mode-locked Ti: sapphire laser,"*Optics Letters*, 24, 411, 1999.

92. D. H. Sutter, G. Steinmeyer, L. Gallmann, N. Matuschek, F. Morier-Genoud, U. Keller et al., "Semiconductor saturable-absorber mirror assisted Kerr-lens mode-locked Ti:sapphire laser producing pulses in the two-cycle regime," *Optics Letters*, 24, 631, 1999.

93. T. Brabec, C. Spielmann, P. Curley, and F. Krausz, "Kerr lens mode locking," *Optics Letters*, 17, 1292, 1992.

94. H. A. Haus, "Mode-locking of lasers," *IEEE Journal of Selected Topics in Quantum Electronics*, 6, 1173, 2000.

95. B. S. Kawasaki, D. C. Johnson, K. Hill, and Y. Fujii, "Bandwidth-limited operation of a mode-locked Brillouin parametric oscillator," *Applied Physics Letters*, 32, 429, 1978.

96. I. Bar-Joseph, A. Friesem, E. Lichtman, and R. Waarts, "Steady and relaxation oscillations of stimulated Brillouin scattering in single-mode optical fibers," *JOSA B*, 2, 1606, 1985.

97. E. Picholle, C. Montes, C. Leycuras, O. Legrand, and J. Botineau, "Observation of dissipative superluminous solitons in a Brillouin fiber ring laser," *Physical Review Letters*, 66, 1454, 1991.

98. T. Mirtchev and N. Minkovski, "Extremely transient pulses from a Brillouin fiber laser," *IEEE Photonics Technology Letters*, 5, 158, 1993.

99. J. Botineau, G. Cheval, and C. Montes, "CW-pumped polarization-maintaining Brillouin fiber ring laser: II. Active mode-locking by phase modulation," *Optics Communications*, 257, 311, 2006.

100. J. Botineau, G. Cheval, and C. Montes, "CW-pumped polarization-maintaining Brillouin fiber ring laser: I. Self-structuration of Brillouin solitons,"*Optics communications*, 257, 319, 2006.

101. S. Randoux and J. Zemmouri, "Polarization dynamics of a Brillouin fiber ring laser," *Physical Review A*, 59, 1644, 1999.

102. C. Montes, A. Mamhoud, and E. Picholle, "Bifurcation in a cw-pumped Brillouin fiber-ring laser: Coherent soliton morphogenesis," *Physical Review A*, 49, 1344, 1994.

103. C. Montes, D. Bahloul, I. Bongrand, J. Botineau, G. Cheval, A. Mamhoud et al., "Self-pulsing and dynamic bistability in cw-pumped Brillouin fiber ring lasers," *JOSA B*, 16, 932, 1999.

104. S. Loranger, V. L. Iezzi, and R. Kashyap, "Demonstration of an ultra-high frequency picosecond pulse generator using an SBS frequency comb and self phase-locking," *Optics Express*, 20, 19455, 2012.

105. E. M. Dianov, S. Isaev, L. S. Kornienko, V. V. Firsov, and Y. P. Yatsenko, "Locking of stimulated Brillouin scattering components in a laser with a waveguide resonator," *Quantum Electronics*, 19, 1, 1989.

106. X. Bao and L. Chen, "Recent progress in Brillouin scattering based fiber sensors," *Sensors*, 11, 4152, 2011.

107. H. Ohno, H. Naruse, M. Kihara, and A. Shimada, "Industrial applications of the BOTDR optical fiber strain sensor," *Optical Fiber Technology*, 7, 45, 2001.

108. H. Ohno, H. Naruse, N. Yasue, Y. Miyajima, H. Uchiyama, Y. Sakairi et al., "Development of highly stable BOTDR strain sensor employing microwave heterodyne detection and tunable electric oscillator," in *Proc. SPIE*, 2001, pp. 74–85.

109. T. Horiguchi and M. Tateda, "Optical-fiber-attenuation investigation using stimulated Brillouin scattering between a pulse and a continuous wave," *Optics Letters*, 14,408, 1989.

110. T. Horiguchi, T. Kurashima, and M. Tateda, "Tensile strain dependence of Brillouin frequency shift in silica optical fibers," *IEEE Photonics Technology Letters*, 1, 107, 1989.

111. T. Kurashima, T. Horiguchi, and M. Tateda, "Distributed-temperature sensing using stimulated Brillouin scattering in optical silica fibers," *Optics Letters*, 15, 1038, 1990.

112. X. Bao, D. J. Webb, and D. A. Jackson, "22-km distributed temperature sensor using Brillouin gain in an optical fiber," *Optics Letters*, 18, 552, 1993.

113. X. Bao, J. Dhliwayo, N. Heron, D. J. Webb, and D. A. Jackson, "Experimental and theoretical studies on a distributed temperature sensor based on Brillouin scattering," *Journal of Lightwave Technology*, 13, 1340, 1995.

114. M. A. Soto, G. Bolognini, F. Di Pasquale, and L. Thévenaz, "Simplex-coded BOTDA fiber sensor with 1 m spatial resolution over a 50 km range," *Optics Letters*, 35, 259, 2010.

115. W. Li, X. Bao, Y. Li, and L. Chen, "Differential pulse-width pair BOTDA for high spatial resolution sensing," *Optics Express*, 16, 21616, 2008.

116. S. M. Maughan, H. H. Kee, and T. P. Newson, "57-km single-ended spontaneous Brillouin-based distributed fiber temperature sensor using microwave coherent detection," *Optics Letters*, 26, 331, 2001.

117. M. N. Alahbabi, Y. T. Cho, and T. P. Newson, "150-km-range distributed temperature sensor based on coherent detection of spontaneous Brillouin backscatter and in-line Raman amplification," *JOSA B*, 22, 1321, 2005.

118. F. Rodríguez-Barrios, S. Martín-López, A. Carrasco-Sanz, P. Corredera, J. D. Ania-Castanon, L. Thévenaz et al., "Distributed Brillouin fiber sensor assisted by first-order Raman amplification," *Journal of Lightwave Technology*, 28, 2162, 2010.

119. M. A. Soto, G. Bolognini, and F. Di Pasquale, "Optimization of long-range BOTDA sensors with high resolution using first-order bi-directional Raman amplification," *Optics Express*, 19, 4444, 2011.

120. A. Zornoza, R. A. Pérez-Herrera, C. Elosúa, S. Diaz, C. Bariain, A. Loaysa et al., "Long-range hybrid network with point and distributed Brillouin sensors using Raman amplification," *Optics Express*, 18, 9531, 2010.

121. L. Zou, X. Bao, Y. Wan, and L. Chen, "Coherent probe-pump-based Brillouin sensor for centimeter-crack detection," *Optics Letters*, 30, 370, 2005.

122. S. Afshar V, G. A. Ferrier, X. Bao, and L. Chen, "Effect of the finite extinction ratio of an electro-optic modulator on the performance of distributed probe-pump Brillouin sensorsystems," *Optics Letters*, 28, 1418, 2003.

123. V. L. Iezzi, S. Loranger, M. Marois, and R. Kashyap, "High-sensitivity temperature sensing using higher-order Stokes stimulated Brillouin scattering in optical fiber," *Optics Letters*, 39, 857, 2014.

16 Nonlinear waves in metamaterials — Forward and backward wave interaction

Andrei I. Maimistov

16.1 INTRODUCTION

Traditionally, only waves in which a phase front and energy flux are propagated in the same direction have been considered [1]. These waves are referred to as *forward waves*. In the case of an inhomogeneous medium, the phase velocity vector and energy flux will sometimes not be collinear. In an extreme situation corresponding to a *backward wave*, the phase velocity and flux are oppositely directed.

Backward waves exist in electric transmission lines, in a reaction-diffusion medium and in acoustics. A new field of application of backward waves has been discovered. That is artificial materials — *metamaterials* — which are characterized by unusual electrodynamic properties [2–7]. First and foremost is negative refraction.

If the phase velocity and energy flux, i.e., the Poynting vector of the incident electromagnetic wave is directed in the same direction but the phase velocity and the Poynting vector of the refracted electromagnetic wave are oppositely directed, then the refraction angle is negative. Snell's formula can be used in this case, if the refractive index is considered a negative index. This phenomenon is referred to as *negative refraction*. Antiparallel orientation of the phase velocity ($\mathbf{v}_{ph}$) and the Poynting vector ($\mathbf{S}$) was first discussed in [8,9]. In [10] it was indicated that antiparallel orientation of $\mathbf{v}_{ph}$ and $\mathbf{S}$ results in negative refraction. Subsequently, this idea was developed by Mandelstam in [11]. It has been predicted that when the real parts of dielectric permittivity and magnetic permeability in the medium simultaneously take on negative values in some frequency range, antiparallel orientation of $\mathbf{v}_{ph}$ and $\mathbf{S}$ occurs [12,13] and the property of negative refraction appears [14]. The existence of media characterized by a negative refractive index (NRI) was demonstrated experimentally first in the microwave and then in the near-infrared ranges. Reviews of the properties of NRI materials are presented in [15–19].

Here we consider examples of electrodynamics and optics where the backward wave interacts with the forward wave in a nonlinear medium. We will

409

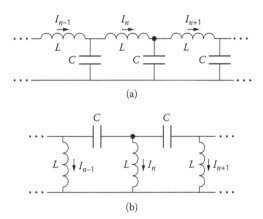

FIGURE 16.1: LC-transmission lines. (a) Transmission line providing forward wave propagation; (b) transmission line providing backward wave propagation.

start with linear waves. Parametric wave interaction and waves in guided systems will be discussed.

16.2 FORWARD AND BACKWARD WAVES

16.2.1 DISCRETE LINEAR MODELS

To illustrate the forward and backward wave concept, it would be useful to consider simple examples of electric transmission lines (Fig 16.1). One unit cell of this line contains capacitor C with charge q_n and current I_n through inductility L. The current I_n flows through capacitor C. $\mathcal{E}_n^L$ is the electromotive force and U_n is the voltage on the capacitor.

$$I_n = \frac{\partial q_n}{\partial t}, \quad U_n = \frac{q_n}{C}, \quad \mathcal{E}_n^L = -\frac{L}{c^2}\frac{\partial I_n}{\partial t}.$$

For the transmission line shown in Fig. 16.1(a), the system of the Kirchhoff equation,

$$I_n - I_{n-1} = I_n = \frac{\partial q_n}{\partial t}, \quad \frac{1}{C}(q_n - q_{n-1}) = -\frac{L}{c^2}\frac{\partial I_n}{\partial t},$$

leads to the wave equation,

$$\frac{\partial^2 I_n}{\partial t^2} = \omega_T^2 \left(I_{n+1} + I_{n-1} - 2I_n \right), \tag{16.1}$$

where $\omega_T = c/\sqrt{LC}$ is the Thompson frequency. From Eq. (16.1), the dispersion relation follows:

$$\omega^2(k) = 4\omega_T^2 \sin^2(ka/2), \tag{16.2}$$

where a is the transmission line period.

For the transmission line that is shown in Fig. 16.1(b), the system of the Kirchhoff equation takes the form

$$C\left(\frac{\partial U_n}{\partial t} - \frac{\partial U_{n+1}}{\partial t}\right) = I_n, \quad \frac{L}{c^2}\left(\frac{\partial I_{n-1}}{\partial t} - \frac{\partial I_n}{\partial t}\right) = U_n.$$

It leads to the wave equation

$$\frac{\partial^2}{\partial t^2}\left(I_{n+1} + I_{n-1} - 2I_n\right) = \omega_T^2 I_n. \tag{16.3}$$

From Equation (16.3), the dispersion relation fallows:

$$\omega^2(k) = \frac{\omega_T^2}{4\sin^2(ka/2)}. \tag{16.4}$$

In the limit $ka \ll 1$ these expressions result in the phase velocities and group velocities of the waves under consideration:

$$v_{ph} = \omega(k)/k = \omega_T a, \qquad v_g = d\omega(k)/dk = \omega_T a,$$
$$v_{ph} = \omega(k)/k = \omega_T/(k^2 a), \quad v_g = d\omega(k)/dk = -\omega_T/(k^2 a).$$

In the first case (Fig.16.1(a)) the phase and group velocities (they are the velocity vector projections in the direction of wave propagation) have the same sign. This corresponds to the forward wave definition. In the second case (Fig.16.1(b)) the phase velocity and group velocity are contra-directed. This corresponds to the backward wave definition.

16.2.2 DISCRETE NONLINEAR MODELS FOR BACKWARD WAVES

Propagation of the backward wave in a nonlinear electric transmission line was considered in [20]. The capacitors in the line were assumed to be nonlinear devices, i.e., capacitance is $C = C(q)$. The voltage on the capacitor takes in the form $W_n = U_n + \gamma U_n^p$, where γ is some constant and $1 < p < 4$. The system of the Kirchhoff equation leads to the wave equation

$$\frac{\partial^2}{\partial t^2}\left(W_{n+1} + W_{n-1} - 2W_n\right) = \omega_T^2 U_n. \tag{16.5}$$

The bright and dark soliton-like solutions were found using the quasi-continuum approximation and setting $p = 3$. In numerical simulations, the stability of these solitons under collisions was demonstrated. Moreover, in the absence of losses they can propagate with nearly any desired velocity.

Propagation of the electromagnetic pulses in magnetic metamaterial has been investigated using the model, where the metamaterial was represented by the array of split-ring resonators (SRR). A one dimensional SRR array was considered in [21,22]. In [23,24] a two-dimensional SRR array was considered. Solitary waves in an array containing two kinds of split-ring resonators were analyzed in [25, 26]. Each of the rings is represented by an LC-circuit coupled with the nearest neighbors because of cross-inductance.

In the case of the one dimensional array of split-ring resonators, the system of the Kirchhoff equation takes the form

$$\frac{\partial q_n}{\partial t} = I_n, \quad \frac{L}{c^2}\frac{\partial I_n}{\partial t} + RI_n + f(q_n) = \frac{L_{cross}}{c^2}\left(\frac{\partial I_{n-1}}{\partial t} + \frac{\partial I_{n+1}}{\partial t}\right), \quad (16.6)$$

where L_{cross} is the cross-inductance coefficient. $f(q_n)$ is the nonlinear capacity voltage of the LC-circuit. In [21, 22] the capacity voltage was defined as $f(q_n) = C_0^{-1}(q_n - \alpha q_n^3 + \beta q_n^5)$. If $R = 0$, Eq. (16.6) lead to

$$\frac{\partial^2}{\partial t^2}\left(\mu q_{n-1} + \mu q_{n+1} - q_n\right) = \omega_T^2 q_n - \omega_T^2(\alpha q_n^3 - \beta q_n^5), \quad (16.7)$$

where $\mu = L_{cross}/L$. In the long wave length limit the dispersion relation for the wave described by this equation has the form

$$\omega^2(k) = \omega_T^2(1 - \mu k^2).$$

Hence, the waves propagating in this one dimensional array of split-ring resonators are backward waves. Discrete breather excitations or intrinsic localized modes were investigated numerically by [21, 22].

A model like the one discussed above was investigated in [27]. The equations of motion are written as

$$q_{n,tt} + \lambda\left(q_{n-1,tt} + q_{n+1,tt}\right) - b(q_n - \alpha q_n^3) - b\Lambda\Omega\sin(\Omega t) = 0, \quad (16.8)$$

where $\Lambda\cos(\Omega t)$ is the normalized electromotive force resulting from a time varying and spatially uniform magnetic field of the form $H = H_0\cos(\Omega t)$, where H_0 is the field amplitude, Ω is the field frequency [21]. The Klein–Gordon equation for this system was derived in the continuum limit. The bright and dark envelope solitons, and bright and dark breather modes, bright and dark pulse solitons are obtained. Comparison of these solutions with appropriate modes in the nonlinear Schrödinger equation approximation was done.

Noteworthy also is the transmission line that was considered in [28]. On the basis of this nonlinear transmission line, the interaction between backward- and forward-propagating solitons was studied analytically and numerically. In the linear regime for certain frequency bands, backward and forward waves can propagate with the same group velocity. In the nonlinear regime, this transmission line supports a variety of backward- and forward-propagating vector solitons of the bright–bright, bright-dark and dark-bright types.

16.3 RESONANT INTERACTION OF FORWARD AND BACKWARD WAVES

Ultrashort pulse propagation in a three-level medium has been studied in many papers (see Chapter 4 in [29]). In the simplest case of this model the resonance levels have V, Λ configurations. It was determined that, if the oscillator forces for every transition in a V or a Λ configuration are equal, then the system of equations governing the evolution of ultrashort pulses in the these cases is completely integrable. The solutions of these equations are solitons. In the general case there is a steady state ultrashort pulse propagating in such a medium without envelope distortion. An ultrashort pulse of this kind was called a *simulton*.

Let the resonant medium consist of three-level atoms with non-degenerated resonance states. We shall consider the collinear propagation of two ultrashort electromagnetic pulses,

$$E_{1,2} = \mathcal{E}_{1,2} e^{-i\omega_{1,2}t + ik_{1,2}z} + c.c.,$$

resonant to adjacent optically permitted transitions $|1\rangle \to |3\rangle$ and $|2\rangle \to |3\rangle$ in the Λ configuration. Let us suppose that $\mathcal{E}_1$ is the envelope of the forward wave and $\mathcal{E}_2$ is the envelope of the backward wave. In this case the normalized Maxwell–Bloch equations under a slowly varying complex envelope approximation can be written as

$$e_{1,\varsigma} + v_1^{-1} e_{1,\tau} = ir_1 \langle \sigma_{31} \rangle, \tag{16.9}$$

$$-e_{2,\varsigma} + v_2^{-1} e_{2,\tau} = ir_2 \langle \sigma_{32} \rangle, \tag{16.10}$$

$$i\sigma_{31,\tau} = (\omega_{31} - \omega_1)\sigma_{31} - e_1 n_1 - e_2 \sigma_{21},$$

$$i\sigma_{32,\tau} = (\omega_{32} - \omega_1)\sigma_{32} - e_2 n_2 - e_1 \sigma_{12},$$

$$i\sigma_{21,\tau} = (\omega_{21} - \omega_1 + \omega_2)\sigma_{21} + e_1 \sigma_{23} - e_2^* \sigma_{31}, \tag{16.11}$$

$$i n_{1,\tau} = 2(e_1 \sigma_{13} - e_1^* \sigma_{31}) + (e_2 \sigma_{23} - e_2^* \sigma_{32}),$$

$$i n_{2,\tau} = (e_1 \sigma_{13} - e_1^* \sigma_{31}) + 2(e_2 \sigma_{23} - e_2^* \sigma_{32}),$$

where σ_{ij} is the slowly varying amplitudes of matrix elements of the density matrix of a three-level atom, and ω_{ij} is the frequency of an optically permitted atomic transition $|j\rangle \to |i\rangle$ [29]. It is suggested that the phase velocities of coupled waves are different but the group velocities are equal. The angular brackets in Eqs. (16.9) and (16.10) represent summation over all the atoms characterized by the frequencies ω_{31} and ω_{32}. The parameters r_1 and r_2 are defined by the expressions

$$r_j = (2\pi i n_A \omega_j / c\hbar) |d_{3j}|^2 \sqrt{\mu(\omega_j)/\varepsilon(\omega_j)}, \quad j = 1,\ 2,$$

where n_A is a three-level atom density, d_{3j} is the matrix element of the dipole transition between resonant states $|j\rangle \to |3\rangle$, $\mu(\omega_j)$ is permeability and $\varepsilon(\omega_j)$ is permittivity at frequency ω_j.

Let us denote a unique independent variable for all dependent variables as $\xi = \tau - \varsigma/V$, where V is the velocity of a steady state pulse

$$V = \frac{v_1 v_2 (r_1 - r_2)}{v_2 r_2 + v_1 r_1}.$$

On the assumption that the absorption line is homogeneously broadened (this is a sharp-line limit of the absorption line) and the exact resonance condition holds, then the system of equations (16.9)–(16.11) can be reduced to the system of equations for real envelopes $a_1(\xi) = |e_1(\xi)|$ and $a_2(\xi) = |e_2(\xi)|$:

$$a_{1,\xi\xi} + 2(a_1^2 + a_2^2)a_1 = \beta n_{10}a_1, \qquad (16.12)$$

$$a_{2,\xi\xi} + 2(a_1^2 + a_2^2)a_2 = \beta n_{20}a_2, \qquad (16.13)$$

where $n_{j0} = n_j$ at $\xi \to -\infty$, and

$$\beta = \frac{v_1 v_2 (r_2 - r_1)}{v_1 + v_2}.$$

This system of equations frequently occurs in nonlinear optics. Using the technique from [30], the solution can be written as

$$a_1(\xi) = \frac{2\sqrt{2}b_1 e^{\theta_1}[1 + e^{2\theta_2 + \Delta_{12}}]}{1 + e^{2\theta_1} + e^{2\theta_2} + e^{2(\theta_1 + \theta_2 + 2\Delta_{12})}}, \qquad (16.14)$$

$$a_2(\xi) = \frac{2\sqrt{2}b_2 e^{\theta_2}[1 + e^{2\theta_1 + \Delta_{12}}]}{1 + e^{2\theta_1} + e^{2\theta_2} + e^{2(\theta_1 + \theta_2 + 2\Delta_{12})}}, \qquad (16.15)$$

where $\theta_j = b_j\sqrt{2}(\xi - \xi_{0j})$, $j = 1, 2$, ξ_{0j} are the integration constants, and

$$b_j = \frac{1}{2}\beta n_{0j}, \quad e^{2\Delta_{12}} = -\frac{b_1 - b_2}{b_1 + b_2}.$$

Assume that all three-level atoms are in the ground state at $\xi \to infty$. Then $n_{10} = 1, n_{20} = 0$. The expressions result in the formula

$$a_1(\xi) = \sqrt{2}b_1 \mathrm{sech}[\sqrt{2}b_1(\xi - \xi_1)], \quad a_2(\xi) = 0.$$

It is the usual 2π-pulse of self-induced transparency.

It should be remarked that cnoidal waves and the bright-dark paired solitary waves can be obtained as special solutions of Eqs. (16.12) and (16.13).

The propagation of two-frequency light pulses through a metamaterial doped with three-level atoms such that the carrier frequencies of the pulses are in resonance with two atomic transitions in the Λ configuration was investigated in [31]. In was suggested that one resonant wave is forward and the other is a backward wave. The nonlinear interaction of these forward- and backward-propagating waves was investigated numerically. Self-trapped waves, counterpropagating radiation waves, and hot spots of medium excitation were found.

In [32] ultrashort pulse propagation in a resonantly absorbing metamaterial was considered. This kind of metamaterial is assumed to be homogeneously doped with two-level atoms. The linear property of host-medium is defined by the electric permittivity and magnetic permeability, which are given according to the lossy Drude model. It was found that an ultrashort laser pulse can evolve into a steady state pulse. This kind of steady state solitary wave was referred to as a backward SIT (self-induced transparency) soliton.

16.4 PARAMETRIC INTERACTION

Parametric interaction of waves is the situation in which the forward (backward) wave with carrier wave frequency ω_p transforms to a backward (forward) wave with another frequency ω_s. Second harmonic generation (SHG) is a special case of three-wave interaction ($\omega_p + \omega_p = \omega_s$) in a $\chi^{(2)}$-medium. Third harmonic generation (THG) is a special case of four-wave interaction ($\omega_p + \omega_p + \omega_p = \omega_3$) in a $\chi^{(3)}$-medium. SHG is one of the first examples of the nonlinear phenomena in which the unusual property of forward and backward waves interaction was identified [33–36].

16.4.1 SECOND-HARMONIC GENERATION

The normalized form of the equations describing the SHG is

$$ie_{1,\zeta} + (s_1/2)e_{1,\tau\tau} - e_2 e_1^* = 0, \tag{16.16}$$
$$ie_{2,\zeta} + i\delta e_{2,\tau} - (s_2/2)e_{2,\tau\tau} - \Delta e_2 + e_1^2/2 = 0. \tag{16.17}$$

Here e_1 and e_2 represent the normalized fundamental and harmonic wave envelopes. δ is normalized group velocity mismatch, Δ is normalized phase mismatch, s_1 and s_2 are parameters of the group-velocity dispersion. The functions $e_{1,2}(\zeta, \tau)$, independent variables ζ, τ and other parameters expressed in terms of the physical values are given in [19].

The process of the SHG in the continuous wave limit was considered in a number of papers [35–37]. The principal result is the existence of some critical value of the mismatch Δ_{cr}. If $\Delta \leq \Delta_{cr}$, the transformation of the fundamental wave into a harmonic wave occurs monotonically. If $\Delta > \Delta_{cr}$, the amplitudes of the interacting waves vary periodically. In the case of conventional harmonic generation the critical value of mismatch is zero.

The evolution of the interacting wave packets of the forward fundamental wave e_1 and the harmonic backward wave e_2 was considered in [38]. It was shown that at high intensities the pulse of second harmonics can be trapped by the fundamental wave pulse and forced to propagate in the same direction. This kind of coupled waves is referred to as a simulton. In the case of $s_1 = s_2 = -1$ the real envelopes of the interacting wave packets are

$$|e_1(\xi)| = \frac{\sqrt{2}(3\Delta - \delta^2)}{6\cosh^2[\sqrt{(3\Delta - \delta^2)/2}\,\xi/3]}, \qquad |e_2(\xi)| = \frac{(3\Delta - \delta^2)}{6\cosh^2[\sqrt{(3\Delta - \delta^2)/2}\,\xi/3]},$$

where $\xi = \tau - \zeta/v_s - \tau_0$, v_s is the simulton group velocity, τ_0 is a position of the simulton amplitude maximum. This solution exists under the condition $\Delta > \delta^2/3$.

Cnoidal waves, bright and dark simultons and two-hump simultons are represented in [38] as examples of coupled steady state waves. The numerical simulation of the SHG and simulton propagation in negative refractive materials with quadratic nonlinearity was done in [39–42]. It was found that under certain conditions the second harmonic pulse can be trapped and dragged along by the fundamental wave pulse.

16.4.2 THIRD-HARMONIC GENERATION

Third-harmonic generation (THG) is due to $\chi^{(3)}$-nonlinearity of material. This is the reason why self-modulation and cross-modulation take place in addition to the transfer of energy from one wave to another. The normalized form of the basic equations is [19, 43, 44]

$$ie_{1,\varsigma} + (s_1/2)e_{1,\tau\tau} - e_3 e_1^{*2} - \alpha_{11}|e_1|^2 e_1 - \alpha_{13}|e_3|^2 e_1 = 0, \tag{16.18}$$
$$ie_{3,\varsigma} + i\delta e_{3,\tau} - (s_3/2)e_{3,\tau\tau} - \Delta e_3 + e_1^3 + \alpha_{31}|e_1|^2 e_3 + \alpha_{33}|e_3|^2 e_3 = 0 \tag{16.19}$$

Here α_{jl} is the self-modulation and cross-modulation coefficient.

Assuming the envelopes of the coupled waves are dependent on $\xi = \tau - \varsigma/v_s$, where v_s is the simulton group velocity and the steady state waves for both frequencies must be propagating as the single one. Furthermore, let α_{jl} be zero. These assumptions allow us to reduce Eqs. (16.18) and (16.19) to a single equation for the real envelope e_1 [19]. If we set dispersion parameters as $s_1 = s_3 = -1$, the bright simulton solution can be found.

$$|e_1(\xi)| = |e_3(\xi)| = \frac{\sqrt{4\Delta - 3\delta^2}}{4\cosh[\sqrt{4\Delta - 3\delta^2}\xi]}. \tag{16.20}$$

This solution exists under the condition $\Delta > 3\delta^2/4$.

Steady state cnoidal waves solutions for Eqs. (16.18) and (16.19) are found in [43], by taking into account all the above-mentioned assumptions.

It is interesting to remark that the walk-off effect, which is accounted for by parameter δ in Eq. (16.19), is liable to frustrate coupling between fundamental and harmonic waves. In this case the waves will propagate independently according to

$$ie_{1,\varsigma} + (s_1/2)e_{1,\tau\tau} - \alpha_{11}|e_1|^2 e_1 = 0,$$
$$ie_{3,\varsigma} + i\delta e_{3,\tau} - (s_3/2)e_{3,\tau\tau} - \Delta e_3 + \alpha_{33}|e_3|^2 e_3 = 0.$$

Either of the two equations is a nonlinear Schrödinger-type equation. Hence, the fundamental wave pulse and/or third-harmonic wave pulse can be transformed into a soliton (or solitons), propagating with its own velocity. That was observed in numerical simulation of the THG in [43].

The case that a fundamental beam is tuned in a positive index region and generates second and/or third harmonics in a negative index region was considered in [42]. Phase matching conditions, phase locking, inhibition of absorption at the harmonic wavelengths, and second- and third-harmonic lenses were investigated.

Investigation of the THG in a continuous wave limit shows [44, 45] that, as with SHG, there exists a critical value of the modulus of phase mismatch that, being a function of self-action effects, separates two generation regimes, monotonic and periodic ones, at which the harmonic amplitude is a monotonic or oscillating function, respectively. This feature does not take place in the case of harmonic generation in a medium possessing the same sign of the refractive index at both frequencies. However, it should be expected during the generation of a wave with any multiple frequency in a nonlinear medium negatively refracting at the fundamental or generated frequency.

16.5 WAVEGUIDE SYSTEMS: COUPLERS, ARRAYS AND BUNDLES

The two closely located waveguides can be coupled due to the tunneling of light from one waveguide to the other. If waveguides are fabricated from materials with a positive refractive index (PRI), then the wave propagation direction is unchanged. The coupler is called a *directional coupler*. One coupler can be coupled with a third waveguide or with another coupler. The procedure can be prolonged step by step. It results in a waveguide system. In the one dimensional case we get a waveguide array. The optical waveguide array provides a convenient setup for experimental investigation of periodic nonlinear systems in one dimension [46]. Nonlinear optical waveguide arrays (NOWA) are a natural generalization of a waveguide array. NOWA with a positive refractive index have many useful applications and are well studied in the literature (see, for example, [47–49]).

Interaction between forward and backward waves is conveniently performed in a waveguide array where the sign of the index of refraction of one of the waveguides is positive and the index of refraction of the neighboring waveguide is negative. This waveguide array is called an *alternating waveguide array*.

16.5.1 ALTERNATING NONLINEAR OPTICAL WAVEGUIDE ZIGZAG ARRAY

Usually the coupling between nearest neighboring waveguides is taken into account. It is a correct approximation for a strong localized electromagnetic wave in a waveguide. However, the coupling between both nearest neighboring waveguides and the next nearest neighboring ones can be introduced by the use of a zigzag arrangement [50,51]. Let ϑ_b be an angle between the lines connecting the centers of neighboring waveguides. In a linear array this angle is π. In a zigzag-like array at $\vartheta_b \approx \pi/2$, the coupling between the nearest neighboring waveguides and the next nearest neighboring ones is approximately the same. Nonlinear optical waveguide zigzag arrays (NOWZA) can be considered a generation of NOWA.

Let us assume that the waveguide marked by index n is characterized by a positive refractive index (PRI), and the nearest neighboring waveguides with indexes $n-1$ and $n+1$ possess a negative refractive index (NRI). If the electromagnetic radiation is localized in each waveguide, the coupled wave theory can be used. In the case of the array which is prepared in the form of a line (i.e., all centers of the waveguide placed on one line), only the interaction between the

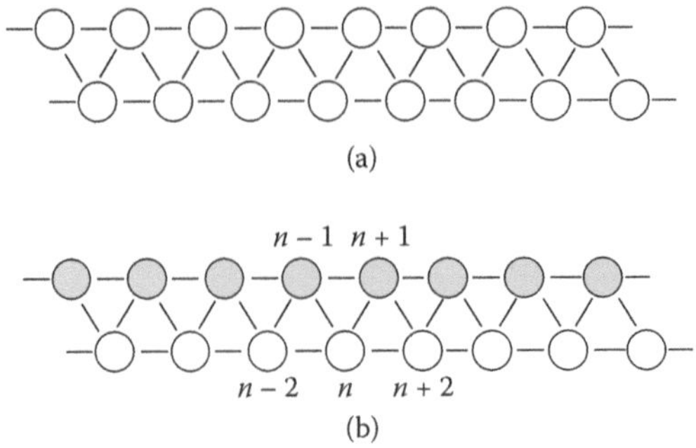

(a)

(b)

FIGURE 16.2: A schematic illustration of a zigzag coupled waveguide array (a) and a zigzag positive-negative coupled waveguide array (b). The white circles are the PRI waveguide; the gray circles are the NRI waveguide.

nearest neighboring waveguides is essential. In this configuration the angles between the lines connecting the waveguides are equal π. However, if the array is deformed in the form of a zigzag, where the angles between the lines connecting the waveguides are equal to approximately $2\pi/3$, interaction between both the nearest neighboring and the next nearest neighboring waveguides will be important. The system of equations describing wave propagation in a nonlinear zigzag waveguide array with alternating signs of refractive index is [52]

$$i(e_{n,\varsigma} + e_{n,\tau}) + c_1(e_{n+1} + e_{n-1}) + c_2(e_{n+2} + e_{n-2}) + r_1|e_n|^2 e_n = 0, \quad (16.21)$$

$$i(-e_{n+1,\varsigma} + e_{n+1,\tau}) + c_1(e_{n+2} + e_n) + c_3(e_{n+3} + e_{n-1}) + r_2|e_n|^2 e_n = 0, \quad (16.22)$$

where e_n is the normalized envelope of the wave localized in the n-th waveguide. Coupling between neighboring PRI and NRI waveguides is defined by parameter c_1. The parameter c_2 (c_3) is the coupling constant between neighboring PRI (NRI) waveguides. The functions $e_n(\varsigma, \tau)$, independent variables ς, τ and other parameters are expressed in terms of the physical values given in [19]. The configuration of these alternating waveguides is called an *alternating nonlinear optical waveguide zigzag array* (ANOWZA).

ANOWZA is a chain of waveguides where the unit cell contains two different waveguides, i.e., PRI and NRI waveguides, per cell. A pair of the two variables $A_n = e_{2n}$, $B_n = e_{2n+1}$ is convenient to describe the alternating nonlinear optical waveguide zigzag array. Instead of Eqs. (16.21) and (16.22) we will consider the following system of equations:

$$i(A_{n,\varsigma} + A_{n,\tau}) + c_1(B_n + B_{n-1}) + c_2(A_{n+1} + A_{n-1}) + r_1|A_n|^2 A_n = 0, \quad (16.23)$$

$$i(-B_{n,\varsigma} + B_{n,\tau}) + c_1(A_n + A_{n+1}) + c_3(B_{n+1} + B_{n-1}) + r_2|B_n|^2 B_n = 0, \quad (16.24)$$

It should be pointed out that now n is the unit cell label.

16.5.2 LINEAR PROPERTIES OF THE ALTERNATING WAVEGUIDE ZIGZAG ARRAY

Let us consider the case of linear waveguides combined into an zigzag alternating waveguide array. This optical system is described by following equations:

$$i\left(A_{n,\varsigma} + A_{n,\tau}\right) + c_1(B_n + B_{n-1}) + c_2(A_{n+1} + A_{n-1}) = 0, \qquad (16.25)$$
$$i\left(-B_{n,\varsigma} + B_{n,\tau}\right) + c_1(A_n + A_{n+1}) + c_3(B_{n+1} + B_{n-1}) = 0. \qquad (16.26)$$

To find the linear wave spectrum we can employ the presentation of the envelopes in the form of harmonic waves,

$$A_n = ae^{-i\omega\tau+iq\varsigma+i\phi n}, \quad B_n = be^{-i\omega\tau+iq\varsigma+i\phi n}.$$

Substitution of this expression in Eqs. (16.25) and (16.26) leads to a system of algebraic linear equations respecting a and b. Nonzero solutions of these equations exist if the determinant

$$\det\begin{pmatrix} q + \omega + \gamma_3 & \gamma_1 \\ \gamma_1^* & \omega - q + \gamma_2 \end{pmatrix}$$

is equal to zero. Here the parameters $\gamma_1 = 2c_1\cos\phi/2\exp(i\phi/2)$, $\gamma_2 = 2c_2\cos\phi$, and $\gamma_3 = 2c_3\cos\phi$ were introduced. This condition results in the equation

$$(\omega + \omega_0)^2 = |\gamma_1|^2 + (q - q_0)^2$$

where

$$2\omega_0 = (\gamma_2 + \gamma_3), \quad 2q_0 = (\gamma_2 - \gamma_3).$$

Thus we obtain the dispersion law for linear waves in linearized ANOWZA:

$$\omega^{(\pm)}(q) = -\omega_0 \pm \sqrt{|\gamma_1|^2 + (q - q_0)^2}. \qquad (16.27)$$

Expression (16.27) shows that (a) the forbidden zone (gap) in the spectrum of the linear waves exists, $\Delta\omega = 2|\gamma_1|$, (b) the spectrum is shifted along both the frequency axis and the wave number axis, (c) the form of the spectrum is like the spectrum for a linear oppositely directional coupler [53, 59]. The gapless spectrum appears only when condition $\phi = \pi$ holds. In this case the radiation propagates along waveguides with the same refractive indexes. The energy flux between neighboring waveguides is zero.

16.5.3 NONLINEAR WAVES IN ANOWZA

The simplest approximation is based on the ansatz

$$A_n(\varsigma, \tau) = a(\varsigma, \tau)e^{i\phi n}, \quad B_n(\varsigma, \tau) = b(\varsigma, \tau)e^{i\phi(n+1)},$$

where a and b are the envelopes of the quasi-harmonic waves. This allows us to reduce Eqs. (16.23) and (16.24) and to obtain the equations

$$i\,(a_{,\varsigma} + a_{,\tau}) + \gamma_1 b + \gamma_2 a + r_1|a|^2 a = 0, \tag{16.28}$$

$$i\,(b_{,\varsigma} - b_{,\tau}) - \gamma_1 a - \gamma_3 b - r_2|b|^2 b = 0. \tag{16.29}$$

At $\phi = \pi$ this system of equations splits into two independent equations. At $\phi = \pi/2$ Eqs. (16.28) and (16.29) are transformed to that which was considered in [58, 59].

Among the different kinds of the waves, steady state waves attract attention. These waves correspond to a solution of the wave equation depending on only one particular variable. Let this variable be

$$\xi = \gamma_1 \frac{\varsigma + \beta\tau}{\sqrt{1 - \beta^2}},$$

where β is the parameter that defines the group velocity. If we introduce new envelopes u_1 and u_2, according to formulae $u_1 = \sqrt{1 + \beta}\,a\ u_2 = \sqrt{1 - \beta}\,b$, then u_1, u_2 are governed by following equations:

$$i\frac{\partial u_1}{\partial \xi} + u_2 + \frac{\gamma_2}{\gamma_1}\sqrt{\frac{1 - \beta}{1 + \beta}}u_1 + \frac{r_1}{\gamma_1(1 + \beta)}\sqrt{\frac{1 - \beta}{1 + \beta}}|u_1|^2 u_1 = 0, \tag{16.30}$$

$$i\frac{\partial u_2}{\partial \xi} - u_1 - \frac{\gamma_3}{\gamma_1}\sqrt{\frac{1 + \beta}{1 - \beta}}u_2 - \frac{r_2}{\gamma_1(1 - \beta)}\sqrt{\frac{1 + \beta}{1 - \beta}}|u_2|^2 u_2 = 0, \tag{16.31}$$

Solutions of Eqs. (16.30) and (16.31) under boundary conditions $a \to 0$, $b \to 0$ at $|\xi| \to \infty$ correspond to solitary waves.

Two kinds of steady state solitary waves were found in [52] under consideration of $r_1 = r$, $r_2 = 0$.

T Soliton-like waves

These waves correspond to the following solutions of Eqs. (16.30) and (16.31):

$$|a(\xi)|^2 = \frac{4\Delta^2}{|\vartheta|(1 + \beta)\{\cosh[2\Delta(\xi - \xi_0)] + \delta/2\}}, \tag{16.32}$$

$$|b(\xi)|^2 = \frac{4\Delta^2}{|\vartheta|(1 - \beta)\{\cosh[2\Delta(\xi - \xi_0)] + \delta/2\}}, \tag{16.33}$$

$$\Phi(\xi) = \Phi(-\infty) + \mathrm{sgn}(\vartheta)\mathcal{S}(\delta/2; X_0), \tag{16.34}$$

where $\Phi = \arg a - \arg b$ is the phase difference of the electric fields of neighboring waveguides. The parameters

$$\vartheta = \frac{r}{\gamma_1(1 + \beta)}\sqrt{\frac{1 - \beta}{1 + \beta}}, \quad \delta = \left(\frac{\gamma_3}{\gamma_1}\sqrt{\frac{1 + \beta}{1 - \beta}} + \frac{\gamma_2}{\gamma_1}\sqrt{\frac{1 - \beta}{1 + \beta}}\right), \quad \Delta^2 = 1 - \frac{\delta^2}{4}$$

are entered. Furthermore, the auxiliary function

$$S(a; X_0) = \arctan \left(\frac{(1 - a^2)^{1/2} e^{X_0}}{1 + a e^{X_0}} \right)$$

and the argument $X_0 = 2\Delta(\xi - \xi_0)$ were introduced.

Due to the fact that at $\xi \to -\infty$ the derivative $\partial|a|/\partial\xi$ is positive, the value of $\Phi(-\infty)$ must equal $\varepsilon\pi/2$, where $\varepsilon = \pm 1$.

It is noteworthy that from the system of Eqs. (16.30) and (16.31) the two integrals of motion follow:

$$|u_1|^2 - |u_2|^2 = C_1 = \text{const},$$

$$(u_1 u_2^* + u_1^* u_2) + \frac{\delta}{2}|u_1|^2 + \frac{\vartheta}{4}|u_1|^4 = C_2 = \text{const}.$$

T **Algebraic solitary waves**

The soliton-like solutions of Eqs. (16.30) and (16.31) describe solitary waves with exponentially decaying wave fronts. However, sometimes solitary waves can be decreasing as $\sim 1/\xi^2$. Algebraic solitary waves correspond to the following solutions of Eqs. (16.30) and (16.31) [52]:

$$a_1^2(\xi) = \frac{8}{\vartheta(1 + \beta_1)\{1 + 4(\xi - \xi_0)^2\}}, \tag{16.35}$$

$$a_2^2(\xi) = \frac{8}{\vartheta(1 - \beta_1)\{1 + 4(\xi - \xi_0)^2\}}. \tag{16.36}$$

Here β_1 corresponds to $\delta = -2$. If δ runs to $+2$, the amplitudes of the waves in Eqs. (16.30) and (16.31) are equal to zero.

It is worth noting that the velocity parameter β is limited due to condition $|\delta| \leq 2$. It should be pointed out that if the next nearest neighboring waveguides are not coupled as in [59], then steady state waves exist for $|\beta| < 1$.

The characteristic coupling constant for ANOWZA can be controlled by the angle between the lines connecting the centers of neighboring waveguides ϑ_b. The numerical simulation of the collision between two steady state solitary waves shows that the complicated picture of the interaction depends on coupling constants which are characterized as coupling between the next nearest neighboring waveguides. Thus the angle ϑ_b is the control parameter for nonlinear waves in ANOWZA.

16.5.4   ALTERNATING NONLINEAR OPTICAL WAVEGUIDE ARRAY

If the angle between the lines connecting the centers of neighboring waveguides in zigzag waveguide array ϑ_b is equal to π, the coupling between the next nearest neighboring ones is terminated. The configuration of these alternating waveguides is called *alternating nonlinear optical waveguide arrays* (ANOWA).

The evolution of the normalized envelope of the wave e_n localized in the n-th waveguide is defined by the following equations [19, 53]:

$$i\left(A_{n,\varsigma}+A_{n,\tau}\right)+c_1(B_n+B_{n-1})+r_1|A_n|^2A_n=0, \quad (16.37)$$
$$i\left(-B_{n,\varsigma}+B_{n,\tau}\right)+c_1(A_{n+1}+A_n)+r_2|B_n|^2B_n=0 \quad (16.38)$$

Coupling between neighboring PRI and NRI waveguides is defined by parameter c_1.

16.5.5 LINEAR PROPERTIES OF THE ALTERNATING WAVEGUIDE ARRAY

Linear properties of the waves in the alternating waveguide array can be considered on the basis of Eqs. (16.37) and (16.38), where parameters r_1 and r_2 are set to zero. It results in the system of Eqs. (16.25) and (16.26), but where $c_2=c_3=0$. Thus the result of Section 16.5.2 can be exploited.

The spectrum of linear waves in an alternating waveguide array results in, from Eq. (16.27),

$$\omega^{(\pm)}(q)=\pm\sqrt{|\gamma_1|^2+q^2}, \quad (16.39)$$

where $|\gamma_1|=2c_1\cos\phi/2$. Here the phase ϕ can be written as $\phi=\kappa d_1$, where d_1 is the distance between the nearest neighboring waveguides and κ is the transverse wave number of the 1D array. Once again, expression (16.39) shows that the forbidden zone (gap) in the spectrum of linear waves exists [53]. The gap width is $\Delta\omega(\kappa)=4c_1|\cos(\kappa d_1/2)|$.

16.5.6 NONLINEAR WAVES IN ANOWA

Nonlinear steady state waves can be described by using the Eqs. (16.32)–(16.34) and setting the parameters of coupling between the next nearest neighboring waveguides $c_2=c_3=0$ equal to zero. It follows that soliton-like solutions take the form

$$|a(\xi)|^2 = \frac{4}{|\vartheta|(1+\beta)\cosh[2(\xi-\xi_0)]}, \quad (16.40)$$
$$|b(\xi)|^2 = \frac{4}{|\vartheta|(1-\beta)\cosh[2(\xi-\xi_0)]}, \quad (16.41)$$
$$\Phi(\xi) = \Phi(-\infty)+\text{sgn}(\vartheta)\arctan e^{2(\xi-\xi_0)} \quad (16.42)$$

Algebraic solutions cannot be obtained.

It should be noted that the system of Eqs. (16.37) and (16.38) has been derived under the assumption of equality of the phase velocity of coupled waves. In general, these equations taking into account phase mismatch $\delta_m=\Delta\beta_{wg}L_c/2$, result in the equations [54]

$$i\left(A_{n,\varsigma}+A_{n,\tau}\right)-\delta_m A_n+c_1(B_n+B_{n-1})+r_1|A_n|^2A_n=0, \quad (16.43)$$
$$i\left(-B_{n,\varsigma}+B_{n,\tau}\right)+\delta_m B_n+c_1(A_{n+1}+A_n)+r_2|B_n|^2B_n=0. \quad (16.44)$$

Here $\Delta\beta_{wg}$ is the difference of the propagating constant and L_c is the coupling length. The soliton-like solution is

$$|a(\xi)|^2 = \frac{4\Delta_m^2}{|\vartheta|(1+\beta)\{\cosh[2\Delta(\xi-\xi_0)]-\mathrm{sgn}(\vartheta)\delta_1\}}, \qquad (16.45)$$

$$|b(\xi)|^2 = \frac{4\Delta^2}{|\vartheta|(1-\beta)\{\cosh[2\Delta_m(\xi-\xi_0)]-\mathrm{sgn}(\vartheta)\delta_1\}}, \qquad (16.46)$$

$$\Phi(\xi) = \Phi(-\infty) + 2\mathrm{sgn}(\vartheta)\arctan\frac{\Delta_m e^{2\Delta_m(\xi-\xi_0)}}{1-\mathrm{sgn}(\vartheta)\delta_1 e^{2\Delta_m(\xi-\xi_0)}}, \quad (16.47)$$

where $\delta_1 = \delta_m(1-\beta^2)^{-1/2}$, $\Delta_m^2 = 1 - \delta_1^2$, and the parameter of nonlinearity is defined by the expression

$$\vartheta = \frac{r_1}{1+\beta}\sqrt{\frac{1-\beta}{1+\beta}} + \frac{r_2}{1-\beta}\sqrt{\frac{1+\beta}{1-\beta}}.$$

Comparison Eqs. (16.45)–(16.47) with Eqs. (16.32) and (16.33) shows that the effect of coupling between the next nearest neighboring waveguides is similar to a phase mismatch effect.

Algebraic solutions can be obtained from Eqs. (16.45)-(16.46) by considering limit $\delta_1 \to \pm\infty$.

16.5.7 SPATIAL DISCRETE SOLITONS IN ANOWA

The existence and properties of discrete solitons in arrays of alternating waveguides with positive and negative refractive indices were studied in [55]. The spatial (i.e., continuous wave) field distribution in the array (A_n, B_n) has been described by the modified version of Eqs. (16.37) and (16.38):

$$iA_{n,\varsigma} + \delta A_n + (B_n + B_{n-1}) + r_1|A_n|^2 A_n = 0, \qquad (16.48)$$
$$iB_{n,\varsigma} - \delta B_n - (A_n + A_{n+1}) - r_2|B_n|^2 B_n = 0. \qquad (16.49)$$

It was supposed that a discrete soliton has the form $A_n(\zeta) = a_n \exp(ik\zeta)$ and $B_n(\zeta) = b_n \exp(ik\zeta)$ where a_n and b_n are real and vanish as $|n| \to \infty$. Amplitudes are determined by the system of algebraic equations

$$(\delta - k)a_n + (b_n + b_{n+1}) + r_1 a_n^3 = 0,$$
$$(\delta + k)b_n + (a_n + a_{n-1}) + r_2 b_n^3 = 0.$$

The spectrum of the linear continues waves is

$$k^2 = \delta^2 - 4\cos^2(\kappa d_1/2),$$

where d_1 is the distance between the nearest neighboring waveguides and κ is the transverse wave number of the 1D array. If $|\delta| > 2$, then the gap $\Delta k =$

$2\sqrt{\delta^2 - 4}$ exists. As is shown in [55], gap solitons exist for waveguides having nonlinearities of different types. When the nonlinearities of all waveguides are focusing, solitons exist if the propagation constant κ lies outside the gap.

Numerical simulation [55] has shown that that there exist soliton families bifurcating from the gap edges of the linear spectrum. The field distribution in the opposite direction of the nonlinear waveguide array reveals nonexponential decay and nonmonotonic dependence of the energy growth in the positive index waveguides on the absorption index.

16.5.8 BUNDLES OF WAVEGUIDES

Let us consider the waveguide array which rests on a cylinder with a large diameter. This device is called the circular waveguide array or waveguide bundle.

A circular waveguide array consisting of coupled waveguides with alternating positive and negative refractive indices will be considered. We assume that the group velocities and the propagation constant s are equal for all waveguides. The total number of waveguides M is equal to $2N$. The waveguide array is said to have N dimers. In the case of $M = 2N + 1$, one waveguide is unpaired, i.e., it is a "defect" waveguide. Circular waveguide arrays are defect free. Circular nonlinear waveguide arrays with alternating waveguides are called *alternating nonlinear optical waveguide bundles* (ANOWB).

The system of governing normalized equations takes the form (like Eqs. (16.37) and (16.38))

$$i\left(A_{n,\varsigma} + A_{n,\tau}\right) + \left(B_n + B_{n-1}\right) + r_1|A_n|^2 A_n = 0, \qquad (16.50)$$
$$i\left(-B_{n,\varsigma} + B_{n,\tau}\right) + \left(A_{n+1} + A_n\right) + r_2|B_n|^2 B_n = 0, \qquad (16.51)$$

with the periodic boundary conditions $A_{n+N} = A_n$ and $B_{n+N} = B_n$.

16.5.9 LINEAR MODES OF AN ALTERNATING WAVEGUIDE BUNDLE

To study the linear modes ANOWB system of Eqs. (16.50) and (16.51), we must consider $r_1 = r_2 = 0$. Substitution of the ansatz

$$A_n = a_n e^{-i\omega\tau + ik\varsigma}, \quad B_n = b_n e^{-i\omega\tau + ik\varsigma}$$

and $r_1 = r_2 = 0$ into the system of equations (16.50) and (16.51) leads to

$$(\omega - k)a_n + (b_n + b_{n+1}) = 0,$$
$$(\omega + k)b_n + (a_n + a_{n-1}) = 0.$$

The periodic boundary conditions $A_{n+N} = A_n$ and $B_{n+N} = B_n$ allow us to use the discrete Fourier transform

$$a_n = \sum_s^N f_s e^{i2\pi ns/N}, \quad b_n = \sum_s^N g_s e^{i2\pi ns/N}.$$

Substitution of this series into the linear system above results in the algebraic system

$$(\omega - k)f_s + g_s e^{i\pi s/N} 2 \cos(\pi s/N) = 0,$$
$$(\omega + k)g_s + f_s e^{-i\pi s/N} 2 \cos(\pi s/N) = 0.$$

The dispersion relation for eigenmodes – eigenvalues – is

$$\begin{vmatrix} (\omega - k) & \Delta_s \\ \Delta_s^* & (\omega + k) \end{vmatrix} = 0,$$

where

$$\Delta_s = e^{i\pi s/N} 2 \cos(\pi s/N).$$

Thus the dispersion relation takes the form

$$\omega_s^2(k) = k^2 + 4 \cos^2(\pi s/N). \tag{16.52}$$

Then some limit cases can be considered. For $N = 1$ (i.e., two waveguides in a bundle), the eigenvalues are

$$\omega_s^2(k) = k^2 + 4 \cos^2(\pi s) = k^2 + 4,$$

where $s = 0$, as $\max |s| = N/2$ or $(N - 1)/2$. Two gap branches exist [59, 64].
For $N = 2$ (i.e., four waveguides in a bundle) the eigenvalues are

$$\omega_0^2(k) = k^2 + 4, \quad \omega_{\pm 1}^2(k) = k^2.$$

Here two gapless modes exist, and two gap modes exist.
For the case of the bundle of six waveguides, $N = 3$. Hence s is $0, 1, 2$. This leads to

$$\begin{aligned} \omega_0^2(k) &= k^2 + 4, \\ \omega_1^2(k) &= k^2 + 4 \cos^2(\pi/N) = k^2 + 1, \\ \omega_2^2(k) &= k^2 + 4 \cos^2(\pi 2/N) = k^2 + 1. \end{aligned}$$

Here three gap modes exist, and there are no gapless modes.
From Eq. (16.52) it follows that gapless modes will appear when $N = 2j$, due to $s = 0, 1, ..., j, ...2j$.
It should be noted that the optical waveguide bundles containing waveguides of like features have only gapless modes.

16.5.10 LINEAR MODES OF THE TWISTED ALTERNATING WAVEGUIDE BUNDLE

In [56] the twisted waveguide bundle was investigated. This system consists of a set of N tunnelling-coupled optical fibers which are twisted along the propagation direction with a twist rate $\epsilon = 2\pi/\Lambda_t$, where Λ_t is the local twist period, which is assumed to be much longer than the light wavelength. This twist generates the additional phase of the coupling constant $\delta \sim \epsilon$. The normalized envelopes of the waves localized in the n-th waveguide (or fiber) are governed by the following system of equations:

$$ie_{n,\zeta} = e_n + \left(e^{-i\delta}e_{n+1} + e^{\delta}e_{n-1}\right). \tag{16.53}$$

It was shown that a fiber twist can be conveniently exploited to control light transfer between two fibers in the array.

The helical spatial twist of the ANOWB can be investigated on the basis the equations [53, 57]

$$i(A_{n,\zeta} + A_{n,\tau}) + \left(B_n + e^{i\delta}B_{n+1}\right) + r_1|A_n|^2 A_n = 0, \tag{16.54}$$
$$i(B_{n,\zeta} - B_{n,\tau}) - \left(e^{-i\delta}A_{n-1} + A_n\right) - r_2|B_n|^2 B_n = 0. \tag{16.55}$$

In the case of linear waves (i.e., Eqs. (16.54) and (16.55) with $r_1 = r_2 = 0$) the dispersion relation takes the form

$$\omega_s^2 = k^2 + 4\cos^2\left(\frac{\pi}{N}s + \frac{\delta}{2}\right). \tag{16.56}$$

Thus, in this model, the spectra of the eigenmodes have gaps, but under some values of parameter δ the gaps can be closed, for example, at

$$\delta/2 = \pi(2j+1)/2 - \pi s/N.$$

In the general case s varies from 0 up to N. Thus the argument of the cosine in Eq. (16.56) varies from $\delta/2$ up to $\delta/2 + \pi$.

16.5.11 NONLINEAR SOLITARY WAVES IN ANOWB

The steady state solitary waves in ANOWB can be found by the same procedure as for ANOWA. But, in the first place, the anzatses

$$A_1 = A, \quad A_2 = \theta A, \ldots, \quad A_{j+1} = \theta^j A,$$

$$B_1 = B, \quad B_2 = \theta B, \ldots, \quad B_{j+1} = \theta^j B$$

are assumed. The condition of periodicity results in

$$A_{N+1} = \theta^N A = A_1 = A, \quad B_{N+1} = \theta^N B = B_1 = B,$$

from which it follows that $\theta^N = 1$ or

$$\theta = \exp\left(i\frac{2\pi l}{N}\right),$$

for any integer number l. Thus,

$$A_j = A \exp\left(i\frac{2\pi l}{N}j\right), \quad B_j = B \exp\left(i\frac{2\pi l}{N}j\right). \tag{16.57}$$

By using these formulae one can find that

$$e^{-i\delta}B_j + e^{i\delta}B_{j-1} = 2\cos\left(\frac{\pi l}{N} - \delta\right) B \exp\left(i\frac{2\pi l}{N}j - i\frac{\pi l}{N}\right),$$

$$e^{i\delta}A_j + e^{-i\delta}A_{j+1} = 2\cos\left(\frac{\pi l}{N} - \delta\right) A \exp\left(i\frac{2\pi l}{N}j + i\frac{\pi l}{N}\right).$$

Substitution of these expressions into Eqs. (16.54) and (16.55) leads to

$$i\left(A_{,\varsigma} + A_{,\tau}\right) - \mu A + C(\delta)\exp\left(-i\pi l/N\right)B + r_1|A|^2 A = 0, \tag{16.58}$$

$$i\left(B_{,\varsigma} - B_{,\tau}\right) - \mu B - C(\delta)\exp\left(i\pi l/N\right)A - r_2|B|^2 B = 0, \tag{16.59}$$

where

$$C(\delta) = 2\cos\left(\pi l/N - \delta\right).$$

The system of Eqs. (16.58) and (16.59) is similar to Eqs. (16.43) and (16.44), where the phase mismatch (i.e., μ) and the coupling constant $C(\delta)$ are taken into account. To obtain a soliton-like solution of the system of Eqs. (16.58) and (16.59) we can use Eqs. (16.45)–(16.47).

16.5.12   OPPOSITELY DIRECTIONAL COUPLERS

The forward-wave-backward-wave interaction can be realized in a nonlinear oppositely directional coupler [58, 59]. This coupler consists of two tightly spaced nonlinear/linear waveguides. The sign of the index of refraction of one of these waveguides is positive and the index of refraction of the other waveguide is negative. In the linear regime two coupled waveguides with opposite signs of the refractive index act as a mirror. This device is called the *oppositely directional coupler* (ODC).

The oppositely directional coupler is a limiting case of the ANOWA, where there are only two coupled waveguides. Hence the base equations in this case can be written as

$$i\left(e_{1,\varsigma} + e_{1,\tau}\right) - \delta_m e_1 + e_2 + r_1|e_1|^2 e_1 = 0, \tag{16.60}$$

$$i\left(-e_{2,\varsigma} + e_{2,\tau}\right) + \delta_m e_2 + e_1 + r_2|e_2|^2 e_2 = 0. \tag{16.61}$$

Soliton-like solutions of Eqs. (16.60) and (16.61) were found in [58]. These solutions are represented by Eqs. (16.45)–(16.47).

Exact solutions in the form of periodical and solitary waves for nonlinear waves described by the system of Eqs. (16.60) and (16.61) are given in [60]. All types of traveling wave solutions were presented. Among them there is the usual type of nonlinear waves: periodical solutions that are cnoidal waves expressed via elliptic functions and solitary waves expressed via hyperbolical functions. Another type is solitary and periodical waves in the form of compactons (a robust solitary wave with compact support beyond which it vanishes identically) [61].

16.5.13   GAP SOLITONS

The usual gap soliton arises from a balance between nonlinearity and grating assisted dispersion in optical fibers [62,63]. Both in the case of ODC and in the cases of ANOWA and ANOWZA, the spectrum of linear waves has a forbidden zone (gap). The opposite directionality of the phase velocity and the energy flow in the NRI waveguide facilitate an effective feedback mechanism that leads to optical bistability [64] and robust solitary wave formation [58,59,65]. Numerical simulation of the the steady state pulse formation and the collisions between the soliton-like waves demonstrated that the pulses are very robust against perturbations [59,65]. This allows the definite conclusion that these steady state pulses are a new kind of gap soliton. It should be noted that this solitary wave exists in a medium without grating.

As usual, the formation of a steady state solitary wave has a threshold character. It has been shown that the gap soliton in ODC formation has a threshold character, too. Numerical simulation of soliton formation from the initially Gaussian envelope pulse $a(\zeta = 0, \tau) = a_0 \exp(-\tau^2)$ allows one to estimate the dependence of the amplitude threshold a_{th} versus the nonlinearity parameter $r : a_{th} = 2.1 \, r^{-1/2}$ [65]. The analytical approximate formula for the amplitude threshold was presented in [53]. The amplitude threshold is

$$a_{th}^2 r = 4\sqrt{(\pi/2)^2 - 1} \approx 4.85.$$

This expression provides a good estimation for the numerical results of [65].

16.5.14   BISTABILITY

Bistability is a phenomenon in which the system exhibits two steady transmission states for the same input intensity. Transmission and reflection coefficients for the nonlinear oppositely directional coupler of finite width in the case of continuum wave radiation were found [64]. Spatial distributions of the electric fields in waveguides are governed by the following equations:

$$ie_{1,\zeta} - \delta_m e_1 + e_2 + r_1 |e_1|^2 e_1 = 0, \tag{16.62}$$

$$ie_{2,\zeta} - \delta_m e_2 - e_1 - r_2 |e_2|^2 e_2 = 0. \tag{16.63}$$

These equations are accompanied by the boundary conditions $e_1(\zeta = 0) = e_0$, $e_2(l) = 0$, where l is the normalized length of the coupler. At condition δ_m,

Eqs. (16.62) and (16.63) can be solved analytically. The phase mismatch affect was taken into account in numerical simulations in [66,67]. It was demonstrated that the transmission (and reflection) coefficient is a multivalued function of input power $|e_0|^2$. It leads to hysteresis of the transmission and reflection coefficients.

The transmission characteristics of the nonlinear ODC are very similar to those of the distributed feedback structures, with an important fundamental difference that bistability in the coupler is facilitated by the effective feedback mechanism originating from the forward wave and backward wave interaction. In the conventional case the feedback mechanism is due to mirrors of a cavity.

16.5.15 MODULATION INSTABILITY

Modulation instability (MI) in nonlinear ODC was investigated in [68]. It was shown that the input power MI threshold exists only in the normal dispersion regime. If the NRI waveguide is manufactured from self-defocusing nonlinear material and the PRI waveguide is produced from self-focusing nonlinear material in the anomalous dispersion regime, then MI occurs only for finite values of input power. Contrary to the MI in the conventional couplers, the increasing input power in ODC may suppress the MI. It is interesting that this feature of MI in ODC is akin to the characteristic of modulation instability in ANOWZA [69].

The influence of saturable nonlinearity on MI was investigated in [70]. It was shown MI can exist in the case of the normal group velocity dispersion regime as well as in the normal group velocity dispersion regime in nonlinear ODC. The saturable nonlinearity can change the number of sidebands or shift the existing sidebands. The maximum value of the increment of MI, as well as its bandwidth, are also affected by saturable nonlinearity.

16.5.16 WAVEGUIDE AMPLIFIER BASED ON ODC

In a conventional distributed feedback laser (DFB laser) or amplifier, directly and oppositely propagating waves propagate in the same waveguide with periodic distribution of the (effective) refractive index and gain medium. Coupling between these waves is due to Bragg resonance. In the ODC, coupling between forward waves and backward waves results from tunnel penetration of the radiation from one waveguide to another. Let us assume that in the ODC, the forward wave propagates in the gain medium, while the backward wave propagates in the absorbing medium. The amplifier and the absorber are spatially separated, which makes this structure different from the DFB laser.

In the case of the linear approximation of the amplifier, the system of equations takes the form

$$i\left(e_{1,\zeta} + e_{1,\tau}\right) - \delta_m e_1 + e_2 + r_1|e_1|^2 e_1 = i\Gamma_1 e_1, \qquad (16.64)$$

$$i\left(-e_{2,\zeta} + e_{2,\tau}\right) + \delta_m e_2 + e_1 + r_2|e_2|^2 e_2 = -i\Gamma_2 e_2, \qquad (16.65)$$

where Γ_1 and Γ_2 are parameters describing dissipation properties of the waveguides. The model of the waveguide amplifier under condition $r_2 = 0$ was considered in [71]. It was shown that compensation and amplification are possible in the linear regime where both $r_2 = 0$ and $r_1 = 0$. However, if $r_1 \neq 0$ nonlinearity becomes dominant in the changing phase difference of the coupled waves because of amplification of the forward wave. This results in weakening of coupling between the forward and backward waves. As a result, the amplifier based on ODC will not be efficient as a loss compensator. The waveguide with a saturable gain may be a more suitable ODC element for compensation of losses and amplification of radiation. In this case the model of the ODC amplifier can be based on the system of equations

$$i\left(e_{1,\zeta} + e_{1,\tau}\right) - \delta_m e_1 + e_2 + r_1|e_1|^2 e_1 = i\Gamma_1 e_1(1 + \varrho|e_1|^2)^{-1}, \quad (16.66)$$
$$i\left(-e_{2,\zeta} + e_{2,\tau}\right) + \delta_m e_2 + e_1 + r_2|e_2|^2 e_2 = -i\Gamma_2 e_2, \quad (16.67)$$

where ϱ is the saturation factor.

If we take into account the evolution of resonant atomic subsystems, for example, the Jaynes–Cummings model, the system of Eqs. (16.66) and (16.67) can be generalized.

CONCLUSION

The interaction of forward waves and backward waves is studied frequently with the use of electric transmission lines. However, the fabrication of new materials — metamaterials — allows us to investigate these waves using the famous nonlinear electromagnetic effects. As the examples of nonlinear phenomena in metamaterials, the parametric interaction of the electromagnetic waves, the solitary waves propagating in a nonlinear oppositely directional coupler and nonlinear waveguide array and waveguide bundles are considered.

There is a wide class of nonlinear phenomena where forward and backward wave coupling exists: nonlinear surface waves at the interfaces of left-handed media [72,73] and nonlinear waves propagating in a waveguide composed of a negative-index medium surrounded by conventional media [74,75]. Nonlinear TE-polarized surface waves localized at the interface between a nonlinear uniform material and a semi-infinite one-dimensional photonic crystal are investigated in [76]. A survey of these phenomena can be found in [19].

ACKNOWLEDGMENTS

I would like to thank I.R. Gabitov, A.S. Desyatnikov, N.M. Litchinitser, E.V. Kazantseva and J.G. Caputo for enlightening discussions. I thank the Department of Mathematics of the University of Arizona, the Laboratoire de Mathematiques, INSA de Rouen and the Nonlinear Physics Center, Australian National University, for support and hospitality. This work was partially supported by NSF grant DMS-0509589, ARO-MURI award 50342-PH-MUR and the State of Arizona (Proposition 301), and by the Russian Foundation for Basic Research (grants No. 09-02-00701-a, 12-02-00561).

REFERENCES

1. G. B. Whitham, *Linear and Nonlinear Waves*, John Wiley & Sons, New York, 1974.
2. H. Chen, B.-I. Wu, J. A. Kong, "Review of electromagnetic theory in left-handed-materials" *J. Electromagn. Waves and Appl.*, 20, 2137 2006.
3. Al. Boltasseva, Vl. M. Shalaev, "Fabrication of optical negative-index metamaterials: Recent advances and outlook," *Metamaterials*, 2, 1 2008.
4. G. V. Eleftheriades, K. G. Balmain, *Negative-Refraction Metamaterials: Fundamental Principles and Applications*, Wiley, New York, 2005.
5. M. A. Noginov, V. A. Podolskiy, *Tutorials in Metamaterials*," Taylor and Francis Group, LLC/CRC Press, Boca Raton, 2012.
6. A. K. Sarychev, V. M. Shalaev, *Electrodynamics of Metamaterials*," World Scientific, Singapore, 2007.
7. W. Cai, V. Shalaev, *Optical Metamaterials: Fundamentals and Applications*," Springer, Dordrecht, 2009.
8. H. Lamb, "On group-velocity," *Proc Proc. London Math. Soc.*, 1, 473, 1904.
9. H. C. Pocklington, "Growth of a wave-group when the group velocity is negative," *Nature*, 71, 607, 1905.
10. A. Schuster, *An Introduction to the Theory of Optics*," London Edward Arnold, 1904.
11. L. I. Mandelstam, "Group velocity in crystalline arrays," *Zh. Eksp. Teor. Fiz.* 15, 475, 1945.
12. D. V. Sivukhin, "The energy of electromagnetic waves in dispersive media," *Opt. Spektrosk*, 3, 308, 1957.
13. V. E. Pafomov, "On transition radiation and the Vavilov-Cherenkov radiation," *Sov.Phys. JETP*, 9, 1321, 1959.
14. V. G. Veselago, "The electrodynamics of substances with simultaneously negative values of ε and μ," *Sov. Phys. Usp.*, 10, 509, 1968.
15. S. A. Ramakrishna, "Physics of negative refractive index materials," *Rep. Prog. Phys.*, 68, 449, 2005.
16. V. Veselago, L. Braginsky, V. Shklover, C. Hafner, "Negative refractive index materials," *J. Computational and Theoretical Nanoscience*, 3, 189, 2006.
17. V. M. Agranovich, Y. N. Gartstein, "Spatial dispersion and negative refraction of light," *Phys. Usp.*, 49, 1029, 2006.
18. N. M. Litchinitser, I. R. Gabitov, A. I. Maimistov, V. Shalaev, "Negative refractive index metamaterials in optics," *Progress in Optics*, 51, 1, 2008.
19. A. I. Maimistov, I. R. Gabitov, "Nonlinear optical effects in artificial materials," *Eur. Phys. J. Special Topics*, 147, 265, 2007.
20. E. Arevalo, "Solitons in anharmonic chains with negative group velocity," *Phys. Rev. E*, 76, 066602, 2007.
21. N. Lazarides, M. Eleftheriou, G. P. Tsironis, "Discrete breathers in nonlinear magnetic metamaterials," *Phys. Rev. Lett.*, 97, 157406, 2006.
22. N. Lazarides, G. P. Tsironis, Y. S. Kivshar, "Surface breathers in discrete magnetic meta-materials," *Phys. Rev. E.*, 77, 065601, 2008.

23. M. Eleftheriou, N. Lazarides, G. P. Tsironis, "Magnetoinductive breathers in metamaterials," *Phys. Rev. E*, 77, 036608, 2008.

24. M. Eleftheriou, N. Lazarides, G. P. Tsironis, Y. S. Kivshar, "Surface magnetoinductive breathers in two-dimensional magnetic metamaterials," *Phys. Rev. E*, 80, 017601, 2009.

25. W. Cui, Y. Zhu, H. Li, S. Liu, "Soliton excitations in a one-dimensional nonlinear diatomic chain of split-ring resonators," *Phys. Rev. E*, 81, 016604, 2010.

26. M. I. Molina, N. Lazarides, G. P. Tsironis, "Bulk and surface magnetoinductive breathers in binary metamaterials," *Phys. Rev. E*, 80, 046605, 2009.

27. P. Giri, K. Choudhary, A. Sen Gupta, A. K. Bandyopadhyay, A. R. McGurn, "Klein-Gordon equation approach to nonlinear split-ring resonator based metamaterials: One-dimensional systems," *Phys. Rev. B*, 84, 155429, 2011.

28. G. P. Veldes, J. Cuevas, P. G. Kevrekidis, D. J. Frantzeskakis, "Coupled backward- and forward-propagating solitons in a composite right/left-handed transmission line," *Phys. Rev. E*, 88, 013203, 2013.

29. A. I. Maimistov, A. M. Basharov, *Nonlinear Optical Waves*, Kluwer Academic Publishers, Dortrecht, Boston, London, 1999.

30. M. V. Tratnik, J. E. Sipe, "Bound solitary waves in a birefringent optical fiber," *Phys. Rev. A*, 38, 2011, 1988.

31. A. O. Korotkevich, K. E. Rasmussen, G. Kovachich, V. Roytburd, A. I. Maimistov, I. R. Gabitov, "Optical pulse dynamics in active metamaterials with positive and negative refractive index," *J. Opt. Soc. Amer. B*, 30, 1077, 2013.

32. J. Zeng, J. Zhou, G. Kurizki, T. Opatrny, "Backward self-induced transparency in metamaterials," *Phys. Rev. A*, 80, 061806, 2009.

33. V. M. Agranovich, Y. R. Shen, R. H. Baughman, A. A. Zakhidov, "Linear and nonlinear wave propagation in negative refraction metamaterials," *Phys. Rev. B*, 69, 165112, 2004.

34. I. V. Shadrivov, A. A. Zharov, Y. S. Kivshar, "Second-harmonic generation in nonlinear left-handed metamaterials," *J. Opt. Soc. Amer.*, B23, 529, 2006.

35. A. K. Popov, V. M. Shalaev, "Negative-index metamaterials: Second-harmonic generation, Manley-Rowe relations and parametric amplification," *Appl. Phys.*, B84, 131, 2006.

36. A. K. Popov, V. V. Slabko, V. M. Shalaev, "Second harmonic generation in left-handed metamaterials," *Laser Phys. Lett.*, 3, 293, 2006.

37. Z. Kudyshev, I. Gabitov, A. Maimistov, "Effect of phase mismatch on second harmonic generation in negative index materials," *Phys. Rev. A*, 87, 063840, 2013.

38. A. I. Maimistov, I. R. Gabitov, E. V. Kazantseva, "Quadratic solitons in media with negative refractive index," *Optics and Spectroscopy*, 102, 90, 2007.

39. M. Scalora, G. D'Aguanno, M. Bloemer, M. Centini, D. de Ceglia, N. Mattiucci, Y. S. Kivshar, "Dynamics of short pulses and phase matched second harmonic generation in negative index materials," *Optics Express*, 14, 4746, 2006.

40. V. Roppo, M. Centini, C. Sibilia, M. Bertolotti, D. de Ceglia, M. Scalora, N. Akozbek, M. J. Bloemer, J. W. Haus, O. G. Kosareva, V. P. Kandidov, "Role of phase matching in pulsed second-harmonic generation: Walk-off and phase-locked twin pulses in negative-index media," *Phys. Rev. A*, 76, 033829, 2007.

41. V. Roppo, M. Centini, D. de Cegliac, M. A. Vicenti, J. Haus, N. Akozbek, J. Mark, M. J. Bloemer, M. Scalora, "Anomalous momentum states, non-specular reflections, and negative refraction of phase-locked, second-harmonic pulses," *Metamaterials*, 2, 135, 2008.

42. V. Roppo, C. Ciraci, C. Cojocaru, M. Scalora, "Second harmonic generation in a generic negative index medium," *J. Opt. Soc. Amer.*, B27, 1671, 2010.

43. S. O. Elyutin, A. I. Maimistov, I. R. Gabitov, "On the third harmonic generation in a medium with negative pump wave refraction," *JETP*, 111, 157, 2010.

44. E. I. Ostroukhova, A. I. Maimistov, "Third harmonic generation in the field of a backward pump wave," *Optics and Spectroscopy*, 112, 255, 2012.

45. E. I. Ostroukhova, A. I. Maimistov, "Spatial distribution of the interacting wavesâÅŹ amplitudes under third harmonic generation in a negative-positive refractive medium," *Optics and Spectroscopy*, 115, 378, 2013.

46. D. N. Christodoulides, F. Lederer, Y. Silberberg, "Discretizing light behavior in linear and nonlinear waveguide lattices," *Nature*, 424, 817, 2003.

47. D. N. Christodoulides, R. I. Joseph, "Discrete self-focusing in nonlinear arrays of coupled waveguides," *Opt. Letts.*, 13, 794, 1988.

48. C. Schmidt-Hattenberger, U. Trutschel, F. Lederer, "Nonlinear switching in multiple-core couplers," *Opt. Letts.*, 16, 294, 1991.

49. S. Darmanyan, I. Relke, F. Lederer, "Instability of continuous waves and rotating solitons in waveguide arrays," *Phys. Rev. E.*, 55, 7662, 1997.

50. N. K. Efremidis, D. N. Christodoulides, "Discrete solitons in nonlinear zigzag optical waveguide arrays with tailored diffraction properties," *Phys. Rev. B*, 65, 056607, 2002.

51. G. Wang, J. P. Huang, K. W. Yu, "Nontrivial Bloch oscillations in waveguide arrays with second-order coupling," *Optics Letters*, 35, 1908, 2010.

52. E. V. Kazantseva, A. I. Maimistov, "Nonlinear waves in an array of zigzag waveguides with alternating positive and negative refractive indices," *Quantum Electronics*, 43, 807, 2013.

53. A. I. Maimistov, E. V. Kazantseva, A. S. Desyatnikov, "Linear and nonlinear properties of the antidirectional coupler. In Coherent optics and optical spectroscopy," *Lect notes, Kazan State University, Kazan, pp. 21–31*, 2012.

54. A. A. Dovgiy, A. I. Maimistov, "Solitary waves in a binary nonlinear waveguide array," *Optics and Spectroscopy*, 116, 626, 2014.

55. D. A. Zezyulin, V. V. Konotop, F. K. Abdullaev, "Discrete solitons in arrays of positive and negative index waveguides," *Optics Letters*, 37 3930, 2012.

56. S. Longhi, "Light transfer control and diffraction management in circular fiber waveguide arrays," *J. Phys. B*, 40, 4477, 2007.

57. A. S. Desyatnikov, A. I. Maimistov, Y. S. Kivshar, "Topological solitons in twisted coupled positive-negative waveguides," *2nd International Workshop "Nonlinear Photonics" NLP*2013*, 2013, Sudak, Crimea, Ukraine.

58. A. I. Maimistov, I. R. Gabitov, N. M. Litchinitser, "Solitary waves in a nonlinear oppositely directed coupler," *Optics and Spectroscopy*, 104, 253, 2008.

59. E. V. Kazantseva, A. I. Maimistov, S. S. Ozhenko, "Solitary electromagnetic wave propagation in the asymmetric oppositely directed coupler," *Phys. Rev.*, A80, 43833, 2009.

60. N. A. Kudryashov, A. I. Maimistov, D. I. Sinelshchikov, "General class of the traveling waves propagating in a nonlinear oppositely-directional coupler," *Phys. Letts.*, A376, 3658, 2012.

61. P. Rosenau, J. Hyman, "Compactons: Solitons with finite wavelength," *Physical Review Letters*, 70, 564, 1993.

62. D. L. Mills, S. E. Trullinger, "Gap solitons in nonlinear periodic structures," *Phys. Rev. B.*, 36, 947, 1987.

63. G. Agrawal, Y. S. Kivshar, *Optical Solitons: From Fibers to Photonic Crystals*, Academic Press, Amsterdam, 2003.

64. N. M. Litchinitser, I. R. Gabitov, A. I. Maimistov, "Optical bistability in a nonlinear optical coupler with a negative index channel," *Phys. Rev. Lett.*, 99, 113902, 2007.

65. M. S. Ryzhov, A. I. Maimistov, "Gap soliton formation in a nonlinear antidirectional coupler," *Quantum Electronics*, 42, 1034, 2012.

66. Z. Kudyshev, G. Venugopal, N. M. Litchinitser, "Generalized analytical solutions for nonlinear positive-negative index couplers," *Phys. Res. International*, 2012, 945807, 2012.

67. G. Venugopal, Z. Kudyshev, N. M. Litchinitser, "Asymmetric positive-negative index nonlinear waveguide couplers," *IEEE Journal of Selected Topics in Quantum Electronics*, 18, 753, 2012.

68. X. Yuanjiang, W. Shuangchun, D. Xiaoyu, F. Dianyuan, "Modulation instability in nonlinear oppositely directed coupler with a negative-index metamaterial channel," *Phys. Rev.*, E82, 056605, 2010.

69. A. A. Dovgiy, "Modulation instability in a zigzag nonlinear waveguide array with alternating positive and negative index waveguides," *Quantum Electron.*, 44, 2014.

70. P. H. Tatsing, A. Mohamadou, C. Bouri, C. G. L. Tiofack, T. C. Kofane, "Modulation instability in nonlinear positive negative index couplers with saturable nonlinearity," *J. Opt. Soc. Amer. B*, 29, 3218, 2012.

71. A. I. Maimistov, E. V. Kazantseva, "A waveguide amplifier based on a counterdirectional coupler," *Optics and Spectroscopy*, 112, 264, 2012.

72. V. M. Agranovich, Y. R. Shen, R. H. Baughman, A. A. Zakhidov, "Optical bulk and surface waves with negative refraction," *J. Luminescence*, 110, 167, 2004.

73. S. A. Darmanyan, M. Neviere, A. A. Zakhidov, "Nonlinear surface waves at the interfaces of left-handed electromagnetic media," *Phys. Rev. E*, 72, 036615, 2005.

74. I. V. Shadrivov, "Nonlinear guided waves and symmetry breaking in left-handed waveguides," *Photonics Nanostruct.: Fundam. Appl.*, 2, 175, 2004.

75. S. A. Darmanyan, A. Kobyakov, D. Q. Chowdhury, "Nonlinear guided waves in a negative-index slab waveguide," *Phys. Lett. A*, 363, 159, 2007.

76. S. R. Entezar, A. Namdar, Z. Eyni, H. Tajalli, "Nonlinear surface waves in one-dimensional photonic crystals containing left-handed metamaterials," *Phys. Rev. A*, 78, 023816, 2008.

17 Optical back propagation for compensation of dispersion and nonlinearity in fiber optic transmission systems

Xiaojun Liang, Jing Shao, and *Shiva Kumar*

17.1 INTRODUCTION

Optical signals propagating in a fiber optic link will be distorted due to fiber dispersion and nonlinearity. Fiber dispersion can be simply compensated using a dispersion compensating fiber (DCF) that has a dispersion coefficient with the opposite sign of that of the transmission fiber (TF). Dispersion compensation can also be achieved by digital signal processing (DSP) using a finite impulse filter (FIR) or fast Fourier transforms (FFTs). However, the compensation of fiber nonlinearity is much more complicated. Lumped compensation of self-phase modulation (SPM) has been investigated [1], but is not applicable to other nonlinear distortions. Recently, a novel method called back propagation has been proposed to provide distributed nonlinearity compensation, which can be implemented in either the optical [2–5] or electrical domain [6,7].

Signal propagation in an optical fiber is governed by the nonlinear Schrödinger equation (NLSE) [8], which can be used to calculate the signal distortions due to fiber dispersion and nonlinearity. The deterministic signal distortions can be compensated by employing a mechanism that inverts the NLSE. The inversion of the NLSE can be implemented in either the electrical or optical domain. In the electrical domain, the distorted signal is propagated backward in a virtual fiber whose loss, dispersion and nonlinear parameters have the opposite sign of those of the TF [6]. In the optical domain, the inversion of the NLSE can be achieved either by using multiple fiber Bragg gratings (FBGs) and highly nonlinear fibers (HNLFs) [2], or by using a dispersion-decreasing fiber (DDF) [5].

Midpoint optical phase conjugation (OPC) [9–12] may be considered as an optical back propagation (OBP) scheme to compensate for fiber dispersive and nonlinear effects. In midpoint OPC, an OPC is placed at the middle of the fiber optic link. The dispersion, nonlinearity and loss profiles of the fibers

435

should be symmetric with respect to the OPC. In erbium doped fiber amplifier (EDFA) systems, the loss profile cannot be symmetric with respect to OPC and hence, the performance improvement obtained using midpoint OPC is marginal for wavelength division multiplexing (WDM) systems. However, using distributed Raman amplifiers, the power profiles can be made symmetric with respect to OPC [10], which significantly enhances the transmission performance. Although midpoint OPC with distributed Raman amplifiers leads to significant benefits in the laboratory, in practical systems, its benefits are expected to be limited due to the fact that the amplifier spacings are not equal, which destroys the symmetry with respect to OPC. Instead, a dedicated OBP module at the receiver is required to compensate for dispersive and nonlinear effects [2–5].

In one of the OBP schemes, multiple FBGs and HNLFs are concatenated in a way analogous to the split-step Fourier scheme (SSFS) to undo fiber impairments. Instead of using uniform step size in back propagation, the transmission performance and/or reach can be significantly enhanced by using optimal step sizes based on the minimum area mismatch (MAM) method [4]. In a real TF, the optical power is decreasing exponentially, and so does the nonlinear phase shift. In back propagation, a reversed loss coefficient is required which indicates an exponentially increasing profile of nonlinear phase shift. This can be realized by designing either an exponentially increasing power profile or a nonlinear coefficient profile. In SSFS-based back propagation, the exponential profile is approximated by a stepwise curve. As a result, the optimal approximation can be obtained using the MAM method which minimizes the area mismatch between the stepwise curve and the exponential profile. In the OBP design, the optimum accumulated dispersion of each section of the FBG and the optimum nonlinear phase shift of each section of the HNLF are calculated by minimizing the area mismatch. The method of Lagrange multipliers is used for optimization. Using the MAM technique, the optical component count in the OBP system can be reduced as compared to the uniform step size, to achieve the same performance.

In SSFS-based OBP, the dispersion and nonlinear effects are compensated in a split-step fashion analogous to SSFS using multiple concatenated FBGs and HNLFs. Although this technique is quite effective for a single channel, for a WDM system, small step size is required and hence the insertion losses due to DCF/FBG and HNLF increase, which limits the transmission performance. A dispersion-decreasing fiber (DDF) with a specific dispersion profile can be used in place of FBGs and HNLFs to fully compensate for fiber dispersive and nonlinear effects [5]. In the DDF, the dispersion coefficient decreases with distance and the walk-off effects also decrease. As a result, the nonlinear effects will increase with distance due to the enhanced interaction among co-propagating pulses. An analytical expression of the required dispersion profile of DDF has been derived [5]. Simulations show that OBP with DDF can increase the transmission performance and reach significantly as compared with DBP.

OBP has the following advantages/disadvantages over digital back propagation (DBP). (i) A very large bandwidth ($\sim$ 4 THz) is available for OBP while the bandwidth of DBP is limited by the bandwidth of the coherent receiver. (ii) DBP requires significant computational resources, especially for a WDM system and hence it is currently limited to off-line signal processing. In contrast, OBP provides compensation in real time. (iii) The number of samples per symbol available for DBP is limited by the sampling rate of the analog-to-digital converter (ADC). Although it is possible to do upsampling on DSP, it leads to additional computational complexity. However, for OBP, the signal processing is done on the analog optical waveform. (iv) OBP requires a real fiber which has loss. So, amplifiers are needed to compensate for fiber loss in the OBP section, which enhances the noise in the system.

17.2   OPTICAL BACK PROPAGATION USING OPTICAL PHASE CON-JUGATION

In this section, we investigate an OBP module that consists of an OPC followed by short lengths of high-dispersion fibers (HDFs) and HNLFs. The HDF provides the accumulated dispersion that is the same as the corresponding TF section and the set of HDF and HNLF provides a nonlinear phase shift that is the same as the corresponding TF section.

The evolution of the optical field envelope in a fiber optic link is described by the NLSE. Ignoring the higher-order dispersion and higher-order/delayed nonlinear effects, the NLSE can be written as

$$i\frac{\partial q}{\partial z} - \frac{\beta_2}{2}\frac{\partial^2 q}{\partial T^2} + \gamma_0 |q|^2 q = -i\frac{\alpha}{2}q, \tag{17.1}$$

where q is the optical field envelope and α, β_2, and γ_0 are the loss, dispersion and nonlinear coefficients, respectively. Using the transformation

$$q(z,T) = \exp[-w(z)/2]u(z,T), \tag{17.2}$$

where $w(z) = \int_0^z \alpha(s)ds$, we obtain the lossless form of the NLSE as

$$\frac{\partial u}{\partial z} = i[D(t) + N(t,z)]u(t,z), \tag{17.3}$$

$$D(t) = -\frac{\beta_2}{2}\frac{\partial^2}{\partial t^2}, N(t,z) = \gamma(z)|u(t,z)|^2, \gamma(z) = \gamma_0 \exp[-w(z)]. \tag{17.4}$$

The solution of Eq. (17.3) for a single span of TF may be written as

$$u(t,L_a) = \exp\left\{i\int_0^{L_a}[D(t) + N(t,z)]dz\right\}u(t,0), \tag{17.5}$$

where L_a is the fiber length. To compensate for the distortion due to fiber dispersion and nonlinear effects, an OBP module is placed at the end of the

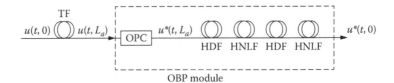

FIGURE 17.1: Schematic of a single-span fiber optic link with OBP.

TF, as shown in Fig. 17.1. Letting the output signal of the TF pass through an OPC, the output of the OPC is

$$u^*(t, L_a) = \exp\left\{-i \int_0^{L_a} [D(t) + N^*(t, z)]dz\right\} u^*(t, 0). \qquad (17.6)$$

If the rest of the OBP module (excluding the OPC) is designed to have channel matrix S,

$$S \equiv \exp\left\{i \int_0^{L_a} [D(t) + N^*(t, z)]dz\right\}, \qquad (17.7)$$

we find that the output of the OBP module will be

$$u_{OBP,out}(t) = Su^*(t, L_a) \qquad (17.8)$$

$$= \exp\left\{i \int_0^{L_a} [D+N^*]dz\right\} \exp\left\{-i \int_0^{L_a} [D+N^*]dz\right\} u^*(t,0) \qquad (17.9)$$

$$= u^*(t, 0). \qquad (17.10)$$

Then the input field can be recovered by performing a complex conjugation in the electrical domain at the receiver.

Equation (17.8) is equivalent to the NLSE

$$\frac{\partial u_b}{\partial z} = i[D(t) + N^*(t, z)]u_b(t, z), \qquad (17.11)$$

with $u_b(t, 0) = u^*(t, L_{tot})$. Hence, the OBP module can be designed by emulating an SSFS with a step size of Δz,

$$S \approx A(t) \cdot B(t, \Delta z) \cdot A(t) \cdot B(t, 2\Delta z) \cdot \dots \cdot A(t) \cdot B(t, L_a), \qquad (17.12)$$

where

$$A(t) = \exp[iD(t) \cdot \Delta z], \qquad (17.13)$$

$$B(t, x) = \exp\left[i \int_x^{x+\Delta z} \gamma(z)|u_b(t, z)|^2 dz\right]. \qquad (17.14)$$

Here, $A(t)$ and $B(t, x)$ are the operators corresponding to the fiber dispersive effect over a length Δz and the nonlinear effect over the interval $[x, x + \Delta z]$.

Typically, $\gamma(z)$ varies more rapidly than $|u_b(t,z)|^2$, so $|u_b(t,z)|^2$ can be approximated to be independent of z. The integral in Eq. (17.14) can be evaluated analytically as

$$B(t,x) = \exp\left[i\gamma_0 \Delta z_{eff} |u_b(t,x)|^2\right]. \tag{17.15}$$

$$\Delta z_{eff} = \frac{1 - \exp(-\alpha \Delta z)}{\alpha}. \tag{17.16}$$

The operator A can be realized using an HDF, as shown in Fig. 17.1, if its nonlinearity is ignored. The operator B is realized using a dispersion-shifted HNLF [13]. The HNLF introduces a nonlinear phase shift, which is the same as that of a corresponding TF of length Δz, if the nonlinearity of the HDF is absent. For example, if $\Delta z = L_a/2$ (as shown in Fig. 17.1), the first HNLF compensates for the nonlinear phase shift in the second half of the TF while the second HNLF compensates for the nonlinear phase shift in the first half of the TF.

The transmission in a HNLF is described by

$$u_{HN,out} = u_{HN,in} \exp\left(i\gamma_{HN} L_{HN,eff} |u_{HN,in}|^2\right), \tag{17.17}$$

$$L_{HN,eff} = \frac{1 - \exp(-\alpha_{HN} L_{HN})}{\alpha_{HN}}, \tag{17.18}$$

where α_{HN}, γ_{HN} and L_{HN} are the loss coefficient, nonlinear coefficient, and length of the HNLF, respectively. Equation (17.17) indicates that, for a fixed nonlinear phase shift, we can use an amplifier to increase the signal power so as to reduce the HNLF length.

So far we considered the compensation of dispersion and nonlinearity of a single-span link. For a general case, Fig. 17.2 shows a fiber optic link consisting of N spans of transmission fibers and the OBP module. In the OBP module, an amplifier is introduced after the OPC in order to reduce the required HNLF length. Also, tiny amplifiers are introduced after the HNLFs to offset the loss due to HDFs and HNLFs.

Let us first consider the case of $\Delta z = L_a$ in the presence of HDF nonlinearity. The set of HDF_j and $HNLF_j$, $j = 1, 2, ..., N$ compensates for the nonlinear phase shift of the corresponding TF_{N-j+1}. Let the nonlinear phase shift of HDF_j, $HNLF_j$, and the TF_{N-j+1} be ϕ_{HDF}, ϕ_{HN}, and ϕ_{TF}, respectively. Then, we have

$$\phi_{TF} = \phi_{HDF} + \phi_{HN}, \tag{17.19}$$

$$\phi_r = \gamma_r P_r L_{r,eff}, \tag{17.20}$$

$$L_{r,eff} = \frac{1 - \exp(-\alpha_r L_r)}{\alpha_r}, \tag{17.21}$$

where $r = $ HDF, HN, TF; and P_r is the launch power to the fiber type r. Using Eq. (17.20) in Eq. (17.19), we find

$$L_{HN,eff} = \frac{\phi_{TF} - \phi_{HDF}}{\gamma_{HN} P_{HN}}. \tag{17.22}$$

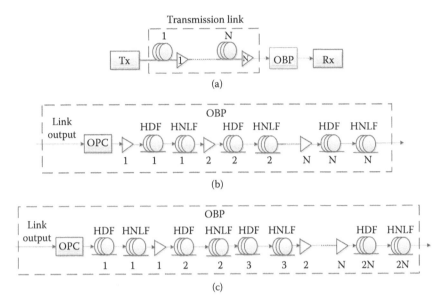

FIGURE 17.2: (a) Schematic of a fiber optic link with OBP; (b) diagram of the OBP with step size $\Delta z = L_a$; (c) diagram of the OBP with step size $\Delta z = L_a/2$.

If the dispersion of the HDF were to be zero, Eq. (17.22) holds true exactly. However, because of the large dispersion of the HDF, Eq. (17.22) is approximate and it should be used as a rough guide to optimize the HNLF length.

For the case of $\Delta z = L_a/2$, the first set of HDF and HNLF in Fig. 17.2(c) compensates for the nonlinear effects of the second half of the last span of the transmission fiber. Since the nonlinear phase shift due to the second half is quite small, we do not really need an amplifier following the OPC. So, the first amplifier is placed after the first set of the HDF and HNLF (see Fig. 17.2(c)). The analytical length of HNLFs can be calculated as before.

It is desirable that the HDF and HNLF have high dispersion and high nonlinearity, respectively, and they are of the shorter lengths so that their insertion losses are minimum. If the standard single-mode fiber (SSMF) is used as the TF, the dispersion of the HDF should be anomalous. However, an HDF with anomalous dispersion is not commercially available. We have used the negative dispersion fiber (NDF) [14] as the TF so that the conventional DCF with high normal dispersion can be used as the HDF.

We simulated a single-channel fiber optic system with OBP at the receiver with the following parameters: symbol rate = 25 Gsymbols/s, modulation = 32 quadrature amplitude modulation (QAM), transmission fiber dispersion = 5 ps²/km, transmission fiber loss = 0.2 dB/km, nonlinear coefficient = 2.2 $W^{-1}km^{-1}$, amplifier spacing = 80 km, spontaneous emission noise factor = 1.5, dispersion of the HDF = 140 ps²/km, loss of the HDF = 0.4 dB/km, nonlinear coefficient of the HDF = 4.4 $W^{-1}km^{-1}$, loss of the HNLF = 0.3 dB/m, and the nonlinear coefficient of the HNLF = 2000 $W^{-1}km^{-1}$. We employ two OBP step size realizations to compensate for the fiber impairment in the transmission link. One is that the step size equals the amplifier spacing L_a; the other is that the step size equals half the amplifier spacing. In the case of $\Delta z = L_a$, we numerically optimized the gain of the first amplifier to obtain the best performance. The optimum gain is found to be 4.8 dB. Then, we optimized the HNLF length numerically to obtain the minimum bit error ratio (BER). We found the optimum HNLF length as 3.1 m at the transmission fiber launch power of 1 dB. The analytical length found using Eq. (17.22) is 3.2 m, in good agreement with numerics. Total loss due to HDF and HNLF is 2.07 dB, which is compensated by the in-line amplifiers in OBP. The amplifiers in OBP have an n_{sp} of 1.5.

In the case of $\Delta z = L_a/2$, the gain of the first amplifier and lengths of HNLFs are numerically optimized. The first amplifier compensates for the loss of the first set of HDF and HNLF, and it gives an extra gain. The optimum extra gain is 3 dB. At 1 dBm transmission fiber launch power, the numerically optimized lengths of $HNLF_1$, $HNLF_2$, $HNLF_3$ are 0.28 m, 10.6 m, and 2.4 m, respectively. The corresponding analytical lengths are 0.28 m, 11.9 m and 1.5 m, respectively.

Figure 17.3 shows the BER as a function of the launch power in different system configurations when the transmission distance is 800 km. The dashed line and the solid line represent the BER of the OBP with step size equaling L_a and $L_a/2$, respectively. In order to compare the OBP with the other schemes, we calculated the BER of the DBP with $\Delta z = L_a$ and the midpoint OPC. In the single-channel simulation of the system based on DBP, eight samples per symbol are used in the transmission link, and after the ADC, two samples

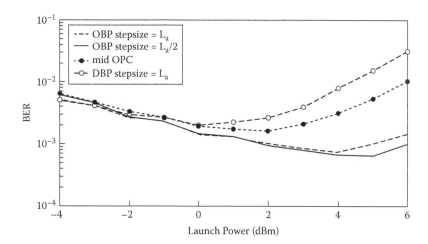

FIGURE 17.3: BER vs launch power. Transmission distance = 800 km.

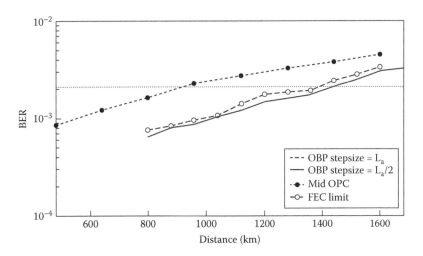

FIGURE 17.4: BER vs transmission distance.

per symbol are used. As shown in Fig. 17.3, the four schemes have almost the same performance if the launch power is less than 0 dBm when the nonlinear effects are small. Beyond 0 dBm launch power, OBP with $\Delta z = L_a/2$ has the best tolerance to nonlinearity. The relatively poor performance of the DBP as compared to the OBP is mainly attributed to the fact that the nonlinear HDF in OBP provides distributed nonlinearity compensation and partly due to the lower number of samples per symbol of DBP. Midpoint OPC does not perform well enough due to the unsymmetrical power profile with respect to the location of the OPC.

Figure 17.4 shows the dependence of the BER on the transmission distance. Each point in Fig. 17.4 is obtained after optimizing the launch power. Using the midpoint OPC, the maximum reach is about 880 km, which can be increased to 1360 km with OBP, $\Delta z = L_a$, and to 1440 km with OBP, $\Delta z = L_a/2$.

17.3 OPTICAL BACK PROPAGATION WITH OPTIMAL STEP SIZE

In this section, we investigate an improved OBP scheme with optimal step size. The OBP module consists of an OPC, FBGs, and HNLFs. Transmission fiber dispersion is compensated by the FBGs and nonlinearity is compensated by HNLFs. Several sections of FBGs and HNLFs are concatenated in a way analogous to the split-step Fourier scheme used for solving the nonlinear Schrödinger equation. The optimum accumulated dispersion of the each section of FBG and the optimum nonlinear phase shift of each section of the HNLF are calculated by minimizing the mismatch between the area under the exponentially increasing nonlinearity profile and its stepwise approximation.

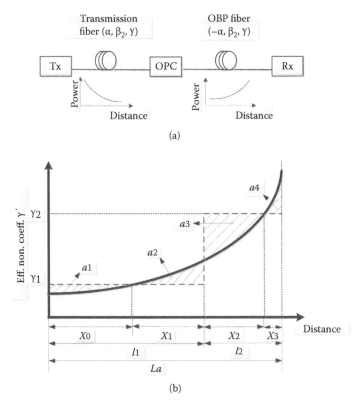

FIGURE 17.5: (a) Diagram of a fiber optic link with OBP; (b) effective nonlinear coefficient and its stepwise approximation for $M = 2$.

This OBP scheme leads to significant performance improvement and/or reach enhancement as compared to uniformly spaced sections, for the given number of sections.

As shown in Section 17.1, the output field of a single-span optical fiber may be written as

$$q(t, L_a) = \exp\left\{ i \int_0^{L_a} [D(t) + N'(q)]dz \right\} q(t, 0), \qquad (17.23)$$

$$D(t) = -\frac{\beta_2}{2}\frac{\partial^2}{\partial t^2}, N'(q) = \gamma_0|q(t, z)|^2 + i\frac{\alpha}{2}. \qquad (17.24)$$

As shown in Fig. 17.5(a), let the output of the transmission fiber be phase-conjugated using an OPC, so that

$$q^*(t, L_a) = \exp\left\{ -i \int_0^{L_a} [D(t) + N'^*(q)]dz \right\} q^*(t, 0). \qquad (17.25)$$

Assume an ideal OBP fiber that is identical to the transmission fiber except that its loss profile (or equivalently, the power profile) is inverted. In other words, the nonlinear operator corresponding to the OBP fiber is $N'^*(q)$. The output of the OBP fiber is

$$q_{OBP,out}(t) = \exp\left\{i \int_0^{L_a} [D(t) + N'^*(q)]dz\right\} q^*(t, L_a)$$

$$= \exp\left\{i \int_0^{L_a} [D + N'^*]dz\right\} \exp\left\{-i \int_0^{L_a} [D + N'^*]dz\right\} q^*(t, 0)$$

$$\tag{17.26}$$

$$= q^*(t, 0).$$

Equation (17.26) is equivalent to

$$\frac{\partial q_b}{\partial z} = i[D(t) + N'^*(q_b)]q_b, \tag{17.27}$$

with $q_b(t, 0) = q^*(t, L_a)$. Using a transformation $q_b = \exp(\alpha z/2)u_b$, Eq. (17.27) can be written as

$$i\frac{\partial u_b}{\partial z} - \frac{\beta_2}{2}\frac{\partial^2 u_b}{\partial t^2} + \gamma'(z)|u_b|^2 u_b = 0, \tag{17.28}$$

$$\gamma'(z) = \gamma_0 \exp(\alpha z). \tag{17.29}$$

If the effective nonlinear coefficient, $\gamma'(z)$, of the OBP fiber increases exponentially with distance (or equivalently, optical power increases exponentially with distance), the output of the OBP fiber would be the conjugate of the transmitter output in the absence of noise. But it is hard to design a fiber with such a property. Instead, we divide the OBP fiber into M sections of length l_j, $j = 1, 2, ..., M$. If the sections are sufficiently small, the propagation in each of the sections is approximated by a single-step SSFS, i.e.,

$$u_b(t, z_{j-1} + l_j) \cong \exp\left(\frac{iD(t)l_j}{2}\right) \exp\left[i \int_{z_{j-1}}^{z_{j-1}+l_j} \gamma'(z)|u_b(t, z)|^2 dz\right]$$

$$\times \exp\left(\frac{iD(t)l_j}{2}\right) u_b(t, z_{j-1}), \tag{17.30}$$

where $z_j = \sum_{k=1}^{j} l_k$, $z_0 = 0$, and $z_M = L_a$. The dispersion operator is realized by an FBG with $\gamma_0 = 0$ and the nonlinear operator is realized by the dispersion-shifted HNLF with $\beta_2 = 0$. Consider the following two cases.

   Case (i) Uniform Spacing:

The OBP fiber has M sections of equal length $l_j = L_a/M$. Each section of the OBP fiber is realized by a combination of FBGs and HNLF. The accumulated dispersion, ξ_j, and nonlinear phase shift, ϕ_j, of the jth section provided by FBGs and HNLF, respectively, are

$$\xi_j = \beta_2 l_j, \quad \phi_j = \gamma_0 L_{eff,j}|q_{b,j}|^2 = \gamma_{HN} L_{HN,eff,j}|q_{b,j}|^2, \tag{17.31}$$

$$L_{eff,j} = \frac{e^{\alpha l_j} - 1}{\alpha}, \quad L_{HN,eff,j} = \frac{1 - e^{-\alpha_{HN} L_{HN,j}}}{\alpha_{HN}}, \tag{17.32}$$

where $q_{b,j}$ is the input field of the jth section, γ_{HN}, α_{HN} and $L_{HN,j}$ are the nonlinear coefficient, loss coefficient and length of the jth HNLF, respectively.

Case (ii) Minimum Area Mismatch (MAM):

As shown in Fig. 17.5(b), the exponentially increasing effective nonlinear coefficient, $\gamma'(z)$, (solid line) can be approximated by a stepwise increasing function (dashed line) such that the sum of the area mismatch between the area under the exponential curve in each section and the area under the corresponding stepwise approximated curve in that section is minimum [15]. Figure 17.5 illustrates the method for the case of $M = 2$. The area mismatch, Δ_1, for the first section is $a_1 + a_2$ where

$$a_1 = \gamma_1 x_0 - \gamma_0 \left(\frac{e^{\alpha x_0} - 1}{\alpha} \right), \quad x_0 = \frac{1}{\alpha} \ln \left(\frac{\gamma_1}{\gamma_0} \right), \tag{17.33}$$

$$a_2 = -\gamma_1 (l_1 - x_0) + \frac{\gamma_0 e^{\alpha l_1} - \gamma_1}{\alpha}. \tag{17.34}$$

The total area mismatch of both sections is $\Delta_1 + \Delta_2 = \sum_{j=1}^{4} a_j$. The section length l_1 and the effective nonlinear coefficients γ_1 and γ_2 are so chosen that $\Delta_1 + \Delta_2$ is minimum under the constraint that the total area under the exponential curve is the same as that under the stepwise curve, i.e.,

$$\gamma_1 l_1 + \gamma_2 (L_a - l_1) = \gamma_0 \left(\frac{e^{\alpha L_a} - 1}{\alpha} \right) \equiv K. \tag{17.35}$$

To solve this optimization problem, we define a Lagrange function [16]

$$\Lambda = \sum_{j=1}^{4} a_j + \lambda[\gamma_1 l_1 + \gamma_2 (L_a - l_1) - K], \tag{17.36}$$

where λ is the Lagrange multiplier. Differentiating Λ with respect to its arguments, and simplifying the resulting equations, we find,

$$\gamma_1 = \gamma_0 e^{\frac{\alpha}{2} l_1 (1 - \lambda)}, \quad \gamma_2 = \gamma_0 e^{\frac{\alpha}{2}[l_1(1+\lambda) + L_a(1-\lambda)]}, \tag{17.37}$$

$$\frac{K - \gamma_1 l_1}{L_a - l_1} - \gamma_0 e^{\alpha l_1} \gamma_1^{(L_a - l_1)/l_1} = 0. \tag{17.38}$$

Equation (17.20) is a transcendental equation for l_1 which needs to be numerically computed. The propagation in each OBP fiber section is approximated by the SSFS. The nonlinear coefficient of the jth HNLF need not be γ_j, but the nonlinear phase shift imparted by the jth HNLF should be

$$\int_{z_{j-1}}^{z_{j-1}+l_j} \gamma'(z)|u_b(t,z)|^2 dz \cong \gamma_j l_j |u_b(t,z_{j-1})|^2, \tag{17.39}$$

$$\gamma_j l_j = \gamma_{HN} L_{HN,eff,j}, \quad j = 1,2,...,M. \tag{17.40}$$

When the number of steps increases, it becomes harder to optimize using the Lagrange multiplier technique. Instead, the steepest descent method can be used to optimize the HNLF lengths. For the case of M sections, the total area mismatch is

$$\chi\left(l_1,l_2,...,l_{M-1},\gamma_1,\gamma_2,...,\gamma_M\right) = \sum_{m=1}^{2 \times M} a_m. \tag{17.41}$$

For convenience, we define

$$x_k = \sum_{j=1}^{k} l_j, \quad x_0 = 0. \tag{17.42}$$

The total area mismatch χ is now a function of x_k and γ_k, $k = 1,2,...,M$. Using the steepest descent method, the derivatives of χ with respect to unknown parameters are found as

$$\frac{\partial \chi}{\partial x_k} = 2\gamma_0 \exp(x_k) - \gamma_k - \gamma_{k+1}, \quad k = 1,2,...,M-1, \tag{17.43}$$

$$\frac{\partial \chi}{\partial \gamma_k} = \frac{2}{\alpha}\left(\ln\frac{\gamma_k}{\gamma_0} + 1\right), \quad k = 1,2,...,M. \tag{17.44}$$

And the parameters are updated in the $(n+1)$-th step using

$$x_k^{(n+1)} = x_k^{(n)} - \frac{\partial \chi}{\partial x_k}\Delta_x, \quad \gamma_k^{(n+1)} = \gamma_k^{(n)} - \frac{\partial \chi}{\partial \gamma_k}\Delta_\gamma, \tag{17.45}$$

where Δ_x and Δ_γ are the iteration step sizes. The number of unknown parameters is $2M-1$. In order to reduce the number of unknown parameters (especially for large M), we fix the value of γ_j by the mean value of the jth section using

$$\gamma_j = \frac{1}{l_j}\int_{z_{j-1}}^{z_j} \gamma_0 \exp(-\alpha z)dz, \tag{17.46}$$

which indicates that the nonlinear phase shift in each section is exactly compensated. That way the number of unknown parameters is reduced roughly by a factor of 2.

Figure 17.6(a) shows a schematic of a fiber optic link consisting of N spans of transmission fiber and the OBP module. The OBP is applied after the whole transmission link. The schematic of the OBP module is shown in Fig. 17.6(b). The output of the OPC passes through a pre-amplifier of gain G_{pre} so that the lengths of HNLFs can be reduced. A bandpass filter BPF1 is introduced after the pre-amplifier to remove the out of band amplified spontaneous emission (ASE) noise. The OBP fiber shown in Fig. 17.5(a) is approximated by M sections consisting of FBGs and HNLFs. To compensate for the losses of FBGs and HNLFs, an amplifier of gain G is used. Since there are N spans in the transmission system, cascaded OBPs, with each OBP consisting of M sections, are required. A bandpass filter BPF2 is introduced after the cascaded OBPs to maximize the signal-to-noise ratio (SNR).

We simulated a single-channel and single-polarization fiber optic system with OBP at the receiver with 25 Gsym/s symbol rate and 32 QAM. Standard single-mode fiber (SSMF) is used as the transmission fiber, with the parameters, $\alpha = 0.046$ km^{-1}, $\beta_2 = -21$ ps^2/km, and $\gamma = 1.1$ W^{-1}km^{-1}. The amplifier spacing L_a is 80 km and the gain of the amplifier is 16 dB. The spontaneous emission noise factor n_{sp} is 1.5 for all the amplifiers (in-line amplifiers as well as OBP amplifiers). Noise loading is done on a per amplifier basis. The parameters for OBP are as follows: loss of FBG = 1.8 dB, nonlinear coefficient of FBG = 0 W^{-1}km^{-1}, loss of the HNLF = 0.3 dB/m, dispersion coefficient of the HNLF = 0 ps^2/km and the nonlinear coefficient of the HNLF = 2000 W^{-1}km^{-1}. The gain of the pre-amplifier G_{pre} is 14 dB. The gain of the amplifiers in the cascaded OBP, G, is 5.7 dB when $M = 2$, which exactly compensates for the losses due to FBGs and HNLFs. In the case of $M = 2$,

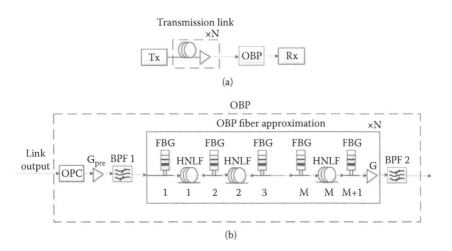

(a)

(b)

FIGURE 17.6: (a) Schematic of a fiber optic link with OBP; (b) diagram of the OBP. BPF, bandpass filter.

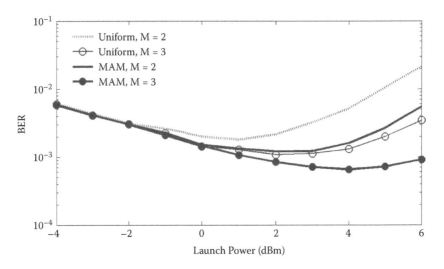

FIGURE 17.7: BER versus launch power for various OBP schemes. Transmission distance = 800 km.

lengths of HNLFs are $L_{HN,1} = 0.2036(0.0936)$ m, $L_{HN,2} = 0.7876(0.9534)$ m for MAM (uniform spacing). Eight and two samples per symbol are used for optical propagation (both forward and backward) and for digital processing, respectively. A coherent receiver is used with the local oscillator laser linewidth (=transmitter laser linewidth) = 22 kHz. A feedforward carrier recovery algorithm is used for phase estimation [17]. Digital filters are not used for dispersion compensation as the OBP compensates for dispersion. Second order Gaussian filters with bandwidths 80 GHz and 50 GHz are used prior to the OBP (BPF1 in Fig. 17.6(b)) and after the OBP (BPF2), respectively. The gain of the pre-amplifier and filter bandwidths is optimized to obtain the minimal BER. The gain of the pre-amplifier is so chosen that the power launched to the first FBG is 14 dB higher than the power launched to the transmission fiber. It is possible to choose a lower gain of the pre-amplifier, and in that case, the lengths of HNLFs become longer and because of the losses in HNLFs, there would be slight performance degradation.

Figure 17.7 shows the BER as a function of the SSMF launch power at 800 km. As can be seen, for the given M, the OBP system with sections designed using the MAM (Case 2) outperforms that designed using the uniformly spaced sections (Case 1). For a 10 span system and $M = 2$, 30 FBGs and 20 HNLFs are required for OBP to compensate for dispersion and nonlinearity of the whole transmission link.

The minimum BER (such as the minimum point in Fig. 17.7) is plotted as a function of the transmission reach in Fig. 17.8. At the forward error collection (FEC) limit of 2.1×10^{-3}, the transmission reach is limited to 560

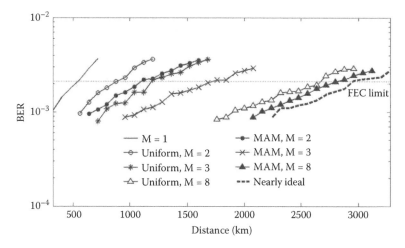

FIGURE 17.8: BER versus transmission distance for various OBP schemes.

km for $M = 1$, which can be increased to 1200 km and 1680 km for $M = 3$ with uniform spacing and $M = 3$ with MAM, respectively. Thus we see that the MAM technique leads to significant reach enhancement as compared to uniform spacing for the given M. If nearly ideal OBP is employed with lossless FBGs and HNLFs and a very small step size of 1 km, transmission reach can be increased to 2960 km. With MAM, $M = 8$, the maximum reach is 2880 km, which is 97% of the reach obtained using the nearly ideal OBP. For the case of nearly ideal OBP, the optimal signal launch power corresponding to the 2960 km transmission link is 6 dBm. If the system were to be linear, there should be no limit on the achievable reach assuming that the launch power to the transmission fiber can be increased arbitrarily. The maximum reach of 2960 km for the nearly ideal case is due to signal-ASE nonlinear interaction. Although receiver-based ideal OBP can compensate for the deterministic (bit pattern dependent) nonlinear effects exactly, signal-ASE nonlinear interaction cannot be compensated for by the receiver-based ideal OBP. In addition, the interaction of polarization mode dispersion (PMD) and nonlinearity could lead to performance degradations which cannot be recovered by the ideal OBP. However, this effect is not considered in our simulation.

Finally, we note that it may be possible to develop a single integrated optic device with alternating sections of Bragg gratings and highly nonlinear waveguides or multiple sections of the gratings written on the highly nonlinear waveguide. For the latter case, optimum section lengths calculated using Eq. (17.38) are still applicable, but the split-step approximation done in Eq. (17.30) is not required, as dispersion and nonlinearity act simultaneously and hence the performance would be better than that shown here for the given M.

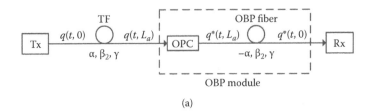

(a)

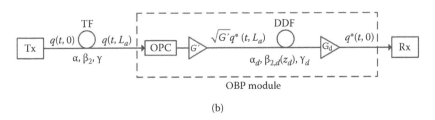

(b)

FIGURE 17.9: A single-span fiber optic system with (a) OBP using an ideal optical back propagation fiber with negative loss coefficient; (b) OBP using a DDF and amplifiers. DDF: dispersion-decreasing fiber.

17.4 IDEAL OPTICAL BACKPROPAGATION USING DISPERSION-DECREASING FIBER

In Sections 17.2 and 17.3, OBP schemes consisting of OPC, HDF/FBG, and HNLF are investigated. DCF/FBG is used to compensate for dispersion, and HNLF is used to compensate for nonlinearity. The dispersion and nonlinear effects are compensated in a split-step fashion analogous to SSFS that is used to solve the NLSE. Although this technique is quite effective for a single channel, for a WDM system, small step size is required and hence the insertion losses due to DCF/FBG and HNLF increase, limiting the transmission performance. In this section, we investigate the possibility of introducing a single optical device which can exactly compensate for dispersion and nonlinearity. A dispersion-decreasing fiber (DDF) with a specific dispersion profile is found to meet our requirements.

It is shown in Section 17.2 and 17.3 that an OPC followed by an ideal OBP fiber with parameters $(-\alpha, \beta_2, \gamma)$ can be used to compensate for the distortions due to propagation in a TF with parameters $(\alpha, \beta_2, \gamma)$, as shown in Fig. 17.9(a). The NLSE for the OBP fiber is

$$\frac{\partial q_b}{\partial z_b} = i[D(t) + N'^*(t, z_b)]q_b(t, z_b), \qquad (17.47)$$

with $q_b(t, 0) = q^*(t, L_a)$, and z_b is the distance in OBP fiber. Using transformations

$$q_b = \sqrt{P_{in}}e^{-\alpha(L_a - z_b)/2}u_b \qquad (17.48)$$

and

$$dz_b' = \beta_2 dz_b, \tag{17.49}$$

Eq. (17.47) can be rewritten as

$$i\frac{\partial u_b}{\partial z_b'} - \frac{1}{2}\frac{\partial^2 u_b}{\partial t^2} + \frac{\gamma P_{in}}{\beta_2}e^{-\alpha(L_a - z_b)}|u_b|^2 u_b = 0, \tag{17.50}$$

where P_{in} is the power launched to the TF. Equation (17.50) describes the field propagation in an ideal fiber with a constant β_2 and a negative loss coefficient (or equivalently the power increasing with distance) that exactly compensates for dispersion and nonlinearity of the TF. However, it is hard to realize such a fiber in practice. For an ideal OBP, we like to have a short length of a fiber (so that its insertion loss is small) which provides the same response as that of the ideal OBP fiber given by Eq. (17.50). Here, we derive an equivalent way of realizing Eq. (17.50) by using amplifiers and a DDF with positive loss coefficient α_d and a dispersion profile $\beta_{2,d}(z_d)$ [see Fig. 17.9(b)].

The optical field envelope in the DDF is described by

$$i\frac{\partial q_b}{\partial z_d} - \frac{\beta_{2,d}(z_d)}{2}\frac{\partial^2 q_b}{\partial t^2} + \gamma_d |q_b|^2 q_b + i\frac{\alpha_d}{2}q_b = 0, \tag{17.51}$$

where α_d and γ_d are the loss and nonlinear coefficients of DDF, respectively; z_d is the distance in the DDF; $q_b(t,0) = \sqrt{G'}q^*(t, L_a)$ and G' is the gain of the amplifier preceding DDF. Using transformations

$$q_b = \sqrt{P_d}e^{-\alpha_d z_d/2}u_b \tag{17.52}$$

and

$$dz_d' = \beta_{2,d}(z_d)dz_d, \tag{17.53}$$

Eq. (17.51) can be rewritten as

$$i\frac{\partial u_b}{\partial z_d'} - \frac{1}{2}\frac{\partial^2 u_b}{\partial t^2} + \frac{\gamma_d P_d e^{-\alpha_d z_d}}{\beta_{2,d}(z_d)}|u_b|^2 u_b = 0, \tag{17.54}$$

where $P_d = GP_{in} = G'e^{-\alpha L_a}P_{in}$ is the input power of the DDF. Equations (17.50) and (17.54) are identical only if

$$dz_b' = dz_d', \tag{17.55}$$

and

$$\frac{\gamma P_{in}}{\beta_2}e^{-\alpha(L_a - z_b)} = \frac{\gamma_d P_d e^{-\alpha_d z_d}}{\beta_{2,d}(z_d)}. \tag{17.56}$$

Substituting Eqs. (17.49) and (17.53) in Eq. (17.55), we find

$$\beta_2\frac{dz_b}{dz_d} = \beta_{2,d}(z_d), \tag{17.57}$$

$$w \equiv \beta_2 z_b = \int_0^{z_d} \beta_{2,d}(z_d) dz_d, \tag{17.58}$$

$$\frac{dw}{dz_d} = \beta_{2,d}(z_d). \tag{17.59}$$

Substituting Eqs. (17.58) and (17.59) in Eq. (17.56), we obtain

$$\frac{dw}{dz_d} e^{\alpha w / \beta_2} = \left(\frac{\gamma_d P_d \beta_2}{\gamma P_{in}} \right) e^{\alpha L_a} e^{-\alpha_d z_d}. \tag{17.60}$$

Integrating Eq. (17.60), we find

$$\frac{1}{\alpha} \left(e^{\frac{\alpha}{\beta_2} w(z_d)} - 1 \right) = \left(\frac{\gamma_d P_d}{\gamma P_{in}} \right) e^{\alpha L_a} \frac{1 - e^{-\alpha_d z_d}}{\alpha_d}. \tag{17.61}$$

Simplifying Eq. (17.61), we obtain

$$w(z_d) = \frac{\beta_2}{\alpha} \ln \left\{ 1 + \frac{\gamma_d G \alpha}{\gamma e^{-\alpha L_a}} \frac{1 - e^{-\alpha_d z_d}}{\alpha_d} \right\}, \tag{17.62}$$

$$\beta_{2,d}(z_d) = \frac{e^{-\alpha_d z_d}}{\frac{\gamma e^{-\alpha L_a}}{\gamma_d G} + \alpha \left(\frac{1 - e^{-\alpha_d z_d}}{\alpha_d} \right)} \beta_2. \tag{17.63}$$

The length of DDF L_d is found as follows. The total accumulated dispersion of the ideal OBP fiber [Fig. 17.9(a)] should be the same as that of the DDF, i.e.,

$$\beta_2 L_a = w(L_d) = \int_0^{L_d} \beta_{2,d}(z_d) dz_d, \tag{17.64}$$

or

$$L_d = -\frac{1}{\alpha_d} \ln \left\{ 1 - \frac{\alpha_d \gamma e^{-\alpha L_a}}{\gamma_d G \alpha} \left(e^{\alpha L_a} - 1 \right) \right\}. \tag{17.65}$$

If the dispersion profile of the DDF is tailored to satisfy Eq. (17.63), the combination of the amplifiers and DDF provides the ideal response described by Eq. (17.50), and hence, signal-signal nonlinear interactions can be exactly compensated. The amplifier with gain $G_d = e^{\alpha_d L_d}$ is introduced after the DDF [see Fig. 17.9(b)] to compensate for the loss of DDF. Figure 17.10 shows the dispersion profiles of DDF that satisfy Eq. (17.63). As can be seen, the relatively shorter length of DDF can compensate for the dispersion and nonlinear effects of the TF.

Figure 17.11 shows the schematic of a WDM fiber optic transmission system consisting of M transmitters, N spans of TFs, the OBP module, and M coherent receivers. The OBP is applied at the end of the transmission link. A pre-amplifier with gain G is introduced so that the required dispersion profile and length of the DDF can be adjusted according to Eqs. (17.63) and (17.65), respectively. A BPF is introduced to remove the out of band ASE

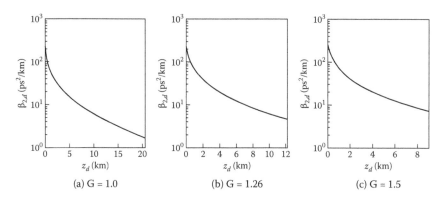

FIGURE 17.10: Dispersion profiles of DDF. TF parameters: $\alpha = 0.2$ dB/km, $\beta_2 = 5$ ps^2/km, $\gamma = 2.2$ W^{-1}km^{-1}, $L_a = 60$km. DDF parameters: $\alpha_d = 0.4$ dB/km, $\gamma_d = 4.86$ W^{-1}km^{-1}. (a) $G = 1.0$: $\beta_{2,d}(0) = 175.1$ ps^2/km, $L_d = 20.5$ km; (b) $G = 1.26$: $\beta_{2,d}(0) = 220.6$ ps^2/km, $L_d = 12.1$ km; (c) $G = 1.5$: $\beta_{2,d}(0) = 262.6$ ps^2/km, $L_d = 9.0$ km.

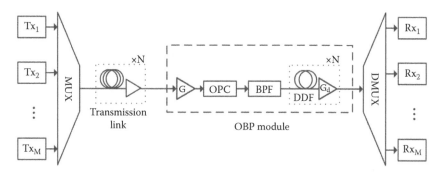

FIGURE 17.11: Schematic diagram of a WDM fiber optic transmission system with OBP. MUX: multiplexer; DMUX: demultiplexer.

noise. During back propagation, amplifiers with gain G_d are used to fully compensate for the loss of each span of DDF.

In DBP, the compensation of fiber dispersion and nonlinearity is implemented in a step-wise manner and the performance is usually limited by the step size, which has to be traded off against computational cost or system complexity. In WDM systems, the required computational load may prevent DBP from real-time implementation. In the OBP with DDF, the compensation of dispersion and nonlinearity is realized by a gradually decreasing dispersion profile, which inherently has a very small step size. The DDF with exponentially dispersion decreasing fibers have been fabricated before [18,19]. The step size of the order of a few meters in DDF can be realized and hence nearly ideal OBP can be realized using DDF. The DDF can be fabricated by tapering the fiber during the drawing process, which alters the waveguide contribution to the dispersion. The maximum dispersion required for OBP fiber is of the same order as the commercially available dispersion compensation fiber and of the same sign.

We simulated a WDM fiber optic transmission system with OBP at the receiver with the following parameters: number of WDM channels = 5, channel spacing = 100 GHz, symbol rate per channel = 25 Gsymbols/s, modulation = 32 QAM, number of symbols simulated = 32768 per channel. The linewidths of the transmitter and local oscillator lasers are 100 kHz each. The dispersion, loss, and nonlinear coefficients of the TF are $\beta_2 = 5$ ps^2/km, $\alpha = 0.2$ dB/km, and $\gamma = 2.2$ W^{-1}km^{-1}, respectively. This type of fiber has been fabricated before and it is known as negative dispersion fiber (NDF). The amplifier spacing is 60 km, and the spontaneous emission noise factor is $n_{sp} = 1.5$. The BPF shown in Fig. 17.11 is a second order Gaussian filter with a full bandwidth of 450 GHz. For the DDF, $\alpha_d = 0.4$ dB/km, $\gamma_d = 4.86$ W^{-1}km^{-1}, and $L_d = 12.1$ km [see Fig. 17.10(b)]. The corresponding amplifier gain for compensating the DDF loss is 4.84 dB. In all the simulations, 32 samples per symbol are used in the transmission link so as to obtain a frequency window covering all the WDM channels. In DBP simulations, two samples per symbol are used after the ADC unless otherwise specified, while in OBP simulations, back propagation is in the optical domain and 32 samples per symbol are used. Using the method of [20], the coupled NLSE is used to compensate for the interchannel nonlinear impairments, ignoring four-wave mixing (FWM). However, the OBP scheme compensates for both cross-phase modulation (XPM) and FWM simultaneously. The central channel is demultiplexed using a second order Gaussian filter with the full bandwidth of 50 GHz. In the coherent receiver, for OBP, two samples per symbol are used after the ADC and phase noise compensation is done using the approach of [17]. A low pass filter (LPF) of bandwidth 25 GHz is used prior to phase noise compensation. For the DBP scheme, the coupled NLSE is solved in the digital domain prior to phase noise compensation. The optical and electrical filter bandwidths are optimized in both the OBP and DBP schemes.

Figure 17.12 shows the BER as a function of the launch power per WDM channel when the transmission distance is 1200 km. The solid curve represents the BER of OBP using DDFs, and the dashed and dotted curves represent the BER of DBP with 3 km and 10 km step sizes, respectively. The DBP step size of the simulated WDM system is limited by the walk-off length [20], which is 3.2 km. We found that there is no obvious performance improvement when a step size smaller than 3 km is chosen for DBP. Also, Fig. 17.12 shows the simulation results of DBP with four samples/symbol ADC sampling rate and DBP with DSP upsampling [6] from two to four samples/symbol. The DBP performance can be improved by increasing the ADC sampling rate or DSP upsampling, at the cost of increased system complexity or computational cost. The OBP outperforms DBP (two samples/symbol, step size = 3 km) by 2.0

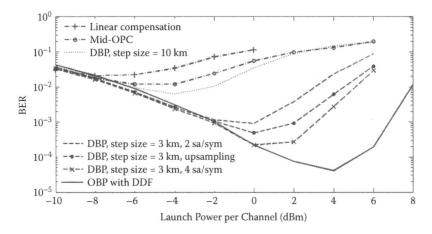

FIGURE 17.12: BER versus launch power per WDM channel. Transmission distance = 1200 km.

dBQ. The relatively poor performance of DBP as compared to OBP is mainly due to the down-sampling penalty and the lack of FWM compensation. The performance of midpoint OPC is worse than DBP, because the power profile is unsymmetrical with respect to the location of OPC. The performance of OBP is worse than that of DBP (with step size = 3 km) when the launch power is less than −2 dBm, which is due to the optical signal to noise ratio (OSNR) penalty resulting from OBP amplifiers. The OSNR penalty due to OBP amplifiers is found to be 0.56 dB. From Fig. 17.12, it can also be seen that the DBP with a step size of 10 km performs worse than the DBP with a step size of 3 km even at lower launch powers (−10 dBm to−6 dBm) due to residual nonlinearity. The curve with + shows the case where no OBP (or DBP) is applied and fiber dispersion and laser phase noise are compensated in the receiver. As can be seen, the performance of this system is much worse than the system with DBP or OBP.

Figure 17.13 shows the minimum BER as a function of transmission distance. The BER_{min} is obtained by optimizing the launch power for each distance. At the BER of 2.1×10^{-3}, the transmission reaches linear compensation only and midpoint OPC are 300 km and 360 km, respectively. For DBP with a 10 km step size and two samples/symbol sampling rate, the reach is 760 km, which can be increased to 1600 km by using a 3 km step size at the cost of more than tripling the computational effort. The transmission reach of OBP with DDF is 2460 km. Although the OBP fully compensates for signal-signal nonlinear interactions, it neither compensates for signal-ASE nonlinear interactions [21, 22] nor mitigates nonlinear PMD [23], which are the limiting factors to enhance the reach in systems based on OBP.

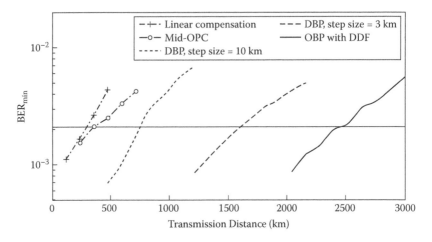

FIGURE 17.13: BER_{min} versus transmission distance.

17.5   CONCLUSION

Backpropagation is an effective technique to compensate for deterministic dispersive and nonlinear effects in optical fibers. Digital back propagation (DBP) is performed in the electrical domain using high-speed digital signal processor (DSP). DBP is effective to compensate for intra-channel nonlinearities in single-channel fiber optic systems, but it requires enormous computational resources in wavelength division multiplexing (WDM) systems and is limited to off-line processing currently. Optical back propagation (OBP) implements signal processing in the optical domain which inherently has large bandwidth and circumvents the heavy computational loads and the limitations of the ADC sampling rate in DBP. OBP is promising for real-time applications. OBP can be realized by concatenating multiple fiber Bragg gratings (FBGs) and highly nonlinear fibers (HNLFs) to emulate the split-step Fourier scheme (SSFS). In such schemes, the exponential profile of the nonlinear coefficient (which is ideal for nonlinearity compensation) is approximated by a stepwise curve. Instead of using a uniform step size in the stepwise curve, a better approximation of the exponential profile can be obtained by using variable step sizes optimized by the minimum area mismatch (MAM) method, which minimizes the area mismatch between the stepwise curve and the exponential profile. Using the MAM technique, the transmission performance is significantly enhanced. Another design of OBP is based on a dispersion-decreasing fiber (DDF). DDF with a specific dispersion profile (given in Eq. (17.63)) is found to be able to fully compensate for the dispersion and nonlinearities of a transmission fiber. The OBP scheme with DDF eliminates the requirement of multiple discrete optical components such as FBGs and HNLFs, and hence the insertion loss will be significantly reduced. Moreover, the OBP scheme with

DDF has a very small step size (on the order of a few meters) and offers almost ideal backpropagation. However, both DBP and OBP neither compensate for signal-amplified spontaneous emission (ASE) noise nonlinear interaction nor mitigate nonlinear polarization mode dispersion (PMD) effects, which are the limiting factors to further enhance the transmission performance.

REFERENCES

1. K. P. Ho and J. M. Kahn, Electronic compensation technique to mitigate nonlinear phase noise, *J. Lightwave Technol.*, 22, 779, 2004.
2. S. Kumar and D. Yang, Optical backpropagation for fiber optic communications using highly nonlinear fibers, *Opt. Lett.*, 36, 1038, 2011.
3. J. Shao and S. Kumar, Optical backpropagation for fiber optic communications using optical phase conjugation at the receiver, *Opt. Lett.*, 37, 3012, 2012.
4. S. Kumar and J. Shao, Optical back propagation with optimal step size for fiber optic transmission systems, *IEEE Photon. Technol. Lett.*, 25, 523, 2013.
5. X. Liang, S. Kumar and J. Shao, Ideal optical backpropagation of scalar NLSE using dispersion-decreasing fibers for WDM transmission, *Opt. Express,* 211, 28668, 2013.
6. X. Li, X. Chen, G. Goldfarb, E. Mateo, I. Kim, F. Yaman and G. Li, Electronic post-compensation of WDM transmission impairments using coherent detection and digital signal processing, *Opt. Express*, 16, 880, 2008.
7. E. Ip and J. M. Kahn, Compensation of dispersion and nonlinear impairments using digital backpropagation, *J. Lightwave Technol.*, 26, 3416, 2008.
8. G. P. Agrawal, *Nonlinear Fiber Optics*, 3rd ed, Academic Press, San Diego, CA, U.S.A, 2001.
9. D. M. Pepper and A. Yariv, Compensation for phase distortions in nonlinear media by phase conjugation, *Opt. Lett.*, 5, 59, 1980.
10. K. Solis-Trapala, T. Inoue, and S. Namiki, Nearly-ideal optical phase conjugation based nonlinear compensation system, in *Optical Fiber Communication Conference*, OSA Technical Digest (Optical Society of America), paper W3F.8, 2014.
11. I. D. Phillips, Exceeding the nonlinear-Shannon limit using Raman laser based amplification and optical phase conjugation, in *Optical Fiber Communication Conference*, OSA Technical Digest (Optical Society of America), paper M3C.1, 2014.
12. M. Morshed, A. J. Lowery and L. B. Du, Reducing nonlinear distortion in optical phase conjugation using a midway phase-shifting filter, in *Optical Fiber Communication Conference*, OSA Technical Digest (Optical Society of America), paper W2A.56, 2014.
13. J. Y. Y. Leong, P. Petropoulos, J. H. V. Price, H. Ebendorff-Heidepriem, S. Asimakis, R. C. Moore, K. E. Frampton, V. Finazzi, X. Feng, T. M. Monro, and D. J. Richardson, High-nonlinear dispersion-shifted lead-silicate holey fibers for efficient $1\text{-}\mu m$ pumped supercontinuum generation, *J. Lightwave Technol.*, 24, 183, 2006.
14. I. Thomkos, D. Chowdhury, J. Conradi, D. Culverhouse, K. Ennser, C. Giroux, B. Hallock, T. Kennedy, A. Kruse, S. Kumar, N. Lascar, I. Roudas, M. Sharma, R. S. Vodhanel and C. C. Wang, Demonstration of negative dispersion fibers for

DWDM metropolitan area networks, *IEEE Selected Topics in J. of Quantum Electronics*, 7, 439, 2001.

15. W. Forysiak, F. M. Knox and N. J. Doran, Average soliton propagation in periodically amplified systems with stepwise dispersion-profile fiber, *Opt. Lett.*, 19, 174, 1994.

16. R. C. Wrede and M. Spiegel, *Advanced Calculus*, 2nd ed., New York: McGraw-Hill, ch. 8, 2002.

17. T. Pfau, S. Hoffmann and R. Noé, Hardware-efficient coherent digital receiver concept with feedforward carrier recovery for M-QAM constellations, *J. Lightwave Technol.*, 27, 989, 2009.

18. S. V. Chernikov, E. M. Dianov, D. J. Richardson, and D. N. Payne, Soliton pulse compression in dispersion-decreasing fiber, *Opt. Lett.*, 18, 476, 1993.

19. A. J. Stentz, R. W. Boyd and A. F. Evans, Dramatically improved transmission of ultrashort solitons through 40 km of dispersion-decreasing fiber, *Opt. Lett.*, 20, 1770, 1995.

20. E. Mateo, L. Zhu, and G. Li, Impact of XPM and FWM on the digital implementation of impairment compensation for WDM transmission using backward propagation, *Opt. Express*, 16, 16124, 2008.

21. J. P. Gordon and L. F. Mollenauer, Phase noise in photonic communications systems using linear amplifiers, *Opt. Lett.*, 15, 1351, 1990.

22. S. Kumar, Effect of dispersion on nonlinear phase noise in optical transmission systems, *Opt. Lett.*, 30, 3278, 2005.

23. C. Xie, Inter-channel nonlinearities in coherent polarization-division-multiplexed quadrature-phase-shift-keying systems, *IEEE Photon. Technol. Lett.*, 21, 274, 2009.

18 Eigenvalue communications in nonlinear fiber channels

Jaroslaw E. Prilepsky and *Sergei K. Turitsyn*

Continuous progress in optical communication technology and corresponding increasing data rates in core fiber communication systems are stimulated by the evergrowing capacity demand due to constantly emerging new bandwidth-hungry services like cloud computing, ultra-high-definition video streams, etc. This demand is pushing the required capacity of optical communication lines close to the theoretical limit of a standard single-mode fiber, which is imposed by Kerr nonlinearity [1–4]. In recent years, there have been extensive efforts in mitigating the detrimental impact of fiber nonlinearity on signal transmission, through various compensation techniques. However, there are still many challenges in applying these methods, because a majority of technologies utilized in the inherently nonlinear fiber communication systems had been originally developed for linear communication channels. Thereby, the application of "linear techniques" in a fiber communication systems is inevitably limited by the nonlinear properties of the fiber medium. The quest for the optimal design of a nonlinear transmission channels, development of nonlinear communication technqiues and the usage of nonlinearity in a "constructive" way have occupied researchers for quite a long time. For instance, the idea of balancing the nonlinear self-phase modulation effect by dispersion (or vice versa) using soliton pulses was proposed by Hasegawa and Tappert in the early 1970s [5], when intensity modulation and direct detection were the main technology trend [6–9]. The relevant advances in this technique include ultra-long wavelengh-division-multiplexing soliton transmission, dispersion-managed solitons and many other interesting methods (see [6–11] and references therein). However, in the past decade traditional soliton approaches lost their appeal due to fast progress in various other (simpler implementation) transmission techniques. Moreover, increasing symbol rates stipulate the use of shorter pulses, which, for solitons with power inversely proportional to pulse width, means increasingly substantial nonlinear interactions and pattern dependent jitter effects. Recently, modification of soliton techniques to the coherent detection system was proposed [12–14], with the first studies indicating a decent potential for this approach. In general, currently there is an evident need for radically different approaches to coding, transmission, and processing of information in fiber communication channels that would take into account the nonlinear properties of optical fiber, allowing one to overcome the limits of linear techniques.

More than 20 years ago Hasegawa and Nyu [15] (see also Chapter 4.4. of the monograph by Hasegawa and Kodama [7]), considering the nonlinear Schrödinger equation (NLSE) as a model for signal propagation in a single-mode fiber, put forward the idea of exploiting the *nonlinear spectrum* of a signal for the purposes of information transmission. Since then, this concept has been known as "eigenvalue communications" because the authors proposed encoding information using special discrete eigenvalues (non-dispersive part of the nonlinear spectrum), which correspond to solitonic degrees of freedom and have no analogues in linear problems. This proposal is an example of a fundamentally nonlinear communication technique based on the unique property of the channel – the *integrability* of the corresponding nonlinear channel model given by the NLSE [16]. Currently, this prefiguring idea of using the nonlinear spectrum in data encoding, processing and transmission processes, is generally understood in a somewhat wider sense, and can be formulated as follows. In linear communication channels, spectral components (modes) defined by the Fourier transform (FT) of the signal propagate without interactions with each other. In certain nonlinear channels (*integrable channels*), such as the one governed by the NLSE, there exist nonlinear modes (nonlinear signal spectrum) that also propagate without interacting with each other and without corresponding nonlinear cross-talk, effectively, in a linear manner. Thus the parameters of these nonlinear modes can be used for encoding and efficient transmission of the information over a nonlinear fiber, the signal propagation inside which can be well modeled (at least in the leading approximation) by an integrable equation, for instance, by the NLSE.

The transition from the true space-time domain into the nonlinear spectral domain and back is achieved by performing the so-called *nonlinear Fourier transform* (NFT) — a technique introduced in the 1970s [16–19]. NFT operations constitute nothing more than the core parts of the general inverse scattering transform (IST) method [20, 21] for the solution of initial-value problems associated with integrable evolutionary equations and, in particular, developed for the the solution of the NLSE by Zakharov and Shabat in [16], and, further, for the so-called Manakov system, relevant to optical signal transmission using polarization degrees of freedom, by Manakov [22]. It should be noted that until recently the potential of the original idea expressed in [15], namely, the usage of the specifically nonlinear quantities from the IST method for the signal processing and transmission, had been largely overlooked by the communications and engineering community, although pulse-to-soliton conversion and soliton evolution problems were being widely studied by physicists in a number of different areas. Only recently this concept has attracted a new wave of attention with application to signals transmission and processing, highlighting the re-emergence of the eigenvalue communications [23–34]. Over the past few years several groups have revisited and extended the original ideas of Hasegava and Nyu in the context of coherent optical communications. The concept itself is being approached from two somewhat "orthogonal and complementary" pathways, neither of which excludes the parallel implementation of the alternative approach. These two main directions in "eigenvalue communication" methodology can be categorized according to what part of the nonlinear spectrum is used for modulation and transmission. The first approach involves the use of discrete eigenvalues and related solitonic quantities for signal transmission [15, 27–29, 34] and processing [26, 30], where the reported spectral efficiency of the method reached a decent value of more than 3 bit/sec/Hz [31]. The second method has been pursued by the Aston group and collaborators: it deals with the modulation of the continuous part of the nonlinear spectrum for signal encoding and efficient transmission in optical fibers [24, 25, 32, 33]. In this chapter we review the present state of this completely new direction in "eigenvalue communications" following recent publications [24, 25, 32, 33].

18.1 INTRODUCTION AND MAIN MODEL DESCRIPTION

Optical fiber systems form the backbone of global telecommunication networks and currently carry the majority of the world's information traffic, with the "fifth generation" of optical transmission systems operating with advanced modulation formats, e.g., orthogonal frequency division multiplexing (OFDM), digital signal processing techniques, etc. Skyrocketing demand for communication speed is exerting great pressure on the networks' infrastructure at every scale, which explains the real motivation behind the overwhelming part of optical communications research. Since the introduction of fiber-optic communications in the late 1970s, many technological advances, such as erbium-doped fiber amplifiers (EDFA), wavelength division multiplexing (WDM), dispersion management, forward error correction, and Raman amplification, have been developed to enable the exponential growth of data traffic [1, 2, 6]. The introduction of advanced modulation formats and digital signal processing for coherent communications led to practical implementation of systems with 100 Gb/sec channel rates. The key to this breakthrough is the possibility of mitigating the most important linear transmission impairments, such as fiber link dispersion and polarization mode dispersion. In coherent fiber optic communication systems, the received optical signal is digitized through high-speed analog-to-digital converters and then processed using digital signal processing (DSP) algorithms. The input signal is then recovered with the accuracy allowed by the channel noise and the transmission effects that are not equalized by the DSP. After the mitigation of linear effects, noise and nonlinear impairments become the key factors in limiting the performance of coherent fiber optic communication systems.

In recent years, a number of techniques have been introduced and studied for surmounting the capacity limit (occurring due to Kerr nonlinearity) through various nonlinearity compensation techniques, including digital back-propagation (DBP) [35], optical phase conjugation [36], and phase-conjugated twin waves [37], to mention a few recent advances. However, there are still many limitations in applying the aforementioned nonlinear compensation methods. A significant step forward would take place if a method could "incorporate" fiber nonlinearity constructively when designing core optical communication coding, transmission, detection, and processing approaches. It actually means that the true capacity limits of nonlinear fiber channels have yet to be found.

In general, the power of a signal transmitted through an optical fiber link is degraded by loss and has to be periodically recovered through optical amplification. In many important practical situations, the averaging of such periodic loss and gain results in an effectively lossless propagation model – the NLSE [6–8, 10, 11], which describes the continuous interplay between dispersion and nonlinearity. Moreover, using technology developed at Aston University, it was demonstrated experimentally that fiber loss can be compensated continuously along a fiber span, leading to effectively quasi-lossless transmission [38–42]. Overall, the NLSE can be considered a principal master model for demonstrating key techniques and approaches in optical fiber communications. Written for a complex slow-varying optical field envelope $q(z, t)$, it reads as (so far we disregard all deviations from the pure integrable case)

$$i \, q_z - \frac{\beta_2}{2} \, q_{tt} + \gamma \, q \, |q|^2 = 0, \tag{18.1}$$

where z stands for the propagation distance and t is the time in the frame co-moving with the group velocity of the envelope. Depending on the sign of the group velocity dispersion coefficient β_2, two physically different situations are generally considered with regard to model (18.1): (i) the case of anomalous dispersion, where the dispersion coefficient $\beta_2 < 0$, resulting in the so-called *focusing* NLSE, and (ii) the normal dispersion case with $\beta_2 > 0$, corresponding to the *defocusing* type of the NLSE (the higher-order dispersion terms are not considered). The instantaneous Kerr nonlinearity coefficient γ is expressed through the nonlinear part of refractive index n_2 and an effective mode area A_{eff}: $\gamma = n_2 \omega_0 / c A_{eff}$, with c being the vacuum speed of light and $\omega_0 = 2\pi\nu_0$ being the carrier frequency of the envelope $q(t, z)$. Further, we will use the explicit form of the NFT operations attributed to the the normalized versions of the NLSE. We normalize time in Eq. (18.1) to the characteristic time related to an input signal T_s, which can be, e.g., the extent of the RZ signal or the characteristic duration of a single information-bearing symbol (the normalization value T_s is rather a matter of convenience), and then use the effective z-scale associated with T_s: $Z_s = T_s^2 / |\beta_2|$. Then, we measure the power of the input in units of $P_0 = (\gamma Z_s)^{-1}$ and normalize the signal amplitude correspondingly. The summary of normalizations is

$$\frac{t}{T_s} \to t, \qquad \frac{z}{Z_s} \to z, \qquad \frac{q}{\sqrt{P_0}} = q \sqrt{\gamma Z_s} \to q. \tag{18.2}$$

For the anomalous dispersion case, the typical value of β_2 is -22 ps^2/km, and for normal dispersion we use the value $\beta_2 = 5$ ps^2/km; the typical value of the Kerr coefficient is $\gamma = 1.27$ (W·km)$^{-1}$.

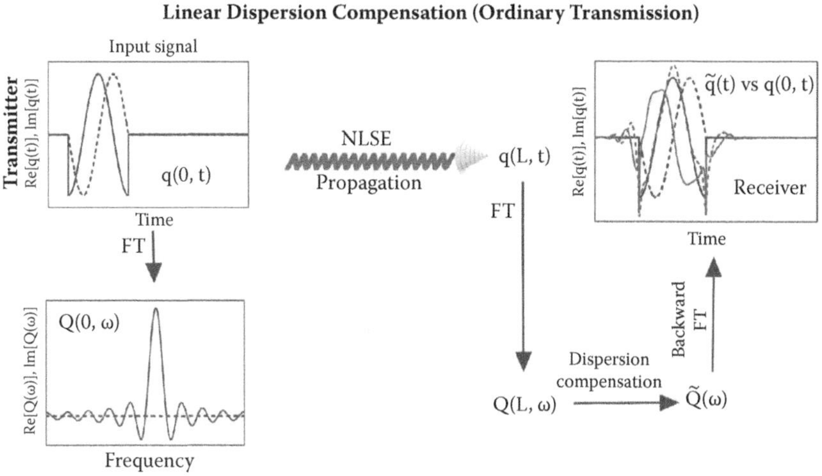

Linear Dispersion Compensation (Ordinary Transmission)

FIGURE 18.1: Flowchart of the linear scheme (chromatic dispersion compensation), valid for both signs of dispersion.

Later we will compare the results of the different NFT-based schemes against the linear dispersion compensation technique, given schematically in Fig. 18.1. For the NLSE, at a very low signal power the digital compensation of dispersion produces perfect recovery of the initial signal (when noise is not taken into account). However, when signal power gets higher, the compensation of dispersion recovers the initial signal, leaving noticeable corruption of the post-processed received waveform (depending on the input power level and propagation distance) even without taking into account the channel noise; see the insets of Fig. 18.1.

In addition to a fiber's dispersion and nonlinearity, the key master model that describes the signal propagation in a fiber link takes into account the effect of noise. The noise term results from amplifier spontaneous emission (ASE) either from the EDFAs (in the path average model) or from the distributed Raman amplification (see for details references and discussion in [1]):

$$iq_z - \frac{\beta_2}{2} q_{tt} + \gamma q |q|^2 = \Gamma(t, z). \tag{18.3}$$

The random complex quantity $\Gamma(t, z)$ describes noisy corruptions due to the ASE: It is generally written as a symmetric additive complex Gaussian white noise (AWGN) with zero average, fully characterized by its autocorrelation intensity:

$$\langle \Gamma(t, z) \bar{\Gamma}(t', z') \rangle = 2D\, \delta(t - t')\, \delta(z - z'),$$

where the overbar stands for the complex conjugation. Reference [1] provides a detailed account of how noise intensity D relates to the parameters of the line.

It is well known that the NLSE (without perturbation) belongs to the class of integrable nonlinear systems [7, 8, 10, 16–21]. In particular, this means that the focusing NLSE possesses a special type of solution: highly robust localized nonlinear waves, called solitons. However, it should be stressed that the methodology of eigenvalue communication is conceptually different from pure soliton-based transmission [12–14], even though the solitonic components can actually be present in the transmitted pulse: The information carriers there are not the soliton waveshapes themselves, but the IST data attributed, in particular (but not necessarily), to the solitonic degrees of freedom. This fact indicates the momentous difference between soliton-based transmission and eigenvalue communication.

18.2   NONLINEAR FOURIER TRANSFORM ASSOCIATED WITH NLSE

One of the particular manifestations of the integrability property is that, given the initial conditions (in the context considered, the waveform of the input signal), we can propagate the signal to a distance $z = L$ in three steps, which have direct analogies with the same stages in the consideration of linear problems, although the immediate implementation of these steps differs significantly from the linear case.

1. The first step is the mapping of the input profile to the spectral domain. For a linear channel, this operation corresponds to the ordinary forward Fourier transform (FT), and we will call this stage the *forward NFT* (FNFT), by analog with the linear situation. For nonlinear propagation, this stage involves solving the specific direct scattering problem associated with an integrable equation and produces a set of *scattering data*, where the particular quantities (continuous spectrum and a set of discrete complex eigenvalues, if the latter exists) are then associated with orthogonal nonlinear "normal modes."

2. The next step is the propagation of the initial spectral distribution (again, a continuous spectrum and complex eigenvalues) to distance L: Here, in both linear and nonlinear cases, the spectrum evolves according to the linear dispersion law. So, a further impetus for an analogy between NFT and its linear counterpart is that the former does to NLSE what the latter does to the linear equations: just as the linear FT changes dispersion to a phase rotation in frequency space so the NFT leads to a trivial phase rotation of the spectral data. This means that the fiber nonlinear transmission effects are effectively included in the NFT.

3. The last stage is the recovery of the solution profile in the space-time domain: It is the backward FT in the linear case. For the nonlinear integrable problem, the backward NFT (BNFT) amounts to the solution of the so-called Gelfand–Levitan–Marchenko equations, and this step accomplishes the finding of a solution (signal profile) at distance L. These three stages, which provide the solution of a nonlinear equation at distance L, constitute the essence of the IST method.

References [7, 16–21] provide numerous details, examples and profound explanation of the IST method, and below we briefly present only some relevant parts of this.

18.2.1 FORWARD NONLINEAR FOURIER TRANSFORM (ZAKHAROV–SHABAT DIRECT SCATTERING PROBLEM) FOR THE FOCUSING NLSE

In this subsection we consider the FNFT attributed to the anomalous dispersion (focusing) case, where the normalized NLSE (18.1) is explicitly rewritten as

$$iq_z + \frac{1}{2}q_{tt} + q|q|^2 = 0. \tag{18.4}$$

The FNFT operation for Eq. (18.4) requires solutions of the so-called Zakharov–Shabat spectral problem (ZSSP), which corresponds to the scattering problem for a non-Hermitian (for the anomalous dispersion) Dirac-type system of equations for two auxiliary functions $\phi_{1,2}(t)$, with the NLSE input waveform $q(0, t) \equiv q(t)$ serving as an effective potential entering the equations

$$\frac{d\phi_1}{dt} = q(t)\phi_2 - i\zeta\phi_1, \qquad \frac{d\phi_2}{dt} = -\bar{q}(t)\phi_1 + i\zeta\phi_2. \tag{18.5}$$

Here, ζ is a (generally complex) spectral parameter, $\zeta = \xi + i\eta$, and the potential $q(t)$ is supposed to decay as $t \to \pm\infty$ (see the specific constraints imposed on $q(t)$ decay in [16–21]).

At the left end $t \to -\infty$ we fix the "initial" condition for the incident wave scattered by the potential $q(t)$ to have the so-called Jost solution $\vec{\Phi}(t, \zeta) = [\phi_1(t, \zeta), \phi_2(t, \zeta)]^T$:

$$\vec{\Phi}(t, \zeta)\Big|_{t\to-\infty} = \begin{pmatrix} 1 \\ 0 \end{pmatrix} \exp(-i\zeta t).$$

With this initial condition, at the right end, $t \to +\infty$, we define two Jost scattering coefficients, $a(\zeta)$ and $b(\zeta)$, constituting the essence of the FNFT:

$$a(\zeta) = \lim_{t\to\infty} \phi_1(t, \zeta), \exp(i\zeta t), \qquad b(\zeta) = \lim_{t\to\infty} \phi_2(t, \zeta) \exp(-i\zeta t), \tag{18.6}$$

with $\phi_{1,2}$ being the corresponding elements of vector $\vec{\Phi}(t, \zeta)$. The (right) reflection coefficient associated with Eq. (18.5) is then defined as

$$\rho(\xi) = \frac{b(\xi)}{a(\xi)} = \lim_{t\to\infty} \frac{\phi_2(\xi, t)}{\phi_1(\xi, t)} \exp(-2i\xi t). \tag{18.7}$$

The FNFT operation corresponds to the mapping of the initial field, $q(0, t) = q(t)$, onto the set of *scattering data*:

$$\Sigma = \left[\rho(\xi), \ \xi \in \mathrm{R}, \quad \left\{ \zeta_n, C_n \equiv \frac{b(\zeta_n)}{a'(\zeta_n)} \right\} \right], \tag{18.8}$$

where index n runs over all discrete eigenvalues of ZSSP, Eq. (18.5) (if these are present). The quantity $\rho(\xi)$ from Eqs. (18.7) and (18.8) defined for real ξ ("frequency") plays the role of continuous nonlinear spectral distribution, while the quantities associated with discrete eigenvalues describe solitonic degrees of freedom and do not have analogs in linear problems. The evolution of the reflection coefficient is given by

$$\rho(L,\xi) = \rho(\xi)\exp(2i\xi^2 L),\qquad(18.9)$$

where $\rho(L,\xi)$ is the value of the coefficient after propagation to the distance L. From Eq. (18.9), one can see that the nonlinear spectrum obeys the linear dispersion law of the NLSE (18.4) if one associates the linear frequency ω with the quantity ξ as $\xi = -\omega/2$. Indeed, from the IST theory, it is known [17] that asymptotically in the linear limit the following formula is valid:

$$\rho(\xi)\Big|_{|q(t)|\to 0} = -\bar{Q}(-2\xi),\qquad(18.10)$$

where $Q(...)$ identifies the linear FT of the signal $q(t)$. In view of Eq. (18.10), it is useful to define the *nonlinear spectral function* (NSF) $N(\omega)$ associated with $\rho(\xi)$ via

$$N(\omega) = -\bar{\rho}(\xi)\Big|_{\xi=-\frac{\omega}{2}}.\qquad(18.11)$$

So, in the linear limit the NSF (18.11) coincides with the linear spectrum of the signal $q(t)$.

Quite often for the sake of computation convenience the *left set* of scattering data is defined for Eq. (18.5); as for the linear FT, one can use different signs in the transform exponent. In particular, the left reflection coefficient on the real axis is given by

$$r(\xi) = \frac{\bar{b}(\xi)}{a(\xi)}.\qquad(18.12)$$

Obviously, the poles of the left reflection coefficient (18.12) coincide with those of the right one, Eq. (18.7), as in both cases these are defined by $a(\zeta) = 0$ in the upper complex half-plane of spectral parameter ζ. The quantities C_n (norming constants) from (18.8) change to $\tilde{C}_n = [b(\zeta_n)a'(\zeta_n)]^{-1}$. Generally, the definitions of the complete set of scattering data through the right and left sets are equivalent, leading to unique recovery of the profile in the time domain, and we refer an interested reader to the work by Ablowitz et al. [17], where the IST method is simultaneously formulated in terms of both left and right sets. One of the distinctions is that the evolution law for $r(L,\xi)$ changes the sign in the exponent

$$r(L,\xi) = r(\xi)\exp(-2i\xi^2 L),\qquad(18.13)$$

and the definition for the NSF in terms of $r(\xi)$ reads

$$N(\omega) = -r(\xi)\Big|_{\xi=-\frac{\omega}{2}}.\qquad(18.14)$$

18.2.2 MODIFICATION OF THE FNFT FOR THE NORMAL DISPERSION CASE

Now consider the case of the normal dispersion NLSE, with the explicit normalized form

$$iq_z - \frac{1}{2}q_{tt} + q|q|^2 = 0. \tag{18.15}$$

The associated ZSSP transforms as follows (cf. Eq.(18.5)):

$$\frac{d\phi_1}{dt} = q(t)\phi_2 - i\xi\phi_1, \qquad \frac{d\phi_2}{dt} = \bar{q}(t)\phi_1 + i\xi\phi_2. \tag{18.16}$$

We have already written this ZSSP specifically for real spectral parameter ξ. The significant difference of the focusing case (18.5) from the defocusing ZSSP (18.16) is that the latter is Hermitian. It signifies that for the normal dispersion one cannot have soliton solutions (complex discrete eigenvalues) emerging from any sufficiently localized input. The right and left reflection coefficients are defined in the same manner as in Subsection 18.2.1. Due to the different sign of the dispersion in Eq. (18.15), we also have the change of sign in the evolution law exponent for the reflection coefficient attributed to Eqs. (18.15) and (18.16):

$$r(L, \xi) = r(\xi) \exp(2i\xi^2 L). \tag{18.17}$$

18.2.3 BACKWARD NONLINEAR FOURIER TRANSFORM (GELFAND–LEVITAN–MARCHENKO EQUATION)

The backward NFT maps the scattering data Σ onto the field $q(t)$: This is achieved via the Gelfand-Levitan-Marchenko equations (GLME) for the unknown functions $K_{1,2}(t, t')$. The general form of the GLME written in terms of the left scattering data reads:

$$\bar{K}_1(t, t') + \int\limits_{-\infty}^{t} dy\, F(t' + y)K_2(t, y) = 0,$$

$$t > t', \tag{18.18}$$

$$\pm\bar{K}_2(t, t') + F(t + t') + \int\limits_{-\infty}^{t} dy\, F(t' + y)K_1(t, y) = 0,$$

where "$+$" corresponds to the focusing and "$-$" to the defocusing NLSE. For the defocusing case ("$-$" sign) the quantity $F(t)$ can contain both contributions from the solitonic and continuous parts:

$$F(t) = -i\sum_k \tilde{C}_k e^{-i\zeta_k t} + \frac{1}{2\pi} \int\limits_{-\infty}^{\infty} d\xi\, r(\xi)\, e^{-i\xi t}. \tag{18.19}$$

Having solved the GLME (18.18) for $K_{1,2}(t, t')$, the solution sought in the space-time domain is recovered as $q(t) = \pm 2\bar{K}_2(t, t)$. However, for the GLME associated with the soliton-free case, considered later in this chapter, this expression reduces to the simple FT of $r(\xi)$,

$$F(t) = \frac{1}{2\pi} \int\limits_{-\infty}^{\infty} d\xi \, r(\xi) \, e^{-i\xi t},$$

which is valid for both signs of the dispersion. When one is interested in the solution $q(L,t)$, the quantity $r(\xi)$ in (18.19) is replaced with $r(L,\xi)$, given either by Eq. (18.13) or (18.17), depending on the sign of the dispersion. So, the resulting solution of the GLME (18.18) becomes the function of L: $K_{1,2}(L; t, t')$.

18.2.4 SOME REMARKS ON NUMERICAL METHODS FOR COMPUTING NFT AND ASSOCIATED COMPLEXITY

Inasmuch as the form of the FNFT and BNFT operations is seemingly different, numerical methods for the solution of the ZSSP, Eqs. (18.5) and (18.16) and for the GLME (18.18) are also distinct. For the latter we note that for our purposes we do not consider the soliton's contribution.

The well-know methods for the solution of ZSSP involve Crank–Nicolson finite-difference discretization, Ablowitz–Ladik discretization (where the resulting differential-difference NLSE form is also integrable), the Boufetta–Osborne method [43,44], where one uses a piecewise-constant approximation of the "potential" $q(t)$, Runge–Kutta integration of the ZSSP [44], the spectral collocation method [26], etc. The methods for the ZSSP are well reviewed in [28,46]. For solitonic quantities, one has to augment the solving routine with a root finding method, such as the Newton–Raphson method, to locate zeros of $a(\xi)$. This step, in general, can bring about an increase in computational complexity.

For the solution of GLME (18.18), there exists a multitude of methods stemming mostly from the Bragg grating s synthesis research, which are applicable in the soliton-free case for both dispersion signs. We mention different peeling algorithms [45] and the Toeplitz matrix based method [33,47]; see also the references in these works.

At this point it is pertinent to discuss the numerical complexity of the NFT as compared to, say, the popular digital back-propagation (DBP) technique for the removal of nonlinear distortions [35]. In the latter one reads the transmitted waveform at the receiver, inserts it as an input for the noiseless NLSE, Eq. (18.1), and then solves it in a backward direction. The numerical solution of the NSLE is usually performed by using the split-step Fourier method [6,35], which requires $\sim N_z M_t \log M_t$ floating-point operations (flops), M_t being the number of discretization (sampling) points in the time domain, and N_z the number of steps in z, which grows with the transmission length and can depend on the pulse power. The transmission techniques that we consider further involve either one or two nonlinear transforms, each of those requiring $\sim M_t^2$ flops with the use of the well-developed methods mentioned above. Even with such an estimate, the complexity of the NIS can be comparable to that of the DBP when $M_t^2 \sim N_z M_t \log M_t$ [25]. However, recent advance in numerical NFT methods indicate that the complexity of the NFT operations can be potentially reduced even further. For the ZSSP, a recent study by Wahls and Poor [46] suggests that the recovery of the *continuous* part of the nonlinear spectrum can be made in only $\sim M_t \log^2 M_t$ flops. For the GLME, in another work by the same authors [48] some arguments in favor of the possibility for the fast BNFT operation with the same order of flops are given. The Toeplitz matrix based GLME solution method [33, 47] can be potentially integrated with the superfast Toeplitz matrix inversion algorithms (see the direct references in [25, 33]), also resulting in overall NFT complexity reductions. Taking these estimations, we believe that NFT-based transmission methods can potentially outperform the DBP and other nonlinearity compensation techniques in terms of numerical complexity for digital signal processing.

18.3   TRANSMISSION USING CONTINUOUS NONLINEAR SPECTRUM — NORMAL DISPERSION CASE

In [24] the case of transmission through the channel described by the normal dispersion NLSE (18.15) was addressed. The normalization parameters were taken as follows: $T_s = 25$ ps, $Z_s = 125$ km. As mentioned in Subsection 18.2.2, for this dispersion sign no solitons can emerge from the input having a localized extent, and this fact greatly simplifies the usage of NFT operations because one does not have to deal with discrete eigenvalues. We call this approach "the straight IST-based method" insofar as in this case the course of actions is completely similar to the linear case; compare Figs. 18.1 and 18.2. In [24] noisy corruptions were not considered and only the proof-of-concept demonstration of how the continuous nonlinear spectrum can be used for transmission was presented. The input pattern used for numerical calculations was built from the sequence of $N = 100$ of pulses and had the following form:

$$q(z = 0, t) = \sum_{k=1}^{N} c_k \, s(t - k \, T), \tag{18.20}$$

with T being the symbol duration ($T = T_s$). As an example, the quadrature phase shift keying modulation (QPSK) of the information coefficients was employed: The absolute value of $c_{\alpha \, k}$ is the same for each coefficient, $|c_k| = c = \text{const.}$, and the phase of each c_k takes four discrete values from the set:

$$\text{Arg}\{c_k\} = 2\pi p/4, \quad \text{with } p = 0 \div 3. \tag{18.21}$$

The whole set of c_k (18.21) can be rotated to an arbitrary angle in the complex plane. So, the constellation diagram of the input signal (the loci of c_k in the complex plane for all carrier numbers k) consists of four points. The value

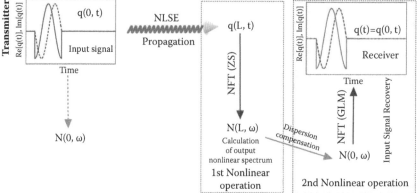

FIGURE 18.2: Flowchart of the transmission scheme for the straight IST-based nonlinearity compensation, utilized in [24] for the focusing NLSE.

$|c_k| = c = 0.5$ was taken, which produced observable nonlinear effects over distances ~ 1000 km. The input carrier pulse shape $f_0(t)$ can have an arbitrary profile, and in [24] a Gaussian pulse shape was chosen,

$$s(t) = \exp\left[-t^2/(2\tau_0^2)\right] \times \exp\left[i\phi\right],$$

where parameter τ_0 is related to the pulse full width at half maximum width through $T_{FWHM} = 1.655\,\tau_0$, and the phases were generated randomly from the QPSK set (18.21).

The transmission scheme considered in [24] is given in Fig. 18.2. The scheme involves two NFT operations, both at the receiver side. First, using the profile at the receiver, one inserts it into the FNFT associated with the focusing NLSE, Eq. (18.16), then unrolls the accumulated dispersion inside the nonlinear spectral domain, and finally recovers the profile using the BNFT operation given by the appropriate GLME (18.18). The results for the application of this scheme vs linear dispersion compensation, Fig. 18.1, to the received waveform emerging from the same input (18.20), were compared at different propagation distances on the eye diagrams, Fig. 18.3 (superposition of waveforms from different slots), and constellation diagrams (indicating the position of coefficients c_k on the complex plane), Fig. 18.4. When the absolute value of coefficients is small (low powers), the reconstructed signals for the linear and nonlinear methods are almost identical; reconstruction via the scheme from Fig. 18.1 is very quick and efficient. However, with an increase of nonlinearity, the FT approach becomes less efficient, whereas the NFT results change very little, mostly due to the increase in NFT computational errors. Figure 18.3 shows the eye diagrams of reconstructed signals for different distances, obtained by using both linear and nonlinear approaches. We can see that the IST (NFT)

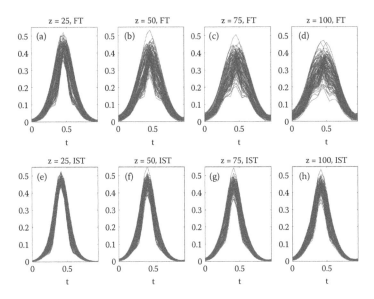

FIGURE 18.3: Eye diagram for the signals reconstructed via FT: (a) at $z = 25$; (b) at $z = 50$; (c) at $z = 75$; (d) at $z = 100$; and via IST: (e) at $z = 25$; (f) at $z = 50$; (g) at $z = 75$; (h) at $z = 100$. Taken from [24].

approach, Fig. 18.3(e–h), provides better results than the compensation via the FT, Fig. 18.3(a–d). With an increase of the propagation distance, the "eye" in Fig. 18.3(a–d) starts "closing," while for the IST approach, the "eye" remains well open. Figure 18.4 shows the corresponding constellation diagrams for reconstructed signals, indicating the same tendency.

So, although nonlinear transmission impairments due to fiber Kerr nonlinearity can be compensated by the DBM method (see Subsection 18.2.4), it requires substantial computational efforts to model reverse signal channel propagation. The key technical difference between compensation of linear channel dispersion and nonlinear effects is that the linear Fourier transform compensates for accumulated channel dispersion analytically, without using any computer time for reverse propagation. The NFT-based scheme illustrated in this section allows one to do the same with nonlinear impairments. Of course, there is a price to pay for such an advantage, meaning that one has to deal with the NFT, instead of performing direct and inverse linear FT, as would be the case in linear channel equalization.

In conclusion, this section simply illustrates the recovery of a nonlinearly distorted signal using NFT-based signal processing. In this technique a propagation part is trivial and technical problems are moved to the receiver side. It has to be noted that the deviations of the waveform obtained with the use of the NFT from the initial ones, observable in Figs. 18.3 (e)–(h) and 18.4 (e)–(h), arise due to computational errors and periodic boundary conditions,

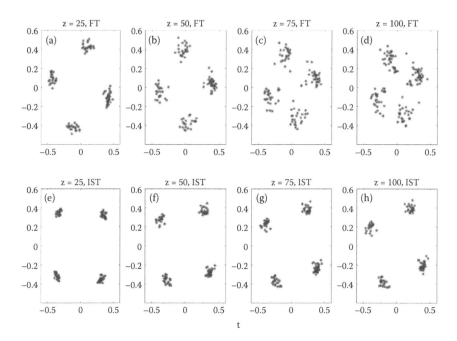

FIGURE 18.4: Constellation diagram for the signals reconstructed via FT: (a) at $z = 25$; (b) at $z = 50$; (c) at $z = 75$; (d) at $z = 100$; and via IST: (e) at $z = 25$; (f) at $z = 50$; (g) at $z = 75$; (h) at $z = 100$. Taken from [24].

used for the integration of the NLSE. When these factors are eliminated (or mitigated), the recovery of the profiles at the receiver side would become perfect.

18.4 METHOD OF NONLINEAR AND LINEAR SPECTRA EQUALIZATION FOR LOW ENERGY SIGNALS: ANOMALOUS DISPERSION

In this section we deal with the anomalous dispersion case and describe a straightforward method of linear and nonlinear spectra equalization [25], applicable in the case of weak power signals $|q(t)| \sim \varepsilon$ having a finite extent $[-T/2, T/2]$ (burst mode transmission). The method provides a good illustration of how the transition from linear to nonlinear quantities occurs. The low power condition implies that no solitons can form from our input, so that the ZSSP (18.5) does not contain discrete eigenspectrum: $\Sigma = [\rho(\xi), \; \xi \in \mathrm{Re}]$, the *complete* set of scattering data (18.8) consists only of the quantity $\rho(\xi)$. The first question with regard to nonlinear spectrum manipulation is how one can encode nonlinear spectral data at the transmitter side and then retrieve them at the receiver. Note that the results given in Subsection 18.4.1 below are general and do not imply any specific modulation format. In fact, the only limitation of the method is the relative smallness of the input signal amplitude multiplied to the pulse duration.

18.4.1   NONLINEAR SPECTRUM EXPANSIONS FOR LOW SIGNAL AMPLITUDE

Using the smallness of ε we can obtain expressions for the nonlinear spectrum using perturbative iterations. We start from Eq. (18.5), looking for the solution expansion in terms of the parameter $\varepsilon\, T$. First, in ZSSP (18.5) one makes the transformation from $\phi_{1,2}$ to slow varying functions $\varphi_{1,2}$ as $\varphi_{1,2} = e^{\mp i\xi t}\phi_{1,2}$. In terms of $\varphi_{1,2}$, the expression for $\rho(\xi)$ (18.7) now changes to

$$\rho(\xi) = \lim_{t \to \infty} \frac{\varphi_2(\xi, t)}{\varphi_1(\xi, t)}. \tag{18.22}$$

Solving the ZSSP recast in terms of $\varphi_{1,2}$ by recursive iterations gives us

$$\varphi_2(t, \xi) = -\int_{-T/2}^{t} dt_1\, e^{-2i\xi t_1}\bar{q}(t_1) + \int_{-T/2}^{t} dt_1 \int_{-T/2}^{t_1} dt_2 \int_{0}^{t_2} dt_3\, e^{2i\xi(t_2-t_1-t_3)}\bar{q}(t_1)q(t_2)\bar{q}(t_3),$$

up to ε^3 (each power of q gives the contribution $\sim \varepsilon$), and

$$\varphi_1(t, \xi) = 1 - \int_{-T/2}^{t} dt_1 \int_{-T/2}^{t_1} dt_2\, e^{2i\xi(t_1-t_2)}q(t_1)\bar{q}(t_2),$$

up to ε^2. The expression for $\rho(\xi)$ takes the form $\rho(\xi) \approx \rho_0(\xi) + \rho_1(\xi)$, where $\rho_0 \sim \varepsilon$ and $\rho_1 \sim \varepsilon^3$ are given as follows (the next term is $\sim \varepsilon^5$):

$$\rho_0(\xi) = -\int_{-T/2}^{T/2} dt_1\, e^{-2i\xi t_1}\bar{q}(t_1), \tag{18.23}$$

$$\rho_1(\xi) = -\int_{-T/2}^{T/2} dt_1 \int_{t_1}^{T/2} dt_2 \int_{-T/2}^{t_2} dt_3\, e^{2i\xi(t_2-t_1-t_3)}\bar{q}(t_1)q(t_2)\bar{q}(t_3). \tag{18.24}$$

Now we "propagate" our $\rho(\xi)$ to the distance L using Eq. (18.9),

$$\rho(L, \xi) = \rho_0(L, \xi) + \rho_1(L, \xi) = \big[\rho_0(\xi) + \rho_1(\xi)\big]e^{2i\xi^2 L}, \tag{18.25}$$

to obtain the expression for nonlinear spectral distribution at $z = L$, where $\rho_0(L, \xi) \sim \varepsilon$ and $\rho_1(L, \xi) \sim \varepsilon^3$.

18.4.2   LINEAR AND NONLINEAR SPECTRA EQUALIZATION USING SIGNAL PRE-DISTORTION

Suppose that at the input $z = 0$, we apply pre-distortion $s(t) \sim \varepsilon^3$ to the initial signal waveform $q(t)$:

$$q_s(t) = q(t) + s(t). \tag{18.26}$$

The idea of the method is to remove the quantity $\rho_1(\xi)$ given by Eq. (18.24) and thus the term $\rho_1(L, \xi)$ from the spectral density at the end point $z = L$ in Eq. (18.25), by using the additional pre-processing given by $s(t)$. When a small quantity $s(t) \sim \varepsilon^3$ is added to the input signal, one gains a correction $\rho_s(\xi) \sim \varepsilon^3$ to the expression for $\rho_1(\xi)$ (see Eq. (18.24)):

$$\rho(\xi) = \rho_0(\xi) + \rho_1(\xi) + \rho_s(\xi) + O(\varepsilon^5),$$

$$\rho_s(\xi) = - \int_{-T/2}^{T/2} dt_1 \, e^{-2i\xi t_1} \bar{s}(t_1). \tag{18.27}$$

For the two terms of the same order, $\rho_1(\xi)$ and $\rho_s(\xi)$, to cancel each other, we choose $s(t)$ in such a way that the following relation is satisfied:

$$\rho_s(\xi) = -\rho_1(\xi). \tag{18.28}$$

Using the definition of the NSF (18.11), we can now obtain the ordinary Fourier spectrum $S(\omega)$ for our correction $s(t)$ as

$$S(\omega) = \bar{\rho}_1(\xi)\Big|_{\xi=-\frac{\omega}{2}}, \tag{18.29}$$

and performing the backward FT of Eq. (18.29), we restore the profile of $s(t)$ in the time domain. Thus, for the pre-distorted signal $q_s(t) = q(t) + s(t)$, with the FT of $s(t)$ given by Eq. (18.29), the addition to the nonlinear spectrum $\sim \varepsilon^3$ disappears altogether and *the nonlinear spectrum associated with $q_s(t)$ coincides with the linear spectrum of initial $q(t)$ up to the terms $\sim \varepsilon^5$*.

The flowchart of the pre-compensation scheme and the signal recovery at distance $z = L$ is given in Fig. 18.5. We note that, aside from recursive Fourier-type integration used to obtain $\rho_1(\xi)$, the scheme involves just one FNFT. So, by means of the pre-compensation described above, one is able to translate the encoded information into the nonlinear spectral domain without using any special formats, and control the accuracy of the data mapping. The transmission itself is effectively performed through the nonlinear spectral domain.

18.4.3 ILLUSTRATION OF THE METHOD

18.4.3.1 Optical frequency division multiplexing (OFDM) modulation

For the illustration of how the current and other methods work, as an example we consider input in the form of the burst-mode (several symbols of) OFDM. Generally, coherent optical OFDM has recently become a popular transmission technique owing to its robustness against chromatic and polarization mode dispersion, efficiency and practicality of implementation; see the monograph [49] and references therein. OFDM is a multi-carrier transmission format where a data stream is carried with many lower-rate tones:

$$q(t) = \sum_{\alpha=-\infty}^{\infty} \sum_{k=0}^{N_{sc}-1} c_{\alpha k} \, s_k(t - \alpha T) \, e^{i \Omega_k t}. \qquad (18.30)$$

Here, $c_{\alpha k}$ is the α-th informational coefficient in the k-th subcarrier, s_k is the waveform of the k-th subcarrier, N_{sc} is the total number of subcarriers, Ω_k is the frequency of the k-th subcarrier, and T is the OFDM symbol (slot) duration. The shape of each subcarrier, $s_k(t)$, is usually a rectangle $\Pi(t)$ of width T and unit height, and such a choice ensures the orthogonality condition

$$\delta_{kl} = \frac{1}{T} \int_0^T s_k(t) \bar{s}_l(t) \, e^{i \, (\Omega_k - \Omega_l) \, t} \, dt,$$

which is met as long as the subcarrier frequencies satisfy $\Omega_k - \Omega_l = (2\pi/T)m$, with an integer m. This means that for linear transmission these orthogonal subcarrier sets, with their frequencies spaced at multiples of the inverse of

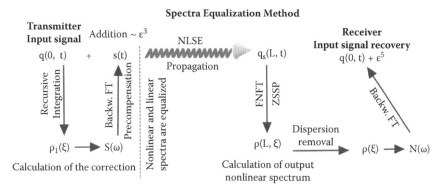

FIGURE 18.5: Flowchart of the pre-compensation scheme for the equalization of the linear and nonlinear spectra up to ε^5 and the subsequent recovery of the informational content at the receiver. Taken from [25].

the symbol rate, $\Omega_k = (2\pi/T)(k-1)$, can be recovered without intercarrier (IC) and intersymbol (IS) interference, in spite of strong spectral signal overlapping. The coefficients $c_{\alpha k}$ are then recovered by a convolution with the appropriate conjugate base function $\bar{q}_{\alpha k} = \Pi(t)(t - \alpha T) e^{-i\Omega_k t}$. The linear spectrum of the OFDM-modulated signal (18.30) is a comb of sinc-like shapes:

$$Q(\omega) = 2 \sum_{\alpha=-\infty}^{\infty} \sum_{k=1}^{N_{sc}} c_{\alpha k} \exp\left[-i\omega\,\alpha\,T + (i/2)(\Omega_k - \omega)T\right] \frac{\sin\left[\frac{(\Omega_k - \omega)\,T}{2}\right]}{\Omega_k - \omega}. \tag{18.31}$$

18.4.3.2 Spectra equalization for OFDM input signals

18.4.3.2.1 Single OFDM tone

First, it is instructive to consider pre-distortion for the simple input in the form of a single OFDM tone $q(0,t) = c\,e^{i\Omega t}$ if $t \in [-\frac{T}{2}, \frac{T}{2}]$, and 0 otherwise, assuming its amplitude $|c| \sim \varepsilon \ll 1$ and using the results of Subsection 18.4.2. The ZSSP for this input profile can be solved analytically [25], so the results of the expansion can be checked directly. The first-order term in the expansion of $\rho(\xi)$ (where the corresponding NSF coincides with the linear spectrum) is given by

$$\rho_0(\xi) = -\bar{c}_k \frac{\sin T(\xi + \Omega/2)}{\xi + \Omega/2}. \tag{18.32}$$

For the nonlinear addition $\sim \varepsilon^3$, we have

$$\rho_1(\xi) = 2\,\bar{c}_k\,|c_k|^2\,\exp\left[i(\xi + \Omega/2)T\right] \frac{\sin T(2\xi + \Omega) - T(2\xi + \Omega)}{(\xi + \Omega/2)^3}. \tag{18.33}$$

Now, one calculates the pre-distortion spectrum $S(\omega)$, inserting (18.33) into Eqs. (18.28) and (18.29). The resulting function $s(t)$ is given in Fig. 18.6(b) for the case $T = 1$, $\varepsilon = 0.5$, $\Omega = 2\pi$ [see the profile of $q(t)$ in Fig. 18.6(a)]. Interestingly, the resulting profile of $s(t)$ is asymmetric with respect to the time axis origin, in contrast the obvious symmetry of the input pulse.

In Fig. 18.6(c), we present a comparison of the absolute errors for the linear scheme in Fig. 18.1 and the error for the scheme from Fig. 18.5. One observes that the resulting error for the equalization method is generally four to five times smaller compared to the case with linear dispersion compensation even for such a low power input. Note that the largest errors for the pre-distorted pulse occurred at the points $t = \pm T/2$, i.e., where there was a sharp change in the input profile. These errors are caused by numerical discretization aliasing.

18.4.3.2.2 Several OFDM symbols

For illustration purposes, in [25] the random QPSK encoding (18.21) of OFDM coefficients was adopted and the following normalizing parameters were used:

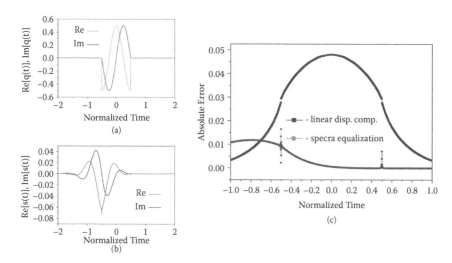

FIGURE 18.6: (a) Real and imaginary parts of the input profile corresponding to a single OFDM tone with $T = 1$. (b) Real and imaginary parts of the corresponding nonlinear pre-distortion profile $s(t) \sim \varepsilon^3$. (c) The absolute errors, obtained by the application of spectra equalization pre-compensation (circles) (see Fig. 18.5) and linear dispersion removal (squares) (see Fig. 18.1) at distance $z = 1$ ($L = 4000$ km). Taken from [25].

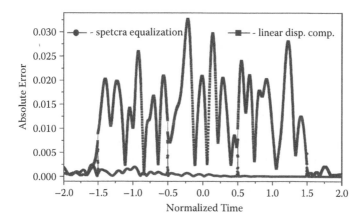

FIGURE 18.7: Absolute errors for the propagation of three slots (normalized duration of each slot $T = 1$) of the 10-mode QPSK-OFDM with $|c_{\alpha k}| = 0.15$, obtained by the application of spectra equalization pre-compensation (circles) (see Fig. 18.5) and linear dispersion removal (squares) (see Fig. 18.1) at distance $z = 1$. Taken from [25].

$T_s = 300$ ps, giving the characteristic z normalization scale as $Z_s \approx 4000$ km. The input pulse was taken as a finite number (a burst) of OFDM symbols (18.30), with 10 tones in each symbol. The ε^3 correction to $\rho(\xi)$ can also be obtained analytically [25] and contains contributions from both the IS and IC interference terms. The absolute errors associated with the propagation of the pre-compensated OFDM vs the errors produced by the linear dispersion compensation (Fig. 18.1) of the same OFDM sequence are summarized in Fig. 18.7. We can see that the pre-compensation error is still much lower than that of the linear method (notice the aliasing contribution at the ends of each slot) in spite of the extended total duration of the burst pulse (for the simulation in Fig. 18.7 three OFDM slots were used). In [25], the robustness of the method from Fig. 18.5 against ASE-induced noise was also checked, and simulations using the NLSE with a noise term (18.3) and realistic noise intensity were performed. It was shown that the presence of ASE does not violate the general performance of the transmission based on nonlinear spectrum evolution. The results for the NFT-based method were better than those for the application of the simple linear scheme from Fig. 18.1.

18.5   NONLINEAR INVERSE SYNTHESIS (NIS) METHOD—ANOMALOUS DISPERSION

18.5.1   GENERAL IDEA OF THE METHOD

In the previous two sections the preliminary "proof-of-concept" demonstration that the non-solitonic part of the nonlinear spectrum can be used for mitigating nonlinear distortions was presented. However, these methods have limited applicability, being confined to the normal dispersion case or to the low-power initial signals. The method described in this section is rather more challenging and practically important, dealing with the focusing NLSE and high-power inputs [32, 33]. Note that, for an arbitrary high-energy input for the focusing NLSE, we usually have the formation of solitons [20, 21], and the eigenvalue spectrum of ZSSP, Eq. (18.5), contains complex eigenvalues corresponding to solitonic degrees of freedom (aside from highly disordered inputs [50, 51], which are less interesting in transmission problems). When the high-power input is randomly coded, the complex eigenvalue portrait of ZSSP can be extremely involved. It means that the implementation of the direct scheme given in Fig. 18.2 brings about considerable difficulties related to finding the location of eigenvalues and the recovery of a profile using the general form of the focusing GLME (18.18) with solitonic components (18.19). To get rid of this issue, we suggest *synthesizing* the profile in the time domain starting from given encoded shapes in the nonlinear spectral domain. In other words, one performs a one-to-one mapping of the linear spectrum $Q(\omega)$ for the known information-bearing signal $q(t)$ to the nonlinear spectrum (NSF) $N(\omega)$, where the latter already corresponds to a new signal $q_{GLM}(t)$: $Q(\omega) \to N(\omega)$. Then the new profile in the time domain $q_{GLM}(t)$ is synthesized using the BNFT, i.e., by solving the GLME (18.18) and using the corresponding $N(\omega)$ (or $r(\xi)$, given by Eq. (18.14)). Notably, for high powers the synthesized signal $q_{GLM}(t)$ can be essentially different from the initial waveform $q(t)$. Such an encoding can explore the advantages of well-developed linear formats, like OFDM, as the propagation of the nonlinear spectrum is linear. The idea itself is similar to that widely used for the inverse syntheses of Bragg grating s [45,47]: One creates the input profile bearing the desired properties, starting from the nonlinear spectral data, and then employs the BNFT, thus synthesizing the profile in the time domain. Because of this similarity we call this method "nonlinear inverse synthesis" (NIS) [32,33]. During the evolution, the spectral data undergo just a trivial phase rotation without nonlinear mode coupling or channel crosstalk, and hence after winding out this "nonlinear dispersion" at the receiver, the initial information can be recovered without nonlinear signal degradation. The scheme of the NIS method is illustrated in Fig. 18.8. The NIS method involves two stages: (i) the BNFT at the transmitter, providing the profile $q_{GLM}(0,t)$ in the time domain corresponding to a desired initial NSF $N(\omega)$; and (ii) the recovery of the reflection coefficient and corresponding NSF at the receiver by the FNFT, i.e., by solving Eq. (18.5), and the consequent dispersion compensation inside the nonlinear spectral domain.

Nonlinear Inverse Synthesis

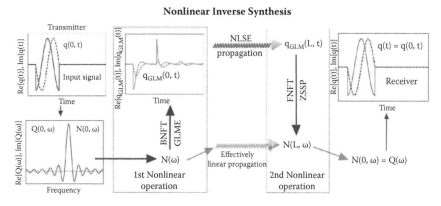

FIGURE 18.8: Flowchart depicting the sequence of operations for the NIS method, the example initial waveform $q(0,t) = e^{2\pi it/T}$ if $t \in [0,T]$, 0 otherwise. The panes display the true profiles for $T = 1$ ns, transmission length $L = 2000$ km (noiseless case). Taken from [32].

18.5.2   ILLUSTRATION OF THE METHOD

18.5.2.1   Synthesis of profiles from some characteristic shapes in the non-linear spectral domain

The first question of interest with respect to NIS method implementation is: What do the elementary base functions look like in the time domain when its spectral shape is used in the nonlinear spectral domain? We present the corresponding results in Fig. 18.9. Figure 18.9(a) and (b) show the results for the sinc base (i.e., it is a rectangle with amplitude c in the ω domain) taken as an NSF. This type of base function is utilized in the Nyquist-format technique, which provides the highest spectral efficiency. As seen from Fig. 18.9(b), the corresponding profiles in the time domain, $q_{GLM}(t)$, are not symmetric, and the asymmetry grows with the increase of amplitude c. In the OFDM scheme the elementary base in the time domain is simply a rectangle, i.e., a sinc-function in the ω-domain. The inverse NFT of the single OFDM spectral tone, $N(\omega) = c\,\mathrm{sinc}(\omega)$, is given in Fig. 18.9(c) and (d) for different values of amplitude c.

We see that for a sufficiently large c the waveform of $q_{GLM}(t)$ is significantly different from a rectangular profile occurring in the linear case: While losing its symmetry, the profile develops an oscillatory advancing tail. The general form of the spectrum of an arbitrary OFDM-encoded data sequence is given by Eq. (18.31). In Figs. 18.9(e) and (f) we present the spectrum of a single OFDM slot containing 10 subcarriers with $c_k = 1$ and a corresponding inverse NFT, $q_{GLM}(t)$. (The general form of the spectrum of an arbitrary OFDM-encoded data sequence is given by Eq. (18.31).) In this case the structure of the advancing tail is more involved and reflects the structure of the pulse itself.

18.5.2.2   NIS for high-efficiency OFDM transmission – Comparison with digital backpropagation

In [33] 56 Gbaud OFDM NIS-based transmission systems (in burst mode) with different modulation formats were studied: In addition to the QPSK, the higher level quadrature amplitude modulations 16QAM (16 possible complex values for c_k), and 64QAM (64 possible values) of the OFDM coefficients were used. The net data rates of these systems, after removing overhead, were 100 Gb/s, 200 Gb/s, and 300 Gb/s, respectively. The guard time duration is chosen as 20 percent longer than the fiber chromatic dispersion induced memory for a 2000 km link. For the OFDM NIS-based system, the total number of tones was 128, where 112 subcarriers were filled with data, while the remaining subcarriers were set to zero. The useful OFDM symbol duration was 2 ns and the cyclic prefix was not used. Each packet data (burst) contained only one OFDM symbol (slot).

The linear spectra of OFDM signals before and after the BNFT are shown in the Fig. 18.10. It can be seen that after the BNFT, the linear spectrum

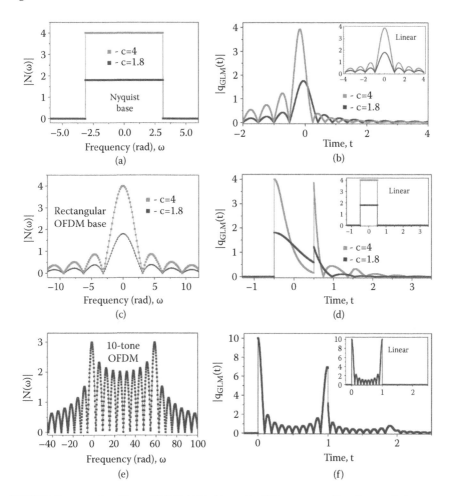

FIGURE 18.9: Profiles of NSF $|N(\omega)|$ with different amplitudes and the corresponding profiles $|q_{GLM}(t)|$ in the time domain, obtained by solving GLME Eq. (18.18). (a), (b) For the rectangle-shaped $N(\omega) = \prod(\omega/2\pi)$; (c), (d) for the sinc-shaped $N(\omega)$ (the base of standard OFDM); (e), (f) $N(\omega)$ and corresponding $q_{GLM}(t)$ for a single slot of 10 OFDM tones, with $c_k = c = 1$. The insets show the corresponding linear FTs of the spectrum. Taken from [32].

of the OFDM signal does not broaden significantly, indicating that the NIS method combined with the OFDM can be effectively applied for a WDM transmission or even multiplexed into superchannels.

In [33] the Q-factor (calculated through the error vector magnitude; see [52]) for the evaluation of the OFDM coefficient deviations was chosen as the performance indicator for the transmission quality assessment. In Fig. 18.11 the information recovery for the NIS method is again compared with the linear dispersion compensation, Fig. 18.1; for the NIS, the Q-factor for the back-to-back (B2B) recovery with no propagation and after the propagation for 2000 km is presented. Noise was not included for the simulations given in Fig. 18.11, so that fiber nonlinearity was the only impairment. This result confirms that the NIS-based approach can perfectly compensate for the deterministic impairment due to fiber nonlinearity, using just a single-tap linear dispersion removal for the nonlinear spectrum at the receiver. However, one can notice that the back-to-back performance of NIS-based systems deteriorates when the input signal power increases. This phenomenon can be explained by the fact that the numerical error of both FNFT and BNFTs grows with the increase of input signal power [33].

Now we compare the performance of the OFDM systems with the use of the NIS and DBP methods [35] (see Subsection 18.2.4) for fiber nonlinearity compensation. For the implementation of DBP, the received signal is first filtered with an eighth order low-pass filter having a bandwidth of 40 GHz. Subsequently, the optical field is reconstructed and the signal is back-propagated with a different number of steps per single span, indicating the numerical complexity of the corresponding DBP realization. In Fig. 18.12 we compare the Q-factors of OFDM systems with NIS and DBP. One can see that the OFDM NIS-based system offers over 3.5 dB advantage over the traditional OFDM system, confirming the effectiveness of the proposed approach for fiber nonlinearity compensation. This performance improvement is comparable with that of DBP with 10 steps per span. The launch power in the NIS-based system is limited to -4 dBm (the optimum launch power), which is mainly due to numerical errors in the NFT operations at the transmitter and receiver.

When combining it with a higher modulation format, such as 16QAM, the OFDM NIS-based approach offers nearly 4 dB advantage over the traditional OFDM scheme; see Fig. 18.13. The transmission bit rate in this case is

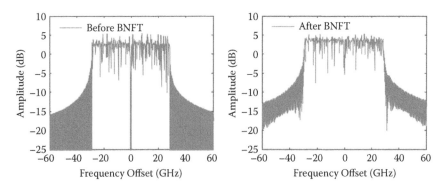

FIGURE 18.10: Linear spectra of 128-tone randomly coded QPSK-OFDM signals before and after BNFT; the launch power is 0 dBm. Taken from [33].

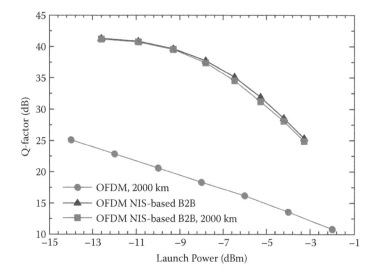

FIGURE 18.11: Q-factor as a function of launch power for linear dispersion re-moval, Fig. 18.1, and for a 100 Gb/s QPSK-OFDM NIS-based system without ASE-induced noise. Taken from [33].

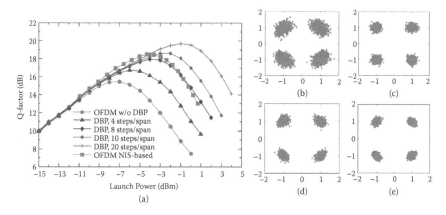

FIGURE 18.12: (a) Performance comparison of 100-Gb/s QPSK-OFDM systems with the NIS vs. the DBP methods for fiber nonlinearity compensation. The receiver filter bandwidth used was 40 GHz; the distance is 2000 km. The right panels show constellation diagrams at the optimum launch powers with and without the NIS and DBP methods for fiber compensation: (b) without NIS and DBP, (c) with the NIS method, (d) DBP with 10 steps/span, (e) DBP with 20 steps/span. Taken from [33].

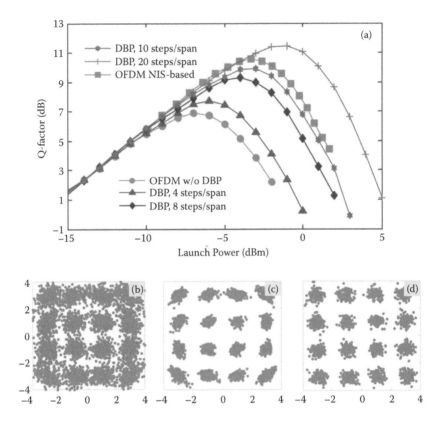

FIGURE 18.13: (a) Performance comparison of the 200-Gb/s 16QAM-OFDM systems with the NIS vs. the DBP methods for fiber nonlinearity compensation. The receiver filter bandwidth used was 40 GHz; the distance is 2000 km. The right panels show constellation diagrams at the optimum launch powers with and without the NIS and DBP methods for fiber compensation: (b) without NIS and DBP, (c) DBP with 20 steps/span, (d) with the NIS method. Taken from [33].

200 Gb/s. It can be seen that for the 16QAM modulation format, the OFDM NIS-based system outperforms the DBP with 10 steps per span. The optimum constellation diagrams for the conventional OFDM system, OFDM NIS-based, and OFDM with 20 steps per span DBP are shown in Fig. 18.13(b)–(d). From this figure one can observe that the NIS method produced fairly clear constellation diagrams. The simulation results for the 300-Gb/s 64QAM OFDM NIS-based system are compared in Fig. 18.14 with the conventional OFDM and OFDM with DBP. It can be seen that for such a high-order modulation format, the OFDM NIS-based system displays almost the same performance as the DBP with 20 steps per span. The performance improvement in compar-

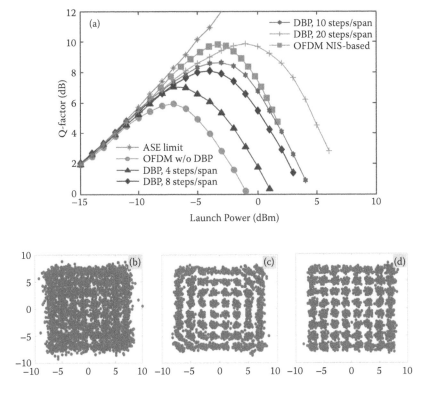

FIGURE 18.14: The same as in Fig. 18.13 but for the 300-Gb/s 64QAM-OFDM systems. Taken from [33].

ison with the conventional OFDM system is about 4.5 dB, which is larger than the values achieved for the QPSK and 16QAM modulation formats. This result indicates that a greater performance advantage of the OFDM NIS-based system over the traditional approaches can be reached for higher-order modulation formats and shows the considerable benefit of the NIS method for fiber nonlinearity compensation for highly spectrally efficient transmission systems. In Fig. 18.14 the curve indicating the ASE transmission limit is also presented: For calculating it, nonlinearity was completely removed. It can be seen that the curve for the NIS-based transmission generally goes above those for the DBP in the noise-dominated region, but it does not intersect the limiting line. This behavior reveals that the NIS-based transmission is less sensitive to noise-induced corruption than the DBP, and the refinements of the NFT processing techniques can improve the NIS performance even further.

To sum up, the NIS method can be successfully combined with the transmission techniques (e.g., OFDM, Nyquist-shaped) having high spectral efficiency and advanced modulation formats, such as QPSK, 16QAM, and 64QAM. This novel transmission scheme suggests encoding the information onto the continuous part of the nonlinear spectrum and requires only single-tap equalization at the receiver to compensate for all the deterministic fiber nonlinearity impairments accumulated along the fiber link. Generally, the NIS concept can be further extended to other optical systems described by different integrable continuous equations. The simulations confirmed the effectiveness of the NIS scheme and showed that an improvement of 4.5 dB can be achieved, which is comparable to the multi-steps per span DBP compensation method. With the utilization of increasingly higher-order modulations (16- and 64QAM), the results for the performance of the NIS-based OFDM system became comparable with the performance of increasingly higher-order DBP compensation (i.e., with progressively more steps per span). This fact reveals that the NIS method efficiency can become strongly competitive and outperform that of the DBP methods for high spectral efficiency formats. In [33] the performance of the Nyquist-shaped modulation format combined with the NIS method was also studied, and it was demonstrated that the OFDM is potentially a more suitable modulation for such systems.

18.6 CONCLUSION

In this chapter we have reviewed recent progress in the promising re-emerging communication technique based on transmission using the continuous part of the nonlinear spectrum associated with an integrable evolutionary equation, in particular, with the NLSE (18.1). This bevy of methods is a ramification the original idea of Hasegawa and Nyu [15], the "eigenvalue communication," where the data are encoded onto the parameters of specifically nonlinear "normal modes" and thus are not affected by nonlinear impairments during transmission. So, the major advantage of using the nonlinear spectral domain for data transmission in a coherent communication channel is a suppression of nonlinear cross-talk insofar as the fiber nonlinearity is effectively included into the digital signal processing based on NFT operations. The resulting channel becomes effectively linear, and the signal propagation boils down to the trivial phase rotation of individual (nonlinear) spectral components. This allows us to employ well-developed modulation formats, recasting them into the nonlinear spectral domain. Also, the robustness of NFT-based transmission in a practical environment that includes ASE noise was demonstrated. For the most advanced and practically attractive recently introduced NIS method (Section 18.5), it was shown that an improvement (in terms of the Q-factor) of 4.5 dB can be achieved, which is comparable to that for the multi-steps per span DBP compensation method. In addition, NFT-based processing can be competitive and even outperform that of the "traditional" digital signal processing methods, like DBP, with the use of recent advances in NFT processing methods [46, 48].

In a recent work of Buelow [34] the important first experimental evidence that nonlinear spectral data can be used for optical transmission was presented. In this experimental work, signal detection based on NFT processing in a coherent receiver was assessed: 16 GBaud binary phase-shift keying (BPSK) signals were transmitted over a few spans of standard single-mode fiber at power levels which induce a strong nonlinear distortion. The detection scheme employed the discrete (multi-soliton) eigenvalues of the ZSSP. The experimental work [34] clearly indicates that NFT-based detection can be applied in practical fiber optic systems.

The concept of eigenvalue communication is re-emerging as a powerful nonlinear digital signal processing technique that paves the way to overcoming current limitations of traditional communications methods in nonlinear fiber channels. Its methodology is not ultimately linked to the NLSE model considered in this chapter but can be further extended to other communication systems described by other integrable evolutionary equations, with the most important example being the Manakov system of equations [22], governing the transmission of a polarization multiplexed optical signal. This means that the technique considered here can be further developed and generalized to the case of polarization division multiplexing systems. The other important direction in the progress of these methods is to develop and optimize modulation formats specifically for coding of the signal inside the nonlinear spectral domain, aiming at an increase of the spectral efficiency and performance improvement with regard to the signal corruptions occurring due to the deviation of the real channel from a purely integrable model. In particular, the experiments of Buelow [34] indicate that the channel performance evaluation with regard to the EDFA-based system with strong attenuation has to be addressed. A no less important area is the further development of superfast numerical methods for NFT-based processing, i.e., the development of fast NFT processing algorithms. Finally, we strongly believe that the two current practical implementations of the "eigenvalue communication," namely, transmission based on the discrete (solitonic) part of the spectrum and transmission techniques using the continuous spectrum part, will eventually merge together, thus emanating into a single, solid, highly efficacious and extremely flexible nonlinear digital signal processing technique. Numerous aspects of the fundamental IST method are available for integration into communication engineering technologies, and we hope that the ideas of the NFT become no less common and routine for optical engineers than the standard linear Fourier operations are now. This requires efforts from different research communities and serves as a remarkable example of the groundbreaking impact that interdisciplinary research can produce.

ACKNOWLEDGMENTS

The work has been supported by EPSRC project UNLOC EP/J017582/1. The support of the Russian Ministry of Education and Science and the European Research Council project ULTRALASER is also acknowledged. We would like to thank all our colleagues who have participated in the research and contributed to the results described in this review chapter.

REFERENCES

1. R. Essiambre, G. Kramer, P. J. Winzer, G. J. Foschini, and B. Goebel. Capacity limits of optical fiber networks. *Journal of Lightwave Technology*, 28(4):662–701, 2010.

2. A. D. Ellis, J. Zhao, and D. Cotter. Approaching the non-linear Shannon limit. *Journal of Lightwave Technology*, 28(4):423–433, 2010.

3. D. J. Richardson. Filling the light pipe. *Science*, 330(6002):327–328, 2010.

4. R. I. Killey and C. Behrens. Shannon's theory in nonlinear systems. *Journal of Modern Optics*, 58(1):1–10, 2011.

5. A. Hasegawa and F. Tappert. Transmission of stationary nonlinear optical pulses in dispersive dielectric fibers. 1. Anomalous dispersion. *Applied Physics Letters*, 23(3):142–144, 1973.

6. G. P. Agrawal. *Fiber-Optic Communication Systems, 4th ed.* John Wiley & Sons, Inc., Hoboken, NJ, 2010.

7. A. Hasegawa and Y. Kodama. *Solitons in Optical Communications*. Clarendon Press, Oxford, 1995.

8. L. F. Mollenauer and J. P. Gordon. *Solitonsin Optical Fibers: Fundamentals and Applications*. Academic Press, San Diego, 2006.

9. M. Nakazawa, H. Kubota, K. Suzuki, and E. Yamada. Recent progress on soliton transmission technology. *Chaos*, 10(3):486–513, 2000.

10. S. K. Turitsyn, B. Bale, and M. P. Fedoruk. Dispersion-managed solitons in fiber systems and lasers. *Physics Reports*, 521(4):135–203, 2012.

11. S. K. Turitsyn, E. G. Shapiro, S. B. Medvedev, M. P. Fedoruk, and V. K. Mezentsev. Physics and mathematics of dispersion-managed optical solitons. *Comptes Rendus Physique* 4(1):145–161, 2003.

12. J. E. Prilepsky, S. A. Derevyanko, and S. K. Turitsyn. Temporal solitonic crystals and non-Hermitian informational lattices. *Physical Review Letters*, 108(187), art. no 183902, 2012.

13. O. Yushko, A. Redyuk, M. Fedoruk et al. Timing and phase jitter suppression in coherent soliton transmission. *Optics Letters*, 39(21):6308–6311, 2014.

14. O. V. Yushko and A. A. Redyuk. Soliton communication lines based on spectrally efficient modulation formats. *Quantum Electronics*, 44(6):606–611, 2014.

15. A. Hasegawa and T. Nyu. Eigenvalue communication. *Journal of Lightwave Technology*, 11(3):395–399, 1993.

16. V. E. Zakharov and A. B. Shabat. Exact theory of two-dimensional self-focusing and one-dimensional self-modulation of waves in nonlinear media. *Soviet Physics-JETP*, 34(1):62–69, 1972.

17. M. J. Ablowitz, D. J. Kaup, A. C. Newell, and H. Segur. The inverse scattering transform-Fourier analysis for nonlinear problems. *Studies in Applied Mathematics*, 53(4):249–315, 1974.

18. A. R. Osborne. The inverse scattering transform: tools for the nonlinear Fourier analysis and filtering of ocean surface waves. *Chaos, Solitons & Fractals*, 5(12):2623–2637, 1995.

19. A. S. Fokas and I. M. Gelfand. Integrability of linear and nonlinear evolution equations and the associated nonlinear Fourier transforms. *Letters in Mathematical Physics*, 32(3):189–210, 1994.

20. V. E. Zakharov, S. V. Manakov, S. P. Novikov, and L. P. Pitaevskii. *Theory of Solitons. The Inverse Scattering Method*. Consultants Bureau, New York, 1984.

21. M. J. Ablowitz and H. Segur. *Solitons and the Inverse Scattering Transform.* SIAM, Philadelphia, 1981.

22. S. V. Manakov. On the theory of two-dimensional stationary self focussing of electromagnetic waves. *Soviet Physics-JETP*, 38(2):248–253, 1974.

23. S. Oda, A. Maruta, and K. Kitayama. All-optical quantization scheme based on fiber nonlinearity. *IEEE Photonics Technology Letters*, 16(2):587–589, 2004.

24. E. G. Turitsyna and S. K. Turitsyn. Digital signal processing based on inverse scattering transform. *Optics Letters*, 38(20):4186–4188, 2013.

25. J. E. Prilepsky, S. A. Derevyanko, and S. K. Turitsyn. Nonlinear spectral management: Linearization of the lossless fiber channel. *Optics Express*, 21(20):24344–24367, 2013.

26. H. Terauchi and A. Maruta. Eigenvalue modulated optical transmission system based on digital coherent technology. In *18th OptoElectronics and Communications Conference held jointly with 2013 International Conference on Photonics in Switching (OECC/PS)*, paper WR2-5. IEICE, 2013.

27. M. I. Yousefi and F. R. Kschischang. Information transmission using the nonlinear Fourier transform, Part I: Mathematical tools. *IEEE Transactions on Information Theory*, 60(7):4312–4328, 2014.

28. M. I. Yousefi and F. R. Kschischang. Information transmission using the nonlinear Fourier transform, Part II: Numerical methods. *IEEE Transactions on Information Theory*, 60(7):4329–4345, 2014.

29. M. I. Yousefi and F. R. Kschischang. Information transmission using the nonlinear Fourier transform, Part III: Spectrum modulation. *IEEE Transactions on Information Theory*, 60(7):4346–4369, 2014.

30. H. Terauchi, Y. Matsuda, A. Toyota, and A. Maruta. Noise tolerance of eigenvalue modulated optical transmission system based on digital coherent technology. In *2014 OptoElectronics and Communication Conference and Australian Conference on Optical Fiber Technology*, pages 778–780. Engineers Australia, 2014.

31. S. Hari, F. Kschischang, and M. Yousefi. Multi-eigenvalue communication via the nonlinear Fourier transform. In *27th Biennial Symposium on Communications (QBSC)*, pages 92–95. IEEE, 2014.

32. J. E. Prilepsky, S. A. Derevyanko, K. J. Blow, I. Gabitov, and S. K. Turitsyn. Nonlinear inverse synthesis and eigenvalue division multiplexing in optical fiber channels. *Physical Review Letters*, 113(1), art. no. 013901, 2014.

33. S. T. Le, J. E. Prilepsky, and S. K. Turitsyn. Nonlinear inverse synthesis for high spectral efficiency transmission in optical fibers. *Optics Express*, 22(22):26720–26741, 2014.

34. H. Buelow. Experimental assessment of nonlinear Fourier transformation based detection under fiber nonlinearity. In *European Conference on Optical Communications*, paper We.2.3.2, 2014.

35. E. Ip and J. M. Kahn. Compensation of dispersion and nonlinear impairments using digital backpropagation. *Journal of Lightwave Technology*, 26(20):3416–3425, 2008.

36. S. L. Jansen, D. Van den Borne, B. Spinnler et al. Optical phase conjugation for ultra long-haul phase-shift-keyed transmission. *Journal of Lightwave Technology*, 24(1):54–64, 2006.

37. X. Liu, A. R. Chraplyvy, P. J. Winzer, R. W. Tkach, and S. Chandrasekhar. Phase-conjugated twin waves for communication beyond the Kerr nonlinearity limit. *Nature Photonics*, 7(7):560–568, 2013.

38. J. D. Ania-Castanon. Quasi-lossless transmission using second-order Raman amplification and fiber Bragg gratings. *Optics Express*, 12(19):4372–4377, 2004.

39. J. D. Ania-Castanon, T. J. Ellingham, R. Ibbotson, X. Chen, L. Zhang, and S. K. Turitsyn. Ultralong Raman fiber lasers as virtually lossless optical media. *Physical Review Letters*, 96(2), art. no. 023902, 2006.

40. T. J. Ellingham, J. D. Ania-Castanon, R. Ibbotson, X. Chen, L. Zhang, and S. K. Turitsyn. Quasi-lossless optical links for broad-band transmission and data processing. *IEEE Photonics Technology Letters*, 18(1):268–270, 2006.

41. J. D. Ania-Castanon, V. Karalekas, P. Harper, and S. K. Turitsyn. Simultaneous spatial and spectral transparency in ultralong fiber lasers. *Physical Review Letters*, 101(12), art. no 123903, 2008.

42. J. D. Ania-Castanon and S. K. Turitsyn. Unrepeated transmission through ultra-long fiber laser cavities. *Optics Communications*, 281(23):5760–5763, 2008.

43. G. Boffetta and A. R. Osborne. Computation of the direct scattering transform for the nonlinear Schrëdinger equation. *Journal of Computational Physics*, 102(2):252–264, 1992.

44. S. Burtsev, R. Camassa, and I. Timofeyev. Numerical algorithms for the direct spectral transform with applications to nonlinear Schrödinger type systems. *Journal of Computational Physics* 147(1):166–186, 1998.

45. A. Buryak, J. Bland-Hawthorn, and V. Steblina. Comparison of inverse scattering algorithms for designing ultrabroadband fiber bragg gratings. *Optics Express*, 17(3):1995–2004, 2009.

46. S. Wahls and H. V. Poor. Introducing the fast nonlinear Fourier transform. In *Proceedings of IEEE International Conference on Acoustics, Speech, and Signal Processing (ICASSP)*, pages 5780–5784. IEEE, 2013.

47. O. V. Belai, L. L. Frumin, E. V. Podivilov, and D. A. Shapiro. Efficient numerical method of the fiber Bragg grating synthesis. *Journal of Optical Society of America B*, 24(7):1451–1457, 2007.

48. S. Wahls and H. V. Poor. Inverse nonlinear Fourier transforms via interpolation: The Ablowitz-Ladik case. In *Proceeding of International Symposium on Mathematical Theory of Networks and Systems (MTNS)*, pages 1848–1855, 2014.

49. S. Shieh and I. Djordjevic. *Orthogonal Frequency Division Multiplexing for Optical Communications*. Academic Press, Amsterdam, 2010.

50. S. A. Derevyanko and J. E. Prilepsky. Random input problem for the nonlinear Schrödinger equation. *Physical Review E*, 78(4), art. no 046610, 2008.

51. S. K. Turitsyn and S. A. Derevyanko. Soliton-based discriminator of noncoherent optical pulses. *Physical Review A*, 78(6), art. no 063819, 2008.

52. S. T. Le, K. J. Blow, V. K. Menzentsev, and S. K. Turitsyn. Comparison of numerical bit error rate estimation methods in 112 Gbs QPSK CO-OFDM transmission. In *39th European Conference and Exhibition on Optical Communication (ECOC)*, paper P4.14. 2013.

19 Digital coherent technology-based eigenvalue modulated optical fiber transmission system

Akihiro Maruta, Yuki Matsuda, Hiroki Terauchi,
and *Akifumi Toyota*

The ideal information carrier is invariable quantity during propagation in nonlinear dispersive fiber. This is the eigenvalue of the associated equation of the nonlinear Schrödinger equation. We introduce an eigenvalue modulated optical fiber transmission system based on digital coherent technology.

19.1 INTRODUCTION

The nonlinear Schrödinger equation (NLSE), which describes the behavior of the complex envelope of an electric field propagating in a nonlinear dispersive fiber, can be solved analytically by using the inverse scattering transform (IST) [1]. In the framework of IST, the eigenvalues of the Dirac-type eigenvalue equation associated with the NLSE are invariables even though the temporal waveforms and frequency spectra dynamically change during propagation in the fiber. Therefore the eigenvalue is a more ideal information carrier than the pulse's amplitude, frequency, and/or phase, which are modulated in conventional formats. Hasegawa and Nyu have proposed eigenvalue communication based on the above mentioned concept [2]. They applied cross-phase modulation induced higher order soliton fission to demodulate embedded multiple eigenvalues, and its feasibility has been experimentally demonstrated [3]. Since the receiver configuration is complicated in this method, its practical implementation is difficult. On the other hand, digital coherent technologies have been recently introduced to optical fiber communication systems [4]. For example, a digital back propagation scheme in which the NLSE is iteratively integrated toward the inverse direction has been proposed to compensate for transmission impairments due to fiber dispersion and nonlinearity [5]. It is a time consuming scheme and seems difficult to implement in a real-time system. To overcome the difficulty, we have proposed an eigenvalue demodulation method based on digital coherent technology [6]. In this chapter, we introduce the eigenvalue modulated optical fiber transmission system based on digital coherent technology. After a detailed explanation of the demodulation method [6], we numerically and experimentally demonstrate the eigenvalue modulated transmission system. We also investigate the noise tolerance of the eigenvalues [7]. In general, noise is added to a signal and the noise induces signal distortion. For eigenvalue modulation, constellation points are located in a plane in which orthogonal axes are real and imaginary parts of eigenvalues. We experimentally add amplified spontaneous emission (ASE) noise to a pulse sequence and study the noise tolerance of the demodulated eigenvalues.

19.2 PRINCIPLE OF EIGENVALUE DEMODULATION

A periodically amplified complex envelope of the electric field propagating in a fiber which exhibits anomalous dispersion, nonlinearity, and loss can be described by the NLSE,

$$i\frac{\partial q}{\partial Z} + \frac{1}{2}\frac{\partial^2 q}{\partial T^2} + |q|^2 q = 0 \,, \tag{19.1}$$

in the framework of the guiding center theory [8]. Here, Z, T, and q are the normalized quantities of propagation distance, time moving with the group velocity, and complex envelope of the electric field, respectively. The eigenvalue equation associated with Eq. (19.1) is given by

$$\begin{cases} i\dfrac{\partial \psi_1}{\partial T} + q\psi_2 = \zeta\psi_1 \,, \\[2mm] -i\dfrac{\partial \psi_2}{\partial T} - q^*\psi_1 = \zeta\psi_2 \,, \end{cases} \tag{19.2}$$

where ζ is the complex eigenvalue and $\psi_\ell(\ell = 1, 2)$ are the eigenfunctions [1]. So far as q is a solution of Eq. (19.1), the eigenvalues ζ of Eq. (19.2) are invariables. To solve Eq. (19.2) based on digital coherent technology, Eq. (19.2) is converted to the following integral equation by using the Fourier transform.

$$\begin{cases} \Omega\widehat{\psi}_1(\Omega) + \dfrac{1}{\sqrt{2\pi}}\displaystyle\int_{-\infty}^{\infty} \widehat{q}(\Omega - \Omega')\widehat{\psi}_2(\Omega')d\Omega' = \zeta\widehat{\psi}_1(\Omega) \,, \\[4mm] -\Omega\widehat{\psi}_2(\Omega) - \dfrac{1}{\sqrt{2\pi}}\displaystyle\int_{-\infty}^{\infty} \widehat{q}^*(\Omega' - \Omega)\widehat{\psi}_1(\Omega')d\Omega' = \zeta\widehat{\psi}_2(\Omega) \,, \end{cases} \tag{19.3}$$

where $\widehat{q}$ and $\widehat{\psi}_\ell$ are the Fourier transform of q and ψ_ℓ, respectively. We discretize Eq. (19.3) in the frequency domain, and the convolution integrals are calculated numerically. Then Eq. (19.3) can be converted to the following eigenvalue problem in matrix form.

$$\begin{bmatrix} \mathbf{A} & \mathbf{B} \\ -\mathbf{B}^* & -\mathbf{A} \end{bmatrix} \begin{Bmatrix} \widehat{\psi}_1 \\ \widehat{\psi}_2 \end{Bmatrix} = \zeta \begin{Bmatrix} \widehat{\psi}_1 \\ \widehat{\psi}_2 \end{Bmatrix} \,, \tag{19.4}$$

where $\widehat{\psi}_\ell$ is the column vector whose elements are $\widehat{\psi}_\ell(\Omega_n)$ $(n = 1, 2, \cdots, N)$, and Ω_n represents the discretized frequency. $\mathbf{B}^*$ represents the conjugate transposed matrix of $\mathbf{B}$. $\mathbf{A}$ and $\mathbf{B}$ are $N \times N$ matrices whose elements are

$$a_{jk} = \begin{cases} \Omega_j & (j = k) \\ 0 & (\text{otherwise}) \end{cases}, \tag{19.5}$$

$$b_{jk} = \begin{cases} \dfrac{\Delta\Omega}{\sqrt{2\pi}}\widehat{q}(\Omega_{n_{jk}}) & (1 \leq n_{jk} \leq N) \\ 0 & (\text{otherwise}) \end{cases}, \tag{19.6}$$

where $n_{jk} = N/2 + j - k + 1$. $\Delta\Omega(= \Omega_{n+1} - \Omega_n)$ is the discretization spacing in the frequency domain. Eq. (19.4) can be easily solved numerically for any complex envelope of electric field q, and it is compatible with signal processing in digital coherent technology.

19.3 NUMERICAL DEMONSTRATION OF EIGENVALUE MODULATED TRANSMISSION

We consider an eigenvalue modulation by the use of a rectangular initial pulse. The amplitude and width of the pulse are 1 and 10, respectively. Figure 19.1 shows the variation of the eigenvalues $\eta = 2\Im[\zeta]$ for the transmission distance of 0 to 2,000 km in a dispersion shifted fiber (DSF) whose dispersion parameter is 1.0 ps/nm/km, the nonlinear coefficient is 0.52 $\times 10^{-9}$/W, and the loss coefficient is 0.24 dB/km. The fiber loss is compensated for by optical amplifiers allocated every 50 km. The observed waveforms at 0 km and 2,000 km are also shown. The three eigenvalues observed are invariable for any Z even though the temporal waveform changes during propagation and they can be used as temporally multiplexed information carriers [9].

19.4 EXPERIMENTAL DEMONSTRATION OF EIGENVALUE MODU-LATED TRANSMISSION

We first demonstrate eigenvalue demodulation from an experimentally observed pulse sequence. Figure 19.2 shows the experimental setup. An NRZ-OOK data signal was generated by modulating continuous wave (CW) light in a lithium niobate intensity modulator (LN-IM) with a 10 Gb/s pseudorandom bit sequence (PRBS) of length $2^{31} - 1$ from a pulse pattern generator (PPG). A generated NRZ-OOK data signal was sampled by a 50 Gs/s digital sampling oscilloscope (DSA) after detection by a photo detector (PD). Figure 19.3 shows the observed pulse sequence consisting of three types of pulses with different pulse widths. We assumed the pulses do not have any frequency chirp, i.e., q in Eq. (19.2) is a real function, and solved Eq. (19.4) for numerical windows shown by dashed rectangulars in Fig. 19.3. The amplitude of the narrowest pulse is calibrated by multiplying a constant number, by which the pulse has only one discrete eigenvalue. Then, seen in Fig. 19.3, three types

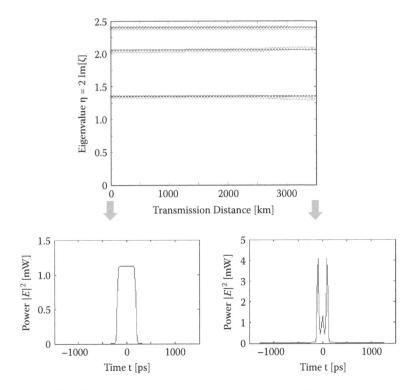

FIGURE 19.1: Variations of the eigenvalues and waveforms.

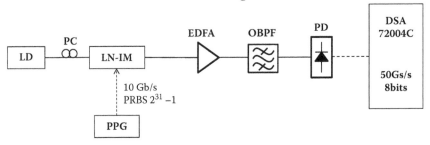

FIGURE 19.2: Experimental setup for eigenvalue modulation.

of pulses have different eigenvalues, and the number of eigenvalues increases when the pulse width becomes wider. In other words, we can say that the eigenvalues can be modulated by modulating the pulse width.

We then conducted transmission experiments of eigenvalue modulated signals and invesigated the change of eigenvalues observed before and after transmission in a fiber. Figure 19.4 shows the experimental setup for eigenvalue transmission. CW light at the wavelength of 1550 nm from laser diode (LD) was modulated by an IQ modulator driven by an RF signal from an arbitrary

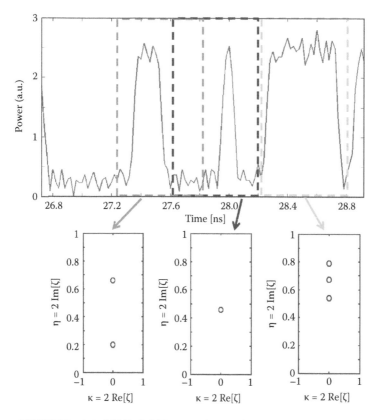

FIGURE 19.3: NRZ-OOK optical signal and its eigenvalues.

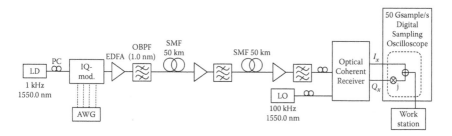

FIGURE 19.4: Experimental setup for eigenvalue transmission.

waveform generator (AWG). The generated optical pulse sequence consists of three kinds of pulses. Pulse width and pulse spacing were set to 200 ps and 800 ps, respectively. The light sources used at the transmitter and the receiver which is used to generate local oscillating (LO) light are semiconductor laser diodes with an external cavity with line widths 1 kHz and 100 kHz, respec-

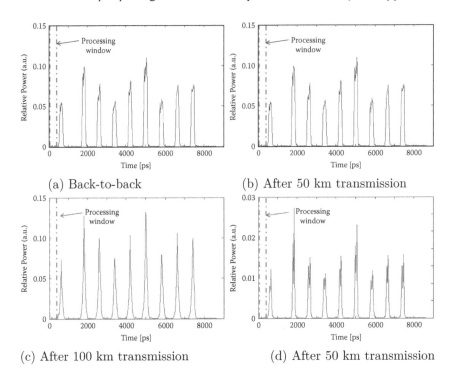

(a) Back-to-back

(b) After 50 km transmission

(c) After 100 km transmission

(d) After 50 km transmission

FIGURE 19.5: Temporal waveform of the received pulse sequence.

tively. In the demonstration, we used a standard single-mode fiber (SMF) as a transmission fiber. The generated pulse sequence was propagated over 50 km and 100 km. For 100 km transmission, the transmitted signal after 50 km propagation was amplified by an erbium doped fiber amplifier (EDFA) and transmitted a further 50 km. Considering the SMF's loss coefficient of 0.2 dB/km, the signal light's average power launched to SMF was set to 2.56 dBm based on the guiding center theory [8]. At the receiver, the signal light was optical-to-electrical (O/E) converted by an optical coherent receiver which consists of a 90° optical hybrid and balanced optical detector and then converted to digital data by a real-time sampling oscilloscope with a sampling rate of 50 GSa/s.

Figure 19.5(a)–(c) shows temporal waveforms of the received pulse sequence for back-to-back, 50 km transmission and 100 km transmission, respectively. In the generated pulse sequence, the amplitude of each pulse was modulated as each pulse has only one eigenvalue and its imaginary part $\eta = 2\text{Im}[\zeta]$ is periodically repeated as "0.2, 0.4, 0.3, 0.2, 0.3, 0.4, 0.2, 0.3, 0.3" when the normalized parameter of time t_0 is chosen as 50 ps. We assume the SMF's dispersion parameter $D = 17$ [ps/nm/km] and nonlinear coefficient $C\gamma = 1.1$ [1/(W · km)]. Comparing the waveforms before and after propagation in SMF, the pulse's

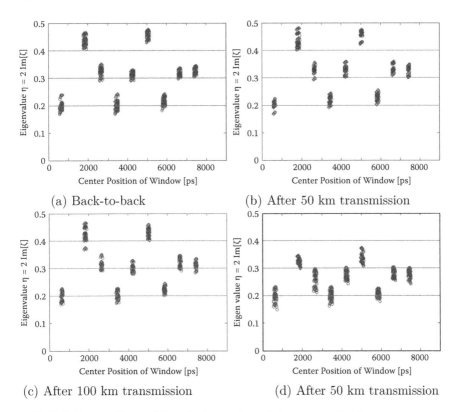

(a) Back-to-back

(b) After 50 km transmission

(c) After 100 km transmission

(d) After 50 km transmission

FIGURE 19.6: Demodulated eigenvalue of the received pulse sequence.

amplitude and width have changed. For demodulating eigenvalues from the received pulse sequence, the eigenvalues are calculated by solving Eq. (19.4) for waveforms within the computational temporal window with gradually shifting the time position of the window. We set the window size as 400 ps, shown by a dotted line in Fig. 19.5. The smallest pulse at the left edge of the sequence was used as the pilot signal. By multiplying a constant by the amplitude of the pulse, we adjusted the scaling with which the averaged eigenvalue η of the pulse is 0.2. In a practical system, the eigenvalue can be demodulated by calibrating the receiver by the use of an appropriate pilot signal. We transmitted waveforms, repeating the pulse sequence shown in Fig. 19.5(a), and overwrote the demodulated eigenvalues over 10 periods in Fig. 19.6(a)–(c). Although the demodulated eigenvalues have spread a bit due to the noise effect, they are close to the theoretical value shown by the dotted line and they did not change before and after transmission. In Fig. 19.5(d) and Fig. 19.6(d), show temporal waveforms of the pulse sequence and demodulated eigenvalues after 50 km propagation when the launched average power of the signal light was increased to 10.3 dBm. By increasing the launched average power, the increased influence of nonlinear effect causes pulse compression, as seen by comparing with Fig.19.5(b). Although the amplitude of the launched optical signal has changed and the corresponding eigenvalue is also changed, calibration had been done such that the eigenvalue of the pilot signal is 0.2 at the receiver. Thus the desired eigenvalue cannot be demodulated at the receiver, as shown in Fig. 19.6(d). So it is important to adjust the appropriate average power launched into the fiber to achieve a desired eigenvalue transmission.

19.5   NOISE TOLERANCE OF EIGENVALUES

Figure 19.7 shows the experimental setup. The pulse sequence was generated by modulating CW light at a wavelength of 1550.0 nm in a dual-parallel Mach–Zehnder modulator with a fixed pattern "010" having a repetition period of 300 ps generated by a 10 GSa/s AWG. ASE noise from EDFA1 was added to the generated pulse sequence. An optical attenuator (ATT) was used to adjust the optical signal to noise ratio (OSNR) of the signal. At the receiver, the noise loaded signal was O/E-converted by an optical coherent receiver consisting of an optical 90°-hybrid and balanced photo detector, and then sampled by a 50 GSa/s digital sampling oscilloscope. Although carrier frequency fluctuation induces a change in the real part of the eigenvalues, the imaginary part is not affected. Therefore, carrier frequency offset (CFO) cancellation is not necessary in an eigenvalue modulated system.

Figures 19.8 and 19.9, respectively, show the temporal waveforms and frequency spectra of the observed pulse sequences for OSNR = 20.4 and 10.1 dB. These spectra were observed just before the optical coherent receiver by using a spectrum analyzer. Each pulse has a different amplitude. The noise level becomes higher when OSNR becomes smaller. The amplitude of the received pulse is calibrated by multiplying a constant number, by which each pulse has only one discrete eigenvalue. We set the size of the processing window as 300 ps and numerically calculate the eigenvalue of each pulse in the window. Due

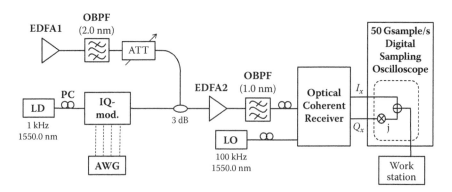

FIGURE 19.7: Experimental setup for noise tolerance evaluation.

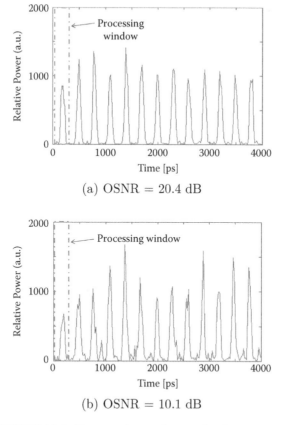

(a) OSNR = 20.4 dB

(b) OSNR = 10.1 dB

FIGURE 19.8: Temporal waveforms of pulse seqences.

to the added ASE noise, demodulated eigenvalues are slightly different even for the same amplitude pulses. Figures 19.10 and 19.11 show constellation points of the eigenvalues and probability mass functions (PMFs) of real part $\kappa = 2\Re[\zeta]$ and imaginary part $\eta = 2\Im[\zeta]$ of the eigenvalues.

The PMFs shown by symbols are in good agreement with those of normal distributions shown by solid lines. Therefore, we can approximate the distribution of the eigenvalues by the normal distribution. Figure 19.12 shows variation coefficients, which are standard deviations σ_κ, σ_η divided by $\bar{\eta}$, that is, the average of η, for various OSNRs. The worse the OSNR is, the bigger the variation coefficients become. So we need to allocate constellation points properly in view of the ASE noise effect for the eigenvalue multiplexed case [9] to distinguish the calculated eigenvalues correctly.

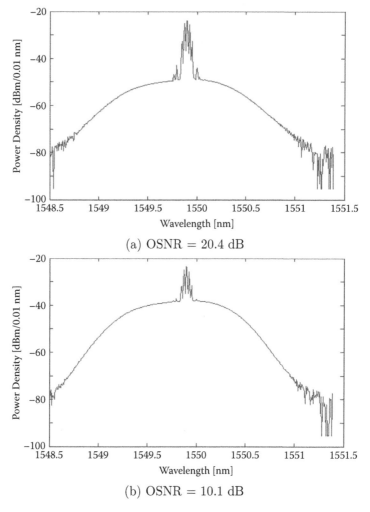

(a) OSNR = 20.4 dB

(b) OSNR = 10.1 dB

FIGURE 19.9: Frequency spectra of pulse sequences.

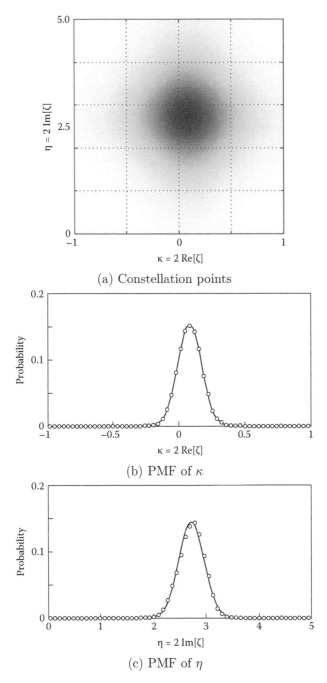

(a) Constellation points

(b) PMF of κ

(c) PMF of η

FIGURE 19.10: Constellation points and probability mass functions of the eigenvalues for OSNR = 20.4 dB.

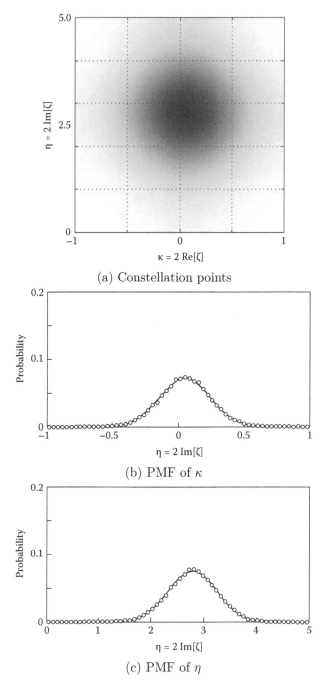

(a) Constellation points

(b) PMF of κ

(c) PMF of η

FIGURE 19.11: Constellation points and probability mass functions of the eigenvalues for OSNR = 10.1 dB.

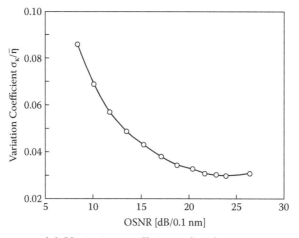

(a) Variation coefficient of real part

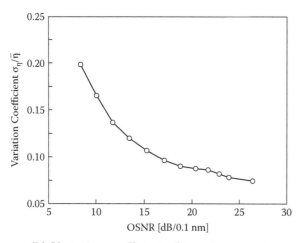

(b) Variation coefficient of imaginary part

FIGURE 19.12: Measured variation coefficients.

19.6 CONCLUSION

We have successfully demonstrated the feasibility of the eigenvalue demodulation scheme based on digital coherent technology. We also experimentally studied the noise tolerance of the demodulated eigenvalues. We showed that the distribution of the eigenvalues can be approximated by the normal distribution. Appropriate eigenvalue allocation in the eigenvalue plane can be then designed with the normally distributed noise model. Since an eigenvalue is an ideal information carrier, the eigenvalue modulated system is a promising candidate for an ultimate optical communication system.

ACKNOWLEDGMENTS

This work was partially supported by a project conducted by NICT, Japan. The authors would like to thank Dr. Yuki Yoshida, Mr. Go Ito, and Mr. Shohei Inudo with Osaka University for their help in using an optical coherent receiver.

REFERENCES

1. V. E. Zakharov and A. B. Shabat, Exact theory of two-dimensional self-focusing and one-dimensional self-modulation of waves in nonlinear media, *Sov. Phys. JETP*, vol. 34, no. 1, pp. 62–69 (1972).
2. A. Hasegawa and T. Nyu, Eigenvalue communication, *IEEE/OSA J. Lightwave Technol.*, vol. 11, no. 3, pp. 395–399 (1993).
3. S. Oda, A. Maruta, and K. Kitayama, All-optical quantization scheme based on fiber nonlinearity, IEEE Photon. Technol. Lett., vol. 16, no. 2, pp. 587–589 (2004).
4. E. Yamazaki, S. Yamanaka, Y. Kisaka, T. Nakagawa, K. Murata, E. Yoshida, T. Sakano, M. Tomizawa, Y. Miyamoto, S. Matsuoka, J. Matsui, A. Shibayama, J. Abe, Y. Nakamura, H. Noguchi, K. Fukuchi, H. Onaka, K. Fukumitsu, K. Komaki, O. Takeuchi, Y. Sakamoto, H. Nakashima, T. Mizuochi, K. Kubo, Y. Miyata, H. Nishimoto, S. Hirano, and K. Onohara, Fast optical channel recovery in field demonstration of 100-Gbit/s Ethernet over OTN using real-time DSP, *Opt. Express*, vol. 19, no. 14, pp. 13179–13184 (2011).
5. X. Li, X. Chen, G. Goldfarb, E. Mateo, I. Kim, F. Yaman, and G. Li, Electronic post-compensation of WDM transmission impairments using coherent detection and digital signal processing, *Opt. Express*, vol. 16, no. 2, pp. 880–888 (2008).
6. H. Terauchi and A. Maruta, Eigenvalue modulated optical transmission system based on digital coherent technology, The 10th Conference on Lasers and Electro-Optics Pacific Rim, and the 18th OptoElectronics and Communications Conference/Photonics in Switching 2013 (CLEO-PR&OECC/PS 2013), Kyoto International Conference Center, Kyoto, Japan, Paper WR2-5 (2013).
7. Hiroki Terauchi, Yuki Matsuda, Akifumi Toyota, and Akihiro Maruta, Noise tolerance of eigenvalue modulated optical transmission system based on digital coherent technology, The 19th OptoElectronics and Communications Conference / Australian Conference on Optical Fiber Technology 2014

(OECC/ACOF 2014), Melbourne Convention and Exhibition Centre, Melbourne, Australia, WEPS2-56, Paper 542 (9 July 2014).

8. A. Hasegawa and Y. Kodama, Guiding-center soliton in optical fibers, *Opt. Lett.*, vol. 15, no. 24, pp.1443–1445 (1990).

9. Yuki Matsuda, Hiroki Terauchi, and Akihiro Maruta, Design of eigenvalue-multiplexed multi-level modulation optical transmission system, The 19th OptoElectronics and Communications Conference/Australian Conference on Optical Fiber Technology 2014 (OECC/ACOF 2014), Melbourne Convention and Exhibition Centre, Melbourne, Australia, Paper TH12B3 (10 July 2014).

20 Quantum field theory analog effects in nonlinear photonic waveguides

Andrea Marini and *Fabio Biancalana*

In this chapter we discuss some recent works and developments on optical simulations and analogies with quantum mechanical/quantum field theory effects. In Sections 20.1–20.3 we focus our attention on optical waveguide arrays (WAs) that allow for the optical simulation of non-relativistic dynamics of quantum particles. They constitute a useful classical laboratory for mimicking quantum effects and can be exploited to analyze fundamental quantum mechanisms with classical tools. In addition, relativistic phenomena of quantum field theory can be optically simulated in binary waveguide arrays (BWAs), as optical propagation in the continuous limit is governed by a (1+1)D Dirac equation. Specifically, we analyze optical analogs of relativistic Dirac solitons (Section 20.1), tachyons (Section 20.2), and neutrinos (Section 20.3). In Section 20.4 we discuss recent developments in the theoretical understanding of novel dispersive waves in nonlinear media, such as negative-frequency resonant radiation (NRR). The role of negative frequencies in nonlinear optics is thoroughly explained and modeled by means of a novel equation for the analytic signal of the optical wave, which inherently accounts for negative-frequency components. Analogies between negative frequencies in nonlinear optics and negative energies in quantum field theory are discussed, and novel frequency conversion applications in nonlinear media are unveiled.

20.1 OPTICAL ANALOG OF RELATIVISTIC DIRAC SOLITONS IN BINARY WAVEGUIDE ARRAYS

20.1.1 INTRODUCTION

Waveguide arrays have been used intensively to simulate the evolution of a nonrelativistic quantum mechanical particle in a periodic potential. Many fundamental phenomena in nonrelativistic classical and quantum mechanics, such as Bloch oscillations [1, 2], Zener tunneling [3, 4], optical dynamical localization [5], and Anderson localization in disordered lattices [6], have been simulated both theoretically and experimentally with waveguide arrays. In a recent study it was shown that, rather surprisingly, most nonlinear fiber optics features (such as resonant radiation and soliton self-wavenumber shift) can also take place in specially excited arrays [7]. Recently, binary waveguide arrays (BWAs) have also been used to mimic relativistic phenomena typical of quantum field theory, such as Klein tunneling [8], the *Zitterbewegung* (trembling motion of a free Dirac electron) [9, 10], Klein paradox [11], and fermion pair production [12], which are all based on the properties of the Dirac equation [13]. Although there is as yet no evidence for fundamental quantum nonlinearities, nonlinear versions of the Dirac equation have been studied for a long time. One of the earlier extensions was provided by Heisenberg [14] in the context of field theory and was motivated by the question of mass. In the quantum mechanical context, nonlinear Dirac equations have been used as effective theories in atomic, nuclear and gravitational physics [15–18] and, more recently, in the study of ultracold atoms [19, 20]. In this regard, BWAs can offer a rather unique model system to simulate nonlinear extensions of the Dirac equation when probed at high light intensities. The discrete gap solitons in BWAs in the *classical* context have been investigated both numerically [21–23] and experimentally [24]. In particular, in [22] soliton profiles with even and odd symmetry were numerically calculated and a scheme with two Gaussian beams, which are tuned to the Bragg angle with opposite inclinations, was proposed to efficiently generate gap solitons. In [24] solitons were experimentally observed when the inclination angle of an input beam is slightly above the Bragg angle.

Inspired by the importance of BWAs as a classical simulator for relativistic quantum phenomena, and also by past achievements in the investigation of discrete gap solitons in BWAs, in this section we present analytical Dirac soliton solutions of the coupled-mode equations (CMEs) in BWAs. Such solutions are found under the condition that the spatial width of solitons should be large enough for the quasicontinuous limit to be valid, as we shall quantify later in the text. We demonstrate that with the soliton solutions found, CMEs can be converted into the nonlinear relativistic 1D Dirac equation , and thus the Dirac soliton solutions can also be analytically constructed and simulated with BWAs. This paves the way for using BWAs to simulate other nonlinear solitonic and nonsolitonic effects of the relativistic Dirac equation.

20.1.2 ANALYTICAL SOLITON SOLUTIONS

Light propagation in a discrete, periodic binary array of Kerr nonlinear waveguides can be described, in the continuous-wave regime, by the following dimensionless CMEs [8, 21]:

$$i\frac{da_n(z)}{dz} = -\kappa[a_{n+1}(z) + a_{n-1}(z)] + (-1)^n \sigma a_n - \gamma|a_n(z)|^2 a_n(z), \quad (20.1)$$

where a_n is the electric field amplitude in the nth waveguide, z is the longitudinal spatial coordinate, 2σ and κ are the propagation mismatch and the coupling coefficient between two adjacent waveguides of the array, respectively, and γ is the nonlinear coefficient of waveguides which is positive for self-focusing, but negative for self-defocusing media. For simplicity, here we suppose all waveguides have the same nonlinear coefficient, but even if these nonlinear coefficients are different (provided they are comparable), then analytical soliton solutions shown later will not be changed, because, as explained later, one component of solitons is much weaker than both unity and the other component, and thus one can eliminate the nonlinear term associated with this weak soliton component. In the dimensionless form, in general, one can normalize variables in the above equation such that γ and κ are equal to unity. However, throughout this chapter we will keep these parameters explicitly in Eq. (20.1). Before proceeding further, it is helpful to analyze the general properties of the general solutions of Eq. (20.1). First of all, let us assume that $(a_{2n}, a_{2n-1})^T = i^{2n}(\varphi_{2n}, \varphi_{2n-1})^T$ is one solution of Eq. (20.1) with φ_{2n} and φ_{2n-1} being appropriate functions. In this case, if we change the sign of γ while keeping the other two parameters constant, one can easily show that a new solution of Eq. (20.1) will be $(a_{2n}, a_{2n-1})^T = i^{2n}(\varphi_{2n-1}^*, \varphi_{2n}^*)^T$, where $*$ denotes the complex conjugation. Second, if the sign of σ is changed while other parameters are kept constant, then a new solution of Eq. (20.1) will be $(a_{2n}, a_{2n-1})^T = i^{2n}(\varphi_{2n-1}, \varphi_{2n})^T$. Of course, when σ changes sign, we still have the same physical system, but with a shift of the wavenumber position n in Eq. (20.1) by one. The above simple rules allow us to quickly find other solutions and their symmetries if one particular solution is known, as will be shown later.

In the specific case when all three parameters γ, κ, and σ are all kept positive, we look for analytical solutions of motionless solitons of Eq. (20.1) in the following form:

$$\begin{bmatrix} a_{2n}(z) \\ a_{2n-1}(z) \end{bmatrix} = \begin{bmatrix} i^{2n}d\frac{2}{n_0}\text{sech}(\frac{2n}{n_0})e^{ifz} \\ -i^{2n-1}b\text{sech}(\frac{2n-1}{n_0})\tanh(\frac{2n-1}{n_0})e^{ifz} \end{bmatrix}, \qquad (20.2)$$

where $n_0 \in \mathbb{R}$ characterizes the beam width (i.e., the average number of waveguides on which the beam extends), and coefficients b, d and f are still unknown. In the system without any loss or gain of energy (i.e., when κ, σ and γ are all real), the coefficient f must also be real, but b and d can be complex in general. Inserting the ansatz (20.2) into Eq. (20.1), assuming a priori that the component a_{2n-1} is much weaker than both unity and the other component a_{2n}, such that one can eliminate the nonlinear term for a_{2n-1}, and also assuming that the quasicontinuous limit is valid (i.e., n_0 is large enough), after some lengthy algebra one gets

$$fd = \kappa bi - \sigma d, \qquad (20.3)$$
$$i\kappa b = 2\gamma|d|^2 d/n_0^2, \qquad (20.4)$$
$$fb = \sigma b + 4d\kappa i/n_0^2. \qquad (20.5)$$

Extracting f and b from Eq. (20.3) and Eq. (20.4), respectively, then inserting them into Eq. (20.5), we will get one quadratic equation for d^2, and thus can find the values for b, d and f. Note that one needs to keep only solutions which satisfy the above assumption that $|a_{2n-1}| \ll |a_{2n}|$. The final solution in the case when $\gamma, \kappa, \sigma > 0$ is

$$
\begin{bmatrix} a_{2n}(z) \\ a_{2n-1}(z) \end{bmatrix} = \begin{bmatrix} i^{2n} \frac{2\kappa}{n_0 \sqrt{\sigma\gamma}} \operatorname{sech}(\frac{2n}{n_0}) e^{iz(\frac{2\kappa^2}{n_0^2\sigma} - \sigma)} \\ i^{2n} \frac{2\kappa^2}{n_0^2\sigma \sqrt{\sigma\gamma}} \operatorname{sech}(\frac{2n-1}{n_0}) \tanh(\frac{2n-1}{n_0}) e^{iz(\frac{2\kappa^2}{n_0^2\sigma} - \sigma)} \end{bmatrix}. \tag{20.6}
$$

It is worth mentioning that the analytical soliton solution in the form of Eq. (20.6) is derived under two conditions: (i) the beam must be large enough such that one can operate in the quasicontinuous limit instead of the discrete one; and (ii) $n_0|\sigma| \gg 2\kappa$. The latter condition is easily satisfied if (i) is held true and if σ is comparable to κ [10]. If condition (ii) is not valid, one can still easily get the analytical solution for b, d and f from Eqs. (20.3)–(20.5), but they are a bit cumbersome and for brevity we do not show it here. The solution in the form of Eq. (20.6) represents a one-parameter family of discrete solitons in BWAs where the beam width parameter n_0 can be arbitrary, provided that $n_0 \gtrsim 4$, a surprisingly small number for the quasicontinuous approximation to be valid.

In Fig. 20.1(a) we plot the soliton profile with even symmetry calculated by using Eq. (20.6) at $z = 0$, with full circles marking the field amplitudes across BWAs, for the parameters given in the caption. Note that the soliton profile in Fig. 20.1(a) consists of two components: one strong component a_{2n} and another much weaker component a_{2n-1} [see also Fig. 20.2(c)]. Once we get the soliton solution in Fig. 20.1(a), we can construct another soliton solution of the same physical system by changing the sign of σ and following the rules

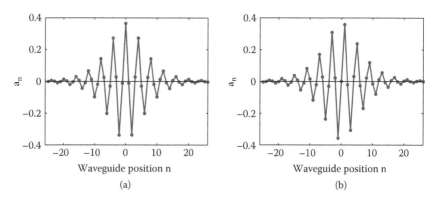

FIGURE 20.1: Discrete soliton profiles (a,b) for even and odd symmetry, respectively. Full circles mark the field amplitudes across the BWA. Parameters in (a): $\kappa = 1$; $\gamma = 1$; $\sigma = 1.2$; and $n_0 = 5$. After getting the even symmetry profile in (a), we construct the odd profile in (b) by switching the sign of σ and following the symmetry trasformations explained in the text.

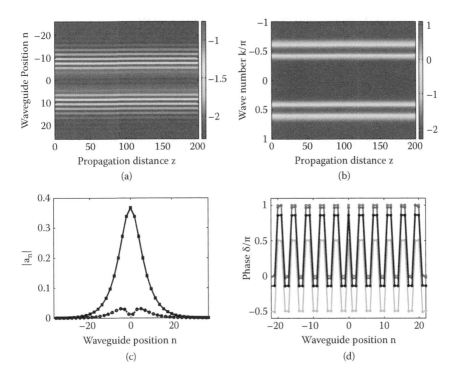

FIGURE 20.2: (a,b) Soliton propagation in the (n, z)-plane (a) and its Fourier transform in the (k, z)-plane (b) with an even symmetry profile at the input. (c) Absolute values of the field amplitudes for intense ($|a_{2n}|$ with solid line and square markers) and weak ($|a_{2n-1}|$ with dashed-dotted curves and round markers) components of a soliton at four different values of $z = 0$ (curves); 50 (curves); 140 (curves); and 200 (curves). The soliton profile is so well preserved that all these curves just stay on top of each other and one can see only the output curves. (d) Phase pattern δ/π of soliton profiles at the above four values of z. Parameters $k = 1$; $\gamma = 1 : \sigma = 1.2$; and $n_0 = 5$. All contour plots are show on a logarithm scale.

explained in the previous section. In that way we obtain the odd symmetry soliton profile depicted in Fig. 20.1(b). It is important to mention that in the case of self-focusing media ($\gamma > 0$), for both even and odd symmetries the strong component is always located at waveguides with a larger propagation constant [channels with $+|\sigma|$ in Eq. (20.1)], whereas the weak component is located at waveguides with a smaller propagation constant [channels with $-|\sigma|$ in Eq. (20.1)]. We are also able to construct the soliton solutions for the self-defocusing media, which also possess soliton solutions with even and odd symmetries. The only difference with the self-focusing media is that now the strong (weak) component is located at waveguides with a smaller (larger) propagation constant.

20.1.3 SOLITON PROPAGATION AND GENERATION

Equation (20.6) and the associated solutions obtained by the above symmetry transformations provide the analytical forms of the two discrete gap soliton branches numerically found in [22]. We note that the propagation constant f of the two solitons, given by $f = -\sigma + 2\kappa^2/(n_0^2\sigma)$, falls in the minigap of the superlattice, near the edge of the lower miniband (because $2\kappa^2/(n_0^2\sigma) \ll \sigma$), and thus they are expected to be stable [22]. As an example, in Fig. 20.2(a) we show the soliton propagation along z as obtained by numerically solving Eq. (20.1) with an input soliton taken from Eq. (20.6) at $z = 0$, demonstrating that the soliton profile is well preserved during propagation. Parameters used for Fig. 20.2 are the same as in Fig. 20.1(a). The evolution of the Fourier transform of the field a_n in Fig. 20.2(a) along z is shown in Fig. 20.2(b), where the wavenumber k represents the phase difference between adjacent waveguides. Due to the periodic nature of BWAs, within the coupled mode approximation, it suffices to investigate k in the first Brillouin zone $-\pi \leq k \leq \pi$ [25]. One very important feature of the wavenumber evolution in Fig. 20.2(b) is the fact that there are two components of wavenumber centered at $k = \pm\pi/2$ which correspond to two Bragg angles [10] with opposite inclinations. These two wavenumber components are generated at the input and preserve their shapes during propagation along z. This feature of k indicates that the soliton operates in the region where CMEs could potentially be converted into the relativistic Dirac equations describing the evolution of a freely moving relativistic particle [9, 10]. We will come back to this important point again later. Figure 20.2(c) shows the two components of the soliton profile at odd and even waveguide position n. The strong component with solid curves and square markers represents the field profile $|a_{2n}|$ at even waveguide positions, whereas the weak component with dashed-dotted curves and round markers represents the field profile $|a_{2n-1}|$ at odd waveguide positions. Field profiles in Fig. 20.2(c) are taken at four values of propagation distance $z = 0$; 50; 140; and 200 — only the black curves are actually visible since the the the profile is perfectly preserved during propagation with a very high precision. The soliton profile also perfectly preserves its phase pattern across the array [Fig. 20.2(d)]. From Eq. (20.6), one can easily see that as the waveguide position variable n runs, the phase pattern of the soliton must be periodic as follows: $\delta_n = ...(\rho, \rho), (\rho + \pi, \rho + \pi), (\rho, \rho)...$ where ρ also changes with z. This pattern is only broken at the soliton center point where the function $tanh$ in Eq. (20.6) changes its sign. This phase pattern is shown in Fig. 20.2(d), where different with meanings as in Fig. 20.2(c) depict patterns at different values of z. The sequence in the phase is important because it allows us to convert Eq. (20.1) into the nonlinear Dirac equation as we shall show shortly. Note that the soliton whose propagation is shown in Fig. 20.2 is the one with even symmetry in Fig. 20.1(a). Our simulations similarly show that the profile of the soliton with odd symmetry in Fig. 20.1(b) is also well preserved during propagation, and we have checked that this true even in the presence of quite a strong numerical noise, demonstrating the robustness and the stability of our solutions.

Although the soliton solutions given by Eq. (20.6) are exact, it is important to consider the possibility of generating the new gap solitons by an input beam with a simpler (and more experimentally accessible) profile. Due to the wavenumber structure shown in Fig. 20.2(b), one can interpret the soliton as a combination of two beams launched under two Bragg angles with opposite tilts $k = \pm \pi/2$, similar to what was suggested in [22]. Here we propose to generate the soliton by an input with a simple phase pattern where the phase difference between adjacent waveguides is equal to $\pi/2$ across the array. The input condition is taken to be $A_n = a_n \exp(in\pi/2)$ where a_n is given by Eq. (20.6) at $z = 0$, but without the term i^{2n}. Note that, since $|a_{2n-1}| \ll |a_{2n}|$, this input condition can be approximately achieved by exciting the BWA with a broad beam tilted at the Bragg angle, with the odd waveguides in the structure being realized at some spatial delay Δz inside the sample (so they are not excited at the input plane); see the scheme shown in Fig. 20.3(f). In the linear regime, the beam broadens and undergoes *Zitterbewegung* [9, 10], whereas in the nonlinear regime soliton formation is expected to take place with suppression of both beam broadening and *Zitterbewegung*. This is clearly shown in Fig. 20.3, which indicates the formation of the soliton during propagation with parameters as in Fig. 20.2. The evolution of field profiles $|a_{2n}|$ and $|a_{2n-1}|$ at even and odd waveguide positions is depicted in Fig. 20.3(a) and 20.3(b), respectively. The evolution of the Fourier transform of the field a_n of Fig. 20.3(a,b) along z is shown in Fig. 20.3(c). One can see that the strong component a_{2n} in Fig. 20.3(a) does not change much during propagation, whereas the weak component a_{2n-1} in Fig. 20.3(b) is dramatically altered during propagation. As seen from Fig. 20.3(b), at the beginning of the propagation the beam undergoes the *Zitterbewegung*. After reaching $z \simeq 70$, the profile $|a_{2n-1}|$ becomes stable. Figure 20.3(d) shows the strong component $|a_{2n}|$ of the soliton profile with solid curves and the weak component $|a_{2n-1}|$ with dashed-dotted curves. As in Fig. 20.2(c,d) field profiles are taken at four values of propagation distance $z = 0$ (curves); 50 (curves); 140 (curves); and 200 (curves). One can also see that the strong component $|a_{2n}|$ is stable, whereas the weak component first gets distorted (see curves), but eventually the output curve (black color) relaxes to the input curve (color). Figure 20.3(e) depicts the phase pattern of the field amplitudes across the array calculated at different z with corresponding as in Fig. 20.3(d). At the input (curve) we have the phase difference equal to $\pi/2$ between adjacent waveguides, but this phase pattern quickly transforms into the phase pattern of the soliton solution given by Eq. (20.6), i.e., $\delta_n = ...(\rho, \rho), (\rho + \pi, \rho + \pi), (\rho, \rho)...$ [see curves in Fig. 20.3(e)]. Therefore, here one can make a local conclusion: a beam with the intensity profiles of the soliton solution given by Eq. (20.6), but with phase difference equal to $\pi/2$ between adjacent waveguides, will first undergo *Zitterbewegung*, but eventually its intensity profile and phase pattern will relax to the ones of the soliton solution given by Eq. (20.6).

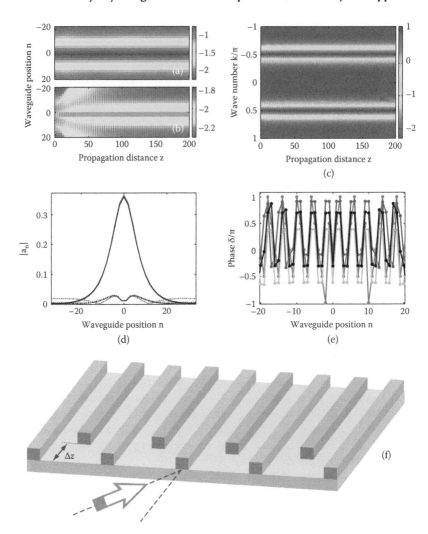

FIGURE 20.3: (a,b) Propagation in the (n, z)-plane of the even and odd components of the beam with initial phase difference equal to $\pi/2$ between adjacent waveguides. (c) Fourier transform of field amplitudes in the (k, z)-plane. (d) Absolute values of the field amplitudes for intense component $|a_{2n}|$ with solid curves and weak component $|a_{2n-1}|$ with dashed-dotted curves at four different values of $z = 0$; 50; 140; and 200. (e) Phase pattern δ/π of field amplitudes for the same values of z as in (d). Colors of curves in (e) have the same meaning as in (d). (f) Scheme of the BWA structure for generating discrete solitons. Parameters: $\kappa = 1$; $\gamma = 1$; $\sigma = 1.2$; and $n_0 = 5$.

20.1.4 DIRAC SOLITONS

As mentioned in the introduction, BWAs have been used to mimic phenomena in both nonrelativistic and relativistic quantum mechanics. To the best of our knowledge, so far all these phenomena which have been simulated by BWAs are linear. In this section we will report on the simulation of nonlinear relativistic Dirac solitons in BWAs. As shown in [9, 10], linear CMEs [Eq. (20.1)] for a beam with phase difference equal to $\pi/2$ can be converted into the linear one-dimensional relativistic Dirac equation (DE). Note that Eq. (20.1) can be converted into the DE only for beams with special phase patterns; for instance, at normal beam incidence Eq. (20.1) cannot be converted into the DE. It turns out that with the soliton solution given by Eq. (20.6), one can also successfully convert Eq. (20.1) into the nonlinear relativistic Dirac equation (NDE). Thus, one can use BWAs to mimic the relativistic Dirac solitons, and soliton solutions in BWAs given by Eq. (20.6) can be used to construct directly the Dirac soliton. Although the solution of Eq. (20.6) does not possess the phase difference equal to $\pi/2$ between adjacent waveguides [see Fig. 20.2(d)], the fact that it exhibits two wavenumbers $k = \pm\pi/2$ [see Fig. 20.2(b)] gives us some hope that the NDE can also be obtained in this case. Indeed, this is the case, as shown below. In general, suppose that $[a_{2n}(z), a_{2n-1}(z)]^T = i^{2n}[g(2n, z), q(2n - 1, z)]^T$, where the two functions g and q are smooth and their derivatives $\partial_n g$ and $\partial_n q$ exist in the quasicontinuous limit [Eq. (20.6) satisfies these requirements]. After setting $\Psi_1(n) = (-1)^n a_{2n}$ and $\Psi_2(n) = i(-1)^n a_{2n-1}$ and following the standard approach developed in [9, 10], we can introduce the continuous transverse coordinate $\xi \leftrightarrow n$ and the two-component spinor $\Psi(\xi, z) = (\Psi_1, \Psi_2)^T$ which satisfies the 1D NDE:

$$i\partial_z \Psi = -i\kappa\alpha\partial_\xi \Psi + \sigma\beta\Psi - \gamma G, \tag{20.7}$$

where the nonlinear terms $G \equiv (|\Psi_1|^2\Psi_1, |\Psi_2|^2\Psi_2)^T$; $\beta = \mathrm{diag}(1, -1)$ is the Pauli matrix σ_z; and α is the Pauli matrix σ_x with diagonal elements equal to zero, but off-diagonal elements equal to unity. Note that Eq. (20.7) is identical to the DE obtained in [9, 10] with the only difference that now we have the nonlinear term G in Eq. (20.7). Similar soliton solutions have been found for the NDE in [26], but with a different and more complicated kind of nonlinearity, in the context of quantum field theory. Note that the nonlinearity that we have in Eq. (20.7) violates Lorentz invariance [27] and is similar to that of the Dirac equations in Bose–Einstein condensates [19]. Using the soliton solution given by Eq. (20.6) and the above relation between a_n and Ψ, one can easily obtain the Dirac soliton solution of Eq. (20.7) as follows:

$$\begin{bmatrix} \Psi_1(\xi, z) \\ \Psi_2(\xi, z) \end{bmatrix} = \begin{bmatrix} \frac{2\kappa}{n_0\sqrt{\sigma\gamma}}\mathrm{sech}\left(\frac{2\xi}{n_0}\right)e^{iz\left(\frac{2\kappa^2}{n_0^2\sigma} - \sigma\right)} \\ i\frac{2\kappa^2}{n_0^2\sigma\sqrt{\sigma\gamma}}\mathrm{sech}\left(\frac{2\xi-1}{n_0}\right)\tanh\left(\frac{2\xi-1}{n_0}\right)e^{iz\left(\frac{2\kappa^2}{n_0^2\sigma} - \sigma\right)} \end{bmatrix}. \tag{20.8}$$

The above solution is obtained for $\sigma > 0$ and $\gamma > 0$. One can use the symmetry properties of Eq. (20.1) to construct other Dirac soliton solutions of

Eq. (20.7), with different sign combinations between σ and γ. The expressions given by Eq. (20.8) give the main result of this section, and the only physically realizable way that we are aware of to produce and observe Dirac solitons with a tabletop experiment.

20.1.5   CONCLUSION

In this section we have provided analytical expressions for the non-moving gap solitons in binary waveguide arrays and shown their connection to Dirac solitons in a nonlinear extension of the relativistic one-dimensional Dirac equation describing the dynamics of a freely moving relativistic particle. Our results suggest that binary waveguide arrays can be used as a classical simulator to investigate relativistic Dirac solitons, enabling us to realize an experimentally accessible model system of quantum nonlinearities that have been so far a subject of speculation in the foundation of quantum field theories. The analysis of analog quantum field theory effects such as those described in this section are applicable to virtually any nonlinear discrete periodic system supporting solitons, therefore making our results very general and of relevance for a number of very diverse communities.

20.2   OPTICAL ANALOG OF SPONTANEOUS SYMMETRY BREAKING AND TACHYON CONDENSATION IN PLASMONIC ARRAYS

Tachyon condensation is a relevant process arising in quantum field theory (QFT) [28]. This is a mechanism in particle physics where the system lowers its energy by spontaneously generating particles. Owing to the complex mass, the tachyonic field is unstable and naturally acquires a non-negative squared mass and a vacuum expectation value by reaching the minimum of a nonlinear Mexican-hat potential. This mechanism is intimately related to the process of *spontaneous symmetry breaking* (SSB), i.e., the spontaneous collapse of a system into solutions that violate one or more symmetries of the governing equations. In other contexts, SSB is inherently related to the existence of Higgs bosons [29], Nambu–Goldstone bosons [30, 31] and fermions [32].

Binary waveguide arrays (BWAs) constitute a useful classical laboratory for the study of QFT phenomena, and can be tailored in order to mimic high-energy mechanisms that can be simulated in a tabletop optical system. In what follows, we focus on a binary amplifying plasmonic array (BAPA). We theoretically investigate optical propagation in a BAPA with alternate couplings, i.e., a layered metal-dielectric stack, sketched in Fig. 20.4. Surface plasmon polaritons (SPPs) propagating at every $y - z$ metal-dielectric interface are weakly coupled to nearest neighbors through alternating positive and negative couplings [33]. Amplification schemes with SPPs have been intensively studied and also demonstrated experimentally [36, 37]. Gain is provided by externally pumped active inclusions embedded in the dielectric layers that can be modeled as two-level atoms.

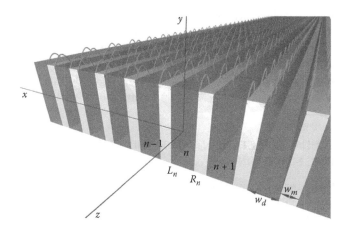

FIGURE 20.4: Layered metal-dielectric stack supporting SPPs at every $z - y$ interface. The dielectric media (dark slabs) embed externally pumped active inclusions, which amplify SPPs propagating along the z-direction. Every nth dielectric slab of width w_d, adjacent to a metallic stripe of thickness w_m, supports SPPs at the left and right interfaces with optical amplitudes L_n, R_n. The system is assumed homogeneous in the y, z directions and infinitely extended in the x, y, z directions.

In the limit of weak nonlinearity and overlap between adjacent SPPs [38], optical propagation in the amplifying plasmonic array sketched in Fig. 20.4 can be modeled by the following pair of coupled mode equations (CMEs) [33]:

$$i\frac{dL_n}{dz} - i\eta L_n + \kappa(R_n - R_{n-1}) + \gamma|L_n|^2 L_n = 0, \qquad (20.9)$$

$$i\frac{dR_n}{dz} - i\eta R_n + \kappa(L_n - L_{n+1}) + \gamma|R_n|^2 R_n = 0, \qquad (20.10)$$

where $\eta = \eta' + i\eta''$, η' is the effective gain parameter, η'' is the linear phase shift induced by two-level atoms, κ, γ are the coupling and nonlinear coefficients and L_n, R_n are the left and right dimensionless field amplitudes at every n-th dielectric slot. The longitudinal coordinate z is normalized to the coupling length z_0, so that $\kappa = 1$ and η, γ are complex dimensionless constants. At $\lambda = 594$ nm, using a silver stripe of width $w_m = 45$ nm and a gaining medium of width $w_d = 260$ nm and $\epsilon_d = 2.13 - 0.05i$, one gets a coupling length of $z_0 \simeq 1\mu$ m, and realistic values for the gain parameter are of the order $|\eta| \simeq 10^{-2}$. A full detailed derivation of Eqs. (20.9) and (20.10) and analytical expressions for the coefficients η, κ, γ are given in [38–40]. Owing to the dual chirality of alternating metal-dielectric interfaces (metal-dielectric and dielectric-metal), the system is inherently binary and every SPP is coupled with left and right adjacent SPPs by means of two different coupling coefficients κ_L, κ_R. However, it is possible to adjust the width of the dielectric slabs (w_d) and metallic stripes

(w_m) in order to achieve the condition $\kappa_L = -\kappa_R = \kappa$ [33]. The nonlinear coefficient is complex $\gamma = \gamma' + i\gamma''$, the real part can be either positive or negative depending on the sign of the detuning, while the imaginary part is always positive $\gamma'' > 0$ and accounts for the nonlinear saturation of gain. Note that Eqs. (20.9) and (20.10) are invariant under reflection in the x-direction $(n \to -n, L_n \to R_{-n}, R_n \to L_{-n})$, due to the inherent chiral symmetry of the total system.

20.2.1   VACUUM EXPECTATION VALUE AND NONLINEAR TACHYON-LIKE DIRAC EQUATION

Owing to the externally pumped active inclusions, small perturbations of the vacuum state $L_n = R_n = 0$ are exponentially amplified at a rate η'. Instability develops until nonlinear effects become important and nonlinear gain saturation comes into play, counterbalancing the linear amplification. Homogeneous nonlinear stationary modes of Eqs. (20.9) and (20.10) can be found by taking the ansatz $L_n = L_0 e^{iqn + i\mu z}$, $R_n = R_0 e^{iqn + i\mu z}$, where q is the transverse momentum and μ is the nonlinear correction to the unperturbed propagation constant β. As a consequence of the dissipative nature of the system, the amplitudes L_0, R_0 do not remain arbitrary and their moduli are fixed to be $A = \sqrt{\eta'/\gamma''}$. The nonlinear correction to the propagation constant is given by $\mu_\pm = \eta'' \pm 2k \sin(q/2) + \gamma'\eta'/\gamma''$. Owing to the inherent alternate coupling of the system, the nonlinear dispersion is characterized by a Dirac diabolical point at $q = 0$. At this special point, the phases of both amplitudes L_0, R_0 remain arbitrary. Conversely, for $q \neq 0$ the mode amplitudes are fixed to $R_0 = \mp i e^{iq/2} L_0$ and only a global phase is left arbitrary. Defining the two-component spinor $\psi = [L_n(z), R_n(z)]^T$, if the transversal patterns of the amplitudes L_n, R_n are smooth, one can take the continuous limit by introducing the continuous spatial coordinate $n \to x$. In this limit, the spinor satisfies the (1+1)D nonlinear tachyon-like Dirac equation

$$i\partial_z\psi - i\eta\psi + i\kappa\hat{\sigma}_y\partial_x\psi + \gamma G(\psi) = 0, \qquad (20.11)$$

where $G(\psi) = (|L|^2 L, |R|^2 R)^T$ is the nonlinear spinorial term and $\hat{\sigma}_y$ is the y-Pauli matrix. Equation (20.11) is analogous to the (1+1)D Thirring model [41] with imaginary mass and nonlinear terms, describing the dynamics of fermionic tachyons. Optical analogs of fermionic tachyons have been recently investigated in optical graphene and in topological insulators [42, 43]. Note that Eq. (20.11) is a *Dirac-like* equation, since the "mass term" $(-i\eta\psi)$ is different from previously studied standard formulations [42] and is responsible for the existence of unstable tachyon-like particles. Owing to amplification, vacuum dynamically acquires a stable expectation value and the ensuing final state is the optical analog of a condensate of stable fermionic particles with non-negative squared mass. In turn, this process is commonly called *tachyon condensation*, e.g., in the context of open string field theories [44].

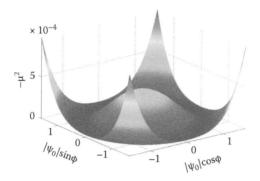

FIGURE 20.5: Optical analog of the Mexican hat potential: $-\mu^2$ is plotted against $|\psi_0|cos\phi$, $|\psi_0|sin\phi$, where $|\psi_0|$ is the field amplitude and ϕ is the relative phase between the spinor components. The plot is made by taking the parameters $\eta' = 0.02$, $\kappa = 1$, $\gamma'' = 0.01i$.

20.2.2   SPONTANEOUS SYMMETRY BREAKING

Note that, analogously to the Nambu–Jona–Lasinio model [30], Thirring models [41], and sine-Gordon models [45], Eq. (20.11) is chirally symmetric since it is left invariant under reflection $x \rightarrow -x$ if the spinor components are transformed as $L(x) \rightarrow R(-x)$, $R(x) \rightarrow L(-x)$. In turn, while the unstable vacuum state $\psi = 0$ is chirally symmetric, the nonlinear homogeneous mode $\psi = \psi_0 e^{i\mu z}$ with finite amplitude ψ_0 and propagation constant μ [where $\mu^2 = -(\eta' - \gamma''|\psi_0|^2)^2$] breaks the chiral symmetry. The optical analog of energy is represented by the propagation constant μ and the system spontaneously evolves to states where $-\mu^2$ is minimum. In Fig. 20.5, we plot $-\mu^2 = (\eta' - \gamma''|\psi_0|^2)^2$ as a function of the mode amplitude $|\psi_0|$ and the relative phase between the spinor components ϕ. We find the characteristic Mexican hat profile, which constitutes the archetypical potential describing spontaneous symmetry breaking in QFT. Tachyon condensation thus drives the physical system to a stable state with broken chiral symmetry where $-\mu^2$ is minimum and particles with non-negative squared mass are generated. These predictions have been confirmed by the direct numerical integration of Eqs. (20.9) and (20.10) using a fourth order Runge–Kutta algorithm. In the panels of Fig. 20.6, we contour plot the modulus of the left optical field $|L_n|$ as a function of the SPP index n and of the propagation direction z for different input conditions. In Fig. 20.6(a), we set as the initial condition a small random perturbation of the vacuum state, which is unstable and dynamically converges to the stable nonlinear homogeneous mode at the Dirac point $q = 0$, which represents the vacuum expectation value. In Fig. 20.6(b), we perturb the homogeneous nonlinear mode of the upper branch at the band edge $q = \pi$ with small random perturbations, finding a modulationally unstable chaotic dynamics that ensues from the presence of dissipative solitons. Owing to the

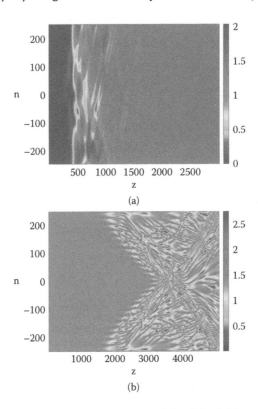

FIGURE 20.6: Propagation contour plots of the left field amplitude $|L_n|$ for two input conditions $L_n(0)$, $R_n(0)$ weakly perturbed with random noise: (a) vacuum state $L_n(0) = R_n(0) = 0$, (b) nonlinear homogeneous mode at the band edge $L_n(0) = R_n(0) = \sqrt{\eta'/\gamma''}e^{i\pi n}$. Numerical integration is taken with the parameters $\eta = 0.01$, $\kappa = 1$ and (a) $\gamma = 0.01i$, and (b) $\gamma = 0.01 + 0.01i$.

instability of the vacuum background, solitons are unstable and behave as strange attractors for the dynamical system [38].

20.3 OPTICAL ANALOG OF NEUTRINO OSCILLATIONS IN BINARY WAVEGUIDE ARRAYS

Neutrinos are uncharged elementary particles with half-integer spin that interact with matter only via weak forces. Neutrinos are created in three different lepton *flavors*: electron, muon and tau neutrinos. *Neutrino oscillation* is a quantum mechanical effect whereby a neutrino created with a specific lepton flavor can later be measured with a different flavor [46]. Neutrino oscillations occur both in vacuum and in matter, as the probability of measuring a neutrino with a specific flavor periodically oscillates during propagation [47–49].

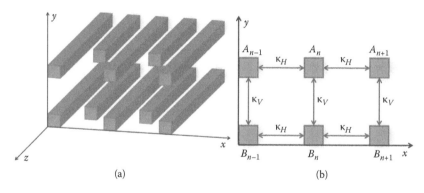

(a)   (b)

FIGURE 20.7: Schematics of two vertically displaced BWAs for the simulation of neutrino oscillation. (a) Three-dimensional sketch of the structure.
Neighboring silica waveguides are offset, while every waveguide shows a longitudinally modulated effective index. Light is trapped by every waveguide in
the transverse x-y directions and propagates in the longitudinal z-direction.
(b) Transverse section of the structure. Optical amplitudes of the upper (A_n)
and lower (B_n) array are coupled to nearest neighbors. Horizontal and vertical coupling constants between nearest neighbor waveguides are denoted by
κ_H, κ_V, respectively.

In recent years, theoretical and experimental investigations of neutrino oscillations have received considerable interest, as the observation of this phenomenon implies that neutrinos have small but finite masses. Neutrino oscillations ensue from a mixture between the flavor and mass eigenstates, since the
three neutrino states weakly interacting with charged leptons are superposi-
tions of three states with definite mass. In the early universe and in collapsing
supernova stars, the density of neutrinos is so large that they strongly interact with each other, many body interactions become important and neutrino
oscillations feedback on themselves [50–53]. In turn, peculiar effects like coherence in collective neutrino oscillations and suppression of self-induced flavor
conversion occur in supernova explosions [54, 55].

In the following we consider a pair of vertically displaced binary waveguide arrays (BWAs) with longitudinally modulated effective refractive index,
sketched in Fig. 20.7. The index modulation is assumed small and sinusoidal,
and can be achieved by longitudinally modulating the core index or the transverse waveguide section [58,59]. We indicate by A_n, B_n the field amplitudes of
the linear fundamental modes of the upper and lower waveguides, which are
weakly affected by the longitudinal modulation of the effective index. Typically, we assume that BWAs are manufactured in silica glass. Taking into
account Kerr nonlinearity, nearest-neighbor evanescent coupling and longitudinal modulation of the effective index, optical propagation is described by

the set of coupled-mode equations (CMEs):

$$i\frac{dA_n}{dz} = V_n(z)A_n - \kappa_H(A_{n+1} + A_{n-1}) - \kappa_V B_n - \gamma|A_n|^2 A_n, \quad (20.12)$$

$$i\frac{dB_n}{dz} = W_n(z)B_n - \kappa_H(B_{n+1} + B_{n-1}) - \kappa_V A_n - \gamma|B_n|^2 B_n, \quad (20.13)$$

where n is an integer labeling the waveguide site, κ_H, κ_V are the horizontal and vertical coupling coefficients between neighboring waveguides, γ is the Kerr nonlinear coefficient, $V_n(z) = (-1)^n m_a + \Delta V_n(z)$, $W_n(z) = -(-1)^n m_b + \Delta W_n(z)$, m_a (m_b) account for the alternating initial offsets of the upper (lower) BWAs, and $\Delta V_n(z), \Delta W_n(z)$ describe the longitudinal modulation of propagation constants of the upper and lower waveguides. In the fast modulation limit $\omega \gg \kappa_H, \kappa_V, |m_{a,b}|$, at leading order we can average Eqs. (20.12) and (20.13) and disregard the rapidly oscillating terms [58, 60]. This yields

$$i\frac{da_n}{dz} = -\kappa(a_{n+1} + a_{n-1}) + (-1)^n m_a a_n + (-1)^n \epsilon b_n - \gamma|a_n|^2 a_n, \quad (20.14)$$

$$i\frac{db_n}{dz} = -\kappa(b_{n+1} + b_{n-1}) - (-1)^n m_b b_n + (-1)^n \epsilon a_n - \gamma|b_n|^2 b_n, \quad (20.15)$$

where $\kappa \simeq 0.761\kappa_H$, $\epsilon \simeq 0.4\kappa_V$ are the effective averaged coupling constants between adjacent waveguides in the horizontal and vertical directions, respectively. Linear modes of Eqs. (20.14) and (20.15) can be calculated by setting $\gamma = 0$ and by taking the ansatz $a_{2n} = a_1 e^{i\beta z + iqn}$, $a_{2n+1} = a_2 e^{i\beta z + iqn}$, $b_{2n} = b_1 e^{i\beta z + iqn}$, $b_{2n+1} = b_2 e^{i\beta z + iqn}$, where q is the transverse quasi-momentum in the x-direction. After substitution of the ansatz above in Eqs. (20.14) and (20.15), one obtains the modes and the linear dispersion relation of the structure.

20.3.1 DIRAC LIMIT: NEUTRINOS

Defining the two-component spinors $A = (A_1 \ A_2)^T = (-1)^n (a_{2n}, i a_{2n-1})^T$ and $B = (B_1, B_2)^T = (-1)^n (i b_{2n-1}, b_{2n})^T$, if the amplitudes A_1, A_2, B_1, B_2 vary slowly with the site index n, one can take the continuous limit by introducing the continuous spatial coordinate $n \to x$ [58]. In this limit, the spinors satisfy two coupled (1+1)D nonlinear Dirac equations for half-spin particles with two different mass eigenstates, i.e., neutrinos:

$$i\partial_z A = -i\kappa\hat{\sigma}_x \partial_x A + m_a \hat{\sigma}_z A + i\epsilon\hat{\sigma}_y B - \gamma G(A), \quad (20.16)$$
$$i\partial_z B = -i\kappa\hat{\sigma}_x \partial_x B + m_b \hat{\sigma}_z B - i\epsilon\hat{\sigma}_y A - \gamma G(B), \quad (20.17)$$

where $G(\psi) \equiv (|\psi_1|^2\psi_1, |\psi_2|^2\psi_2)^T$ is the nonlinear spinorial term and $\hat{\sigma}_x$, $\hat{\sigma}_y$, $\hat{\sigma}_z$ are the x, y, z Pauli matrices. In the Dirac limit, the array alternating offsets m_a, m_b play the role of the neutrino masses, while the coupling coefficients κ, ϵ play the role of the speed of light in vacuum and of the charged-current electroweak interactions. Remarkably, the linear parts of Eqs. (20.16) and (20.17) are identical to models routinely used in particle physics for describing neutrino oscillations in matter [49]. Nonlinear terms are usually disregarded as neutrinos interact weakly in standard conditions of matter densities. However, recent studies demonstrate that nonlinearity plays an important role in supernova stars and in the early universe, where matter density is enormous [50, 53–55]. The linear supermodes $|\psi_+\rangle$, $|\psi_-\rangle$ of Eqs. (20.16) and (20.17) (calculated by setting $\gamma = 0$) represent the instantaneous mass neutrino eigenstates in matter [49], which can be expressed as linear superpositions of the mass eigenstates $|M_a\rangle$, $|M_b\rangle$, i.e., the linear modes of Eqs. (20.16) and (20.17) in the uncoupled limit $\epsilon \to 0$ (limit of non-interacting neutrinos, e.g., in vacuum):

$$|\psi_+\rangle = \sin\Theta|M_a\rangle - \cos\Theta|M_b\rangle, \qquad (20.18)$$
$$|\psi_-\rangle = \cos\Theta|M_a\rangle + \sin\Theta|M_b\rangle, \qquad (20.19)$$

where $\Theta = \tan^{-1}[2\epsilon/(m_a + m_b)]/2$ is the mixing angle. Note that Eqs. (20.18) and (20.19) coincide with the neutrino mixing matrix in matter [49]. The supermodes $|\psi_+\rangle$, $|\psi_-\rangle$ can be calculated with the ansatz $A = \mathcal{A}e^{iEz+ipx}$, $B = \mathcal{B}e^{iEz+ipx}$, which yields the dispersion relation $E_\pm^2 = \kappa^2 p^2 + \epsilon^2 + (m_a^2 + m_b^2)/2 \pm (m_b - m_a)\sqrt{\epsilon^2 + (m_a + m_b)^2/4}$.

20.3.2 NEUTRINO OSCILLATIONS

Equations (20.16) and (20.17), analogously to models routinely used in particle physics [49], describe neutrino oscillations in matter. Neutrinos are created in weak processes in their flavor eigenstates. As they propagate, the quantum mechanical phases of the mass states flow at different rates owing to the diverse neutrino masses. Analogously to Eqs. (20.16) and 20.17), neutrino flavors can also be expressed in terms of mass states through the Pontecorvo–Maki–Nakagawa–Sakata (PMNS) matrix formulation of flavor mixing [56, 57]. In turn, neutrino oscillations in vacuum are trivial, in the sense that they are simply given by the beating between mass states with different energies. In matter, neutrino oscillations are due to electroweak interactions, which in our optical setup are accounted for by the coupling coefficient ϵ. For instance, considering a neutrino that at $z = 0$ is in the electron flavor state, and initially neglecting nonlinear effects, the probability of measuring the neutrino with electron and muon flavors oscillates in z: $P(A) = \sin^2(2\Theta)\sin^2[(E_+ - E_-)z/2]$, $P(B) = 1 - \sin^2(2\Theta)\sin^2[(E_+ - E_-)z/2]$. In massive supernova stars, matter density is so high that electroweak interactions become nonlinear, affecting neutrino oscillations. We have numerically solved Eqs. (20.16) and (20.17) for homogeneous waves with null transverse momentum using a fourth order Runge–Kutta algorithm. Results of numerical simulations for several values of the nonlinear coefficient γ are shown in Fig. 20.8(a) and (b). The probabilities $P(A), P(B)$ of measuring neutrinos with flavors A, B oscillate in z both

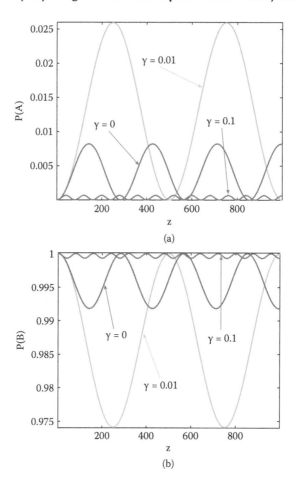

FIGURE 20.8: Probabilities $P(A)$ [panel(a)] and $P(B)$ [panel (b)] of measuring the neutrino with flavors A and B as functions of the propagation length for several nonlinear coefficients $\gamma = 0, 0.01, 0.1$, corresponding to the curves. The other parameter values are $\kappa = 1$, $\epsilon = 0.001$, $m_a = 0.01$, $m_b = 0.012$.

in the linear and nonlinear regimes. Note that the effect of nonlinearity on the amplitude and the period of the oscillations is nontrivial. Indeed, as the nonlinear coefficient γ increases, the oscillation amplitude and period initially increase for $\gamma = 0.01$. Conversely, oscillations are quenched for the higher nonlinear coefficient $\gamma = 0.1$.

In order to grasp a better understanding of the nonlinear quenching of neutrino oscillation, we have calculated the homogeneous nonlinear mode families of Eqs. (20.16) and (20.17) by using the Newton–Raphson method. We

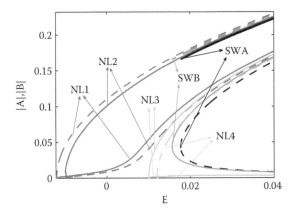

FIGURE 20.9: Homogeneous nonlinear mode families not undergoing oscillation. Full (dashed) curves show the modulus of the spinor amplitudes $|A|$ ($|B|$) as functions of the energy E. In our calculations we have used the parameters $\kappa = 1$, $\gamma = 1$, $m_a = 0.01$, $m_b = 0.012$.

focused our attention on nonlinear modes with null transverse momentum, and numerically found several neutrino families, which are depicted in Fig. 20.9. The families $NL1, NL2$ bifurcate from the linear antineutrino bands with negative energy, while other families represent neutrinos with positive energies. The nonlinear families $NL3, NL4$ exist for any neutrino amplitude as they bifurcate from the linear neutrino bands, while the switching families SWA, SWB ensue after a certain power threshold. In turn, the nonlinear quenching of neutrino oscillations comes from the excitation of the switching families SWA, SWB, which arise when nonlinearity is sufficiently large.

20.4 NEGATIVE FREQUENCIES IN NONLINEAR OPTICS

20.4.1 INTRODUCTION

The study of supercontinuum generation (SCG), i.e., the explosive broadening of the spectrum of an intense and short input pulse due to nonlinear effects in a medium, typically an optical fiber or a bulk crystal, has been an active area of research since its first discovery in 1970 [61], due to the many applications in metrology and device characterization [62,63].

Constructing a theory of SCG has proved to be crucial in order to understand and control the dynamics of pulses in optical fibers [64]. Such a theory is based on the so-called generalized nonlinear Schrödinger equation (GNLSE), an enhanced version of the integrable nonlinear Schrödinger equation [62]. The GNLSE, based on the concept of slowly varying envelope approximation (SVEA) of the electric field, is paradigmatic in nonlinear optics, and has been extremely successful in explaining most of the features of SCG [62,63]. One of the most successful predictions was the emission of dispersive waves from optical solitons, which are phase-matched at specific wavelengths, usually referred to as resonant radiation (RR) or Cherenkov radiation [65–68]. RR contributes significantly to the formation of SCG spectra and can have many applications, especially when using photonic crystal fibers [64, 69, 70].

It is known that the conventional GNLSE fails when considering few- or single-cycle pulses. Many approaches for deriving accurate propagation equations with minimal approximations have been successfully developed for this situation over the years [71, 72] which are capable of dealing with broadband spectral evolution, Raman effect and the inclusion of backward waves. However, recent experiments have revealed that new resonant frequencies [referred to as negative-frequency resonant radiation (NRR)] can be emitted by solitons, which are not predicted by any GNLSE formulation [73–75]. Such frequencies can be numerically predicted by using the full Maxwell equations [solved with the finite difference time domain (FDTD, [76]) or the pseudo-spectral spatial domain (PSSD, [77]) techniques], or alternatively by the so-called unidirectional pulse propagation equation (UPPE, [78]), which includes only forward propagating waves but uses the full oscillating electric field, while the phase-matching condition for NRR formation has been derived heuristically [73, 75]. NRR has been attributed to the presence of negative frequency components in the UPPE, which are absent in the GNLSE due to SVEA. However, this claim sparked some controversy in the community [79], due to a lack of a solid theoretical support that could confirm or disprove the given interpretation. For example, this radiation could be confused with that generated by backward waves or by the conventional four-wave mixing (FWM) between the soliton and co-propagating radiation as in [80]. It is also interesting to notice that, despite the fact that negative frequencies are routinely used in quantum optics [81] (where they are associated with the photon creation operator), quantum field theory [82] and water waves [83], in nonlinear optics there is still some resistance to accepting this concept.

In this section we introduce a new equation for a properly defined pulse envelope that is able to capture the surprising and peculiar interaction between positive and negative frequency components during the propagation of an ultrashort pulse. Such an interaction is able to generate phase-matched dispersive waves that would not exist in any model based on the conventional envelope defined when deriving the NLSE, currently referred to as NRR in the literature. We demonstrate that our new equation can be efficiently solved numerically and gives an analytical insight into the very nature of ultrashort pulse propagation in any dielectric medium. Moreover, in this section we also show that there are some serious deficiencies in the universally adopted equation based on the GNLSE, since the latter neglects the contribution of the cross-phase modulation between the positive and negative frequency parts of the spectrum, which gives rise to new and unexpected nonlinear phenomena that have been previously overlooked.

The structure of this section is the following. In Section 20.4.2 we discuss the physical relevance and the existence of positive and negative frequency states, first in linear optics, and then in nonlinear optics. We support our claims with simple yet robust mathematical arguments, and we discuss the necessity for including negative energy states in the dispersion relation. In Section 20.4.3 we introduce our notation for the analytic signal, and we derive our new equation based on the envelope of the analytic signal, starting from the UPPE. The transparency of this equation when compared to the UPPE allows us to identify the term responsible for the NRR. In Section 20.4.4 we derive the phase-matching conditions of the radiated frequencies, and we discover new phase-matched frequencies that were previously unknown. In Section 20.4.5 we present detailed numerical simulations supporting our theory, and in particular we focus our attention on the dynamics of the NRR. Discussion and conclusion are given in Section 20.4.6.

20.4.2   EXISTENCE AND REALITY OF NEGATIVE FREQUENCIES IN OPTICS

In this section we discuss the somewhat disputed physical reality and the interpretation of negative frequencies in optics. The logical conclusion is that both positive and negative frequencies must be treated on equal footing and must be included in any meaningful formulation of a fully consistent pulse propagation equation.

Let us discuss light propagation in the vacuum case for simplicity. The linear curl Maxwell equations can be written (in the Heaviside–Lorentz unit system) as

$$\nabla \times \mathbf{E} = -\frac{1}{c}\partial_t \mathbf{B}, \qquad (20.20)$$

$$\nabla \times \mathbf{B} = \frac{1}{c}\partial_t \mathbf{E}, \qquad (20.21)$$

where $\mathbf{E}$ and $\mathbf{B}$ are the electric and magnetic field vectors, and c is the speed of light in vacuum. In order to analyze the frequency content of these equations, we note that Eqs. (20.20) and (20.21) are exactly equivalent to a linear relativistic Dirac equation. In order to see this we follow [84], and we define a Dirac spinor as $\psi = (E_y + iE_x, -iE_z, -B_x + iB_y, B_z)^T$. This is only one possible choice out of a total of eight equivalent vectors [84]. With this definition, we can write Eqs. (20.20) and (20.21) in the following form:

$$\gamma^\mu \partial_\mu \psi = 0, \qquad (20.22)$$

where γ^μ are the 4×4 Dirac matrices and $\partial_\mu \equiv (\frac{1}{c}\partial_t, \nabla)$ is the derivative four-vector, with $\mu = \{0, 1, 2, 3\}$. Equation (20.22) is exactly the massless Dirac equation that is encountered in the theory of relativistic quantum mechanics of spin-1/2 particles. This equation is fully Lorentz-invariant, as it should be since Maxwell equations have also this property. The two degrees of freedom that correspond to the two states of spin in quantum mechanics, here represent the two possible circular polarizations of light. Alternative (but equivalent) ways to write a Dirac-like equation for Maxwell's equations make use, for instance, of the Riemann–Silberstein vector [85], and are treated in some textbooks as a mere curiosity [86].

From Eq. (20.22) one derives the wave equation, $i\partial_t\psi = -ic\alpha \cdot \nabla\psi$, where $\alpha^j \equiv \gamma^0\gamma^j = \begin{pmatrix} 0 & \sigma_j \\ \sigma_j & 0 \end{pmatrix}$, and $j = \{1, 2, 3\}$. The Hamiltonian associated with this wave equation is thus $\hat{H} = -ic\alpha \cdot \nabla$. Eigenvalues of the Hamiltonian are found by solving the secular equation $\det(c\alpha \cdot \mathbf{k} - \lambda) = 0$. By using the well-known relation $(\alpha \cdot \mathbf{k})^2 = |\mathbf{k}|^2$, we finally obtain $\lambda_\pm = \pm|\mathbf{k}|c$. This brings us to the core conclusion of this section: Eq. (20.22) exhibits two frequency eigenvalues with opposite sign, irrespective of the direction of propagation of the wave. This is an unavoidable consequence of the fact that Maxwell's equations are relativistic and can be cast in a form identical to the massless Dirac equation, at least in the vacuum case. These two eigenvalues correspond to positive and negative energy states: in full analogy with the classical solutions of the Dirac equation, negative energy states are not unphysical solutions that should be discarded. On the contrary, one must include them *on equal footing* with the positive energy states, in order to preserve the internal consistency of Maxwell's equations.

In Fig. 20.10 we show a full plot of the dispersion for plane waves for bulk silica. There are four quadrants, containing all the combinations of forward/backward propagation and positive/negative energy states. It is customary in nonlinear optics, but certainly not complete, to consider only the upper-rightmost quadrant with positive frequencies and forward propagation when neglecting the backward waves, for example, in the UPPE formulation. This is incorrect and incomplete, since forward waves are described also by the bottom-leftmost quadrant of Fig. 20.10, which contains negative frequency states. Note that the forward propagation constant must satisfy (for lossless media) the relation $\beta(-\omega) = -\beta(\omega)$, and thus it must be an odd function with respect to the axis $\omega = 0$. As a consequence, the real refractive index must satisfy $n(\omega) = n(-\omega)$, and it is thus an even function of the frequency.

In a medium different from vacuum, Maxwell's equations cannot be written in the Dirac form of Eq. (20.22), since the presence of the medium introduces a preferred reference frame, breaking the Lorentz invariance. However, in this case, Maxwell's equations can be cast in a form similar to the Schrödinger equation by using 6×6 matrices; see [87]. The associated Hamiltonian operator has positive and negative eigenvalues, exactly as in the vacuum case [87]. This result can be directly generalized to the nonlinear case by using perturbation theory, assuming that the nonlinear polarization does not perturb considerably the linear modes.

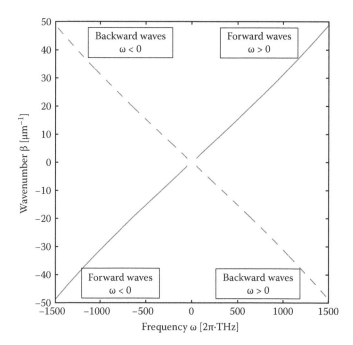

FIGURE 20.10: Dispersion curves for bulk silica. All possible combinations of forward/backward waves and positive/negative energy states appear in this plot.

20.4.3   DERIVATION OF THE ENVELOPE EQUATION FOR THE ANALYTIC SIGNAL FROM THE UNIDIRECTIONAL PULSE PROPAGATION EQUATION

We now introduce some important definitions that will be crucial for the following discussion, mainly following [88–91].

The real electric field propagating in the fiber is denoted by $E(z,t)$, where z is the propagation direction and t is the time variable. The Fourier transform of the electric field is denoted by $E_\omega(z) \equiv \mathcal{F}[E(z,t)] = \int_{-\infty}^{+\infty} E(z,t)e^{i\omega t}dt$. The *analytic signal* of the electric field, i.e., the positive frequency part of the field, which is a complex function, is defined as $\mathcal{E}(z,t) \equiv \pi^{-1}\int_0^\infty E_\omega(z)e^{-i\omega t}d\omega$. The analytic signal can also be defined alternatively by using the Hilbert transform: $\mathcal{E}(z,t) = E(z,t) - i\mathcal{H}[E(z,t)]$, where $\mathcal{H}[E(z,t)] \equiv \pi^{-1}\mathbb{P}\int_{-\infty}^{+\infty} dt' E(z,t')/(t - t')$, and the simbol $\mathbb{P}\int_{-\infty}^{+\infty}$ indicates that the integral must be taken in the sense of the Cauchy principal value. With these definitions, the Fourier transform of the electric field can be written as the sum $E_\omega = [\mathcal{E}_\omega + (\mathcal{E}_{-\omega})^*]/2$ since only the positive (or negative) frequency part of the spectrum carries information, while for the same reason the electric field itself is real and is given by $E(z,t) = [\mathcal{E}(z,t) + \mathcal{E}^*(z,t)]/2$. The analytic signal satisfies the following requirements:

$\mathcal{E}_{\omega>0} = 2E_\omega$, $\mathcal{E}_{\omega<0} = 0$ and $\mathcal{E}_{\omega=0} = E_{\omega=0}$. Note that $(\mathcal{E}^*)_\omega$ and $(\mathcal{E}_\omega)^*$ are different in general and must be distinguished.

It can be proved that the fields $\mathcal{E}$ and $\mathcal{E}^*$ are the classical analogs of the annihilation and creation operators a and $a^\dagger$ used after quantization of the electromagnetic field; see, e.g., [88,89]. This fact is quite understandable since a and $a^\dagger$ are related to the positive and negative energy parts of the electric field, which in quantum optics correspond to absorption and emission of a photon [82]. Since the concept of discrete absorption or emission is extraneous to classical electromagnetism, in a pre-second quantization context one is forced to talk about "conjugated" fields or "negative frequency" fields, as we shall do in the following.

Operators a and $a^\dagger$ are not linked to observables in quantum optics, since they are not Hermitian. Only the real electric field, which is a Hermitian combination of a and $a^\dagger$, is an observable quantity [82]. Due to the requirement that the electric field be real at all moments of time and at all locations in space, the positive and negative energy content of any pulse in optics and in quantum optics must be absolutely identical. This means that, since photons and antiphotons are equivalent, i.e., they carry the same information, there are no quantum numbers that can distinguish between them, and this results in a perfectly real-valued electric field [85]. An interesting, albeit somewhat trivial, corollary is that (contrary to common belief) a measurement can only detect the positive and negative frequency content of a pulse *simultaneously*, and never one or the other individually. For instance, a real-valued cosine wave is the combination of exponentials with both positive and negative energies. The exponential form is the correct basis that has to be used, since only exponentials are eigenfunctions of the Hamiltonian $\hat{H}$ written in the previous section, which is a first order operator.

The starting point of our discussion is the so-called unidirectional pulse propagation equation (UPPE) [78], which is a reduction of Maxwell's equations that accounts only for the forward propagating part of the electric field:

$$i\frac{\partial E_\omega}{\partial z} + \beta(\omega)E_\omega + \frac{\omega}{2cn(\omega)}P_{\mathrm{NL},\omega} = 0, \qquad (20.23)$$

where $\beta(\omega)$ is the full propagation constant of the medium, c is the speed of light in vacuum, $n(\omega)$ is the linear refractive index, and $P_{\mathrm{NL},\omega}(z) \equiv \chi^{(3)}\mathcal{F}[E(z,t)^3]_\omega$ is the nonlinear Kerr polarization. Particular care must be devoted to the definition of the complex envelope, since we do not want to put any limitation the frequency extent of the signals. This aspect is overlooked in the literature, and it is taken for granted that the frequency bandwidth of the envelope is narrow with respect to the carrier wave.

The key element that we introduce here is that only a proper definition of the envelope is able to capture the correct coupling between the positive and the negative frequency parts of the spectrum. The "envelope" we introduce here is based on the analytic signal and is defined as

$$A(z,t) \equiv \mathcal{E}(z,t)e^{-i\beta_0 z + i\omega_0 t}, \tag{20.24}$$

i.e., the frequency components of the analytic signal are "shifted" by an amount $-\omega_0$. By doing this, we shift the carrier frequency of the analytic signal to zero, so that we deal with frequency detuning $\Delta\omega$ from ω_0, and not with absolute frequencies, in analogy to the conventional definition of envelope given in many textbooks [63]. However, *there is a key difference* between the conventional definition of envelope (see, e.g., [63]) and Eq. (20.24): the former is adequate only if the spectral extension of the pulse evolution is much smaller than the pulse central frequency, $|\Delta\omega| \equiv |\omega - \omega_0| \ll \omega_0$, i.e., only under SVEA conditions, while the envelope of the analytic signal $A(z,t)$ considered here does not suffer from this limitation, and so $\mathrm{supp}\{A_{\Delta\omega}(z)\} = [-\omega_0, +\infty)$. By clearly dividing the envelope associated with the positive frequency components from that associated with the negative frequency components, we will be able to write the envelope equation that correctly describes the dynamics of pulses *of arbitrary duration and spectral extension*, taking into account the peculiar and non-trivial interaction between positive and negative frequencies that arises due to the nonlinear polarization.

With the above definitions, the nonlinear polarization is now written as

$$\begin{aligned} P_{\mathrm{NL}}(z,t) = \frac{\chi^{(3)}}{8} & \left[A^3 e^{-3i\omega_0 t + 3i\beta_0 z} + A^{*3} e^{3i\omega_0 t - 3i\beta_0 z} \right] \\ & + \frac{\chi^{(3)}}{8} \left[3|A|^2 A e^{-i\omega_0 t + i\beta_0 z} + 3|A|^2 A^* e^{i\omega_0 t - i\beta_0 z} \right]. \end{aligned} \tag{20.25}$$

Due to our definition of A, the first (second) term in the square brackets contains only positive (negative) frequencies, and they are responsible for third harmonic generation (THG). The third and fourth terms *contain both positive and negative frequencies*, because the Fourier transform of $|A|^2$ has a frequency support (i.e., a domain of existence) that extends from $-\infty$ to $+\infty$. In fact, $\mathcal{F}[|A|^2]_{\Delta\omega}$ is the convolution between $A_{\Delta\omega}$, whose support is $[-\omega_0, +\infty)$, and $A^*_{\Delta\omega}$, whose support is $(-\infty, +\omega_0]$. By applying the Titchmarsh convolution theorem (i.e., the support of the convolution is contained in the sum of the supports of its individual terms [92]), it immediately follows that $\mathrm{supp}\{\mathcal{F}[|A|^2]_{\Delta\omega}\} \subseteq (-\infty, +\infty)$. This means that, although in the absence of nonlinearities positive and negative frequencies live a completely separate existence, *in the presence of nonlinear terms they can interact nonlinearly*. Such an interaction is also present in the traditional Kerr term $|A|^2 A$. However, new nonlinear effects will only be visible numerically and experimentally in the presence of resonant processes, such as the emission of RR and NRR from solitons. If we denote with $\mathcal{P}_{\mathrm{NL}}(z,t)$ the analytic signal for the nonlinear polarization, then its envelope $A_p(z,t) = \mathcal{P}_{\mathrm{NL}} e^{-i\beta_0 z + i\omega_0 t}$ can be expressed as

$$A_p(z,t) = \frac{3\chi^{(3)}}{4} \left[|A|^2 A + |A|^2 A^* e^{2i\omega_0 t - 2i\beta_0 z} + \frac{1}{3} A^3 e^{-2i\omega_0 t + 2i\beta_0 z} \right]_+ \quad (20.26)$$

The subscript $+$ prescribes that only positive frequencies must be taken (i.e., $\Delta\omega > -\omega_0$) and is a shorthand notation to indicate the positive frequency *spectral filtering* involved in the analytic signal, and operated in the time domain by the Hilbert transform, which is crucial in our formulation. The first and third terms in Eq. (20.26) are the conventional Kerr term and the THG term, respectively. The second term, which we call a *conjugated Kerr term*, is the new feature of our formulation and it emerges as a consequence of the analytic signal envelope.

Finally, with all the above ingredients, one can write an equation for the analytic signal envelope A which contains only positive frequencies:

$$i\partial_\xi A + \hat{D}(i\partial_\tau)A + \gamma\hat{S}(i\partial_\tau)\left[|A|^2 A + |A|^2 A^* e^{2i\omega_0\tau + 2i\Delta k\xi}\right]_+$$

$$+\gamma\hat{S}(i\partial_\tau)\left[\frac{1}{3}A^3 e^{-2i\omega_0\tau - 2i\Delta k\xi}\right]_+ = 0, \quad (20.27)$$

where $\Delta k \equiv (\beta_1\omega_0 - \beta_0)$ (this is a central quantity in this work), $\xi \equiv z$ and $\tau \equiv t - \beta_1 z$ are the new space-time variables in the co-moving frame, the dispersive operator $\hat{D}(i\partial_\tau) \equiv \sum_{m=2}^\infty \beta_m(i\partial_\tau)^m/m!$, γ is the nonlinear coefficient of the medium, and $\hat{S}(i\partial_\tau)$ is the operator accounting for the dispersion of the nonlinearity [provided in the Fourier space by the factor $\omega/n(\omega)$ in Eq. (20.23)], which is necessary to include since the equations are broadband and SVEA is not used. For our purposes, and without loss of generality (our results are valid for any form of this operator, provided it includes at least the first order expansion term), it will be sufficient to perform the traditionally adopted approximation $\hat{S}(i\partial_\tau) \simeq 1 + i\partial_\tau/\omega_0$. Note that the field A feels a dispersion given by $D(\Delta\omega) = \sum_{m=2}^\infty \beta_m\Delta\omega^m/m!$ (where $\Delta\omega$ is the detuning from ω_0) and a positive nonlinearity, while the field A^* feels a different, "conjugate" dispersion $-D(-\Delta\omega) \neq D(\Delta\omega)$ and a negative nonlinearity, and both fields are forward-propagating.

Equation (20.27) is the central result of this section. Since A and A^* carry the same amount of information, it is sufficient to consider a single equation only: indeed, the dynamics around the positive carrier frequency (ω_0) must be the mirror image of the dynamics around the negative carrier frequency ($-\omega_0$), due to the requirement that the electric field E must be real. The two modes A and A^* do not see each other in the absence of nonlinearity, but they mutually exchange energy when the nonlinear terms are included, thus generating new frequencies. Since the interaction modifies the phase, new resonant nonlinear effects occur. It is possible to prove that in Eq. (20.27) the energy is perfectly conserved, i.e., $\partial_\xi \int_{-\infty}^{+\infty} |A(\xi,\tau)|^2 d\tau = 0$, due to the detailed balance of the energy flow from A to A^* and back. It is interesting to note that the presence of the shock operator and the THG term are *essential* for energy conservation, which establishes *a deep and previously unnoticed connection* between the shock operator, THG and negative frequencies. In the absence of THG terms, Eq. (20.27) would exhibit a small non-conservation of energy proportional to the missing THG energy. Another crucial point to notice is that, although Eq. (20.27) looks at first sight as a kind of GNLSE written for an envelope, this equation is in its physical content *completely equivalent to the UPPE* [Eq. (20.23)]: the analytic signal and the filtering procedure are used to completely separate positive from negative frequency parts in the UPPE, and the introduction of the envelope of the analytic signal is used only to give Eq. (20.27) a formal resemblance to the GNLSE, but our equation is not restricted by any of the limitations of the latter.

20.4.4 PHASE-MATCHING CONDITIONS BETWEEN SOLITON AND RADIA-TION

In order to derive phase-matching conditions between a soliton and its reso-nant radiations , we follow a standard procedure described in [80]. We first pose $A(\xi, \tau) = F(\tau)e^{iq\xi} + g(\xi, \tau)$, where $F(\tau)$ is the (purely real) envelope of the optical soliton, q is the nonlinear mismatch and g is a small amplitude dispersive wave. After substitution into Eq. (20.27), and by taking only the fundamental and first order terms, one obtains (neglecting the shock term for simplicity, without loss of generality)

$$(i\partial_\xi + \hat{D})g + \gamma F^2 g^* e^{2iq\xi} + 2\gamma F^2 g = - \left(\hat{D} + \frac{1}{2}\beta_2\partial_\tau^2 \right) Fe^{iq\xi}$$

$$-\gamma F^3 e^{2i\omega_0\tau + 2i\Delta k\xi - iq\xi} - \frac{1}{3}\gamma F^3 e^{-2i\omega_0\tau - 2i\Delta k\xi + 3iq\xi}. \qquad (20.28)$$

The phase-matching conditions derived from Eq. (20.28) are then easily found:

$$D(\Delta\omega) = q, \qquad (20.29)$$
$$D(\Delta\omega) = 2\Delta k - q, \qquad (20.30)$$
$$D(\Delta\omega) = -2\Delta k + 3q. \qquad (20.31)$$

Solving Eqs. (20.29) and (20.31) for $\Delta\omega$ will provide all the phase-matched frequencies. In particular, Eq. (20.29) is very well known [65, 67] and corre-sponds to the positive-frequency RR, while Eq. (20.30), found experimentally in [73] and heuristically in [73,75], corresponds to the negative-frequency RR. Equation (20.31) represents the phase-matching condition of the non-solitonic radiation due to THG (which we call *third harmonic resonant radiation*, or TH-RR), and is also a new unexpected feature of our model, which is vindi-cated by our numerical simulations. A curious and unexpected feature of the TH-RR radiation is that, even though this is due to the interaction between the soliton and its third harmonic waves, it is strongly detuned to long wave-lengths, and is thus very feeble and unobservable in bulk crystals, but it is possible that in small-core waveguides such radiation could become experimentally accessible.

Equations (20.30) and (20.31) are impossible to find by using a single GNLSE based on SVEA and thus correspond to new features of our envelope model Eq. (20.27). In all the transparent bulk crystals we have examined, $\Delta k > 0$ and $2\Delta k \gg q$, thus the NRR is usually strongly blueshifted with respect to the RR. This fact is ultimately due to the structure of the Lorentz oscillator theory. However, this restriction seems not to be fundamental for transparent waveguides and there might well be waveguide structures with a range of frequencies for which $\Delta k < 0$, so that they would be able to exhibit redshifted NRR. This interesting and potentially important question is left for future investigations.

The last important thing to mention regarding NRR generation is that there is an intimate relation between the NRR and the FWM between the soliton and THG, which can contribute to its amplitude at the second order of perturbation theory. In fact, one can prove by using the theory reported in [80] that the FWM between the soliton and THG gives *exactly* the phase-matching condition Eq. (20.30). Thus, the formation of the NRR is due to two different contributions, one (of first order) coming directly from the conjugated Kerr term, and the other one (of second order) coming indirectly from the process of FWM of soliton and CW waves, but the amplitudes of these two contributions could be quite different since they are of different order, as we will see below. This interesting "coincidence" has deep roots in the structure itself of Eq. (20.27), and in particular in the energy conservation law.

Figure 20.11(a) shows the phase-matching curve $D(\Delta\omega)$ versus pump frequency (normalized to β_0 and ω_0, respectively), together with its intersections with q, $2\Delta k - q$ and $-2\Delta k + 3q$, which give, respectively, the RR, NRR and TH-RR frequencies. Figure 20.11(b) shows the value of $2\Delta k$ versus pump wavelength, showing that in bulk silica there is an optimal pump wavelength (in the normal dispersion regime) for which the NRR would be closer to the pump frequency, and thus would have an unusually large amplitude.

The above procedure is able to give only the exact frequency position of each resonant radiation that is emitted by the soliton. However, the procedure is not able to provide a correct value for the amplitude of the NRR, for the following reason. When we wrote Eq. (20.28), we assumed implicitly that an analytical solution for the soliton field F is known. Although it is reasonable to assume, at least for pulses that are not sub-cycle, that the conventional Schrödinger soliton is a good approximation, this cannot be completely correct since a true soliton solution of Eq. (20.27) must include the conjugated Kerr term and the THG term as well. Such a solution is not currently available (we leave this problem to future investigations, with a few indications in the literature on how to do that [93, 94]), and thus we are temporarily forced to

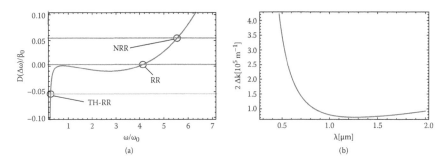

FIGURE 20.11: (a) Phase-matching curve (normalized to β_0) derived by using Eqs. (20.29) and (20.30) in bulk silica. q/β_0, $(2\Delta k - q)/\beta_0$ and $(-2\Delta k + 3q)/\beta_0$ are indicated by the first, second, and third horizontal lines, respectively. RR, NRR and TH-RR frequencies are indicated by circles. (b) Δk for bulk silica vs. pump wavelength.

use the conjugated Kerr term as a source term in Eq. (20.28), which gives the correct phase-matching condition but cannot give the correct amplitude, since the amplitude of the NRR must be proportional to the third order dispersion β_3, as happens for the normal RR. We can, however, always find the emitted NRR amplitude numerically, as done in Fig. 20.13.

20.4.5   NUMERICAL SIMULATIONS

In this section we support the above theory with numerical simulations performed by integrating Eq. (20.27). In Fig. 20.12(a) we show the spectral evolution of a 15 fs sech pulse, with peak intensity 1.4 TW/cm^2 propagating in bulk silica, for a pump wavelength $\lambda_0 = 2\ \mu$m, obtained by solving Eq. (20.27) when the THG term is neglected. Both RR and NRR emissions are visible. Vertical black dashed lines indicate the predictions given by Eqs. (20.29) and (20.30); see also Fig. 20.11(a). Figure 20.12(b) shows the same as Fig. 20.12(a), omitting also the second nonlinear term inside the square brackets in Eq. (20.27). No NRR radiation is generated in this case, showing that such radiation is indeed coming from the interaction between the positive and the negative frequency spectral components. Figure 20.12(c) shows the same simulation as in Fig. 20.12(a) but switching off the shock term, i.e. $\hat{S}(i\partial_\tau) = 1$, and for a peak intensity of 2.6 TW/cm^2. One can see that both RR and NRR are visible, conclusively proving that NRR is not due to the shock effect, even though the shock helps to further broaden the spectrum and thus to feed the soliton tail that excites the NRR. Finally, Fig. 20.12(d) shows the evolution of the pulse by solving the full-field UPPE, Eq. (20.23), which also shows evidence of small THG. Exactly the same figure is obtained by solving Eq. (20.23), showing that our envelope model based on the analytic signal is indeed correct.

Figure 20.13 shows the comparison between the amplitudes of the generated

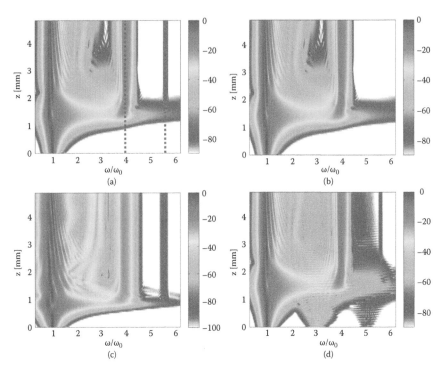

FIGURE 20.12: (a) Contour plot of the spectral evolution of a short sech pulse in bulk silica, obtained by direct simulation of Eq. (20.27), when THG is neglected. The pulse is pumped at $\lambda_0 = 2$ μm, with a peak intensity of 1.4 TW/cm^2 and a duration $t_0 = 15$ fsec. The formation of RR and NRR is clearly visible. Vertical black dashed lines indicate the position of the radiation as predicted by Eqs. (20.29) and (20.30), compare with Fig. 20.11(a). (b) Same as (a) when also switching off the second nonlinear term inside the square brackets of Eq. (20.27), i.e., the conjugated Kerr term. The NRR line has completely disappeared. (c) Same as (a) but when switching off the shock operator, and for a peak intensity 2.6 TW/cm^2. (d) Results obtained with the UPPE of Eq. (20.23), using the same parameters as in (a). All plots are in logarithmic scale.

NRR when switching on and off the various terms of Eq. (20.27). Parameters are the same as in Fig. 20.12, and the spectra are recorded after $z = 5$ mm of propagation. One can notice that, for the chosen parameters, the conjugated term alone overestimates the radiation amplitude, while the THG alone underestimates it with respect to the case when both terms are maintained. Figure 20.13 first of all proves that the FWM between the soliton and THG on one hand, and the contribution of the conjugated term on the other hand give exactly the same phase-matching point for the NRR, as we have pre-

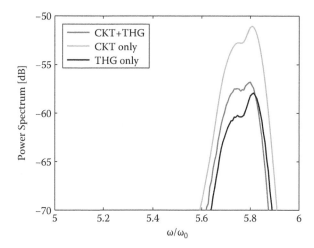

FIGURE 20.13: Comparison between the amplitudes of the generated NRR when the conjugated Kerr term (CKT) is present but THG is absent (top line), when THG is present but the conjugated Kerr term is absent (bottom line), and in the case when both terms are present (middle line). Equation (20.27) has been used in the simulation, and parameters are the same as in Fig. 20.12. Spectra are recorded after $z = 5$ mm of propagation, and the vertical scale is logarithmic.

dicted in the previous section. Moreover, since, when omitting the THG or the conjugated Kerr term energy is not conserved, the radiation amplitude is not correctly predicted in these cases, and the only consistent way to simulate correctly the problem is to consider all the terms in Eq. (20.27), which strictly conserves energy.

20.4.6   DISCUSSION AND CONCLUSION

In conclusion, we have derived an equation that correctly describes the non-linear interaction between the positive and the negative frequency parts of the spectrum of optical pulses. The key concept is that the envelope function is now defined in terms of the analytic signal of the electric field, therefore clearly dividing the dynamics of the negative and positive frequency parts of the spectrum, and avoiding SVEA altogether, while still retaining an envelope formulation. The interaction between positive and negative frequencies is due to the presence of a cross-phase-modulation-like term in the nonlinear polarization, the role of which we have elucidated here for the first time. By using the new equation we have analytically derived the phase-matching conditions between a soliton and the positive- and negative-frequency resonant radiation emitted by it. Our theory opens up a new realm in nonlinear optics and in other areas that are described by NLSE-like equations (for instance,

BEC, plasmas, etc.), since it proves that conventional treatments based on GNLSE are deficient, due to the lack of the negative frequency terms. These interactions are of course present in the UPPE and in the nonlinear Maxwell's equations, which are, however, less transparent and less suitable for analytical treatment than Eq. (20.27). Exciting future perspectives are represented by the inclusion of the Raman nonlinearity, which could provide additional unexplored nonlinear effects that are not captured by conventional the GNLSE based on SVEA.

REFERENCES

1. T. Pertsch, P. Dannberg, W. Elflein, A. Bräuer, and F. Lederer, "Optical Bloch oscillations in temperature tuned waveguide arrays," *Phys. Rev. Lett.*, 83, 4752, 1999.
2. R. Morandotti, U. Peschel, J. S. Aitchison, H. S. Eisenberg, and Y. Silberberg, "Experimental observation of linear and nonlinear optical bloch oscillations," *Phys. Rev. Lett.*, 83, 4756, 1999.
3. Mher Ghulinyan, Claudio J. Oton, Zeno Gaburro, Lorenzo Pavesi, Costanza Toninelli, and Diederik S. Wiersma, "Zener tunneling of light waves in an optical superlattice," *Phys. Rev. Lett.*, 94, 127401, 2005.
4. Henrike Trompeter, Thomas Pertsch, Falk Lederer, Dirk Michaelis, Ulrich Streppel, Andreas Bräuer, and Ulf Peschel, "Visual observation of zener tunneling," *Phys. Rev. Lett.*, 96, 023901, 2006.
5. S. Longhi, M. Marangoni, M. Lobino, R. Ramponi, P. Laporta, E. Cianci, and V. Foglietti, "Observation of dynamic localization in periodically curved waveguide arrays," *Phys. Rev. Lett.*, 96, 243901, 2006.
6. Yoav Lahini, Assaf Avidan, Francesca Pozzi, Marc Sorel, Roberto Morandotti, Demetrios N. Christodoulides, and Yaron Silberberg, "Anderson localization and nonlinearity in one-dimensional disordered photonic lattices," *Phys. Rev. Lett.*, 100, 013906, 2008.
7. T. X. Tran and F. Biancalana, "Diffractive resonant radiation emitted by spatial solitons in waveguide arrays," *Phys. Rev. Lett.*, 110, 113903, 2013.
8. S. Longhi, "Klein tunneling in binary photonic superlattices," *Phys. Rev. B*, 81, 075102, 2010.
9. S. Longhi, "Photonic analog of Zitterbewegung in binary waveguide arrays," *Opt. Lett.*, 35, 235, 2010.
10. Felix Dreisow, Matthias Heinrich, Robert Keil, Andreas Tünnermann, Stefan Nolte, Stefano Longhi, and Alexander Szameit, "Classical simulation of relativistic Zitterbewegung in photonic lattices," *Phys. Rev. Lett.*, 105, 143902, 2010.
11. F. Dreisow, R. Keil, A. Tünnermann, S. Nolte, S. Longhi, and A. Szameit, "Klein tunneling of light in waveguide superlattices ," *Eur. Phys. Lett.*, 97, 10008, 2012.
12. S. Longhi, "Classical simulation of relativistic quantum mechanics in periodic optical structures," *Appl. Phys. B*, 104, 453, 2011.
13. J. M. Zeuner, N. K. Efremidis, R. Keil, F. Dreisow, D. N. Christodoulides, A. Tünnermann, S. Nolte, and A. Szameit, "Optical analogues for massless

dirac particles and conical diffraction in one dimension," *Phys. Rev. Lett.*, 109, 023602, 2012.

14. W. Heisenberg, "Quantum theory of fields and elementary particles," *Rev. Mod. Phys.*, 29, 269, 1957.

15. D. C. Ionescu, J. Reinhardt, B. Müller, W. Greiner, and G. Soff, "Nonlinear extensions of the Dirac equation and their implications in QED," *Phys. Rev. A*, 38, 616, 1988.

16. Antonio Zecca, "Dirac equation in space-time with torsion," *Internat. J. Theoret. Phys*, 41, 421, 2002.

17. M.J. Esteban and E. Sere, "An overview on linear and nonlinear Dirac equations," *Discrete Contin. Dyn. Syst.*, 8, 381, 2002.

18. I. Bialynicki-Birula and J. Mycielski, "Nonlinear wave mechanics," *Ann. Phys.*, 100, 62, 1976.

19. L. H. Haddad and L. D. Carr, "The nonlinear Dirac equation in Bose-Einstein condensates: Foundation and symmetries," *Physica D*, 238, 1413, 2009.

20. L.H. Haddad and L.D. Carr, "Relativistic linear stability equations for the nonlinear Dirac equation in Bose-Einstein condensates," *EPL*, 94, 56002, 2011.

21. A. A. Sukhorukov and Y. S. Kivshar, "Discrete gap solitons in modulated waveguide arrays," *Opt. Lett.*, 27, 2112, 2002.

22. A. A. Sukhorukov and Y. S. Kivshar, "Generation and stability of discrete gap solitons," *Opt. Lett.*, 28, 2345, 2003.

23. M. Conforti, C. De Angelis, and T. R. Akylas, "Energy localization and transport in binary waveguide arrays," *Phys. Rev. A*, 83, 043822, 2011.

24. Roberto Morandotti, Daniel Mandelik, Yaron Silberberg, J. Stewart Aitchison, Marc Sorel, Demetrios N. Christodoulides, Andrey A. Sukhorukov, and Yuri S. Kivshar, "Observation of discrete gap solitons in binary waveguide arrays," *Opt. Lett.*, 29, 2890, 2004.

25. Falk Lederer, George I. Stegeman, Demetri N. Christodoulides, Gaetano Assanto, Moti Segev, Yaron Silberberg, "Discrete solitons in optics," *Phys. Reports*, 463, 1, 2008.

26. Y Nogami, F M Toyama, and Z Zhao, "Nonlinear Dirac soliton in an external field," *J. Phys. A: Math. Gen.*, 28, 1413, 1995.

27. Nonlinear extensions of the Dirac equation that violate Lorentz invariance have been so far the subject of conjecture. See, for instance, Rajesh R. Parwani, "An information-theoretic link between spacetime symmetries and quantum linearity," *Ann. Phys.*, 315, 419 (2005); W. K. Ng and R.R. Parwani, SIGMA, 5, 023 (2009).

28. G. Feinberg, "Possibility of faster-than-light particles," *Phys. Rev.*, 159, 1089, 1967.

29. P. W. Higgs, "Broken symmetries and the masses of gauge bosons," *Phys. Rev. Lett.*, 13, 508, 1964.

30. Yoichiro Nambu, "Quasi-particles and gauge invariance in the theory of superconductivity," *Phys. Rev.*, 117, 648, 1960.

31. Jeffrey Goldstone, Abdus Salam, and Steven Weinberg "Broken symmetries," *Phys. Rev.*, 127, 965, 1962.

32. A. Salam and J. Strathdee, "On goldstone fermions," *Phys. Lett. B*, 49, 465, 1974.

33. A. Marini, A. V. Gorbach, and D. V. Skryabin, "Coupled-mode approach to surface plasmon polaritons in nonlinear periodic structures," *Opt. Lett.*, 35, 3532, 2010.

34. Nikolaos K. Efremidis, Peng Zhang, Zhigang Chen, Demetrios N. Christodoulides, Christian E. Rüter, and Detlef Kip, "Wave propagation in waveguide arrays with alternating positive and negative couplings," *Phys. Rev. A*, 81, 053817, 2010.

35. A. Szameit, Y. V. Kartashov, F. Dreisow, M. Heinrich, T. Pertsch, S. Nolte, A. Tünnermann, V. A. Vysloukh, F. Lederer, and L. Torner, "Inhibition of light tunneling in waveguide arrays," *Phys. Rev. Lett.*, 102, 153901, 2009.

36. A. Marini, A. V. Gorbach, D. V. Skryabin, and A. V. Zayats, "Amplification of surface plasmon polaritons in the presence of nonlinearity and spectral signatures of threshold crossover," *Opt Lett.*, 34, 2864, 2009.

37. P. Berini and I. De Leon, "Surface plasmonâĂŞpolariton amplifiers and lasers," *Nat. Phot.*, 6, 16, 2011.

38. A. Marini, Tr. X. Tran, S. Roy, S. Longhi, and F. Biancalana, "Optical analog of spontaneous symmetry breaking induced by tachyon condensation in amplifying plasmonic arrays," *Phys. Rev. A*, 89, 023840, 2014.

39. A. Marini and D. V. Skryabin, "Ginzburg-Landau equation bound to the metal-dielectric interface and transverse nonlinear optics with amplified plasmon polaritons," *Phys. Rev. A*, 81, 033850, 2010.

40. D. V. Skryabin, A. V. Gorbach, and A. Marini, "Surface-induced nonlinearity enhancement of TM modes in planar subwavelength waveguides," *JOSA B*, 28, 109, 2011.

41. Walter E Thirring, "A soluble relativistic field theory," *Ann. of Phys.*, 3, 91, 1958.

42. Alexander Szameit, Mikael C. Rechtsman, Omri Bahat-Treidel, and Mordechai Segev, "PT-symmetry in honeycomb photonic lattices," *Phys. Rev. A*, 84, 021806(R), 2011.

43. Vadim M. Apalkov and Tapash Chakraborty, "Superluminal tachyon-like excitations of Dirac fermions in a topological insulator junction," *Europhys. Lett.*, 100, 17002, 2012.

44. Simeon Hellerman and Martin Schnabl, "Light-like tachyon condensation in open string field theory," *JHEP*, 2013.

45. Sidney Coleman, "More about the massive Schwinger model," *Ann. Phys.*, 101, 239, 1976.

46. B. Pontecorvo, "Mesonium and antimesonium," *Sov. Phys. JETP*, 6, 429, 1957.

47. L. Wolfenstein, "Neutrino oscillations in matter," *Phys. Rev. D*, 17, 2369, 1978; S. P. Mikheyev and A. Yu. Smirnov, *Sov. J. Nucl. Phys.*, 42, 913, 1985.

48. T. K. Kuo and James Pantaleone, "Neutrino oscillations in matter," *Rev. Mod. Phys.*, 61, 937, 1989.

49. M. C. Gonzalez-Garcia and Y. Nir, "Neutrino masses and mixing: evidence and implications," *Rev. Mod. Phys.*, 75, 345, 2003.

50. James Pantaleone, "Neutrino oscillations at high densities," *Phys. Lett. B*, 287, 128, 1992.

51. Alessandro Strumia and Francesco Vissani, "Neutrino masses and mixings and...," *arXiv.org/abs/hep-ph/0606054*, 2010.

52. Huaiyu Duan, George M. Fuller, and Yong-Zhong Qian, "Collective neutrino oscillations," *Annu. Rev. Nucl. Part. Sci.*, 60, 569, 2010.

53. A. B. Balantekin, "Flavor oscillations in core-collapse supernovae," *Nuclear Physics B (Proc. Suppl.)*, 235–236, 388, 2013.

54. Georg G. Raffelt, "N-mode coherence in collective neutrino oscillations," *Phys. Rev. D*, 83, 105022, 2011.

55. Srdjan Sarikas, Georg G. Raffelt, Lorenz Hüdepohl, and Hans-Thomas Janka, "Suppression of self-induced flavor conversion in the supernova accretion phase," *Phys. Rev. Lett.*, 108, 061101, 2012.

56. Z. Maki, M. Nakagawa, and S. Sakata, "Remarks on the unified model of elementary particles," *Progress of Theoretical Physics*, 28, 870, 1962.

57. B. Pontecorvo, "Neutrino experiments and the problem of conservation of leptonic charge," *Sov. Phys. JETP*, 26, 984, 1968.

58. Andrea Marini, Stefano Longhi, and Fabio Biancalana, "Optical simulation of neutrino oscillations in binary waveguide arrays," *Phys. Rev. Lett.*, 113, 150401, 2014.

59. A. Szameit, Y. V. Kartashov, F. Dreisow, M. Heinrich, T. Pertsch, S. Nolte, A. Tünnermann, V. A. Vysloukh, F. Lederer, and L. Torner, "Inhibition of light tunneling in waveguide arrays," *Phys. Rev. Lett.*, 102, 153901, 2009.

60. Stefano Longhi, "Coherent control of tunneling in driven tight-binding chains: Perturbative analysis," *S. Longhi, Phys. Rev. B*, 77, 195326, 2008.

61. R. R. Alfano and S. L. Shapiro, "Observation of self-phase modulation and small-scale filaments in crystals and glasses," *Phys. Rev. Lett.*, 24, 592, 1970.

62. J. M. Dudley, G. Genty, and S. Coen, "Supercontinuum generation in photonic crystal fiber," *Rev. Mod. Phys.*, 78, 1135, 2006.

63. G. P. Agrawal, *Nonlinear Fiber Optics*, Academic Press, 2007, 4th edition.

64. P. St. J. Russell, "Photonic crystal fibers," *Science*, 299, 358, 2003.

65. N. Akhmediev and M. Karlsson, "Cherenkov radiation emitted by solitons in optical fibers," *Phys. Rev. A*, 51, 2602, 1995.

66. A. V. Husakou and J. Herrmann, "Supercontinuum generation of higher-order solitons by fission in photonic crystal fibers," *Phys. Rev. Lett.*, 87, 203901, 2001.

67. F. Biancalana, D. V. Skryabin, and A. V. Yulin, "Theory of the soliton self-frequency shift compensation by the resonant radiation in photonic crystal fibers," *Phys. Rev. E*, 70, 016615, 2004.

68. D. V. Skryabin, F. Luan, J. C. Knight, and P. St. J. Russell, "Soliton self-frequency shift cancellation in photonic crystal fibers," *Science*, 301, 1705, 2003.

69. P. H'olzer, W. Chang, J. C. Travers, A. Nazarkin, J. Nold, N. Y. Joly, M. F. Saleh, F. Biancalana, and P. St. J. Russell, "Femtosecond nonlinear fiber optics in the ionization regime," *Phys. Rev. Lett.*, 107, 203901, 2011.

70. M. F. Saleh, W. Chang, P. H'olzer, A. Nazarkin, J. C. Travers, N. Y. Joly, P. St.J. Russell, and F. Biancalana, "Theory of photoionization-induced blueshift of ultrashort solitons in gas-filled hollow-core photonic crystal fibers," *Phys. Rev. Lett.*, 107, 203902, 2011.

71. P. Kinsler, "Optical pulse propagation with minimal approximations," *Phys. Rev. A*, 81, 013819, 2010.

72. G. Genty, P. Kinsler, B. Kibler, and J. M. Dudley, "Nonlinear envelope equation modeling of sub-cycle dynamics and harmonic generation in nonlinear waveguides," *Opt. Express*, 15, 5382, 2007.

73. E. Rubino, J. McLenaghan, S. C. Kehr, F. Belgiorno, D. Townsend, S. Rohr, C. E. Kuklewicz, U. Leonhardt, F. K\u{o}nig, and D. Faccio, "Negative-frequency resonant radiation," *Phys. Rev. Lett.*, 108, 253901, 2012.

74. E. Rubino, A. Lotti, F. Belgiorno, S. L. Cacciatori, A. Couairon, U. Leonhardt, and D. Faccio, "Soliton-induced relativistic-scattering and amplification," *Scientific Reports*, 2, 932, 2012.

75. M. Conforti, N. Westerberg, F. Baronio, S. Trillo, and Daniele Faccio, "Negative-frequency dispersive wave generation in quadratic media," *Phys. Rev. A*, 88, 013829, 2013.

76. A. Taflove and S. C. Hagness, *Computational Electrodynamics: The Finite-Difference Time-Domain Method*, Artech House Publishers, 2005, 3rd edition.

77. J. C. A. Tyrrell, P. Kinsler, and G. H. C. New, "Pseudospectral spatial-domain: A new method for nonlinear pulse propagation in the few-cycle regime with arbitrary dispersion," *J. Mod. Opt.*, 52, 973, 2005.

78. M. Kolesik, P. Townsend Whalen, and J. V. Moloney, "Theory and simulation of ultrafast intense pulse propagation in extended media," *IEEE J. Sel. Top. Quantum Electron.*, 18, 494, 2012.

79. F. Biancalana, "Negative frequencies get real," *Physics*, 5, 68, 2012.

80. D. V. Skryabin and A. V. Yulin, "Theory of generation of new frequencies by mixing of solitons and dispersive waves in optical fibers," *Phys. Rev. E*, 72, 016619, 2005.

81. D. F. Walls and G. J. Milburn, *Quantum Optics*, Springer-Verlag, 2010.

82. F. Mandl and G. Shaw, *Quantum Field Theory*, John Wiley and Sons, 1996.

83. G. Rousseaux, C. Mathis, P. Maïssa, T. G. Philbin, and U. Leonhardt, "Observation of negative-frequency waves in a water tank: A classical analogue to the Hawking effect?" *New J. Phys.*, 10, 053015, 2008.

84. V. M. Simulik, "Connection between the symmetry properties of the Dirac and Maxwell equations. Conservation laws," *Theor. and Math. Phys.*, 87, 386, 1991.

85. J. H. Eberly, L. Mandel, and E. Wolf (Eds.), *Coherence and Quantum Optics VII*, Springer, 1996, p. 313.

86. J. J. Sakurai, *Advanced Quantum Mechanics*, Addison-Wesley Publishing, 1967, p. 169.

87. V. Grigoriev and F. Biancalana, "Coupled-mode theory for on-channel nonlinear microcavities," *J. Opt. Soc. B*, 28, 2165, 2011.

88. Sh. Amiranashvili, U. Bandelow, and N. Akhmediev, "Few-cycle optical solitary waves in nonlinear dispersive media," *Phys. Rev A*, 87, 013805, 2013.

89. Sh. Amiranashvili and A. Demircan, "Ultrashort optical pulse propagation in terms of analytic signal," *Adv. Opt. Technol.*, 2011, 989515, 2011.

90. M. Conforti, F. Baronio, and C. De Angelis, "Ultrabroadband optical phenomena in quadratic nonlinear media," *IEEE Photonics*, 2, 600, 2010.

91. M. Conforti, F. Baronio, and C. De Angelis, "Nonlinear envelope equation for broadband optical pulses in quadratic media," *Phys. Rev. A*, 81, 053841, 2010.

92. E. C. Titchmarsh, "The zeros of certain integral functions," *Proc. of the London Math. Soc.*, 25, 283, 1926.

93. R. S. Tasgal, Y. B. Band, and B. Malomed, "Gap solitons in a medium with third-harmonic generation," *Phys. Rev. E*, 72, 016624, 2005.

94. V. Cao Long, P. P. Goldstein, and M. Trippenbach, "On existence of solitons for the 3rd harmonic of a light beam in planar waveguides," *Act. Phys. Pol. A*, 105, 437, 2004.

Index

2CNLS, see Two coupled nonlinear
    Schrödinger (2CNLS)
    (Manakov) equations

A

Ablowitz–Kaup–Newell–Segur
    (AKNS) hierarchies,
    145–177
Ablowitz–Ladik (AL) equation,
    79–93
  closed form, 82
  discretization, 468
  discretization effects, 85–89
  and nonlinear Schrödinger
    equation (NLSE), 79–80
  rogue wave hierarchy, 89–92
  theory, 80–83
Accelerating soliton, 149–151
Acousto-optics modulator (AOM),
    379, 380
Acousto-optics shaper, 341
Active mode-locking, 379
Adaptation concept, 176–177
Additive complex Gaussian white
    noise, 463
Agitated breather, 175
Air holes in fibers, 306
Airy beams, 153
Airy function, 151–153
Akhmediev breathers, 189
AKNS, see Ablowitz-Kaup-Newell-
    Segur (AKNS)
    hierarchies
AL, see Ablowitz-Ladik (AL)
    equation
Algebraic solitary waves, 421
All-optical data-processing, 199
All-optical switching, 98
Alternating nonlinear optical
    waveguide array

(ANOWA), 421–424
Alternating nonlinear optical
    waveguide bundles
    (ANOWB), 424–425
Alternating nonlinear optical
    waveguide zigzag array
    (ANOWZA), 418–421
Alternating waveguide array, 417
Alternating waveguide bundle,
    424–425
Alternating waveguide zigzag array,
    419
Ambient pressure effects, 367
Amplification span, 190
Amplification to compensate for loss,
    462
Amplified spontaneous emission
    (ASE), 447
Amplifier spontaneous emission
    (ASE), 463
Amplitude-shift keying, 368
Analog effects, 507–538
Analytical soliton solutions, 508–511
Anderson localization, 507
ANOWA, see Alternating nonlinear
    optical waveguide array
    (ANOWA)
ANOWB, see Alternating nonlinear
    optical waveguide bundles
    (ANOWB)
ANOWZA, see Alternating nonlinear
    optical waveguide zigzag
    array (ANOWZA)
Antiresonance mode, 306
Anti-Stokes emission, 375–378; see
    also Stokes emission process
Anti-Stokes intensity variation with
    temperature, 398
Anti-VK criteria, 215–216
AOM, see Acousto-optics modulator
    (AOM)

Arrayed waveguide grating (FBG), 337

Arrhenius-like relaxation function, 366–367

ASE, see Amplified spontaneous emission (ASE); Amplifier spontaneous emission (ASE)

Asymptotic attracting state, 327–329

Asymptotic parabolic regime, 330

Asymptotic state, 330

Atom optics, 253–275

Attractors, 201, 235–236, 330

Auto-Bäcklund transformation, 162, 169

Autonomous systems, 145–146

B

Back propagation, 435, 456

Backward and forward wave interaction, 409–430

Backward NFT (BNFT), 464, 467

Backward pitchfork bifurcation, 209–210

Backward propagation in optical resonators, 32

Backward pulse generation, 32

Backward waves
    definition, 409
    model, 410–412

Bandgaps, 280
    linear stability analysis, 287

Bandwidth window limitation, 98

BAPA, see Binary amplifying plasmonic array (BAPA)

Bathythermographs, 367

BECs, see Bose-Einstein condensates (BECs)

BER, see Bit error ration (BER)

Bessel basis, 32–33

Bessel function basis, 48

BG, see Bragg grating (BG)

BGS, see Brillouin gain spectrum (BGS)

Bidirectional pulse propagation equation, 39

Bifurcations in laser locking, 236

Bifurcations in locking Brillouin fiber ring laser, 383

Binary amplifying plasmonic array (BAPA), 516*8

Binary arrays, 68–73

Binary phase-shift keying (BPSK), 487

Binary waveguide arrays (BWAs), 507, 508
    periodic nature, 512

Birefringence, 331

Bistability, 428–429

Bit error rates increased by dispersion, 24

Bit error ration (BER), 441, 448–449

Bit rate
    definition, 7
    limited by pulse dispersion, 7
    maximum in single-mode fibers, 11

Bloch equations, 31–32

Bloch waves, 284, 287, 288–294

Bloch oscillations, 507

BNFT, see Backward NFT (BNFT)

Boomeron model, 147

Bose-Einstein condensates (BECs), 146, 253–275
    at absolute zero, 254–256
    behavior of, 253–254
    lens transformation, 148–149
    matter-wave solitons in, 151
    matter-wave solitons in multi-component, 266–275
    matter wave switching, 270–275

BOTDA, see Brillouin optical time domain analysis (BOTDA)

BOTDR, see Brillouin optical time domain reflectometry (BOTDR)

Bottle-neck problems, 62

Boundary regime propagation, 31, 32

projection method for, 41–45
BP, see Brillouin pump (BP)
BPSK, see Binary phase-shift keying (BPSK)
Bragg angles, 512
Bragg grating (BG)
    lossy, 202
    for pulse shaping, 337
Bragg grating gap solitons, 219–225
Bragg grating solitons, 63
Bragg induced dispersion, 62
Bragg planes in photonic crystal fibers, 306
Bragg resonance, 429
Bragg solitary waves in FBG, 63
Bragg solitons, 62
Bragg-type resonance, 76
Breathers, 79
    agitated, 175
    Akhmediev, 189
    interactions between, 214
    Satsuma–Yajima, 175
    stable formation, 211
Bright–bright solitons, 270–275
Bright matter-wave solitons, 253–254
Bright NLSE nonautonomous solitons, 169–172
Bright simultons, 415–416
Bright soliton-like solutions, 411–412
Brillouin coupling, 352
Brillouin distributed measurement of strain, 394–395
Brillouin fiber laser, 371
Brillouin frequency comb, 378
Brillouin frequency shift, 355
Brillouin gain, 357–360, 368
Brillouin gain spectrum (BGS), 357–360
Brillouin laser
    distributed feedback (DFB) laser, 373–374
    mode-locked, 378–384
    self-phase-locked, 384–388
Brillouin linewidth, 363–367

Brillouin mitigation, 367–368
Brillouin optical time domain analysis (BOTDA), 390, 392–394
Brillouin optical time domain reflectometry (BOTDR), 390–392
Brillouin power threshold, 360–363
    modulation via temperature or strain, 368
Brillouin pump (BP), 375
Brillouin regimes in cavities, 383
Brillouin scattering (BS), 351–401;
    see also Stimulated Brillouin scattering (SBS)
    history of, 351–352
    sensor applications, 388–400
    spectral broadening, 355
    spontaneous, 352–355
Brillouin sensing, 395–397
Brillouin Stokes light, 372
Brillouin-Stokes waves, 363
Brillouin strain temperature dependence, 363–367
Brillouin zone, 282, 512

C

Canonical soliton, 145–146
Capacity expansion in fiber communications, 459
Carrier frequency offset (CFO) cancellation, 498
Cauchy problem, 31
    one-dimensional formulation, 33–34
Cavity types for fiber lasers, 371
CCD, see Coupling coefficient dispersion (CCD)
Center frequency, 16
CFO, see Carrier frequency offset (CFO) cancellation
CGLEs, see Complex Ginzburg–Landau equations (CGLEs);

Ginzburg–Landau
equations (CGLEs)
Chalcogenide fiber, 360
in Fabry–Perot configuration,
388
laser, 374, 378
Chaotic bunches, 243–246
Chaotic pulse bunches, 243–248
Chen and Liu accelerating soliton,
149–151
Cherenkov radiation, see Resonant
radiation (RR)
Chirp coefficient, 331
Chirped de Broglie wave, 155
Chirped hyperbolic pulses, 329
Chirped pulse amplification (CPA),
337–338
Chirping
highly linear, 340
linear, 329
multiple may cancel, 16
negative, 340–341
source of, 18, 19
Circular waveguide array, 424
CME, see Couple mode equations
(CME)
CMEs, see Coupled-mode equations
(CMEs)
CMOS, see Complementary
metal-oxide-semiconductor
(CMOS) technology
Cnoidal structures, 79
Cnoidal waves in coupled steady
state, 415–416
CNS, see coupled nonlinear
Schrödinger equation
(CNS)
Coherent detection system, 459
Coherent states, 153–157
Cold solitons, 189–197; see also
Hypothermic solitons
Collision of pulses, 20
Colored nonautonomous solitons,
172–175

Compact chaotic bunches, 245–246
Compactons, definition, 428
Complementary
metal-oxide-semiconductor
(CMOS) technology, 374
Complex Ginzburg-Landau
equations (CGLEs), 199
cubic-quintic (CQ), 201
Complimentary pathways, 460
COMSOL Multiphysics, 313
Condensed soliton phase, 242
Conjugated fields, 530
Conjugated Kerr term, 532
"Connection in time", 151
Conservative systems, 64
Continuous wave (cw)
instability development, 133
symmetric/antisymmetric,
110–111
Continuous wave (cw) probe beam,
20
Continuous wave (cw) SBS lasers,
371–373
Convolution theorem, 531
Corning LEAF, 13
Counter-propagating wave envelope,
61
Coupled circuits, 65–68
Coupled mode dynamics, 61–76
Coupled-mode equations (CMEs),
508
Coupled nano-resonators, 65–68
Coupled nonlinear Schrödinger
equation (CNS or CNLSE),
31
derivation, 33
for slowly varying envelope, 105
Coupled steady state waves, 415–417
Coupled system of equations, 61
Couple mode equations (CME), 61
for gap solitons, 223–225
Coupling coefficient dispersion
(CCD), 110–111
Coupling coefficient effects, 133–143

Couterpropagating pulse
    polarizations, 32
CPA, see Chirped pulse
    amplification (CPA)
Crank-Nicolson scheme, 298–300
    finite-difference discretization,
    468
Critical coupling coefficient
    dispersion, 111
    in anomalous dispersion regime,
    116–118
    interplay with SNL, 116–118
Cross-phase modulation (XPM), 13,
    18–21
    definition, 18–19
    induces phase shifts, 20
    induction of modulational
    instability, 134
    modulational instability (MI)
    and, 98–99
    by parabolic pulse, 340
Cubic nonlinearity, 31, 199
Cubic-quintic (CQ) CGLE, 201
Cut-off wavelength, 6
CW or cw, see Continuous wave (cw)

D

Darboux scheme, 83–85
Dark-bright solitons, 268–270
Dark matter-wave solitons, 253–254
Dark NLSE nonautonomous solitons,
    169–172
Dark simultons, 415–416
Dark soliton-like solutions, 411–412
Darwin wave packet, 149–151
DBP, see Digital back propagation
    (DBP)
DCF, see Dispersion-compensating
    fiber (DCF)
DDF, see Dispersion-decreasing fiber
    (DDF)
De Broglie wavelength solitonic
    analogue, 151
De Broglie wave-particle duality, 154

Defect states, 287
Defocusing nonlinearity variational
    approach, 291–294
Delay role in modulational
    instability, 111
Dense wavelength division
    multiplexing (DWDM), 29
Depressed clad designs, 12
Detection of positive and negative
    frequency, 530
DFB, see Distributed feedback
    (DFB) lasers; Distributed
    feedback fiber laser (DFB)
DFT, see dispersive Fourier
    transform technique (DFT)
Dielectric cylindric waveguide
    projection method, 45–54
Dielectric structures, 305–306
Dielectric susceptibility coefficient,
    42, 54
Dielectric susceptibility of third
    order, 45
Diffraction and spatial dissipative
    solitons (SDSs), 199
Digital back propagation (DBP)
    nonlinearity compensation
    technique, 461–462
    vs. optical back propagation,
    437, 441–442, 455
    vs. optical frequency division
    multiplexing (OFDM)
    modulation, 480–486
Digital coherent technology, 491–504
Digital signal processing (DSP), 435
Dimer chain, 68
Dimers, 76
Dirac electron motion, 508
Dirac equations
    nonlinear relativistic (NDE), 515
    nonlinear versions, 508
    soliton generation and
    propagation, 512–514
    tachyon-like, 518

Dirac law of mathematical beauty, 146–147
Dirac-like equation, 518, 528
Dirac limit for neutrinos, 522–523
Dirac matrices, 527
Dirac point, 71
Dirac solitons, 507
Dirac solitons in binary waveguide arrays, 515–516
Dirac solution solitons, 508
Dirac-type eigenvalue equation, 491
Directional coupler, 63, 417
    oppositely directional, 427–428
Discontinuous periodic lattice wave propagation, 79
Discrete traveling waves, 76
Discretization
    effects, 85–89
    for solution, 80–83
Dispersion; see also specific types
    cancelled by soliton nonlinearity, 16
    equation depends on form of, 39–41
    form dictates equation type, 39–41
    higher order, 10–11
    increases bit error rates, 24
    and maximum bit rate in single mode fibers, 11
    nonlinear, 70
    in photonic crystal fibers, 310
    second order, 16
Dispersion coefficient, 8, 99
Dispersion-compensating fiber (DCF), 12–13
    in laser systems, 239–240
    use of, 435
Dispersion-decreasing fiber (DDF), 331, 436
    with optical back propagation (OBP), 450–455
    use of, 435

Dispersion-managed solitons, 189, 459
Dispersion management, 24
    modulational instability (MI) for, 98
Dispersion power penalty, 11
Dispersion relation, 31
Dispersion spectrum, 9
Dispersion tailored fibers and parabolic pulse dynamics, 332
Dispersive Fourier transform technique (DFT) and dual-resonant radiation, 246–248
Dispersive wave and dual-resonant radiation, 317–320
Dissipative rogue waves, 243–245
Dissipative soliton paradigm, 235
Dissipative solitons pinned to hot spots, 205–214
Distributed feedback (DFB) lasers, 352
Distributed feedback fiber laser (DFB), 373–374, 429–430
Distributed measurement of strain, 394–395
Distributed Raman amplifiers, 436
Distributed sensing, 394–395, 398
Dopants
    effect on fiber properties, 358–359
    inhomogeneous, 202
    properties influence nonlinear response, 112
    rare earth, 332
Doped dispersion decreasing fiber, 125–128
Doping alters wavelength dependence, 8
Double-peak solitons, 216–219
DOWN phase, 123
DPM, see Dynamical projecting method (DPM)

Drude model, 414
Dual chirality, 517
Dual core coaxial designs, 12
Dual radiation generation, 305–320
Dual-resonant radiation in
    microstructured optical
    fibers (MOFs), 317–320
Duffing oscillator, 158
DWDM, see Dense wavelength
    division multiplexing
    (DWDM)
Dynamical projecting method
    (DPM)
    development of, 33
    purpose of, 31

E

EDF, see Erbium doped fiber (EDF)
    laser
EDFA, see Erbium doped fiber
    amplifiers (EDFA)
Effective fiber length, 14
Effective index, 5
    wavelength dependence, 8–9
Effective length, 363
    definition, 20
Eigenvalue as ideal information
    carrier, 491
Eigenvalue communications, 459–486
    applications of, 487
    definition, 460
    in nonlinear fiber channels,
        459–487
    vs. pure soliton-based
        transmission, 464
Eigenvalue demodulation method,
    491–503
Eigenvalue modulated optical fiber
    transmission system,
    491–503
Eigenvalue modulated transmission,
    491–504
    experimental demonstration,
        493–498

numerical demonstration, 493
Eigenvalue modulation experimental
    setup, 494
Eigenvalue noise tolerance, 498–503
Elastically interacting wave packets,
    155
Electric field distribution in
    transverse, 16
Electro-optic modulator, 379
Electrostriction, 368
    definition, 355
Encoding techniques, 368
Envelope equation for analytic
    signal, 529–533
Envelope soliton, 17
Erbium doped fiber (EDF) laser,
    239–240
Erbium doped fiber amplifiers
    (EDFA), 335
    comb generation using, 375–378
    for long distance
        communication, 17
    for optical amplification, 12
    properties, 360
    use of, 436
Ermakov equation, 153–154
Extended chaotic bunches, 243–245

F

Fabry-Perot (FP) cavity, 371–372
    formation of, 375
Fast dynamic span, 190
Fast response system, 103–104
FBG, see Arrayed waveguide grating
    (FBG); Fiber Bragg grating
    (FBG)
FDTD, see Finite difference time
    domain (FDTD)
Feedforward carrier, 448
FEM, see Finite element method
    (FEM) solver
Fermionic particles, 518
Fermion pair production, 508
Fermi–Pasta–Ulam recurrence, 79

Feshback resonance, 146
Fiber amplifiers, 75
    similaritons in, 334–335
Fiber array wave propagation, 79
Fiber-based cavities, 332–333
Fiber Bragg grating (FBG), 61
    building of, 65
    long period, 75
    quasi-distributed sensing, 398
    use of, 435
Fiber laser cavities, 236–237
Fiber lasers and mode locking,
    235–249
Fiber loops, 281
    as mirrors, 375–376
Fiber optic parameter amplifiers
    (FOPA), 25–27
Fiber-optic waveguides (FOWGs)
    containing solitons, 190
Fiber properties; see also specific
    fiber types
    and Brillouin strain, 363–367
    and frequency shifts, 358
Fiber Raman amplifiers, 17
Fibers with air holes, 306
Fifth generation of optical
    transmission systems, 461
Finite difference time domain
    (FDTD), 526
Finite element method (FEM)
    solver, 313
Finite impulse filter (FIR), 435
Fluoride fiber, 358
FNFT, see Forward NFT (FNFT)
Focusing NLSE, 462
FOPA, see Fiber optic parameter
    amplifiers (FOPA)
Forbidden zone, 419, 422, 428
Forced harmonic oscillator, 147
Forward and backward wave
    interaction, 409–430
Forward NFT (FNFT), 464
    for normal dispersion case, 467

Forward pitchfork bifurcation,
    209–210
Forward wave model, 410–411
Forward waves
    definition, 409
    interaction with backward
        waves, 412–414
Four coupled NLS equations,
    133–143
Fourth order dispersion (FOD), 100,
    308
Four wave mixing (FWM), 9, 13,
    21–25, 305, 526
    cross talk caused by, 24
    degenerate, 22, 27
FOWGs, see fiber-optic waveguides
    (FOWGs)
FP, see Fabry-Perot (FP)
Freak waves, 244
Free space wave number of
    transverse field, 2
Frequencies unstable, 108–109
Frequency comb, 98
Frequency gap, 65
Frequency jitter, 393–394
Frequency-shift keying, 368
Fundamental mode, 5
    field distribution, 6
    spot size, 6–7
FWM, see Four wave mixing (FWM)

G

Gain effects at higher order, 324
Gains and losses balanced, 64
Gamow tunneling effect, 151
Gap solitary waves in FBG, 63
Gap solitons, 63, 219–225, 288–290,
    428
Gaussian potential
    soliton dynamical evolution and,
        298–299
    stability analysis, 295–296

Geiger–Nuttall law, 151
Gelfand–Levitan–Marchenko
    equations (GLME), 464,
    467–468
Generalized non-linear Schrödinger
    equation (GNLSE), 193,
    525–527
    UPPE equivalent, 532–533
Germanium oxide effect on fiber
    properties, 358
Gibbs-like phenomena, 92
Ginzburg-Landau equations
    (CGLEs), 199, 236; see also
    Complex Ginzburg-Landau
    equations (CGLEs)
GLME, see Gelfand–Levitan–
    Marchenko equations
    (GLME)
Global attractor, 330
GNLSE, see Generalized non-linear
    Schrödinger equation
    (GNLSE)
Gold-tipped fiber, 385
Gordon models, 519
Gouy phase shift, 341
GP, see Gross–PItaevskii (GP)
Green's function, 154
Grid spacing for discretization, 80–83
Gross-PItaevskii (GP) equation,
    254–256, 267
    ruling equation, 280
Group velocity, 16
Group velocity dispersion (GVD), 8,
    98
    anomalous, 8, 107–108
    linearizes accumulated phase,
        330–331
    in photonic crystal fibers,
        309–311
Guided modes, 3–5
Guiding-center solitons, 189, 195–196
Guiding center theory, 496
GVD, see Group velocity dispersion
    (GVD)

H

Hasegawa and Tappert form, 170,
    172–175
Hasegawa-Kumar quasi-solitons, 157
Heaviside-Lorentz unit system, 527
Helmholtz equation, paraxial, 279
Hermite accelerating wave packets,
    151–153
Hermite-Gaussian modes, 153
Hidden group symmetry, 147
Hidden symmetries, 147
Hidden symmetry, 156–157, 176–177
Higher order dispersion (HOD)
    effects, 10–11, 98–99
Higher-order effects on similaritons,
    331
Higher order gain effects, 324
Highly coherent continuums, 339
Highly nonlinear fiber (HNLF),
    324–325, 339
Hirota equation, 160–162
Hirota's bilinear method, 265
HNLF, see Highly nonlinear fiber
    (HNLF)
HOD, see Higher order dispersion
    (HOD) effects
Holey fibers, 25
Hondros–Debye basis, 32–33
    for cylindrical waveguide, 46
Honeycomb arrays, 71
Honeycomb lattices, 76
Hopf-type bifurcations, 211, 236
Hot solitons, 189–197; see also
    Hyperthermic solitons
Hot spot (HS) amplifying region,
    201–202
    description, 202–203
    double hot spot model, 211–213
    pinned solitons associated,
        199–229
HS, see Hot spot (HS) amplifying
    region
HT&T, see Husimi–Taniuti and
    Talanov transformation;

Husimi–Taniuti transformation; Talanov transformation
Hisami parametric oscillator, 147
Husimi–Taniuti and Talanov transformation, 149, 157
Husimi–Taniuti transformation, 147–149
Hybrid fiber configurations, 332
Hybrid solitons, 189–197
Hydrostatic pressure effect on SBS frequency shift, 367
Hyperbolic pulses, 329
Hyperthermic solitons, 190, 192
Hypothermic solitons, 193

I

Ideal solitons, see Isothermic solitons
Idempotent operators, 33
Idler waves, 25
Incoherent tunneling, 366–367
Increasing bit-rates issues, 62
Index, effective, 8–9
Index of refraction, complex, 64
In-phase quadrature (IQ) modulator, 494–495
Instability band at normal dispersion regime, 112
Instantaneous frequency, 19
Integrability of nonlinear channel model, 460
Integrability property, 464
Integrable channels, 460
Intensity modulation converted from phase modulation, 21
Intermodal dispersion in multimoded fiber, 7
Intersymbol interference, 11
Intra-pulse Raman scattering, 324
Inverse scattering transform (IST) method, 147, 460
  reduction procedure, 160
Inversion symmetry and nonlinear effects, 13

IQ, see In-phase quadrature (IQ) modulator
Isothermic solitons, 192
IST, see Inverse scattering transform (IST) method

J

Jaynes–Cummings model, 430

K

KdV, see Korteweg–de Vries (KdV) equation
Kelly sidebands, 194
Kennard states, 155
Kerr-based phase modulation, 310
Kerr case vs. SNL case, 115–116
Kerr coefficient, 122
Kerr effect, 61
  instantaneous, 111
Kerr nonlinearity, 33, 98, 462
Kerr nonlinear waveguides, 508–509
Kerr response
  fails for ultra-short pulses, 99
  of highly nonlinear fiber, 339–340
Kerr terms conjugated, 532
Kirchhoff equation, 412
Klein-Gordon equation, 412
Klein paradox, 508
Klein tunneling, 508
Korteweg-de Vries (KdV) equation, 147, 164–165
Kronecker symbols, 204

L

Lagrange multipliers method, 436
Large effective area fiber (LEAF), 13
Laser diode (LD)
  semiconductor with external cavity, 495–496
  spectral width, 7–8
  use of, 8
Laser tuning, 385–387

Lattice, discontinuous periodic, 79
Lattice waveguides, 204
Lax equation, 168–169
Lax operator method, 157–160
Lax pair matrices, 169
Layered metal-dielectric stack,
        516–518
LCOS, see Liquid crystal on silicon
        (LCOS)
LD, see Laser diode (LD)
LEAF, see Large effective area fiber
        (LEAF)
LED, see light emitting diode (LED)
Lens transformation, 148–149
Leptons, 520
Light emitting diode (LED) spectral
        width, 7
Lighthill criterion, 97
Linear bandgaps, 287
Linear chirp, 329
Linear curl Maxwell equations, 527
Linearly polarized (LP) modes, 2
Linear pulse shaping, 337
Linear resonant coupling, 61
Linear spectra equalization, 472–473
Linear stability analysis (LSA), 100,
        136–137
        of PT-symmetric solitons,
            294–297
Linear transmission impairments,
        461
Liquid analogy in soliton rain
        dynamics, 242–243
Liquid crystal on silicon (LCOS),
        337, 339
        pulse shaping, 340–341
Lithium niobate intensity modulator
        (LN-IM), 493
LMF, see Loop mirror filter (LMF)
LN-IM, see Lithium niobate intensity
        modulator (LN-IM)
LO, see Local oscillating (LO) light
Localized gain pump, 199–229
Local oscillating (LO) light, 495–496

Long haul fiber optic
        communications limitations,
            98
Long-term far-field evolution, 326
Loop mirror filter (LMF), 376
Loss compensator, 430
Losses and gains balanced, 64
Lossy Drude model, 414
Lossy lattice, 225–229
LP, see Linearly polarized (LP)
        modes
Lumped compensation, 435

M

Mach–Zehnder modulator, 498
MAM, see Minimum area mismatch
        (MAM) method
Manakov model, 254, 268–269
Manakov system, 460
Massive Thirring Model (MTM), 63;
        see also Thirring models
Material dispersion, 7–8
Material dispersion wavelength zero
        point, 8
Material properties
        equation, 8
        influence nonlinear response, 112
Mathematical beauty, 146–147
Matter-wave solitons, 146, 253–275
        applications of, 253–254
        bright and dark, 256–266
        in multi-component BECs,
            266–275
Maxwell-Bloch equations, 413
Maxwell equations linear curl, 527
Mehler kernel, 154
Metamaterials, 39
        definition, 409
        doped, 414
        nonlinear waves in, 409–430
Mexican hat potential, 519
MFD, see Mode field diameter
        (MFD)
MI, see Modulational instability (MI)

Microstructured optical fibers (MOFs), 307
  dispersion and, 316–317
Minimum area mismatch (MAM) method, 436, 445–446
Mirrors for comb generation, 375–378
Misfit parameter, 325
Mitigating transmission impairments, 461
Modal field in a step index fiber, 3
Mode effective area, 14
Mode field diameter (MFD), 6
Mode-locked Brillouin laser, 378–384
Mode-locked fiber lasers, 235–249
Mode-locked laser dynamics interest, 249
Mode-locked lasers
  bifurcations in locking, 236
  usefulness of, 235
Mode-locking principle, 379
Mode-locking transition, 240–242
Mode-locking with fiber lasers, 235–249
Modes, 1–3
  defining equation, 2
  four-fold degenerate, 3
  two-fold degenerate, 3
Modulational instability (MI), 79, 97–129, 133–143, 429
  applications of, 98
  critical CCD and, 111
  definition, 97
  effect of, 98
  and group velocity dispersion (GVD), 98–99
  Lighthill criterion, 97
  parametric regime, 101
  phenomenon in multiple environments, 133
  as power gain, 137–140
  relaxation influence on response and, 107–108
  in relaxing system, 105–109
  role of delay in, 111

  semiconductor doped dispersion decreasing (SD-DDF), 125–128
  in two core system, 110–114
  two state in saturable nonlinear system, 122–125
  walk-off effect and, 105–107
MOFs, see Microstructured optical fibers (MOFs)
MTM, see Massive Thirring Model (MTM)
Multimoded fiber intermodal dispersion, 7
Multimode waves, 31
Multiple scale method, 31–57
Multi-soliton (N-soliton) dynamics, 260–266, 268–269
Multi-Stokes comb generation, 352
Multi-Stokes order comb, 374–378

N

Nambu–Jona–Lasinio model, 519
Nano-resonators, 65–68
NDE, see Nonlinear relativistic Dirac equations (NDE)
Negative dispersion fiber (NDF), 440
Negative frequencies, 525–537
  in nonlinear optics, 526–528
  role of, 507
Negative frequency fields, 530
Negative-frequency resonant radiation (NRR), 507
  relationships with, 534
Negative refraction, definition, 409
Negative refractive index (NRI), 409, 417–418
Neutrino oscillations in binary waveguide arrays, 520–522
Neutrinos, 507
  Dirac limit, 522–523
  oscillations, 523–525
Newton–Raphson method, 468, 524–525

NFT, see Nonlinear Fourier
    transform (NFT)
NIS, see Nonlinear inverse synthesis
    (NIS) method
NLSE, see Nonlinear Schrödinger
    equation (NLSE or NS)
Noise-like pulses, 245–246
Nonautonomous Hirota equation,
    160–162
Nonautonomous solitons, 145–177
Nonautonomous systems, 145–146
Nonisospectral IST, 167
Nonlinear dispersion, 70
Nonlinear Duffing oscillator, 158
Nonlinear effects, 13
    dependent on power
        propagating, 14
Nonlinear fiber optics introduction,
    1–29
Nonlinear Fourier transform (NFT),
    460
    associated with NLSE, 464–465
    backward, 464, 467
    forward, 464–466
    for normal dispersion case, 467
    numerical methods for, 468–469
Nonlinear inverse synthesis (NIS)
        method, 469, 478–486
    applications of, 485–486
Nonlinearity
    cancelled by dispersion, 16
    compensation techniques,
        461–462
    in dielectric cylindric waveguide,
        54–56
    of material, 416
    second order, 16
Nonlinearity of material, 416
Nonlinear optical effects caused by
        optical intensities, 1
Nonlinear optical waveguide arrays
    (NOWA), 417
Nonlinear optical waveguide zigzag
    arrays (NOWZA), 417–418

Nonlinear oscillation theory, 153–154
Nonlinear phase shifts, 79
Nonlinear polarization evolution
    (NPE), 239
Nonlinear relativistic Dirac
    equations (NDE), 515
Nonlinear response saturation, 99
Nonlinear Schrödinger equation
    (NLSE or NS), 16–17, 31,
    61
    and Ablowitz-Ladik (AL)
        equation, 79–80
    derivation, 33
    discretization effects, 85–89
    fields of application, 79
    inversion of, 435
Nonlinear science
    across fields, 145
    overlaps, 61
Nonlinear spectrum
    for communication, 460, 469–472
    equalization, 472–473
    expansions, 473
Nonlinear steady state waves,
    422–423
Nonlinear waveguides with
    similaritons, 333–334
Nonlinear wave phenomena, 61
Nonperiodic bunches, 237
Non-radiating hyperthermic solitons,
    192
Non-return to zero (NRZ) pulses, 20
Nonsingular perturbation theory,
    39–41
Nontrivial group symmetry, 147
Non-zero dispersion shifted fibers, 9
Normalized propagation constant,
    3–4
NOWA, see Nonlinear optical
    waveguide arrays (NOWA)
NOWZA, see Nonlinear optical
    waveguide zigzag arrays
    (NOWZA)

NPE, see Nonlinear polarization evolution (NPE)

NRI, see Negative refractive index (NRI)

NRR, see Negative-frequency resonant radiation (NRR)

NRZ, see Non-return to zero (NRZ) pulses

NS, see Nonlinear Schrödinger equation (NLSE or NS)

N-soliton dynamics, see Multi-soliton (N-soliton) dynamics

Nyquist pulses, 340

O

OBP, see Optical back propagation (OBP)

Oceanic parallels, see Rogue waves

Oceanographic optical fiber bathythermographs, 367

ODC, see Oppositely directional coupler (ODC)

OFDM, see Optical frequency division multiplexing (OFDM) modulation; Orthogonal frequency division multiplexing (OFDM)

OI, see Oscillatory instability (OI)

OMF, see Optimum frequency (OMF); Optimum modulation frequency (OMF)

One-mode fiber description, 31

OPC, see Optical phase conjugation (OPC)

Open string field theories, 518

Opposite directed waves, 33

Oppositely directional coupler (ODC), 427–428
    waveguide amplifier using, 429–430

Optical amplification

erbium doped fiber amplifiers (EDFA), 17. see also Erbium doped fiber amplifiers (EDFA)

fiber Raman amplifiers, 17

Optical back propagation (OBP), 435–457
    module, 437–442
    optimal step size, 442–450
    using dispersion-decreasing fiber, 450–455
    using optical phase conjugation, 437–442
    vs. digital back propagation, 437, 441–442, 455

Optical dynamical localization, 507

Optical fiber
    as concept testbed, 323
    development, 305

Optical frequency division multiplexing (OFDM) modulation, 475–478
    vs. digital back propagation, 480–486

Optical phase conjugation (OPC)
    nonlinearity compensation technique, 461–462
    optical back propagation (OBP) and, 437–442
    use of, 435–436

Optical solitary modes, 199–229

Optics parallels, 279

Optimal step size for optical back propagation, 442–450

Optimum frequency (OMF), 103–104

Optimum modulation frequency (OMF), 111

Orthogonal frequency division multiplexing (OFDM), 461

Orthogonal pathways, 460

Oscillating wave group, 154

Oscillatory instability (OI), 280, 298–299

Oscillatory regime, 383

Oxide glasses three states, 366

P

Parabolic amplification, 337–338
Parabolic intensity profile, 327
Parabolic pulses
    applications, 337–341
    profiles, 329–331
    propagation, 323–324
    shape generation, 334–337
Parabolic similaritons, 323–341
    properties, 329–331
Parabolic waveform effects, 327–329
Parametric amplification pairs, 26
Parametric interaction of waves,
    415–417
Paraxial Helmholtz equation, 279
Parity-Time (PT) optical systems,
    64–68
Partially mode-locked laser dynamics
    interest, 249
Particle conservation, 283–284
Passive mode-locking, 379, 381
Passive segments, 334
PBG, see Photonic bandgap (PBG)
PBS, see Polarizing beam splitter
    (PBS)
PCFs, see Photonic crystal fiber
    (PCF)
PCs, see Polarization controllers
    (PCs)
Peregrine breather, see Peregrine
    soliton
Peregrine soliton; see also Rogue
    waves
    Ablowitz-Ladik equation analog,
        81–82
    solutions, 83
Pereira–Stenflo soliton, 200
Periodic bunches, 237
Periodic dielectric structures,
    305–306
Periodic potential, 287–290

soliton dynamical evolution and,
    299–300
stability analysis, 297
variational approach, 291–294
Periodic power transfer, 134
Permittivity function, 39–40
Perturbation theory region, 366–367
Phase-conjugated twin waves
    nonlinearity compensation
        technique, 461–462
Phase insensitive amplifier, 27
Phase matching condition, 102
    and dual-resonant radiation,
        317–320
    between soliton and radiation,
        533–535
Phase matching numerical
    demonstration, 535–537
Phase mismatch, 415–417, 422
Phase modulation conversion to
    intensity modulation, 21
Phase-shift keying, 368
Phase shifts
    induced by cross-phase
        modulation (XPM), 20
    nonlinear, 79
Phase transition occurrence, 65
Phase velocity, 409
Photonic bandgap (PBG), 306
    fibers, 25
    structure, 305–306
Photonic crystal fiber (PCF), 25,
    73–75, 305
    development of, 306
    solid core, 307, 308–309
    suspended core, 311–313
    suspended core properties,
        313–317
Pinned solitons
    equation for, 219
    self-trapping and stability,
        210–211
    zero solution, 209–210

Plasmonic array, see Binary
   amplifying plasmonic array
   (BAPA)
Plasmonic waves, 199
PMD, see Polarization mode
   dispersion (PMD)
PMF, see Polarization maintain fiber
   (PMF)
PMFs, see Probability mass
   functions (PMFs)
PMNS, see Pontecorvo–Maki–
   Nakagawa–Sakata (PMNS)
   matrix
Poisson equation screened, 281
Polarization controllers (PCs),
   239–240
   in Sagnac loop, 375
Polarization effects, 369–371
Polarization for improved sensing,
   398
Polarization maintain fiber (PMF),
   376–378, 381
Polarization mode dispersion
   (PMD), 7, 449
Polarization mode interactions,
   38–39
Polarized modes, 2
Polarizing beam splitter (PBS),
   239–240
Pontecorvo-Maki-Nakagawa-Sakata
   (PMNS) matrix, 523–525
Positive refractive index (PRI),
   417–418
Power threshold, see Brillouin power
   threshold
PPG, see Pulse pattern generator
   (PPG)
PRBS, see Pseudorandom bit
   sequence (PRBS)
PRI, see Positive refractive index
   (PRI)
Probability mass functions (PMFs),
   499
Projecting operators method, 31–57

Projection method
   for boundary regime
      propagation, 41–45
   for dielectric cylindric
      waveguide, 45–54
   for one-dimensional Cauchy
      problem, 34–37
Projection operator method for
   Cauchy problem, 32
Propagation constant, 16
   normalized propagation
      constant, 3–4
Properties and fundamental mode, 6
Pseudo energy, 191
Pseudo power, 191
Pseudorandom bit sequence (PRBS),
   493
Pseudo-spectral method, 284
Pseudo-spectral spatial domain
   (PSSD), 526
PSSD, see Pseudo-spectral spatial
   domain (PSSD)
PT, see Parity-Time
PT breaking, 280
PT linear modes, 281–282
PT-symmetric dipole
   solitons pinned to, 215–219
PT-symmetric solitons, 279–301
PT-symmetric system, 200, 203–204
   stability analysis, 294–297
Pulse amplification, 337–338
Pulse broadening, 8
Pulse collision, 20
Pulsed Brillouin soliton regime, 383
Pulse dispersion, 7
Pulse dynamics, 31–57
Pulse interactions, 19–20
Pulse pattern generator (PPG), 493
Pulse propagation equation
   derivation, 32
Pulse pump, 19–20
Pulse shaping schemes, 337
Pulse signal, 19–20
Pump laser modulation, 368

Pump power effects, 360
Pump-probe technique, 358
Pump pulse, 19–20

Q

Q-factor, 481–482
QFT, see Quantum field theory (QFT)
QPSK, see Quadrature phase shift keying modulation (QPSK)
Quadrature amplitude modulation (QAM), 440–441
Quadrature phase shift keying modulation (QPSK), 469, 470
    random encoding, 476–477
    single tone, 476–477
Quantum field theory (QFT), 516
    analog effects, 507–538
Quantum mechanical harmonic oscillator, 147
Quasi-distributed sensing, 398
Quasi-monochromatic wave packets, 79
Quasi-monochromatic weakly nonlinear wave phenomena, 61

R

Radiating hyperthermic solitons, 192
Radiating solitons, 194
Radiation modes, 3
Radiation waves, 189
Raindrop analogy in soliton rain dynamics, 242
Raman amplification, 340
Raman amplifiers, 17, 336
    distributed, 436
    properties, 360
Raman band
    at dispersionless limit, 112
    due to delayed nonlinear response, 111

self-phase matched, 111
Raman colored optical solitons, 173
Raman coupling, 352
Raman effect self-induced, 173
Raman instability, 103–105
Raman oscillator with extended gain fiber, 333
Raman response of silica, 331
Raman scattering; see also Stimulated Raman scattering (SRS)
    distributed sensing and, 398
    intra-pulse, 324
Ramsauer–Townsend effect, 151
Rare earth dopants, 332
Ray dispersion, 7
Rayleigh distance, 280, 295, 298
Rayleigh scattering, 352
    distributed sensing and, 398
RDS, see Relative dispersion slope (RDS)
Refractive index, 279
    intensity dependence of, 18
    intensity dependent, 13–14
    and wavelength, 8
Refractive index distribution in step index fiber, 1–2
Refractive index profile and waveguide dispersion, 9
Relative dispersion slope (RDS), 12–13
Relaxation dynamics region, 366–367
Relaxation influence on response, 107–108
Relaxing nonlinear system, 102
Relaxing system and vector MI, 105–109
Resonance mode, 306
Resonant coupling mechanism, 61
Resonant radiation (RR), 507, 526
Return to zero (RZ) pulses, 20
Return to zero (RZ) signal, 462
Riemann-Silberstein vector, 528
Ring cavity, 371

Ritz optimization, 291
Robustness of laser profile, 236
Rogue waves, 76, 79, 243–245
    in Ablowitz-Ladik (AL)
        equation, 89–92
    forms of, 83
    hierarchy, 80
    Peregrine soliton, 323
    spectra affected by
        discretization, 87
    triplets, 83–85, 89
RR, see Resonant radiation (RR)
Ruby laser model-locked, 378–379
RZ, see return to zero (RZ)

S

Sagnac loop interferometer (SLI),
    375, 385
Satsuma–Yajima breather, 175
Saturable nonlinearity (SNL)
    case vs. Kerr case, 115–116
    interplay with CCD, 116–118
Saturable nonlinear media, 99
Saturation role, 103
SBS, see Stimulated Brillouin
        scattering (SBS)
SC, see Supercontinuum (SC)
Scalar wave approximation, 2
Scattering data, 464, 465
SCG, see Supercontinuum generation
        (SCG)
"Schrödinger cat male and female
    states", 155
Schrödinger equation
    hidden symmetry, 147
    nonlinear, 16–17, 31. see also
        Nonlinear Schrödinger
        equation (NLSE or NS)
Screened Poisson equation, 281
SD-DDF, see Semiconductor doped
        dispersion decreasing
        (SD-DDF)
SDF, see Semiconductor doped fiber
        (SDF)

SDSs, see Spatial dissipative solitons
    (SDSs)
Secant pulses, 329
Sech pulse, 535
sech-type envelope, 151
Second harmonic generation (SHG),
    415–416
Second order dispersion (SOD), 16,
    101
Second order nonlinearity, 16
Segmented fibers and Brillouin gain,
    368
Self-accelerating optical beams, 153
Self-defocusing medium, 207–208
Self-defocusing regime, 226–228
Self-focusing (SF) nonlinearity and
        spatial dissipative solitons
        (SDSs), 199
Self-focusing medium, 208–209
Self-focusing regime, 228
Self-induced Raman effect, 173
Self-induced transparency, 62
Self-phase matched Raman band,
    111
Self-phase modulation (SPM), 13,
    15, 305
    due to intensity dependent
        refractive index, 13–14
    frequency generation, 18
    use of, 435
Self-steepening, 324
Sellmeier equation, 8, 310, 316–317
Semiconductor doped dispersion
        decreasing (SD-DDF), 99,
        125–128
Semiconductor doped fiber (SDF),
        125–128
Semi-discrete systems, 76
Sensing techniques, 397–400
Sensor application of Brillouin
        scattering, 388–400
Separatrix, 201
Serkin–Hasegawa (SH) theorems,
        168–169

SEWA, see Slowly evolving wave approximation (SEWA)
SF, see self-focusing (SF); Suspension factor (SF)
SH, see Serkin-Hasegawa (SH) theorems
SHG, see Second harmonic generation (SHG)
Shifts in Darboux scheme, 85
Shock effect, 535
Shock operator, 532–533
Shock wave modes, 76
Short pulse (SP) equation, 32
Short-pulse dynamics, 324–329
Signal carrier solitons, 199
Signal pre-distortion, 473, 474
Signal pulse, 19–20
Significant wave height (SWH), 246; see also Zakharov–Shabat (ZS) formalism
Silica in conventional optical fibers, 8
Silica optical fibers
    higher order nonlinearity in, 13
    Raman response, 331
Similariton laser, 333
Similaritons, 323–341
    definition, 329
    higher-order effects, 331
    parabolic, 333–334
    performance effects, 332
Simultaneous detection of positive and negative frequency, 530
Simultons, 413
    bright and dark, 415–416
Sine-Gordon models, 519
Single mode fiber (SMF), 5–6
    pulse dispersion causes, 7
    SMF-28, 360–361
    spectral variations in, 9
Single-mode twin core liquid filled photonic crystal fiber, 114–121
Single-peak solitons, 216–219

Single soliton dynamics, 256–260, 268
SIT soliton, 414
SLI, see Sagnac loop interferometer (SLI)
Slightly perturbed states, 189–190
Slow dynamic span, 190
Slowing light, 62
Slowly evolving wave approximation (SEWA), 31
Slowly varying envelope approximation (SVEA), 31, 525–527
Slow response system, 103–104
SMF, see Single mode fiber (SMF)
SMF-28 fiber, 360–361
SNL, see Saturable nonlinearity (SNL)
SOD, see Second order dispersion (SOD)
Solitary waves, 63
Soliton adaptation law, 168–169
Soliton adaptation to external potentials, 174–175
Soliton binding energy, 151
Soliton energy accumulation, 151
Solitonic analogue
    of de Broglie wavelength, 151
    of the Ramsauer Townsend effect, 151
Solitonic degrees of freedom, 460
Soliton lasers, 146
Soliton-like compression, 324
Soliton-like solutions, 411–412
Soliton-like waves, 420–421
Soliton paradigm, 145
Soliton pulse formation, 17
Soliton rain dynamics, 238–243
Solitons; see also Bragg grating solitons; Gap solitons; specific types
    adaptation law, 168–169
    analytical solutions, 508–511
    applications of, 133

attractive interaction, 256–257
autonomous, 145–146
bright, 169–172
bright matter-wave, 253–254
canonical soliton, 145–146
cold, 190
colored nonautonomous,
    172–175
condensed phase, 242–243
dark, 169–172
dark matter-wave, 253–254
definition, 16
dispersion-managed, 189
dissipative paradigm, 235
double-peak, 216–219
dynamical evolution, 298–300
envelope, 17
evolution of energies in
    perturbed states, 196–197
gap, 288–290, 428
Geiger-Nuttall law, 151
generation, 512–514
guideline for short pulses, 235
guiding-center, 189, 195–196
Hasegawa-Kumar quasi-solitons,
    157
hot, 190
hybrid, 190
hyperthermic, 190, 192
hypothermic, 190, 193
ideal, 190
isothermic, 190, 192
of Korteweg-de Vries equation,
    164–165
matter-wave, 146, 253–254
matter-wave in multi-component
    BECs, 266–275
multi-soliton (N-soliton)
    dynamics, 260–266
nonautonomous, 145–146,
    165–167
Peregrine, 80
Pereira-Stenflo, 200
perturbed types, 191–192

propagation, 512–514
radiating, 194
relationship with NRR, 534
repulsive interaction, 256–257
robust, 190
self-wavenumber shift, 507
in semiconductor doped fiber,
    126–128
shape-changing, 270–275
single dynamics, 256–260
single-peak, 216–219
SIT soliton, 414
spatial discrete, 423–424
spatial dissipative (SDSs), 199
stable, 200–201
types distinguished by position
    of eigenvalue, 287
ultra-long wavelength-division-
    multiplexing transmission,
    459
Soliton self-frequency shift, 173
Soliton symmetries, 151
Soliton tunneling, 151
Source spectral width, 7–8
SP, see Short pulse (SP) equation;
    stationary pulse (SP)
Spatial dissipative solitons (SDSs)
    chirped de Broglie wave, 200
    definition, 199
Spectral broadening
    causes supercontinuum
        generation, 307
    due to cross-phase modulation,
        18
    due to self-phase modulation
        (SPM), 18
Spectral compression application of
    parabolic pulses, 340–341
Spectral density, 87
Spectral filtering, 532
Spectral rogue waves, 246–248
Spectral self-imaging, 340
Spectral variations of fiber, 9
Spectral width of a source, 7–8

Spectronic nature of nonlinear structures, 326
Spinor components, 519
Spinorial term, 518
Split-ring resonators (SRR), 411–412
Split-step Fourier method, 126, 140
Split-step Fourier scheme (SSFS), 436, 468
SPM, see Self-phase modulation (SPM)
Spontaneous Brillouin scattering, 352–355
Spontaneous symmetry breaking, 516–520
Spontaneous symmetry breaking (SSB), 516–520
Spot size
    dependency on wavelength, 6
    of fundamental mode, 6–7
Squeezed states, 147, 153–157
Squeezions, 157
SRR, see Split-ring resonators (SRR)
SRS, see Stimulated Raman scattering (SRS)
SSB, see Spontaneous symmetry breaking (SSB)
SSFS, see Split-step Fourier scheme (SSFS)
SSMF, see standard single-mode fiber (SSMF)
Stable focus dynamical attractors, 235–236
Staggering transformation, 225–226
Stair in radiating solitons, 194
Standard single-mode fiber (SSMF)
    as transmission fiber, 440
Stationary Brillouin mirror regime, 383
Stationary pulse (SP) ,definition, 190
Step index fiber
    modes, 1–3
    waveguide dispersion in, 9
Stimulated Brillouin scattering (SBS), 13, 305, 355–357

comb, 378
history of, 351–352
laser mode-locking, 380
vs. stimulated Raman scattering, 355
Stimulated Raman scattering (SRS), 13, 305
    vs. stimulated Brillouin scattering, 355
Stokes comb, 374–378
Stokes emission process, 352
Stokes wave, 355–357
    amplification span, 393
    and BGS, 362–363
    propagation, 375
Straight IST-based method, 469
Strain effect on Brillouin scattering, 388–390
Strain measurement, 394–395
Strain sensing vs. temperature sensing, 395–396
Sturm–Liuoville problem, 45
Subcritical pitchfork bifurcation, 209–210
Supercontinuum (SC)
    broadband, 27
    generation, 27, 98
Supercontinuum generation (SCG), 98, 525–527
    caused by spectral broadening, 307
    dispersion and, 316–317
Supercritical pitchfork bifurcation, 209–210
Suspended core photonic crystal fibers, 305–320
Suspension factor (SF), 307, 313–314
    zero dispersion wavelength and, 317
SVEA, see Slowly varying envelope approximation (SVEA)
SW ,derivation, 33
SWH, see Significant wave height (SWH)

Symmetric/antisymmetric case, 115
Symmetric/antisymmetric states unaffected by CCD, 134–135
Symmetry breaking, 282

T

Tachyon condensation, 518, 519
Tachyon-like Dirac equations, 518
Tachyons, 507
    condensation, 516–520
Talanov transformation, 147–149
Tangent pulses, 329
Tappert transformation, 150
Temperature dependence
    of Brillouin scattering, 388–390
    and Brillouin strain, 363–367
Temperature sensing vs. strain sensing, 395–396
Temporal triangular intensity profile, 331
TF, see Transmission fiber (TF)
Themode, 73
Thermal distortions limit power output, 75
THG, see Third harmonic generation (THG)
Third harmonic generation (THG), 415, 416–417, 531
    relationships with, 534
Third harmonic resonant radiation (TH-RR), 533–534
Third-order dispersion (TOD), 324
Thirring models, 74, 518, 519
Thompson frequency, 410
Threshold factor, 371
TH-RR, see Third harmonic resonant radiation (TH-RR)
Time-dependent trap, 260
    frequency, 256
Time-independent trap, 259–260
Time-lens effects, 340
TIR, see Total internal reflection (TIR)

Titanium-sapphire laser, 244
Titchmarsh convolution theorem, 531
TOD, see Third-order dispersion (TOD)
Toeplitz matrix based method, 468
Total dispersion, 9–11
Total internal reflection (TIR), 306–307
    in photonic crystal fibers, 308–309
Transmission fiber (TF), 435
Transversal eigenfunction subspaces, 31
Transversal mode, 31
Transverse electric field distribution, 16
Transverse electric field patterns low-order guided modes, 5
Transverse field
    configurations during propagation, 2
    dependency on spatial coordinates and time, 2–3
    profile of modes, 5
Trap frequency, 258–259
    time-dependent, 256
Triangular intensity profile, 331
Tsunami wave studies, 244
Tunneling-coupled optical fibers, 425–426
Twisted alternating waveguide bundle, 425–426
Two color light propagation, 99
Two core fibers periodic power transfer, 134
Two coupled nonlinear Schrödinger (2CNLS) (Manakov) equations, 267, 271
Two-hump simultons, 415–416
Two-soliton dynamics, 268–269

U

Ultrafast all-optical signal processing, 339–340

Ultra-long wavelength-division-multiplexing soliton transmission, 459
Ultrashort pulse
    due to modulational instability, 97
    generation, 18, 337–338
    simulton, 413
    study of, 31–32
Unidirectional pulse propagation equation (UPPE), 526–527
    derivation of envelope equation, 529–533
    GNLSE equivalent, 532–533
Unidirectional wave approximation, 31
Universal curves, 4
Unsaturated nonlinear gain, 226–227
Unstable frequencies, 108–109, 112
Upgrading/updating fiber installations very expensive, 12
UPPE, see Unidirectional pulse propagation equation (UPPE)
UP phase, 123

V

Vacuum expectation value, 518
Vakhitov–Kolokolov criteria, 215–216
Van der Pol oscillators coupled, 65
Vapor cloud analogy in soliton rain dynamics, 242
Variational approach, 291–294
vc-cmKdV, see Korteweg-de Vries (KdV) equation
VECSELs, see Vertical external-cavity surface-emitting lasers (VECSELs)
Vertical external-cavity surface-emitting lasers (VECSELs), 334–335
Virtual mirrors, 378

Vortice modes, 76

W

Walk off, 18–19
Walk off effect, 18
    modulational instability (MI) and, 105–107
Walk off length ,definition, 20
Warm spot (WS), 204
Wave analogy in soliton rain dynamics, 243–245
Waveguide arrays (WAs), 417
    binary, 507
Waveguide bundles, 424–427
Waveguide dispersion, 8–9
    refractive index profile and, 9
    source of, 8–9
Waveguide parameter in a step index fiber, 3–4
Waveguide systems, 417–430
Wavelength and refractive index, 8
Wavelength division multiplexed systems
    back propagation in, 452–455
    dispersion compensation, 12–13
Wavelength division multiplexing (WDM)
    cross talk among channels, 24
    dense (DWDM), 29
    improvement of, 436
    systems, 98
    ultra-long, 459
Wavelength gap, 65
Wave number free space, 2
Wave packets elastically interacting, 155
Wave-particle duality, 154
Wave propagation numerical simulations, 140–142
WDM, see Wavelength Division Multiplexing (WDM)
Weakly guiding approximation, 2
Wentzel–Kramers–Brillouin (WKB) limit, 327

White-light laser, 98
WKB, see
    Wentzel–Kramers–Brillouin
    (WKB) limit
Wronskian, 161
WS, see Warm spot (WS)
W-type fiber, 12

X

$X(3)$-nonlinearity of material, 416
XPM, see Cross-phase modulation
    (XPM)
x-polarized modes, 2

Y

Y-Pauli matrix, 518
y-polarized modes, 2
Ytterbium-doped fiber amplifier,
    334–335
Yukawa concept, 146
Yukawa equation, 281

X

Zakharov–Shabat (ZS)
    direct scattering problem,
    464–466

spectral problem (ZSSP), 465
Zakharov-Shabat (ZS) formalism,
    160
Zakharov-Shabat (ZS) spectral
    problem, 465
ZBLAN fluoride fiber, 358–360, 362
ZDW, see Zero dispersion
    wavelength (ZDW)
Zener tunneling, 507
Zero dispersion second order effects
    absent, 10
Zero dispersion shifted fibers, 9
Zero dispersion wavelength (ZDW)
    multiple, 310
    in photonic crystal fibers, 310
    of saturable nonlinear media,
        99–100
    shifting of, 9
    significant parameter, 9
    and suspension factor, 317
Zero material dispersion wavelength,
    8
Zigzag arrays, 76, 417–418
Zitterbewegung, 508, 513
ZS, see Zakharov-Shabat (ZS)
ZSSP, see Zakharov-Shabat (ZS)
    spectral problem